Electronic Structure and the Propertie

Electronic Structure
and the Properties of Solids

THE PHYSICS OF THE CHEMICAL BOND

Walter A. Harrison
STANFORD UNIVERSITY

W. H. FREEMAN AND COMPANY, SAN FRANCISCO

SPONSORING EDITOR: Peter Renz
MANUSCRIPT EDITOR: Kirk Sargent
PRODUCTION COORDINATOR: Fran Mitchell
ILLUSTRATION COORDINATOR: Batyah Janowski
ILLUSTRATORS: Evan Gillespie, Tim Keenan, John and Jean Foster
DESIGNER: Marie Carluccio
COMPOSITOR: Santype International Limited
PRINTER AND BINDER: The Maple-Vail Book Manufacturing Group

Library of Congress Cataloging in Publication Data

Harrison, Walter Ashley, 1930–
    Electronic structure and the properties of solids.

    Bibliography: p.
    Includes index.
    1. Electronic structure. 2. Chemical bonds.
3. Solid state physics. 4. Solid state chemistry.
I. Title.
QC176.8.E4H37   530.4′1   79-17364
ISBN 0-7167-1000-5

9 8 7 6 5 4 3 2 1

To my wife, Lucky,
and to my sons, Rick, John, Bill, and Bob

Preface

IN THE PAST FEW YEARS the understanding of the electronic structure of solids has become sufficient that it can now be used as the basis for direct prediction of the entire range of dielectric and bonding properties, that is, for the prediction of properties of solids in terms of their chemical composition. Before that, good theories of generic properties had been available (for example, the free-electron theory of metals), but these theories required adjustment of parameters for each material. It had also been possible to interpolate properties among similar materials (as with ionicity theory) or to make detailed prediction of isolated properties (such as the energy bands for perfect crystals). The newer predictions have ranged from Augmented Plane Wave (APW) or multiple-scattering techniques for calculating total energies in perfect crystals, possible with full-scale computers, to elementary calculations of defect structures, which can be done with linear combinations of atomic orbitals (LCAO theory) or pseudopotentials on hand-held calculators. The latter, simpler category is of such importance in the design of materials and in the interpretation of experiments that there is need for a comprehensive text on these methods. This book has been written to meet that need.

The Solid State Table of the Elements, folded into the book near the back cover, exemplifies the unified view of electronic structure which is sought, and its relation to the properties of solids. The table contains the parameters needed to calculate nearly any property of any solid, using a hand-held calculator; these are parameters such as the LCAO matrix elements and pseudopotential core radii, in terms of which elementary descriptions of the electronic structure can be given. The approach used throughout this book has been to simplify the description of *sufficiently so*

the electronic structure of solids ~~enough~~ that not only electronic states but also the entire range of properties of those solids can be calculated. This is always possible; the only questions are: how difficult is the calculation, and how accurate are the results? For determining the energy bands of the perfect crystal, the simplified approach does not offer a competitive alternative to more traditional techniques; therefore, accurate band calculations are used as input information— just as experimental results are used—in establishing understanding, tests, and parameters. It is only with great difficulty that these band-calculational techniques can be extended beyond the energy bands of the perfect crystal. On the other hand, the simplified approaches explained in this book, though they give only tolerable descriptions of the bands, can easily be applied to the entire range of dielectric, transport, and bonding properties of imperfect as well as perfect solids. In most cases, they give analytic forms for the results which are easily evaluated with a hand-held electronic calculator.

Linear combinations of atomic orbitals are used as a basis for studying covalent and ionic solids; for metals the basis consists of plane waves. Both bases are related, however, and the relations between the parameters of the two systems are identified in the text. The essential point is not which basis is used for expansion: either basis can give an arbitrarily accurate description if carried far enough. The point is that isolating the essential aspects within either framework, and then discarding or (correcting for) the less essential aspects, provides the possibility for making simple numerical estimates. It is also at the root of what is meant by "learning the physics of the system" (or "learning the chemistry of the system," if one is of that background.)

Use of LCAO and plane wave bases does not necessarily make the parts of the text where they are used independent, since we continually draw on insight from both outlooks. The most striking case of this is an analysis in Chapter 2 in which the requirement that energy bands be consistent for both bases provides formulae for the interatomic matrix elements used in the LCAO studies of *sp*-bonded solids. This remarkable result was obtained only in late 1978 by Sverre Froyen and me, and it provided a theoretical basis for what had been empirical formulae when the text was first drafted. The development came in time to be included as a fundamental part of the exposition; it followed on the heels of the much more intricate formulation of the corresponding LCAO matrix elements in transition metals and transition metal compounds, which is described in Chapter 20.

Neither of these developments has yet appeared in the physics journals. Indeed, because the theoretical approaches have been developing so rapidly, several studies contained here are original with this book. The analysis of angular forces in ionic crystals—the chemical grip—is one such case, and there are a number of others. I think of the subject as new; the text could not have been written a few years ago and certainly some changes would be made if it were to be written a few years from now. However, I believe that the main features of the theory will not change, as the general theory of pseudopotentials has not changed fundamentally since the writing of *Pseudopotentials in the Theory of Metals* at the very inception of that field. In any case, the subject is much too important to wait for exposition until every avenue has been explored.

The text itself is designed for a senior or first-year graduate course. It grew out of a one-quarter course in solid state chemistry offered as a sequel to a one-quarter solid state physics course taught at the level of Kittel's *Introduction to Solid State Physics*. A single quarter is a very short time for either course. The two courses, though separate, were complementary, and were appropriate for students of physics, applied physics, chemistry, chemical engineering, materials science, and electrical engineering.

Serving so broad an audience has dictated a simplified analysis that depends on three approximations: a one-electron framework, simple approximate interatomic matrix elements, and empty-core pseudopotentials. Refinement of these methods is not difficult, and is in fact carried out in a series of appendixes. The text begins with an introduction to the quantum mechanics needed in the text. An introductory course in quantum mechanics can be considered a prerequisite. What is reviewed here will not be adequate for a reader with no background in quantum theory, but should aid readers with limited background.

The problems at the ends of chapters are an important aspect of the book. They clearly show that the calculations for systems and properties of genuine and current interest are actually quite elementary. A set of problem solutions, and comments on teaching the material, are contained in a teacher's guide that can be obtained from the publisher.

I anticipate that some users will object that much of the material covered in this book is so recent it is not possible to feel as comfortable in teaching it as in teaching a more settled field such as solid state physics. I believe, however, that the subject dealt with here is so important, particularly now that techniques such as molecular beam epitaxy enable one to produce almost any material one designs, that no modern solid state scientist should be trained without a working knowledge of the kind of solid state chemistry described in this text.

Walter A. Harrison
June 1979

Contents

Electronic Structure and the Properties of Solids

PART I

ELECTRON
STATES

IN THIS PART of the book, we shall attempt to describe solids in the simplest
meaningful framework. Chapter 1 contains a simple, brief statement of the
quantum-mechanical framework needed for all subsequent discussions. Prior
knowledge of quantum mechanics is desirable. However, for review, the premises
upon which we will proceed are outlined here. This is followed by a brief description of electronic structure and bonding in atoms and small molecules, which
includes only those aspects that will be directly relevant to discussions of solids.
Chapter 2 treats the electronic structure of solids by extending the framework
established in Chapter 1. At the end of Chapter 2, values for the interatomic
matrix elements and term values are introduced. These appear also in a Solid
State Table of the Elements at the back of the book. These will be used extensively
to calculate properties of covalent and ionic solids.

The summaries at the beginnings of all chapters are intended to give readers
a concise overview of the topics dealt with in each chapter. The summaries will
also enable readers to select between familiar and unfamiliar material.

CHAPTER 1

The Quantum-Mechanical Basis

SUMMARY

This chapter introduces the quantum mechanics required for the analyses in this text. The state of an electron is represented by a wave function ψ. Each observable is represented by an operator O. Quantum theory asserts that the average of many measurements of an observable on electrons in a certain state is given in terms of these by $\int \psi^* O \psi d^3 r$. The quantization of energy follows, as does the determination of states from a Hamiltonian matrix and the perturbative solution. The Pauli principle and the time-dependence of the state are given as separate assertions.

In the one-electron approximation, electron orbitals in atoms may be classified according to angular momentum. Orbitals with zero, one, two, and three units of angular momentum are called s, p, d, and f orbitals, respectively. Electrons in the last unfilled shell of s and p electron orbitals are called valence electrons. The principal periods of the periodic table contain atoms with differing numbers of valence electrons in the same shell, and the properties of the atom depend mainly upon its valence, equal to the number of valence electrons. Transition elements, having different numbers of d orbitals or f orbitals filled, are found between the principal periods.

When atoms are brought together to form molecules, the atomic states become combined (that is, mathematically, they are represented by linear combinations of atomic orbitals, or LCAO's) and their energies are shifted. The combinations of valence atomic orbitals with lowered energy are called bond orbitals, and their occupation by electrons bonds the molecules together. Bond orbitals are symmetric or nonpolar when identical atoms bond but become asymmetric or polar if the atoms are different. Simple calculations of the energy levels are made for a series of nonpolar diatomic molecules.

1-A   Quantum Mechanics

For the purpose of our discussion, let us assume that only electrons have important quantum-mechanical behavior. Five assertions about quantum mechanics will enable us to discuss properties of electrons. Along with these assertions, we shall make one or two clarifying remarks and state a few consequences.

Our first assertion is that

(a) *Each electron is represented by a wave function*, designated as $\psi(\mathbf{r})$. A wave function can have both real and imaginary parts. A parallel statement for light would be that each photon can be represented by an electric field $\mathscr{E}(\mathbf{r}, t)$. To say that an electron is *represented* by a wave function means that specification of the wave function gives all the information that can exist for that electron except information about the electron spin, which will be explained later, before assertion (d). In a mathematical sense, representation of each electron in terms of its own wave function is called a one-electron approximation.

(b) *Physical observables are represented by linear operators on the wave function.* The operators corresponding to the two fundamental observables, position and momentum, are

$$\left. \begin{array}{l} \text{position} \leftrightarrow \mathbf{r}, \\[2mm] \text{momentum} \leftrightarrow \mathbf{p} = \dfrac{\hbar}{i} \nabla, \end{array} \right\} \tag{1-1}$$

where $\hbar$ is Planck's constant. An analogous representation in the physics of light is of the observable, frequency of light; the operator representing the observable is proportional to the derivative (operating on the electric field) with respect to time, $\partial/\partial t$. The operator $\mathbf{r}$ in Eq. (1-1) means simply multiplication (of the wave function) by position $\mathbf{r}$. Operators for other observables can be obtained from Eq. (1-1) by substituting these operators in the classical expressions for other observables. For example, potential energy is represented by a multiplication by $V(\mathbf{r})$. Kinetic energy is represented by $p^2/2m = -(\hbar^2/2m)\nabla^2$. A particularly important observable is electron energy, which can be represented by a Hamiltonian operator:

$$H = \frac{-\hbar^2}{2m} \nabla^2 + V(\mathbf{r}). \tag{1-2}$$

The way we use a wave function of an electron and the operator representing an observable is stated in a third assertion:

(c) *The average value of measurements of an observable O, for an electron with wave function ψ, is*

$$\boxed{\langle O \rangle = \frac{\int \psi^*(\mathbf{r})O\psi(\mathbf{r})d^3r}{\int \psi^*(\mathbf{r})\psi(\mathbf{r})d^3r}.} \tag{1-3}$$

(If ψ depends on time, then so also will $\langle O \rangle$.) Even though the wave function describes an electron fully, different values can be obtained from a particular measurement of some observable. The average value of many measurements of the observable O for the same ψ is written in Eq. (1-3) as $\langle O \rangle$. The integral in the numerator on the right side of the equation is a special case of a *matrix element*; in general the wave function appearing to the left of the operator may be different from the wave function to the right of it. In such a case, the Dirac notation for the matrix element is

$$\langle \psi_1 | O | \psi_2 \rangle \equiv \int \psi_1^*(\mathbf{r}) O \psi_2(\mathbf{r}) d^3 r. \tag{1-4}$$

In a similar way the denominator on the right side of Eq. (1-3) can be shortened to $\langle \psi | \psi \rangle$. The angular brackets are also used separately. The *bra* $\langle 1 |$ or $\langle \psi_1 |$ means $\psi_1(\mathbf{r})^*$; the *ket* $| 2 \rangle$ or $| \psi_2 \rangle$ means $\psi_2(\mathbf{r})$. (These terms come from splitting the word "bracket.") When they are combined face to face, as in Eq. (1-4), an integration should be performed.

Eq. (1-3) is the principal assertion of the quantum mechanics needed in this book. Assertions (a) and (b) simply define wave functions and operators, but assertion (c) makes a connection with experiment. It follows from Eq. (1-3), for example, that the probability of finding an electron in a small region of space, $d^3 r$, is $\psi^*(\mathbf{r}) \psi(\mathbf{r}) d^3 r$. Thus $\psi^* \psi$ is the probability density for the electron.

It follows also from Eq. (1-3) that there exist electron states having discrete or definite values for energy (or, states with discrete values for any other observable). This can be proved by construction. Since any measured quantity must be real, Eq. (1-3) suggests that the operator O is Hermitian. We know from mathematics that it is possible to construct *eigenstates* of any Hermitian operator. However, for the Hamiltonian operator, which is a Hermitian operator, eigenstates are obtained as solutions of a differential equation, the *time-independent Schroedinger equation*,

$$H\psi(\mathbf{r}) = E\psi(\mathbf{r}), \tag{1-5}$$

where E is the eigenvalue. It is known also that the existence of boundary conditions (such as the condition that the wave functions vanish outside a given region of space) will restrict the solutions to a discrete set of eigenvalues E, and that these different eigenstates can be taken to be *orthogonal* to each other. It is important to recognize that eigenstates are wave functions which an electron may or may not have. If an electron has a certain eigenstate, it is said that the corresponding state is *occupied* by the electron. However, the various states exist whether or not they are occupied.

We see immediately that a measurement of the energy of an electron represented by an eigenstate will always give the value E for that eigenstate, since the average value of the mean-squared deviation from that value is zero:

$$\langle (H - E)^2 \rangle = \frac{\langle \psi | H^2 - 2EH + E^2 | \psi \rangle}{\langle \psi | \psi \rangle} = \frac{\langle \psi | E^2 - 2E^2 + E^2 | \psi \rangle}{\langle \psi | \psi \rangle} = 0. \quad (1\text{-}6)$$

We have used the eigenvalue equation, Eq. (1-5), to write $H|\psi\rangle = E|\psi\rangle$. The electron energy eigenstates, or *energy levels*, will be fundamental in many of the discussions in the book. In most cases we shall discuss that state of some entire system which is of minimum energy, that is, the *ground state*, in which, therefore, each electron is represented by an energy eigenstate corresponding to the lowest available energy level.

In solving problems in this book, we shall not obtain wave functions by solving differential equations such as Eq. (1-5), but shall instead assume that the wave functions that interest us can be written in terms of a small number of known functions. For example, to obtain the wave function ψ for one electron in a diatomic molecule, we can make a linear combination of wave functions ψ_1 and ψ_2, where 1 and 2 designate energy eigenstates for electrons in the separate atoms that make up the molecule. Thus,

$$\psi(\mathbf{r}) = u_1 \psi_1(\mathbf{r}) + u_2 \psi_2(\mathbf{r}), \quad (1\text{-}7)$$

where u_1 and u_2 are constants. The average energy, or *energy expectation value*, for such an electron is given by

$$\frac{\langle \psi | H | \psi \rangle}{\langle \psi | \psi \rangle}$$

$$= \frac{u_1^* u_1 \langle \psi_1 | H | \psi_1 \rangle + u_1^* u_2 \langle \psi_1 | H | \psi_2 \rangle + u_2^* u_1 \langle \psi_2 | H | \psi_1 \rangle + u_2^* u_2 \langle \psi_2 | H | \psi_2 \rangle}{u_1^* u_1 \langle \psi_1 | \psi_1 \rangle + u_1^* u_2 \langle \psi_1 | \psi_2 \rangle + u_2^* u_1 \langle \psi_2 | \psi_1 \rangle + u_2^* u_2 \langle \psi_2 | \psi_2 \rangle}.$$

$$(1\text{-}8)$$

The states comprising the set (here, represented by $|\psi_1\rangle$ and $|\psi_2\rangle$) in which the wave function is expanded are called *basis states*. It is customary to choose the scale of the basis states such that they are *normalized*; that is, $\langle \psi_1 | \psi_1 \rangle = \langle \psi_2 | \psi_2 \rangle = 1$. Moreover, we shall assume that the basis states are orthogonal: $\langle \psi_1 | \psi_2 \rangle = 0$. This may in fact not be true, and in Appendix B we carry out a derivation of the energy expectation value while retaining overlaps in $\langle \psi_1 | \psi_2 \rangle$. It will be seen in Appendix B that the corrections can largely be absorbed in the parameters of the theory. In the interests of conceptual simplicity, overlaps are omitted in the main text, though their effect is indicated at the few places where they are of consequence.

We can use the notation $H_{ij} = \langle \psi_i | H | \psi_j \rangle$; then Eq. (1-8) becomes

$$\frac{\langle \psi | H | \psi \rangle}{\langle \psi | \psi \rangle} = \frac{u_1^* u_1 H_{11} + u_1^* u_2 H_{12} + u_2^* u_1 H_{21} + u_2^* u_2 H_{22}}{u_1^* u_1 + u_2^* u_2}. \qquad (1\text{-}9)$$

(Actually, by Hermiticity, $H_{21} = H_{12}^*$, but that fact is not needed here.)

Eq. (1-7) describes only an approximate energy eigenstate, since the two terms on the right side are ordinarily not adequate for exact description. However, within this approximation, the best estimate of the lowest energy eigenvalue can be obtained by minimizing the entire expression (which we call E) on the right in Eq. (1-9) with respect to u_1 and u_2. In particular, setting the partial derivatives of that expression, with respect to u_1^* and u_2^*, equal to zero leads to the two equations

$$\left. \begin{array}{l} H_{11} u_1 + H_{12} u_2 = E u_1; \\ H_{21} u_1 + H_{22} u_2 = E u_2. \end{array} \right\} \qquad (1\text{-}10)$$

(In taking these partial derivatives we have treated u_1, u_1^*, u_2, and u_2^* as independent. It can be shown that this is valid, but the proof will not be given here.) Solving Eqs. (1-10) gives two values of E. The lower value is the energy expectation value of the lowest energy state, called the **bonding state**. It is

$$E_b = \frac{H_{11} + H_{22}}{2} - \sqrt{\left(\frac{H_{11} - H_{22}}{2}\right)^2 + H_{12} H_{21}}. \qquad (1\text{-}11)$$

An electron in a bonding state has energy lowered by the proximity of the two atoms of a diatomic molecule; the lowered energy helps hold the atoms together in a **bond**. The second solution to Eqs. (1-10) gives the energy of another state, also in the form of Eq. (1-7) but with different u_1 and u_2. This second state is called the **antibonding state**. Its wave function is orthogonal to that of the bonding state; its energy is given by

$$E_a = \frac{H_{11} + H_{22}}{2} + \sqrt{\left(\frac{H_{11} - H_{22}}{2}\right)^2 + H_{12} H_{21}}. \qquad (1\text{-}12)$$

We may substitute either of these energies, E_b or E_a, back into Eqs. (1-10) to obtain values for u_1 and u_2 for each of the two states, and therefore, also the form of the wave function for an electron in either state.

A particularly significant, simple approximation can be made in Eqs. (1-11) or (1-12) when the matrix element H_{12} is much smaller than the magnitude of the difference $|H_{11} - H_{22}|$. Then, Eq. (1-11) or Eq. (1-12) can be expanded in the **perturbation** H_{12} (and H_{21}) to obtain

$$E^1 \approx H_{11} + \frac{H_{12} H_{21}}{H_{11} - H_{22}}, \qquad (1\text{-}13)$$

for the energy of a state near H_{11}; a similar expression may be obtained for an energy near H_{22}. These results are part of *perturbation theory*. The corresponding result when many terms, rather than only two, are required in the expansion of the wave function is

$$E^i \approx H_{ii} + \sum_j \frac{H_{ij}H_{ji}}{H_{ii} - H_{jj}}. \qquad (1\text{-}14)$$

Similarly, for the state with energy near H_{11}, the coefficient u_2 obtained by solving Eq. (1-10) is

$$u_2 = \frac{E - H_{11}}{H_{12}} u_1 \approx \frac{H_{21}u_1}{H_{11} - H_{22}}. \qquad (1\text{-}15)$$

The last step uses Eq. (1-13). When H_{21} is small, u_2 is small, and the term $u_2\psi_2(\mathbf{r})$ in Eq. (1-7) is the correction to the *unperturbed* state, $\psi_1(\mathbf{r})$, obtained by perturbation theory. The wave function can be written to first order in the perturbation, divided by $H_{11} - H_{22}$, and generalized to a coupling with many terms as

$$\psi(\mathbf{r}) = \psi_i(\mathbf{r}) + \sum_j \frac{\psi_j(\mathbf{r})H_{ji}}{H_{ii} - H_{jj}}. \qquad (1\text{-}16)$$

The perturbation-theoretic expressions for the electron energy, Eq. (1-14), and wave function, Eq. (1-16), will be useful at many places in this text.

All of the discussion to this point has concerned the spatial wave function $\psi(\mathbf{r})$ of an electron. An electron also has spin. For any $\psi(\mathbf{r})$ there are two possible spin states. Thus, assertion (a) set forth earlier should be amended to say that an electron is described by its spatial wave function and its spin state. The term "state" is commonly used to refer to only the spatial wave function, when electron spin is not of interest. It is also frequently used to encompass both wave function and electron spin.

In almost all systems discussed in this book, there will be more than one electron. The individual electron states in the systems and the occupation of those states by electrons will be treated separately. The two aspects cannot be entirely separated because the electrons interact with each other. At various points we shall need to discuss the effects of these interactions.

In discussing electron occupation of states we shall require an additional assertion—the *Pauli principle*:

(d) *Only two electrons can occupy a single spatial state; these electrons must be of opposite spin.* Because of the discreteness of the energy eigenstates discussed above, we can use the Pauli principle to specify how states are filled with electrons to attain a system of lowest energy.

Because we shall discuss states of minimum energy, we shall not ordinarily be interested in how the wave function changes with time. For the few cases in which that information is wanted, a fifth assertion applies:

(e) *The time evolution of the wave function is given by the Schroedinger equation,*

$$ih\frac{\partial \psi}{\partial t} = H\psi. \tag{1-17}$$

This assertion is not independent of assertion (c); nevertheless, it is convenient to separate them.

At some places, particularly in the discussion of angular momentum in the next section, consequences of these five assertions will be needed which are not immediately obvious. These consequences will be stated explicitly in the context in which they arise.

1-B   Electronic Structure of Atoms

Because the potential energy $V(\mathbf{r})$ of an electron in a free atom is spherically symmetric (or at least we assume it to be), we can expect the angular momentum of an orbiting electron not to change with time. In the quantum-mechanical context this means that electron energy eigenstates can also be chosen to be angular momentum eigenstates. It is convenient to state the result in terms of the square of the magnitude of the angular momentum, L^2, which takes on the discrete values

$$L^2 = l(l + 1)\hbar^2, \tag{1-18}$$

where l is an integer greater than or equal to 0. For each value of l there are $2l + 1$ different orthogonal eigenstates; that is, the component of angular momentum along any given direction can take on the values $m\hbar$, with $m = -l, -l+1, \ldots, l-1, l$.

The spatial wave functions representing these states are called orbitals since we can imagine the corresponding classical (that is, not quantum-mechanical) electron orbits as having fixed energy and fixed angular momentum around a given axis. The term *orbital* will be used to refer specifically to the spatial wave function of an electron in an atom or molecule. We will also use the term orbital for electron wave functions representing chemical bonds where the corresponding electron orbits would not be so simple.

The $2l + 1$ orthogonal eigenstates with different m values all have the same energy, because the potential $V(\mathbf{r})$ is spherically symmetric and the energy does not depend upon the orientation of the angular momentum. States of the same energy are said to be *degenerate*. The angular momentum properties follow from assertions (a), (b), and (c) in Section 1-A but are not derived here. The concept of angular momentum is convenient since it makes it possible to classify all energy eigenstates by means of two quantum numbers, the integers l and m.

In the common terminology for states of small angular momentum, the first four—of smallest angular momentum—are

$$l = 0: \quad s \text{ state,}$$

$$l = 1: \quad p \text{ state,}$$

$$l = 2: \quad d \text{ state,}$$

$$l = 3: \quad f \text{ state.}$$

The first three letters, s, p, and d, were first used nearly a century ago to describe characteristic features of spectroscopic lines and stand for "sharp," "principal," and "diffuse."

For any given value of l and m there are many different energy eigenstates; these are numbered by a third integer, n, in order of increasing energy, starting with $n = l + 1$. This starting point is chosen since, for the hydrogen atom, states of different l but the same n are degenerate; that is, $E = \langle n, l, m | H | n, l, m \rangle$ depends only on the quantum number n. Thus n is called the **principal quantum number**. Only for the hydrogen atom, where the potential is simply $-e^2/r$, does the energy depend on n alone. However, the same numbering system is universally used for all other atoms too.

In each state specified by n, l, and m, two electrons can be accommodated, with opposite spins, according to the Pauli principle. These atomic states are the building blocks for description of the electron energies in small molecules, and in solids, as well as in individual atoms.

The **s orbitals** have vanishing angular momentum; $l = 0$ (and $m = 0$, since $|m| \leq l$). The wave function for an s orbital is spherically symmetric, and it is depicted in diagrams as a circle with a dot representing the nucleus at the center (Fig. 1-1). The lowest energy state, $n = 1$, is called a $1s$ state. Its wave function decreases monotonically with distance from the nucleus. The wave function of the next state, the $2s$ state, drops to zero, becomes negative, and then decays upward to zero. Each subsequent s orbital has an additional node. (Such forms are in fact necessary if the orbitals are to be orthogonal to each other.)

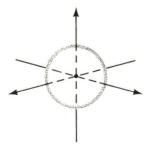

FIGURE 1-1

This depiction of an s orbital will be used frequently in this book.

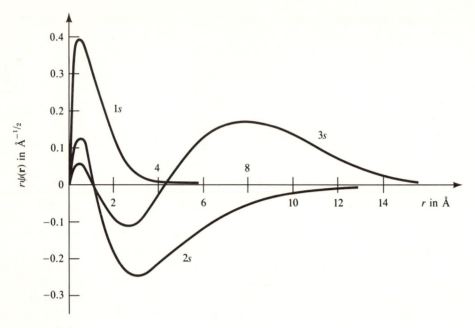

FIGURE 1-2
The three *s* states of lowest energy for atomic hydrogen. The orbitals, multiplied by *r*, are plotted as a function of distance from the nucleus.

A plot of the first three *s* orbitals for a hydrogen atom is given in Fig. 1-2.

The ***p orbitals*** have one unit of angular momentum, $l = 1$; there are three orbitals corresponding to $m = -1$, $m = 0$, and $m = 1$. (See Fig. 1-3.) Any orbital, including those of the *p* series, can be written as a product of a function of radial distance from the nucleus and one of the spherical harmonics Y_l^m, which are functions of angle only (this is explained in Schiff, 1968, p. 79):

$$\psi_{nlm}(\mathbf{r}) = R_{nl}(r)Y_l^m(\theta, \varphi). \tag{1-19}$$

For a given *l*, the radial function is independent of *m*. For *s* orbitals, the spherical

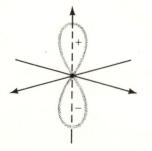

FIGURE 1-3

This *p*-orbital depiction will be used frequently in the book.

harmonic is $Y_0^0 = (4\pi)^{-1/2}$. For p orbitals, the spherical harmonics are

$$Y_1^{-1} = (3/8\pi)^{1/2} \sin \theta e^{-i\varphi},$$
$$Y_1^0 = (3/4\pi)^{1/2} \cos \theta,$$

and

$$Y_1^1 = (3/8\pi)^{1/2} \sin \theta e^{i\varphi}.$$

In solid state physics it is frequently more convenient to take linear combinations of the spherical harmonics to obtain angular dependences proportional to the component of radial distance from the nucleus along one of the three orthogonal axes x, y, or z. In this way, the three independent p orbitals may be written

$$\psi_{n1m}(\mathbf{r}) = \left(\frac{3}{4\pi}\right)^{1/2} R_{n1}(r) \begin{cases} x/r \\ y/r \\ z/r \end{cases} \tag{1-20}$$

These forms, used by Slater and Koster (1954), will be used extensively in this text. For each n when $l = 1$, there are three p orbitals oriented along the three Cartesian axes. Diagrams such as those shown in Fig. 1-4 illustrate the three angular forms.

Except for the different orientations, the orbitals look the same. The wave function is zero in an entire plane perpendicular to the axis of orientation and, at a given radius, the wave function is positive on one side and negative on the other. There are various other ways to visualize such orbitals. Three are compared in Fig. 1-5; Fig. 1-5,c is simplest and most common and, except for the sign of the wave function, is the same as the orbital shown at the left in Fig. 1-4.

The *d orbitals* have two units of angular momentum, $l = 2$, and therefore five m values: $m = -2$, $m = -1$, $m = 0$, $m = 1$, and $m = 2$. They can be conveniently

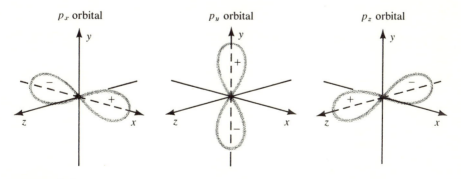

FIGURE 1-4

Three p orbitals, each directed along a different Cartesian axis.

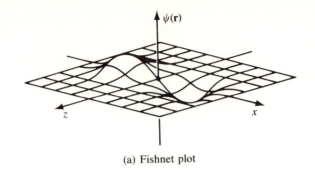

(a) Fishnet plot

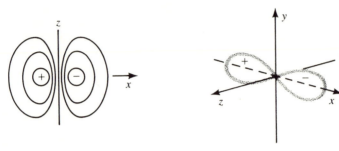

(b) Contour plot

(c) Schematic representation

FIGURE 1-5

Three ways of representing atomic p orbitals.

represented in terms of Cartesian coordinates in the form

$$\psi_{n2m}(\mathbf{r}) = \left(\frac{15}{4\pi}\right)^{1/2} R_{n2}(r) \begin{cases} yz/r^2 \\ zx/r^2 \\ xy/r^2 \\ (x^2 - y^2)/(2r^2) \\ (3z^2 - r^2)/(2r^2 3^{1/2}) \end{cases} \tag{1-21}$$

Fig. 1.6 corresponds to the third angular form listed in Eq. (1-21).

A very important feature of d orbitals is that they are concentrated much more closely at the nucleus than are s and p orbitals. The physical origin of this can be

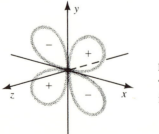

FIGURE 1-6

The d orbital corresponding to the xy/r^2 form in Eq. (1-21).

understood in terms of the $n = 3$ state of hydrogen. The $3s$, $3p$, and $3d$ states all have the same energy, but of these three, the d state corresponds classically to an orbit that is circular. At lesser angular momentum, a classical orbit of the same energy reaches further into space; this corresponds to the great spatial extent of the p orbital. The s state, which corresponds classically to an electron vibrating radially through the nucleus, stretches even further from the nucleus. Therefore, d states tend to be influenced much less by neighboring atoms than are s and p states of similar energy.

We shall have little occasion to discuss f orbitals, though they are important in studying properties of the rare-earth metals. The f orbitals are even more strongly concentrated near the nucleus and isolated from neighboring atoms than are d orbitals.

Let us now discuss the electronic states in the hydrogen atom. As indicated, the energy of an electronic state for hydrogen depends only upon the principal quantum number n. In this book, atomic energy eigenvalues, or other eigenvalues measured from the same zero of energy, will be designated by ε rather than E. For hydrogen,

$$\varepsilon_{\mathrm{H}} = -\frac{e^2}{2a_0\,n^2} = -\frac{me^4}{2\hbar^2 n^2} = -\frac{1}{n^2} \times 13.6 \text{ eV}, \tag{1-22}$$

where a_0 is the Bohr radius, 0.529 Å, e is the magnitude of the electron charge, m is the electron mass, n is the principal quantum number, and the unit of energy is the electron volt (eV).

A sketch of the energies of the states of hydrogen, the energy levels, is given in Fig. 1-7. In the ground state of the hydrogen atom, a single electron occupies the $1s$ orbital. All of the other states, having higher energies, represent excited states of the system. The electron can be transferred from the ground state to an excited state by exposing it to light of angular frequency $\omega = \Delta E/\hbar$, where ΔE is the energy difference between the two levels. Indeed, the most direct experimental study of energy levels of atoms (also called **term values**) in excited states is based upon spectroscopic analysis of the corresponding light absorption and emission lines.

To understand the electron states systematically in elements other than hydrogen, imagine that the charge of the hydrogen nucleus is increased element by element and, thereby, the atomic number, Z, is steadily increased. At the same time, imagine that an electron is added each time the nuclear charge is increased by one unit e. As the nuclear charge increases, the entire set of states drops in energy, relative to hydrogen. In all atoms but hydrogen, s-state energies are lower than p-state energies of the same principal quantum number. In Fig. 1-8 is shown the relative variation in energy of occupied $1s$, $2s$, $2p$, $3s$, $3p$, $3d$, $4s$, and $4p$ orbitals as the atomic number (equal to the number of protons in the nucleus) increases.

In lithium, atomic number 3, the $1s$ level has dropped to a very low energy and is occupied by two electrons. The $1s$ orbital is considered part of the **atomic core** of lithium; a single electron occupies a $2s$ orbital. In the lithium row, all elements, to neon, $Z = 10$, have a "lithium core"; the energy levels in successive atoms

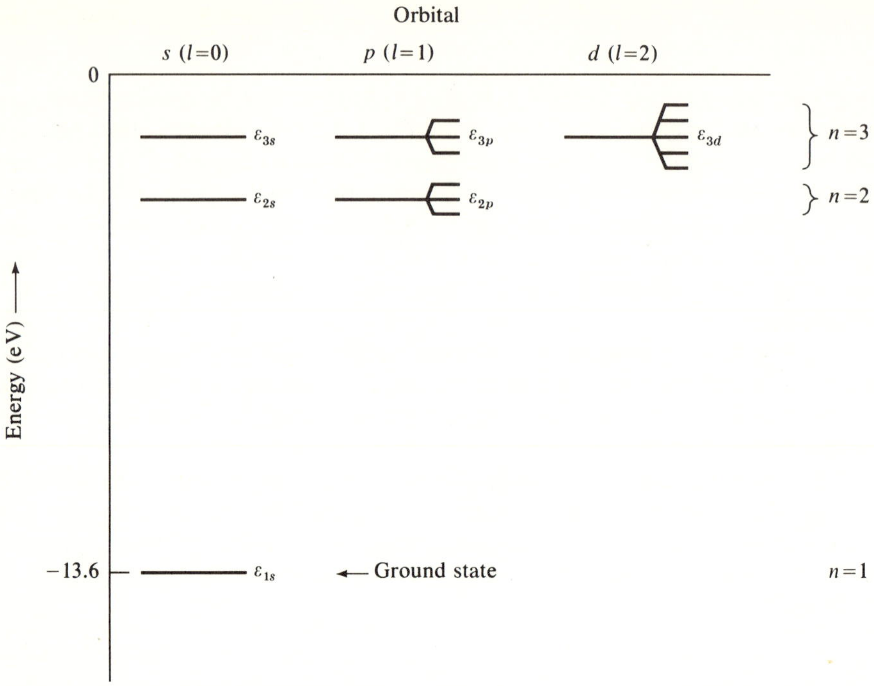

Orbital

s ($l=0$) p ($l=1$) d ($l=2$)

0

ε_{3s} ε_{3p} ε_{3d} $\left.\right\} n=3$

ε_{2s} ε_{2p} $\left.\right\} n=2$

Energy (eV) ⟶

−13.6 ε_{1s} ⟵ Ground state $n=1$

FIGURE 1-7

Energy-level diagram for atomic hydrogen. The lines are branched at the right to show how many orbitals each line represents.

continue to drop in energy and *sp splitting* (the difference in energy between levels, or $\varepsilon_{2p} - \varepsilon_{2s}$) increases. At neon, both 2s and 2p orbitals have become filled; starting with the next element, sodium, they become part of the atomic core, since, at sodium, filling of the 3s orbital begins, to be followed by filling of the 3p orbitals. The filling of successive levels is the essence of periodic variation in the properties of elements as the atomic number increases. The levels are filled in each subsequent row of the periodic table the same way they are filled in the lithium row, but the number of states in the atomic core is larger in lower rows of the table.

In the potassium row, the unoccupied 3d level begins to be filled; its energy has dropped more slowly than that of the 3s and 3p levels, but it becomes filled before the 4p level begins to fill; then in the ground state of scandium the 3d level becomes occupied with one electron. Elements in which some d states are occupied are called *transition metals*. The 3d states have become completely filled when copper, atomic number 29, is reached. The 3d states become part of the atomic core as Z increases further, and the series Cu, Zn, Ga, ..., gains electrons in an order similar to that of the series Na, Mg, Al, ...

Almost all of the properties of elements are determined by the occupied levels of highest energy; the electrons filling the s and p levels in each row (and sometimes those filling d levels) are traditionally called *valence electrons* and determine

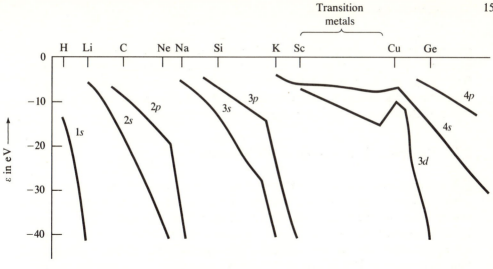

FIGURE 1-8

Trends in the variation of atomic term values, in electron volts (data from Herman and Skillman, 1963). The ε_s and ε_p values are the same as those listed also in the Solid State Table at the back of the book. The $3d$ and $4s$ energy plots were made smooth in the middle of the transition series for clarity.

1 H																1 H	2 He
3 Li	4 Be											5 B	6 C	7 N	8 O	9 F	10 Ne
11 Na	12 Mg											13 Al	14 Si	15 P	16 S	17 Cl	18 Ar
19 K	20 Ca	21 Sc	22 Ti	23 V	24 Cr	25 Mn	26 Fe	27 Co	28 Ni	29 Cu	30 Zn	31 Ga	32 Ge	33 As	34 Se	35 Br	36 Kr
37 Rb	38 Sr	39 Y	40 Zr	41 Nb	42 Mo	43 Tc	44 Ru	45 Rh	46 Pd	47 Ag	48 Cd	49 In	50 Sn	51 Sb	52 Te	53 I	54 Xe
55 Cs	56 Ba	57 *La	72 Hf	73 Ta	74 W	75 Re	76 Os	77 Ir	78 Pt	79 Au	80 Hg	81 Tl	82 Pb	83 Bi	84 Po	85 At	86 Rn
87 Fr	88 Ra	89 †Ac	104 —	105 —													

*Lanthanum Series

58 Ce	59 Pr	60 Nd	61 Pm	62 Sm	63 Eu	64 Gd	65 Tb	66 Dy	67 Ho	68 Er	69 Tm	70 Yb	71 Lu

†Actinium Series

90 Th	91 Pa	92 U	93 Np	94 Pu	95 Am	96 Cm	97 Bk	98 Cf	99 Es	100 Fm	101 Md	102 No	103 Lw

FIGURE 1-9

Periodic chart of the elements.

chemical properties. They also have excited states available to them within a few electron volts. Since these energy differences correspond to electromagnetic frequencies in the optical range, the valence electrons determine the optical properties of the elements. The periodic table (Fig. 1-9) summarizes the successive filling of electronic levels as the atomic number increases.

1-C Electronic Structure of Small Molecules

We have seen how to enumerate the electron states of single atoms. If we consider several isolated atoms as a system, the composite list of electron states for the system would simply be the collection of all states from all atoms. If the atoms are brought together closely enough that the wave functions of one atom overlap the wave functions of another, the energies of the states will change, but in all cases the number of states will be conserved. *No states disappear or are created.* If the sum of the energies of the occupied states decreases as the atoms are brought together, a molecule is said to be bound. An additional energy must be supplied to separate the atoms. (It should be noted that other terms influence the total energy of a system, and all influences must be considered in evaluating bonding energy. We shall return to this later.)

It turns out that the energy of *occupied* electronic states in small molecules, and indeed in solids, which have large numbers of atoms, can be rather well approximated with **linear combinations of atomic orbitals** (or LCAO's). Making such an approximation constitutes a very great simplification in the problem of determining molecular energies since, instead of unknown *functions*, only unknown *coefficients* appear in the linear combination. The LCAO description of the occupied molecular orbitals is much more accurate if the atomic orbitals upon which the approximation is based differ somewhat from those of the isolated constituent atoms; this complication will not arise in this book since ultimately our calculations will be in terms of matrix elements, not in terms of the orbitals themselves. The smaller the number of atomic orbitals used, the greater will be the simplification, but the poorer will be the accuracy. For our discussion of solids, a set of orbitals will be chosen that is small enough to enable calculation of a wide range of properties simply. For calculations of properties depending only upon occupied states, the accuracy will be quite good, but for excited states—those electron states which are unoccupied in the ground state of the system—the properties are not accurately calculated. We can make the same choice of orbitals in diatomic molecules that will turn out to be appropriate for solids.

In describing states of the small molecule (as well as the solid) the first step is to enumerate each of the electronic states in the atom that will be used in the mathematical expansion of the electron states in the molecule. These become our basis states. We let the index $\alpha = 1, 2, 3, \ldots, n$ run from one up to the number of states that are used. Then the molecular state may be written (with the notation discussed in Section 1-A) as

$$|\psi\rangle = \sum_{\alpha} u_{\alpha}|\alpha\rangle, \qquad (1\text{-}23)$$

where the u_α are the coefficients that must be determined. The orbitals $|\alpha\rangle$ representing the basis states are selected to be normalized, $\langle\alpha|\alpha\rangle = 1$. We also take them (as in Section 1-A) to be orthogonal to each other; $\langle\beta|\alpha\rangle = 0$ if $\beta \neq \alpha$.

Next, we must find the coefficients u_α of Eq. (1-23) for the electron state of lowest energy, by doing a variational calculation as indicated in Section 1-A. That is, we evaluate the variation

$$\delta\frac{\langle\psi|H|\psi\rangle}{\langle\psi|\psi\rangle} = \delta\frac{\sum\limits_{\alpha\beta} u_\beta^* u_\alpha\langle\beta|H|\alpha\rangle}{\sum\limits_{\alpha} u_\alpha^* u_\alpha} = 0. \tag{1-24}$$

In obtaining the second form, we allow the u_α to be complex, though ordinarily for our purposes this would not be essential. We also make use of the linearity of the Hamiltonian operator to separate the various terms in the expectation value of the Hamiltonian. In particular, if we require that variations with respect to a particular u_β^* be zero (as in Eq. 1-10), we obtain

$$\frac{\sum\limits_{\alpha} u_\alpha\langle\beta|H|\alpha\rangle}{\sum\limits_{\alpha} u_\alpha^* u_\alpha} - \frac{\sum\limits_{\alpha\beta'} u_{\beta'}^* u_\alpha\langle\beta'|H|\alpha\rangle u_\beta}{\left(\sum\limits_{\alpha} u_\alpha^* u_\alpha\right)^2} = 0, \tag{1-25}$$

or more simply,

$$\sum_{\alpha}\langle\beta|H|\alpha\rangle u_\alpha - Eu_\beta = 0, \tag{1-26}$$

with $E = \langle\psi|H|\psi\rangle/\langle\psi|\psi\rangle$. (Later, specific eigenvalues will be written as ε's with appropriate subscripts.) There is one such equation for each β corresponding to a basis state.

We have obtained a set of simultaneous linear algebraic equations with unknown coefficients u_α. Their solution gives as many eigenvalues E as there are equations. The lowest E corresponds to the lowest electron state; the next lowest, to the lowest electron state having a wave function orthogonal to that of the first, and so on. The solution of these equations gives the u_α which, with Eq. (1-23), give wave functions for the one-electron energy eigenstates directly. The eigenvalues themselves can also be obtained directly from the *secular equation*, familiar from ordinary algebra. The secular determinant vanishes,

$$\det(H_{\beta\alpha} - E\delta_{\beta\alpha}) = 0, \tag{1-27}$$

where "det" means "determinant of" and $\delta_{\beta\alpha}$ is the unit matrix. We have made one further simplification of the notation in writing $H_{\beta\alpha} = \langle\beta|H|\alpha\rangle$. We shall see in Section 2-D how simple estimates of these matrix elements can be made. Then, from Eqs. (1-26) and (1-27), we can obtain the energies and the states themselves.

Let us use the foregoing method to describe the states in a small molecule. The hydrogen molecule, with two electrons, is a simple case and is more closely related to the systems we shall be considering than the simpler hydrogen molecular ion, H_2^+. For the hydrogen molecule, we use two orbitals, $|1\rangle$ and $|2\rangle$, which represent 1s states on atoms 1 and 2 respectively. Eq. (1-26) then becomes

$$\begin{aligned}(\varepsilon_s - E)u_1 - V_2 u_2 &= 0; \\ -V_2 u_1 + (\varepsilon_s - E)u_2 &= 0,\end{aligned}$$ (1-28)

where we have made the natural definition of the 1s energy $\varepsilon_s = \langle 1|H|1\rangle = \langle 2|H|2\rangle$. The energy ε_s is slightly different from what it would be in a free atom, first, because an electron associated with atom 1 has a potential energy lowered by the presence of the second atom, and second, because the energy may be lowered as a result of the choice of a 1s function slightly different from that of the free atom. We have defined a matrix element $V_2 = -H_{12} = -H_{21}$ to correspond to the notation we shall use later. The matrix element V_2 is called a *covalent energy*, and is defined to be greater than zero; V_2 will generally be used for interatomic matrix elements, in this case between s orbitals. All the wave function coefficients are taken to be real in this case; we may always choose real coefficients but in solids will find it convenient to use complex coefficients. Eq. (1-28) is easily solved to obtain a low-energy solution, the *bonding state*, with energy

$$\varepsilon_b = \varepsilon_s - V_2,$$ (1-29)

as well as a high-energy solution, the *antibonding state*, with

$$\varepsilon_a = \varepsilon_s + V_2.$$ (1-30)

Substituting the eigenvalues given in Eqs. (1-29) and (1-30) back into Eq. (1-28) gives coefficients u_1 and u_2. For the bonding state, $u_1 = u_2 = 2^{-1/2}$, and for the antibonding state, $u_1 = -u_2 = 2^{-1/2}$. The conventional depiction of these *bond orbitals* and *antibond orbitals* is illustrated in Fig. 1-10,a.

Notice that the use of orthogonal eigenfunctions for the two atomic states (taking the overlap $\langle 1|2\rangle = 0$) is not consistent with Fig. 1-10,b, in which a clear nonzero overlap is shown. The derivation made in Appendix B allows for a nonzero overlap and shows that part of its effect can be absorbed by a modification of the value of V_2 and the other part can be absorbed in a central-force overlap interaction between the atoms, which is discussed in Chapter 7. Here, for the hydrogen molecule, the lowering of the energy of the molecule, in comparison to separated atoms, is only approximately accounted for by Eq. (1-29). If one wishes to describe the total energy as a function of the separation between atoms, one cannot simply add the energy of the two electrons in the bonding state. The central-force corrections required by this overlap, as well as other terms, must all be included.

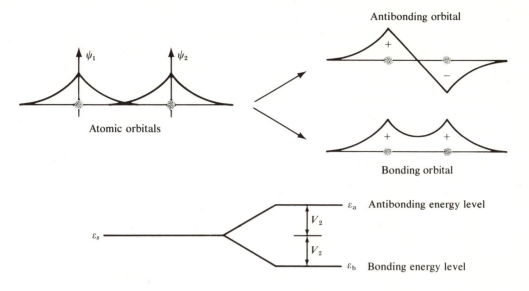

(a) Homopolar diatomic molecule

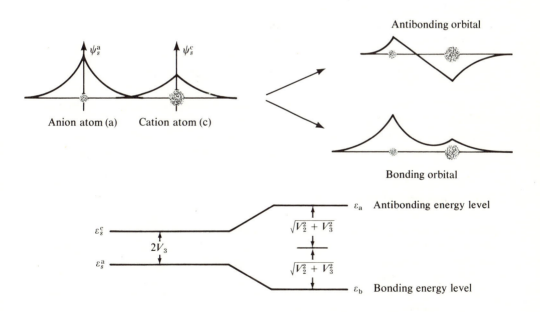

(b) Heteropolar diatomic molecule

FIGURE 1-10

The formation of bonding and antibonding combinations of atomic orbitals in diatomic molecules, and the corresponding energy-level diagrams.

Although it is possible to understand the hydrogen molecule in terms of the ideas we have discussed, hydrogen has only limited relevance to the problems we will be considering. In fact, it is not the most satisfactory way to describe the hydrogen molecule itself. In the equilibrium configuration for hydrogen, the two protons are so close together that a much better model is one in which the two protons are thought of as being superimposed; that is, we consider the nucleus to be that of the helium atom. Once this is understood, one can make corrections for the fact that in hydrogen the two protons are actually separated. Such an approach is more in tune with the spirit of this text: we will always seek the simplest description appropriate to the system we are interested in, and make corrections afterward. It has been argued that this *united atom approach*, treating H_2 as a correction applied to He, is inappropriate when the protons are far apart. That is indeed true, but we are ultimately interested in H_2 at equilibrium spacing. We will therefore simply restate our results for H_2 in the terminology to be used later and move on.

We found that hydrogen $1s$ levels are split into bonding and antibonding levels when the two atoms form the molecule. The separation of those two levels is $2V_2$, where V_2 is the covalent energy. To find the total energy of this system it is necessary to add a number of corrections to the simple sum of energies of the electrons. It will be convenient to postpone consideration of such corrections until systematic treatment in Chapter 7.

Hydrogen is a very special case also when it is a part of other molecules. We saw that in the lithium row and in the sodium row of the periodic table both a valence s state and a valence p state are present. We will see that when these atoms form molecules, the bond orbitals are mixtures of both s and p orbitals. There is no valence p state in hydrogen, and its behavior is quite different. In many ways the hydrogen proton may be regarded as a loose positive charge that keeps a molecule neutral rather than as an atom that forms a bond in the same sense that heavier atoms do. Thus we can think of methane, CH_4, as "neon" with four protons split off from the nucleus, just as we can think of H_2 as "helium" with a split nucleus.

1-D   The Simple Polar Bond

In the H_2 molecule just discussed, the two hydrogen atoms brought together were identical, and their two energies ε_s were the same. We shall often be interested in systems in which the diagonal energies H_{11} and H_{22} (that is, diagonal elements of the Hamiltonian matrix) are different; such molecules are said to have a *heteropolar* or simply *polar bond*. Let us use, as an example, the molecule LiH. We expect the linear combinations to be those of the hydrogen $1s$ orbitals and lithium $2s$ orbitals, though as we indicated at the end of the preceding section, special considerations govern molecules involving hydrogen.

In calculating the energy of heteropolar bonds, Eqs. (1-28) must be modified so that ε_s is replaced by two different energies, ε_s^1 for the low-energy state (for the energy of the anion) and ε_s^2 for the high-energy state (for the energy of the cation).

Eqs. (1-28) then become

$$(\varepsilon_s^1 - E)u_1 - V_2 u_2 = 0; \\ -V_2 u_1 + (\varepsilon_s^2 - E)u_2 = 0. \Big\}$$

(1-31)

The value of one half of the anion–cation energy-difference is the **polar energy**:

$$V_3 = \frac{\varepsilon_s^2 - \varepsilon_s^1}{2}.$$

(1-32)

It is convenient to define the average of the cation and anion energy, written as

$$\bar{\varepsilon} = \frac{\varepsilon_s^1 + \varepsilon_s^2}{2}.$$

(1-33)

Then Eqs. (1-31) become

$$(\bar{\varepsilon} - V_3 - E)u_1 - V_2 u_2 = 0; \\ -V_2 u_1 + (\bar{\varepsilon} + V_3 - E)u_2 = 0. \Big\}$$

(1-34)

The solution of Eqs. (3-34) is trivial:

$$\varepsilon_b = \bar{\varepsilon} - (V_2^2 + V_3^2)^{1/2}; \\ \varepsilon_a = \bar{\varepsilon} + (V_2^2 + V_3^2)^{1/2}; \Big\}$$

(1-35)

ε_b and ε_a are bonding and antibonding energies, respectively. The splitting of these levels is shown in Fig. 1-10,b. In looking at the energy-level diagram of that figure, imagine that the interaction between the two atomic levels, represented by V_2, pushes the levels apart. This is the qualitative result that follows also from the perturbation-theoretic expression, Eq. (1-14).

It is also shown in the figure that the charge density associated with the bonding state shifts to the low-energy side of the molecule (the direction of the anion). This means that the molecule has an electric dipole; the molecule is said to have a polar bond. Polarity of bonding is an important concept in solids and it is desirable to introduce the notion here briefly; it will be examined later, more fully, in discussion of solids. To describe polarity mathematically, first we obtain u_1 and u_2 values for the bonding state by substituting ε_b for the energy E in Eqs. (1-34), the first equation of which can then be rewritten as

$$u_1 = \frac{V_2}{\sqrt{V_2^2 + V_3^2} - V_3} u_2.$$

(1-36)

Second, if the individual atomic wave functions do not overlap, the probability of finding the electron on atom 1 will be $u_1^2/(u_1^2 + u_2^2)$ and the probability of finding it on atom 2 will be $u_2^2/(u_1^2 + u_2^2)$. This follows from the average-value theorem, Eq. (1-3). Manipulation of Eq. (1-36) leads to the result that the probability of the electron appearing on atom 1 is $(1 + \alpha_p)/2$ and the probability of finding it on atom 2 is $(1 - \alpha_p)/2$, where α_p is the *polarity* defined by

$$\alpha_p = V_3/(V_2^2 + V_3^2)^{1/2}. \tag{1-37}$$

We can expect the dipole of the bond to be proportional to $u_1^2 - u_2^2 = \alpha_p$. The polarity of the bond and the resulting dipole are central to an understanding of partially covalent solids.

Another useful concept is the complementary quantity, *covalency*, defined by

$$\alpha_c = V_2/(V_2^2 + V_3^2)^{1/2}. \tag{1-38}$$

1-E   Diatomic Molecules

In Section 1-C we noted that molecular hydrogen is unique in that a single atomic state, the $1s$ state, dominates its bonding properties. In the bonding of other diatomic molecules, valence s states and p states are important, and this will be true also in solids. Only aspects of diatomic molecules that have direct relevance to solids will be taken up here. A more complete discussion can be found in Slater (1968) or Coulson (1970).

Homopolar Bonds

Specific examples of homopolar diatomic molecules are Li_2, Be_2, B_2, C_2, N_2, O_2, and F_2, though, as seen in Fig. 1-8, variation in energy of the s and p electron states is very much the same in other series of the periodic table as it is for these elements. Four valence states for each atom must be considered—a single s state and three p states. It might seem at first that the mathematical expansion of each molecular electronic state would require a linear combination of all of these valence states; however, the matrix elements between some sets of orbitals can be seen by symmetry to vanish, and the problem of determining the states separates into two simpler problems. Fig. 1-11 indicates schematically which orbitals are coupled. The matrix elements between other orbitals than those indicated by a connecting line are zero.

The p_y orbitals of atoms 1 and 2 are coupled only to each other. They form simple bonding and antibonding combinations just as in the hydrogen molecule. In a similar way, the p_z orbitals form bonding and antibonding combinations. The four resulting p-orbital combinations are called π *states*, by analogy with p states, because each has one unit of angular momentum around the molecular axis. The π states are also frequently distinguished by a g, for *gerade* (German for "even"), or

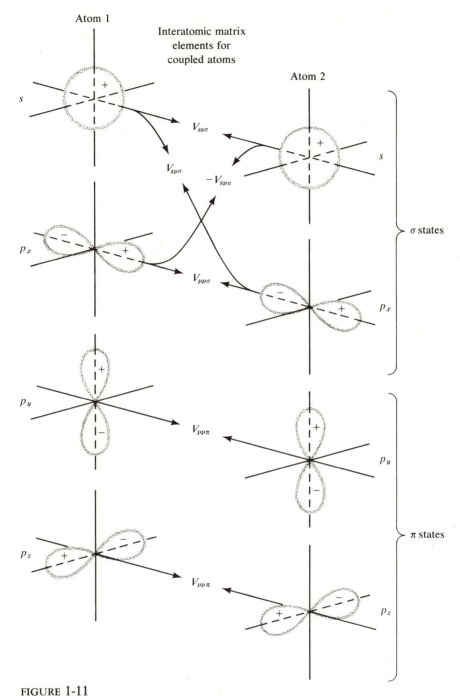

FIGURE 1-11

The coupling of atomic orbitals in lithium-row diatomic molecules, and the resultant bond designations (at right).

u, for **ungerade** ("odd"), depending on whether the wave function of the orbital is even or odd when inverted through a point midway between the atoms. For π orbitals, the bonding combination is ungerade (π_u); a π orbital that is gerade (π_g) is zero on the plane bisecting the bond.

A feature of homopolar diatomic molecules is that *s* states and p_x states are also coupled, and all four states are required in the expansion of the corresponding molecular orbitals, called **σ states**. The bonding combination for σ orbitals is gerade (σ_g). The *s* and *p* states are **hybridized** in the molecule. (The σ-orbital combinations have no angular momentum around the molecular axis.) However, it is not necessary to solve four simultaneous equations; instead, construct gerade and ungerade combinations of *s* states and of *p* states. There are no matrix elements of the Hamiltonian between the gerade and ungerade combinations, so the calculation of states again reduces to the solution of quadratic equations, as in the case of the hydrogen molecule. Notice that the two pairs of coupled *s* and *p* states have matrix elements of opposite sign $(V_{sp\sigma}, -V_{sp\sigma})$ because of the difference in the sign of the *p* lobe in the two cases. The general convention for signs will be specified in Section 2-D.

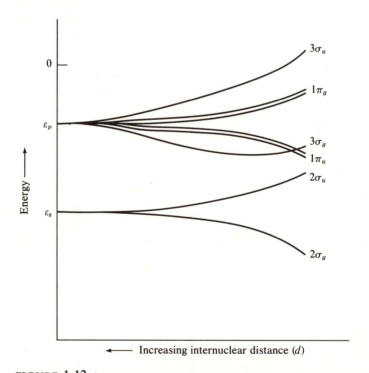

FIGURE 1-12

The development of molecular energy levels as a pair of lithium-row atoms is brought together (that is, internuclear distance *d* decreases from left to right).

Let us trace the changes in energy that occur as a pair of identical atoms from the lithium row come together. Qualitatively these changes are the same for any of the elements and they are illustrated schematically in Fig. 1-12. On the left, corresponding to large separations of the atoms, the energy levels have simply the atomic energies ε_s (one s orbital for each atom) and ε_p (three p orbitals for each atom, p_x, p_y, and p_z). As the atoms are brought together, the electron levels split (one energy going down and the other, up) and bonding and antibonding pairs are formed. The π orbitals oriented along the y-axis have the same energies as those oriented along the z-axis. The bonding and antibonding combinations for these are indicated by $1\pi_u$ and $1\pi_g$, respectively. The number one indicates the first combination of that symmetry in order of increasing energy. Each corresponds to two orbitals and is drawn with double lines. At large separation the σ orbitals are, to a good approximation, a bonding combination of s states and an antibonding combination of s states, and a bonding combination of p_x states and an antibonding combination of p_x states, in order of increasing energy. The energies of the intermediate levels, indicated by $2\sigma_u$ and $3\sigma_g$ in the figure, become comparable and should be thought of as bonding and antibonding combinations of sp-hybrids, mixtures of s states and p states. Their ordering is as shown, and is the same for all the diatomic molecules of the lithium row (Slater, 1968, pp. 451 and 452).

A particularly significant aspect of the energy levels seems to apply to all of these simple diatomic molecules: the energy of the low-lying antibonding state $2\sigma_u$ is never greater than that of either of the two high-energy bonding states $3\sigma_g$ and $1\pi_u$. (The latter two can occur in either order, as suggested in the figure.) Such crossings of bonding and antibonding levels *do* occur in solids and are an essential feature of the electronic structure of what are called covalent solids.

The Occupation of Levels

As indicated in Section 1-A, the energy of electron states and their occupation by electrons are quite separate topics. For example, it is possible to specify the energy values at an observed spacing, as in Fig. 1-12, and then to assign to them, in order of increasing energy, whatever electrons are available, ignoring any effect that an electron in one level may have on an electron in another level. More precisely, the energy of a state in any system is defined to be the negative of the energy required to move a single electron from the designated state to an infinitely distant location, without changing the number of electrons in the other states. Most theoretical calculations of energy levels determine what that energy is for each state, since this information is closely related to a wide variety of properties. When we calculate the total energy of solids, we will consider corrections to the sum of these energies; for the present, it is satisfactory to think of these energy levels as remaining fixed in energy as electrons are added to them.

If two atoms forming a diatomic molecule are both lithium, there are only two valence electrons, which would be put in the $2\sigma_g$ bonding state; the qualitative picture of electronic structure and binding of Li_2 is exactly the same for H_2; the levels deriving from the valence p state of lithium may be disregarded. If the molecule were Be_2, there would be four electrons in the molecule; two would occupy the $2\sigma_g$ bonding state, and the other two would occupy the $2\sigma_u$ antibonding state. The greater energy of the antibonding electrons (in comparison to the atomic levels) would tend to cancel the energy of the bonding electrons, and hence, bonding would not be expected. In fact, Be_2 is not found in nature. As the atomic number of the constituents increases, bonding and antibonding states are filled in succession. F_2 would have enough electrons to fill all but the highest antibonding state, $3\sigma_u$. A pair of neon atoms would have enough electrons to fill all bonding and antibonding states and, like Be_2, would not be bound at all.

In O_2, when the last levels to be filled are degenerate, a special situation occurs. Only two electrons occupy the $1\pi_g$ state though there are states to accommodate four. There are different ways the state could be filled, and *Hund's rule* tells us which arrangement will have lowest energy. It states that *when there is orbital degeneracy, the electrons will be arranged to maximize the total spin*. This means that each electron added to a set of degenerate levels will have the same (parallel) spin, if possible, as the electron which preceded it. The physical origin of this rule is the fact that two electrons of the same spin can never be found at precisely the same place, for basically the same reason that leads to the Pauli principle. Thus electrons of the same spin avoid each other, and the repulsive Coulomb interaction energy between them is smaller than for electrons of opposite spin. The corresponding lowering in energy per electron for parallel-spin electrons, compared to antiparallel-spin electrons, is called *exchange energy*. It tends to be small enough that it is dominant only when there is orbital degeneracy, as in the case of O_2, or very near orbital-degeneracy. The dominance of exchange energy is the origin of the spin alignment in ferromagnetic metals. (A more complete discussion of exchange energy appears in Appendixes A and C.)

In O_2, the two degenerate $1\pi_g$ states take one electron in a p_y state and one in a p_z state. As a result, the charge density around the O_2 molecule has cylindrical symmetry, though there is a net spin from the two electrons. In contrast, if both electrons were in p_y states, they would necessarily also have opposite spin. This would lead to a flattened charge distribution around the molecule. Hund's rule tells us that the former arrangement has lower energy because of the exchange energy.

In the same sense that H_2 is like He (as mentioned at the end of Section 1-C), the molecule C_2H_4 is like O_2, except that the two hydrogen protons are outside the carbon nucleus rather than inside. The number of electrons is the same in both C_2H_4 and O_2 and essentially the same classification of electron levels can be made. However, if the protons in C_2H_4 are all placed in the same plane, the $1\pi_u$ state oriented in that plane will have lower energy than that oriented perpendicular to the plane. The orbital energy will then be lowered if the first orbital is occupied with electrons with both spins. This planar form in fact gives the stable ground-state arrangement of nuclei and electrons in ethylene. If it were possible to increase the exchange energy it would eventually become energetically favorable to occupy one p_y state and one p_z state of parallel spin. Then the electron density would be cylindrically symmetric as in oxygen, and the protons would rotate into perpendicular planes in order to attain lower Coulomb interaction energy. C_2H_4 illustrates several points of interest. First, any elimination of orbital degeneracy will tend to override the influence of exchange energy. Second, atoms (in this case, protons) can arrange themselves in such a way as to eliminate degeneracy; this creates an asymmetric electron density that stabilizes the new arrangement. Through this self-consistent, cooperative arrangement, electrons and atoms minimize their mutual energy. This same cooperative action is often responsible for the spatial arrangement of atoms in solids. Once that arrangement is specified in solids, a particular conception of the electronic structure becomes appropriate, just as in the case of C_2H_4. Furthermore, that conception can be quite different from solid to solid, depending on which stable configuration of atoms is present.

To make the discussion of the electronic structure of diatomic molecules quantitative, it is necessary to have values for the various matrix elements. It will be found that for solids, a reasonably good approximation of the interatomic matrix elements can be obtained from the formula $V_{ij\alpha} = \eta_{ij\alpha} \hbar^2/(md^2)$, where d is the internuclear distance and values for $\eta_{ij\alpha}$ are four universal constants for $ss\sigma$, $sp\sigma$, $pp\sigma$, and $pp\pi$ matrix elements, as given in the next chapter (Table 2-1). Furthermore, atomic term values (given in Table 2-2) can be used for ε_p and ε_s. Applying such an approximation to the well-understood diatomic molecules will not reveal anything about those molecules, but can tell something about the reliability of the approximations that will be used in the study of solids. The necessary quadratic equations can be solved to obtain the molecular orbital energies in terms of the matrix elements and values for all matrix elements can be obtained from Tables 2-1 and 2-2. This gives the one-electron energies listed in Table 1-1, where the bond lengths (distance between the two nuclei) are also listed. For comparison with these values, results of full-scale self-consistent molecular orbital calculations are listed in parentheses. The solid state matrix elements give a very good semi-quantitative account of the occupied states (which lie below the shaded area) for the entire range of homopolar molecules; there are major errors only for the $3\sigma_g$ levels in O_2 and F_2. The empty levels above (shaded) are not well given. Neither will the empty levels be as well given as the occupied ones in the description of solids in terms of simple LCAO theory. This degree of success in applying solid state matrix elements outside the realm of solids, to diatomic molecules, gives confidence in their application in a wide range of solid state problems.

Heteropolar Bonds

Bonding of diatomic molecules in which the constituent atoms are different can be analyzed very directly, and only one or two points need be made. The π states in heteropolar diatomic bonding are calculated just as the simple polar bond was. In each case only one orbital on each atom is involved. A polarity can be assigned to these bonds, just as it was in Section 1-D.

TABLE 1-1

One-electron energies in homopolar diatomic molecules, as obtained by using solid state matrix elements. Values in parentheses are from accurate molecular orbital calculations. Shading denotes empty orbitals. Energies are in eV.

		Li_2	Be_2	B_2	C_2	N_2	O_2	F_2
	$d(\text{Å})$	2.67	2.0	1.59	1.24	1.09	1.22	1.42
Molecular orbital	$3\sigma_u$	2.2	1.2	5.4	11.0	14.2	5.7	−3.0
		—	—	—	—	—	(+20.0)	(+9.3)
	$1\pi_g$	−1.0	−2.7	−4.2	−5.0	−6.3	−10.0	−13.9
		(+4.4)	—	—	(−7.2)	(−8.2)	(−10.7)	(−12.9)
	$3\sigma_g$	−4.1	−5.4	−11.0	−15.6	−19.0	−23.7	−26.0
		(+2.9)	(+1.5)	(+0.2)	(−0.6)	(−15.1)	(−15.1)	(−14.9)
	$1\pi_u$	−2.8	−5.7	−9.1	−13.0	−16.7	−18.3	−20.2
		(+1.7)	—	(−9.2)	(−11.4)	(−14.8)	(−15.0)	(−16.5)
	$2\sigma_u$	−4.6	−7.4	−10.6	−14.5	−20.3	−25.2	−32.3
		(+0.8)	(−6.0)	(−9.5)	(−13.1)	(−19.4)	(−26.6)	(−37.0)
	$2\sigma_g$	−8.3	−13.1	−22.1	−33.9	−43.9	−43.4	−44.3
		(−4.9)	(−11.6)	(−18.4)	(−28.0)	(−38.6)	(−41.3)	(−44.2)

SOURCES of data in parentheses: Li_2, Be_2, C_2, N_2, and F_2 from Ransil (1960); B_2 from Padgett and Griffing (1959); O_2 from Kotani, Mizuno, Kayama, and Ishiguro (1957); all reported in Slater (1968).

There is, however, a complication in the treatment of the σ bonds. Because the states are no longer purely gerade and ungerade, the four simultaneous equations cannot be reduced to two sets of two. In a diatomic molecule this would not be much of a complication, but it is very serious in solids. Fortunately, for many solids containing σ bonds, hybrid basis states can be made from s and p states, and these can be treated approximately as independent pairs, which reduces the problem to that of finding two unknowns for each bond. In other cases, solutions can be approximated by use of perturbation theory. The approximations that are appropriate in solids will often be very different from those appropriate for diatomic molecules. Therefore, we will not discuss the special case of σ-bonded heteropolar molecule.

PROBLEM 1-1 *Elementary quantum mechanics*

An electron in a hydrogen atom has a potential energy, $-e^2/r$. The wave function for the lowest energy state is

$$\psi(\mathbf{r}) = Ae^{-r/a_0}$$

where a_0 is the Bohr radius, $a_0 = \hbar^2/me^2$, and A is a real constant.
(a) Obtain A such that the wave function is normalized, $\langle \psi | \psi \rangle = 1$.

(b) Obtain the expectation value of the potential energy, $\langle \psi | V | \psi \rangle$.

(c) Calculation of the expectation value of the kinetic energy,

$$\text{K.E.} = \langle \psi | \frac{-h^2}{2m} \nabla^2 | \psi \rangle,$$

is trickier because of the infinite curvature at $r = 0$. By partial integration in Eq. (1-3), an equivalent form is obtained:

$$\text{K.E.} = \frac{\frac{h^2}{2m} \int \nabla \psi^* \cdot \nabla \psi d^3 r}{\int \psi^* \psi d^3 r}.$$

Evaluate this expression to obtain K.E.

(d) Verify that the expectation value of the total energy, $\langle \psi | V | \psi \rangle + \text{K.E.}$ is a minimum with respect to variation of a_0. Thus a variational solution of the form $e^{-\mu r}$ would have given the correct wave function.

(e) Verify that this $\psi(\mathbf{r})$ is a solution of Eq. (1-5).

PROBLEM 1-2   *Atomic orbitals*

The hydrogen 2s and 2p orbitals can be written

$$\psi_{2s}(\mathbf{r}) = \left(\frac{1}{32\pi a_0^3} \right)^{1/2} \left(2 - \frac{r}{a_0} \right) e^{-r/2a_0},$$

and

$$\psi_{2p}(\mathbf{r}) = \left(\frac{1}{32\pi a_0^3} \right)^{1/2} \frac{x}{a_0} e^{-r/2a_0}$$

(see Schiff, 1968, p. 94), and p orbitals can also be written with x replaced by y and by z. All four hydrogen orbitals have the same energy, $-e^2/(8a_0)$.

Approximate the lithium 2s and 2p orbitals by the same functions and approximate the lithium potential by $-e^2/r + v_{\text{core}}(r)$, where

$$v_{\text{core}} = \begin{cases} -2e^2/a_0 & \text{for } r \leq a_0 \\ 0 & \text{for } r > a_0 \end{cases}$$

Calculate the expectation value of the energy of the 2s and 2p orbitals. The easiest way may be to calculate corrections to the $-e^2/(8a_0)$ value.

This gives the correct qualitative picture of the lithium valence states but is quantitatively inaccurate. Good quantitative results can be obtained by using forms such as are shown above and varying the parameters in the exponents. Such variational forms are called "Slater orbitals."

PROBLEM 1-3   *Diatomic molecules*

For C_2, obtain the σ states for the homopolar diatomic molecule (see Fig. 1-11), by using the matrix elements from the Solid State Table, at the back of the book, or from Tables 2-1 and 2-2, in Chapter 2. Writing

$$|\psi\rangle = u_1|s_1\rangle + u_2|s_2\rangle + u_3|p_{x1}\rangle + u_4|p_{x2}\rangle,$$

the equations analogous to Eq. (2-2) become

$$(\varepsilon_s - E)u_1 + V_{ss\sigma}u_2 + 0 + V_{sp\sigma}u_4 = 0;$$

$$V_{ss\sigma}u_1 + (\varepsilon_s - E)u_2 - V_{sp\sigma}u_3 + 0 = 0;$$

$$0 - V_{sp\sigma}u_2 + (\varepsilon_p - E)u_3 + V_{pp\sigma}u_4 = 0;$$

$$V_{sp\sigma}u_1 + \cdots$$

Solutions will be even or odd, by symmetry, so there can be solutions with $u_2 = u_1$ and $u_4 = -u_3$, and the above reduce to two equations in two unknowns. Solve them for E. Then, solve again with $u_2 = -u_1$ and $u_4 = u_3$.

Confirm the values of these energies as given in Table 1-1 for C_2.

The lowest state contains comparable contributions from the s and p orbitals. What is the fraction of s character, that is, $(u_1^2 + u_2^2)/(u_1^2 + u_2^2 + u_3^2 + u_4^2)$?

CHAPTER 2

Electronic Structure of Solids

SUMMARY

In solids, atomic valence levels broaden into bands comprising as many states as there are atoms in the solid. Electrons in these band states are mobile, each electron state being characterized by a momentum $\mathbf{p}$ or wave number $\mathbf{k} = \mathbf{p}/\hbar$ that is restricted to a Brillouin Zone. If each atom in the solid has only four neighboring atoms, the atomic valence orbitals can be combined to form bond orbitals between each set of neighbors, and two electrons per bond can stabilize such an arrangement of atoms. In such covalent structures, bands of states based upon the bond orbitals will be fully occupied by electrons but other bands will be empty. The bonds may be symmetric or polar. The covalent structure will not be stable if there are not two electrons per bond, if the bond energy is too small, or if the bond is too polar. Under these circumstances the lattice will tend to collapse to a denser structure. It may be an ionic crystal, which is a particularly stable arrangement, if by redistributing the electrons it can leave every atomic shell full or empty. Otherwise it will be metallic, having bands of states that are only partially occupied.

If the electron states are represented by linear combinations of atomic orbitals, the electron energy bands are found to depend on a set of orbital energies and interatomic matrix elements. Fitting these to accurate bands suggests that atomic term values suffice for the orbital energies and that nearest-neighbor interatomic matrix elements scale with bond-length d from system to system as d^{-2}. This form, and approximate coefficients, all follow from the observation that the bands are also approximately given by a free-electron approximation. Atomic term values and coefficients determining interatomic matrix elements are listed in the Solid State Table and will be used in the study of covalent and ionic solids.

In this chapter we give a very brief description of solids, which is the principal subject of the book. The main goal is to fit solids into the context of atoms and molecules. In addition, we shall carefully formulate the energy band in the simplest possible case and study the behavior of electrons in energy bands.

2-A   Energy Bands

When many atoms are brought together to form a solid, the number of electron states is conserved, just as in the formation of diatomic molecules. Likewise, as in diatomic molecules, the one-electron states for the solid can, to a reasonable approximation, be written as LCAO's. However, in solids, the number of basis states is great. A solid cube one centimeter on an edge may contain 10^{23} atoms, and for each, there is an atomic s orbital and three p orbitals. At first glance it might seem that such a problem, involving some 4×10^{23} equations, could not be attacked. However, the simplicity of the crystalline solid system allows us to proceed effectively and accurately. As the atoms are brought together, the atomic energy levels split into **bands**, which are analogous to the states illustrated for diatomic molecules in Fig. 1-12. The difference is that rather than splitting into a single bonding and a single antibonding state, the atomic levels split into an entire band of states distributed between extreme bonding and antibonding limits.

To see how this occurs, let us consider the simplest interesting case, that of cesium chloride. The structure of CsCl is shown in Fig. 2-1,a. The chlorine atoms, represented by open circles, appear on the corners of a cube, and this cubic array is repeated throughout the entire crystal. At the center of each cube is a cesium atom (at the body-center position in the cube). Cesium chloride is very polar, so the occupied orbitals lie almost entirely upon the chlorine atoms. As a first approximation we can say that the cesium atom has given up a valence electron to

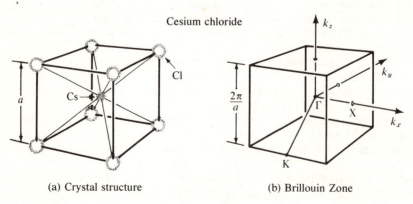

(a) Crystal structure   (b) Brillouin Zone

FIGURE 2-1

(a) A unit cube of the cesium chloride crystal structure, and (b) the corresponding Brillouin Zone in wave number space.

fill the shell of the chlorine atom, which becomes a charged atom, called an *ion*. Thus we take chlorine $3s$ orbitals and $3p$ orbitals as the basis states for describing the occupied states. Furthermore, the chlorine ions are spaced far enough apart that the s and p states can be considered separately, as was true at large inter-nuclear distance d in Fig. 1-12. Let us consider first the electron states in the crystal that are based upon the chlorine atomic $3s$ orbitals.

We define an index i that numbers all of the chlorine ions in the crystal. The chlorine atomic s state for each ion is written $|s_i\rangle$. We can approximate a crystal-line state by

$$|\psi\rangle = \sum_i u_i |s_i\rangle. \tag{2-1}$$

The variational calculation then leads immediately to a set of equations, in anal-ogy to Eq. (1-26):

$$\sum_i H_{ji} u_i - E u_j = 0. \tag{2-2}$$

It is convenient at this stage to avoid the complications that arise from con-sideration of the crystalline surface, by introducing *periodic boundary conditions*. Imagine a crystal of chlorine ions that is N_1 ions long in the x-direction, N_2 long in the y-direction, and N_3 long in the z-direction. The right surface of the crystal is connected to the left, the top to the bottom, and the front to the back. This is difficult to imagine in three dimensions, but in one dimension such a structure corresponds to a ring of ions rather than a straight segment with two ends. Closing the ring adds an H_{ij} matrix element coupling the states on the end ions. Periodic boundary conditions greatly simplify the problem mathematically; the only error that is introduced is the neglect of the effect of surfaces, which is beyond the scope of the discussion here.

The approximate description of the crystalline state, Eq. (2-1), contains a basis set of $N_p = N_1 N_2 N_3$ states (for the N_p pairs of ions), and there are N_p solutions of Eq. (2-2). These solutions can be written down directly and verified by substitu-tion into Eq. (2-2). To do this we define a *wave number* that will be associated with each state:

$$\mathbf{k} = \left(\frac{n_1 \hat{\mathbf{x}}}{N_1} + \frac{n_2 \hat{\mathbf{y}}}{N_2} + \frac{n_3 \hat{\mathbf{z}}}{N_3} \right) \frac{2\pi}{a}, \tag{2-3}$$

where n_1, n_2, and n_3 are integers such that $-N_1/2 \leq n_1 < N_1/2, \ldots$, and $\hat{\mathbf{x}}, \hat{\mathbf{y}}$, and $\hat{\mathbf{z}}$ are units vectors in the three perpendicular directions, as indicated in Fig. 2-1,b. Then for each $\mathbf{k}$ allowed by Eq. (2-3), we can write the coefficient u_j in the form

$$u_j(\mathbf{k}) = \frac{e^{i\mathbf{k} \cdot \mathbf{r}_j}}{\sqrt{N_p}} = \frac{e^{2\pi i [(n_1 m_1/N_1) + (n_2 m_2/N_2) + (n_3 m_3/N_3)]}}{\sqrt{N_1 N_2 N_3}}. \tag{2-4}$$

Here the $r_j = (m_1 \hat{x} + m_2 \hat{y} + m_3 \hat{z})a$ are the positions of the ions. We see immediately that there are as many values of $\mathbf{k}$ as there are chlorine ions; these correspond to the conservation of chlorine electron states. We also see that the wave functions for states of different $\mathbf{k}$ are orthogonal to each other. Values for $\mathbf{k}$ run almost continuously over a cubic region of wave number space, $-\pi/a \leq k_x < \pi/a$, $-\pi/a \leq k_y < \pi/a$, and $-\pi/a \leq k_z < \pi/a$. This domain of $\mathbf{k}$ is called a **Brillouin Zone**. (The shape of the Brillouin Zone, here cubic, depends upon the crystal structure.) For a macroscopic crystal the N_i are very large, and the change in wave number for unit change in n_i is very tiny. Eq. (2-4) is an exact solution of Eq. (2-2); however, we will show it for only the simplest approximation, namely, for the assumption that the $|s_i\rangle$ are sufficiently localized that we can neglect the matrix element $H_{ji} = \langle s_j | H | s_i \rangle$ unless (1) two states in question are the same ($i = j$) or (2) they are from nearest-neighbor chlorine ions. For these two cases, the magnitudes of the matrix elements are, in analogy with the molecular case,

$$H_{jj} = \varepsilon_s;$$
$$H_{ji} = -V_2.$$

In cesium chloride the main contribution to V_2 comes from cesium ion states acting as intermediaries in a form that can be obtained from perturbation theory. We need not be further concerned here with the origin of V_2. (We shall discuss the ionic crystal matrix elements in Chapter 14.) For a particular value of j in Eq. (2-2), there are only seven values of i that contribute to the sum: $i = j$ numbered as 0, and the six nearest-neighbor chlorine s states. The solution (valid for any j) is

$$E(\mathbf{k}) = \varepsilon_s - V_2 \sum_{i=1}^{6} e^{i\mathbf{k} \cdot \mathbf{r}_i}. \tag{2-5}$$

This energy varies with the wave number over the entire Brillouin Zone of Fig. 2-1,b. The results are customarily displayed graphically along certain lines within that Brillouin Zone. For example, Fig. 2-2,a shows a variation along the lines ΓX and ΓK of Fig. 2-1,b.

The calculation of bands based on p states proceeds in much the same way. In particular, if we make the simplest possible assumption—that each p_x orbital is coupled by a matrix element V_2' only to the p_x orbitals on the nearest neighbors in the x-direction and to no other p orbitals, and similarly for the p_y and p_z orbitals— then the calculation can be separated for the three types of states. (Otherwise it would be necessary to solve three simultaneous equations together.) For the states based upon the p_x orbitals,

$$E = \varepsilon_p + 2V_2' \cos k_x a. \tag{2-6}$$

For p_y orbitals and p_z orbitals, the second term is $2V_2' \cos k_y a$ and $2V_2' \cos k_z a$, respectively. The three corresponding p bands are also shown in Fig. 2-2,a. In later

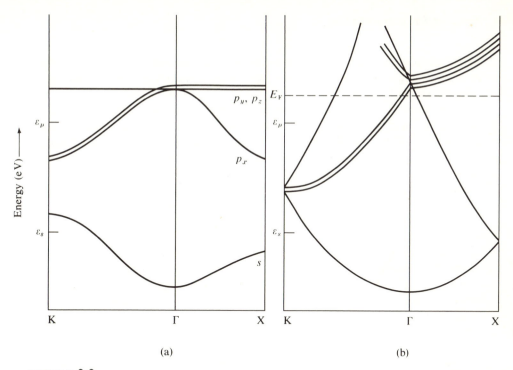

FIGURE 2-2

Valence energy bands for cesium chloride in (a) the simplest LCAO approximation and (b) in the nearly-free-electron limit. Parameters ε_p, ε_s, v_2, etc., are chosen for convenience and are not realistic.

discussions we shall see that by the addition of matrix elements between orbitals that are more distant it is possible to obtain as accurate a description of the true bands as we like; for the present, crude approximations are sufficient to illustrate the method.

Can we construct other bands, for other orbitals, such as the cesium s orbital? It turns out that states that are not occupied in the ground state of the crystal are frequently not well described in the simplest LCAO descriptions, but an approximate description can be made in the same way.

How would the simple bands change if we could somehow slowly eliminate the strong atomic potentials that give rise to the atomic states upon which the bands are based? The answer is given in Fig. 2-2,b. The gaps between bands decrease, including the gap between the cesium bands (not shown in Fig. 2-2,a) and the chlorine bands. The lowest bands have a recognizable similarity to each other in these two extreme limits. The limit shown in Fig. 2-2,b is in fact the limit as the electrons become completely free; the lowest band there is given by the equation for free-electron kinetic energy, $E = \hbar^2 k^2 / 2m$. The other bands in Fig. 2-2,b are also free-electron bands but are centered at different wave numbers (e.g., as $E = \hbar^2 (\mathbf{k} - \mathbf{q})^2 / 2m$), in keeping with the choice to represent all states by wave numbers in the Brillouin Zone. Such free-electron descriptions will be appropriate later when we discuss metals; for cesium chloride, these descriptions are not so far from LCAO descriptions as one might have thought, and in fact the similarity will provide us, in Section 2-D, with approximate values for interatomic matrix elements such as V_2 and V_2'.

Since there are as many states in each band as there are chlorine ions in the crystal, the four bands of Fig. 2-2,a, allowing both spins in each spatial state, can accommodate the seven chlorine electrons and one cesium electron. All states will be filled. This is the characteristic feature of an insulator; the state of the system cannot be changed without exciting an electron with several electron volts of energy, thus transferring it to one of the empty bands of greater energy. For that reason, light with frequency less than the difference between bands, divided by $\hbar$, cannot be absorbed, and the crystal will be transparent. Similarly, currents cannot be induced by small applied voltages. This absence of electrical conductivity results from the full bands, not from any localization of the electrons at atoms or in bonds. It is important to recognize that bands exist in crystals and that the electrons are in states of the *crystal* just as, in the molecule O_2, electrons form bonding and antibonding molecular states, rather than atomic states at the individual atoms.

If, on the other hand, the bands of cesium chloride were as in Fig. 2-2,b, the eight electrons of each chlorine–cesium atom pair would fill the states only to the energy E_F shown in the figure; this is called the **Fermi energy**. Each band would only be partly filled, a feature that, as we shall see, is characteristic of a metal.

2-B   Electron Dynamics

In circumstances where the electron energy bands are neither completely full nor completely empty, the behavior of individual electrons in the bands will be of interest. This is not the principal area of concern in this text, but it is important to understand electron dynamics because this provides the link between the band properties and electronic properties of solids.

Consider a Brillouin Zone, such as that defined for CsCl, and an energy band $E(\mathbf{k})$, defined within that zone. Further, imagine a single electron within that band. If its wave function is an energy eigenstate, the time-dependent Schroedinger equation, Eq. (1-17), tells us that

$$\psi_k(\mathbf{r}, t) = \psi_k(\mathbf{r}, 0)e^{-iE(\mathbf{k})t/\hbar}. \tag{2-7}$$

The magnitude of the wave function and therefore also the probability density at any point do not change with time. To discuss electron dynamics we must consider linear combinations of energy eigenstates of different energy. The convenient choice is a **wave packet**. In particular, we construct a packet, using states with wave numbers near $\mathbf{k}_0$ and parallel to it in the Brillouin Zone:

$$\psi(\mathbf{r}, t) \propto \sum_k e^{-\alpha|\mathbf{k} - \mathbf{k}_0|^2}\psi_k(\mathbf{r}, 0)e^{-iE(\mathbf{k})t/\hbar}. \tag{2-8}$$

Taking the form of ψ_k from Eqs. (2-1) and (2-3), and treating $\mathbf{k} - \mathbf{k}_0$ as small, a little algebra shows that at $t = 0$, Eq. (2-8) corresponds to the state ψ_{k_0} modulated by a gaussian peak centered at $\mathbf{r} = 0$. Furthermore, writing $E(\mathbf{k}) = E(\mathbf{k}_0) + (dE/d\mathbf{k}) \cdot (\mathbf{k} - \mathbf{k}_0)$, we may see that the center of the gaussian moves with a velocity

$$\dot{\mathbf{r}} = \frac{1}{\hbar} \frac{\partial E(\mathbf{k})}{\partial \mathbf{k}}. \tag{2-9}$$

Thus it is natural to associate this velocity with an electron in the state $\psi_{\mathbf{k}_0}$. Indeed, the relation is consistent with the expectation value of the current operator obtained for that state.

We are also interested in the effects of small applied fields: imagine the electron wave packet described above, but now allow a weak, slowly varying potential $V(\mathbf{r})$ to be present. The packet will work against this potential at the rate $\mathbf{v} \cdot dV/d\mathbf{r}$. This energy can only come from the band energy of the electron, through a change, with time, of the central wave number $\mathbf{k}_0$ of the packet:

$$\frac{dE(\mathbf{k})}{d\mathbf{k}} \cdot \frac{d\mathbf{k}}{dt} = -\mathbf{v} \cdot \frac{dV}{d\mathbf{r}} = -\frac{1}{\hbar} \frac{dE(\mathbf{k})}{d\mathbf{k}} \cdot \frac{dV}{d\mathbf{r}}. \tag{2-10}$$

This is consistent with the relation

$$\hbar \frac{d\mathbf{k}}{dt} = -\frac{dV}{d\mathbf{r}}. \tag{2-11}$$

This can, in fact, be generalized to magnetic forces by replacing $-dV/d\mathbf{r}$ by the Lorenz force, $-e[-\nabla\varphi + (\mathbf{v}/c) \times \mathbf{H}]$.

Eqs. (2-9) and (2-11) completely describe the dynamics of electrons in bands wherever it is possible to think in terms of wave packets; that is, whenever the fields are slowly varying relative to interatomic spacings. Notice that if we think of $\hbar\mathbf{k}$ as the canonical momentum, then the band energy, written in terms of $\mathbf{p} = \hbar\mathbf{k}$, plus the potential energy, $V(\mathbf{r})$, play precisely the role of the classical Hamiltonian, since with these definitions, Eqs. (2-9) and (2-11), are precisely Hamilton's equations. Thus, in terms of the energy bands $E(\mathbf{k})$, we may proceed directly by using kinetic theory to examine the transport properties of solids, without thinking again of the microscopic theory that led to those bands. We may go even further and use this classical Hamiltonian to discuss a wave function for the packet itself, just as we constructed wave functions for electrons in Chapter 1. This enables us to treat band electrons bound to impurities in the solid with methods similar to those used to treat electrons bound to free atoms; however, it is imperative to keep in mind that the approximations are good only when the resulting wave functions vary slowly with position, and therefore their usefulness would be restricted to weakly bound impurity states.

Let us note some qualitative aspects of electron dynamics. If the bands are narrow in energy, electron velocities will be small and electrons will behave like heavy particles. These qualities are observed in insulator valence bands and in transition-metal d bands. In simple metals and semiconductors the bands tend to be broader and the electrons are more mobile; in metals the electrons typically behave as free particles with masses near the true electron mass.

One question that might be asked is: what happens when an electron is accelerated into the Brillouin Zone surface? The answer is that it jumps across the zone and appears on the opposite face. It is not difficult to see from Eq. (2-3) that if, for example, m_1 is changed by N_1 (corresponding to going from a wave number on one zone face to a wave number on the opposite face) the phase factors change by $e^{2\pi i}$; the states are therefore identical. In general, equivalent states are found on opposite zone faces, and an electron accelerated into one face will appear at the opposite face and continue to change its wave number according to Eq. (2-11).

2-C Characteristic Solid Types

Before discussing in detail the various categories of solids, it is helpful to survey them in general terms. This is conveniently done by conceptually constructing the semiconductor silicon from free atoms. In the course of this, it will become apparent how the metallicity of a semiconductor varies with row number in the periodic table. With the general model as a basis we can also construct compounds of increasing polarity, starting with silicon or germanium and moving outward in the same row of the periodic table. Metallicity and polarity are the two principal trends shown by compounds and will provide a suitable framework for the main body of our discussions.

Imagine silicon atoms arranged as in a diamond crystal structure but widely spaced. This structure will be discussed in the next chapter; a two-dimensional analogue of it is shown in Fig. 2-3. At large internuclear distance, two electrons are on each individual atom in s states and two are in p states. As the atoms are brought together, the atomic states broaden into bands, as we have indicated. (There are complications, unimportant here, if one goes beyond a one-electron picture.) The s bands are completely full, whereas the p bands can accommodate six electrons per atom and are only one third full. This partial filling of bands is characteristic of a metal. As the atoms are brought still closer together, the broadening bands finally reach each other, as shown in Fig. 2-3, and a new gap opens up with four bands below and four above. The bonding bands below (called *valence bands*) are completely full and the antibonding bands above (called *conduction bands*) are completely empty; now the system is that of an insulator or, when the gap is small, of a semiconductor. In Chapter 1, it was noted that a crossing of bonding and antibonding states does not occur in the simple diatomic molecules, but that it can in larger molecules and in solids, as shown here.

The qualitative change in properties associated with such crossing is one of the most important concepts necessary for an understanding of chemical bonding, yet

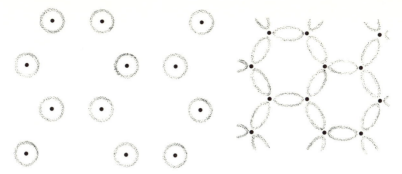

Atoms (large internuclear distance)   Bonds (smaller internuclear distance)

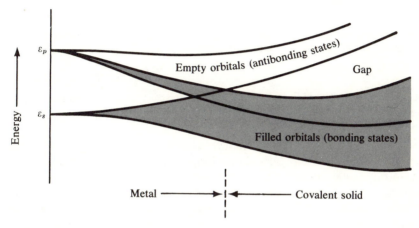

FIGURE 2-3

The formation of bands in a homopolar tetrahedral semiconductor as the atoms are brought together. Internuclear distance decreases to the right.

it has not been widely examined until recently. Particular attention has been brought by Woodward and Hoffmann (1971) in their discussion of reactions between molecules. In that context, Woodward and Hoffmann found that when bonding and antibonding states are equally occupied, as in Be_2, discussed earlier, no bonding energy is gained and the atoms repel each other. Only when the atoms are close enough that upper bonding levels can surpass or cross the energy of the lower antibonding levels above can bonding result. In some such cases (not Be_2) a stably bonded system can be formed, but an energy barrier must be overcome in order to cause the atoms to bond. Reactions in which energy barriers must be overcome are called "symmetry forbidden reactions." (See Woodward and Hoffmann, 1971, p. 10ff, for a discussion of $2C_2H_4 \rightarrow C_4H_8$.) The barrier remains, in fact, when there is no symmetry. In silicon, illustrated in Fig. 2-3, the crossing occurs because high symmetry is assumed to exist in the atomic arrangement. Because of this symmetry, the matrix elements of the Hamiltonian are zero between wave functions of states that are dropping in energy and those that are rising (ultimately to cross each other). If, instead, the silicon atoms were to come

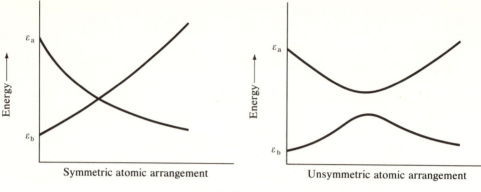

Symmetric atomic arrangement   Unsymmetric atomic arrangement

◄────── Increasing interatomic distance (d)

FIGURE 2-4

The variation of energy of two levels which cross, as a function of atomic spacing d, in a symmetric situation, but do not cross when there is not sufficient symmetry.

together as a distorted lattice with no symmetry, the corresponding matrix elements of the Hamiltonian would not be zero, and decreasing and increasing energy levels would not cross (see Fig. 2-4).

In an arrangement of high symmetry, a plotting of total energy as a function of d may show a cusp in the region where electrons switch from bonding to antibonding states; a clear and abrupt qualitative change in behavior coincides with this cusp region. In an unsymmetric arrangement, change in total energy as a function of d is gradual but at small or at large internuclear distances, energies are indistinguishable from those observed in symmetric arrangements. Thus, though the crossing is artificial (and dependent on path), the qualitative difference, which we associate with **covalent bonding**, is not. For this reason, it is absolutely essential to know on which side of a diagram such as Fig. 2-3 or Fig. 2-4 a particular system lies. For example, in covalent silicon, bonding–antibonding splitting is the large term and the sp splitting is the small one. That statement explains why there is a gap between occupied states and unoccupied states, which makes covalent silicon a semiconductor, and knowing this guides us in numerical approximations. Similarly, in metals, bonding–antibonding splitting is the small term and the sp splitting the large term; this explains why it is a metal and guides our numerical approximations in metals.

If we wished to make full, accurate machine calculations we would never need to make this distinction; we could simply look at the results of the full calculation to check for the presence of an energy gap. Instead, our methods are designed to result in intuitive understanding and approximate calculations of properties, which will allow us to guess trends without calculations in some cases, and which will allow us to treat complicated compounds that would otherwise be intractable by full, accurate calculation in other cases.

The diagram at the bottom of Fig. 2-3 was drawn to represent silicon but also, surprisingly, illustrates the homopolar series of semiconductors C, Si, Ge, and Sn. The internuclear distance is smallest in diamond, corresponding to the largest gap, far to the right in the figure. The internuclear distance becomes larger element by element down the series, corresponding to progression leftward in the figure to tin, for which the gap is zero. (Notice that in a plot of the bands, as in Fig. 2-2, the gap can vary with wave number. In tin it vanishes at only one wave number, as will be seen in Chapter 6, in Fig. 6-10.) Nonetheless we must regard each of these semiconductors—even tin—as a covalent solid in which the dominant energy is the bonding–antibonding splitting. We can define a "metallicity" that increases from C to Sn, reflecting a decreasing ratio of bonding–antibonding splitting to sp splitting; nevertheless, if the structure is tetrahedral, the bonding–antibonding splitting has won the contest and the system is covalent.

The discussion of Fig. 2-3 fits well with the LCAO description but the degree to which a solid is covalent or metallic is independent of which basis states are used in the calculation. Most of the analysis of covalent solids that will be made here will be based upon linear combinations of atomic orbitals, but we also wish to understand them in terms of free-electron-like behavior. (These two extreme approaches are illustrated for cesium chloride in Fig. 2-2.) Free-electron-like behavior is treated in Chapter 18, where two physical parameters will be designated, one of which dominates in the covalent solid and one of which dominates in the metallic solid. It can be useful here to see how these parameters correspond to the concepts discussed so far.

In Fig. 2-2, the width of the bands, approximately $\varepsilon_p - \varepsilon_s$, corresponds to the kinetic energy, E_F, of the highest filled states. The bonding–antibonding splitting similarly corresponds to the residual splitting between bands which was suppressed completely in Fig. 2-2,b. For metals, this residual splitting is described by a *pseudopotential*. In metals, the small parameter is the pseudopotential divided by the Fermi energy (corresponding to the ratio of bonding–antibonding splitting to sp splitting, or the reciprocal of the metallicity). In the covalent solids, on the other hand, we would say that the pseudopotential is the dominant aspect of the problem and the kinetic energy can be treated as the small correction. In fact, in Chapter 18 the pseudopotential approach will be applied to simple tetrahedral solids; there, treating kinetic energies as small compared to the pseudopotential leads to a simple description of the covalent bond in which a one-to-one correspondence can be obtained between matrix elements of the pseudopotential (that is, between plane waves) and matrix elements of the Hamiltonian between atomic states. The correspondence between these two opposite approaches is even more remarkable than the similarity between the LCAO and free-electron bands in Fig. 2-2, though it is the latter similarity which will provide us with LCAO matrix elements.

Now, as an introduction to polar semiconductors, let us follow the variation of electronic structure, beginning with an elemental semiconductor and moving to more polar solids. For this, germanium is a better starting point than silicon, and in order of increasing polarity the series is Ge, GaAs, ZnSe, and CuBr. The total number of electrons in each of these solids is the same (they are *isoelectronic*) and the structure is the same for all; they differ in that the nuclear charge increases on one of the atoms (the anion) and decreases on the other (the cation). The qualita-

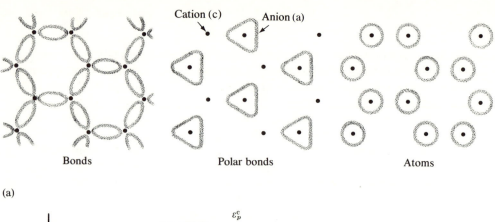

Cation (c)   Anion (a)

Bonds   Polar bonds   Atoms

(a)

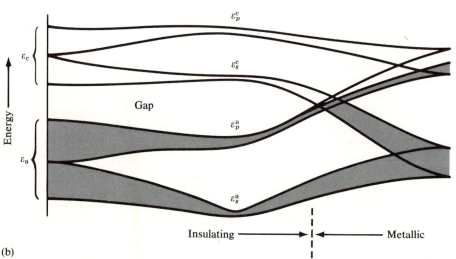

ε_p^c

ε_c

ε_s^c

Gap

ε_p^a

ε_a

ε_s^a

Energy ⟶

Insulating ⟶   ⟵ Metallic

(b)

FIGURE 2-5

Change in the bands as a homopolar semiconductor is made increasingly polar, and then as the two atom types are made more alike without broadening the levels.

tive variation in electronic structure in this series is illustrated in Fig. 2-5,a. Bear in mind that even in nonpolar solids there are two types of atomic sites, one to the right and one to the left of the horizontal bonds in the figure. In polar solids the nuclear charge on the atom to the right is increased, compound by compound. This will tend to displace the bond charges (electron density) toward the atom with higher nuclear charge (center diagram in Fig. 2-5,a) and, in fact, the corresponding transfer of charge in most cases is even larger than the change in nuclear charge, so the atom with greater nuclear charge should be thought of as negative; hence, the term *anion* is used to denote the nonmetallic atom. At high polarities most of the electronic charge may be thought of as residing on the nonmetallic atom, as shown.

The most noticeable change in the energy bands of Fig. 2-5,b, as polarity increases, is the opening up of a gap between the valence bands as shown. There is also a widening of the gap between valence and conduction bands and some broadening of the valence band. In extremely polar solids, at the center of the figure, the valence band, to a first approximation, has split into an anion s band and three narrow anion p bands. The conduction bands in this model—the unoccupied bands—also split into s bands and p bands, but in a real crystal of high polarity, the bands for unoccupied orbitals remain very broad and even free-electron-like.

We can complete the sequence of changes in the model shown in Fig. 2-5 by pulling the atoms apart to obtain isolated free atom energies. Perhaps the simplest path is that shown on the right side of Fig. 2-5, where the metallic and nonmetallic atoms become more alike and where the individual energy bands remain narrow. Where the levels cross, electrons of the anion fill available orbitals of the cation; the crossing results in a reduction of the atomic charges to zero.

By comparing Fig. 2-5 with Fig. 2-3, we can see that there is no discontinuous change in the qualitative nature of the electronic structure in going from homopolar to highly polar solids of the same crystal structure (Fig. 2-5), but that discontinuity is encountered in going from the atomic electronic structure to the covalent one (Fig. 2-3). Properties vary smoothly with polarity over the entire range. This feature has been apparent for a long time and led Pauling to define *ionicity* in terms of energies of formation in order to provide a scale for the trend (Pauling, 1960). Coulson et al. (1962) redefined ionicity in terms of an LCAO description much like the one we shall use in Chapter 3. Phillips (1970) gave still a third definition in terms of the dielectric constant. The formula for polarity of a simple bond, introduced in Eq. (1-37), is essentially equivalent to the ionicity defined by Coulson, but the ionicities defined by Pauling and by Phillips are to a first approximation proportional to the square of that polarity. We will use the term polarity to describe a variation in electronic structure in covalent solids, and the particular values defined by Eq. (1-37) will directly enter the calculation of some properties. We do *not* use polarity to interpolate properties from one material to another. However, such interpolative approaches are commonly used, and degree of ionicity or polarity is frequently used to rationalize trends in properties. Therefore it is best to examine that approach briefly. The distinction between these two approaches is subtle but of fundamental importance.

We have seen that there are trends with polarity and with metallicity among the tetrahedral solids. One of the trends is the decrease, with increasing metallicity and increasing polarity, of the angular rigidity that stabilizes the open tetrahedral structure. Thus, if either increases too far, the structure collapses to form a close-packed structure. When this happens, the new system has a qualitatively different electronic structure, and different concepts and approximations become appropriate. We may think of this as analogous to a phase diagram, as illustrated in Fig. 2-6. If a combination of atoms (e.g., lithium and flourine) is too polar, a close-packed rocksalt structure is formed. LiF is an *ionic crystal* and most frequently the best initial approximation to the electronic structure is based on independent ions, which we used in the discussion of the cesium chloride energy bands. Ionic solids can be distinguished from covalent solids by their characteristic crystalline structures, a topic that will be taken up later.

When the metallicity is too great, a close-packed structure again becomes more

44

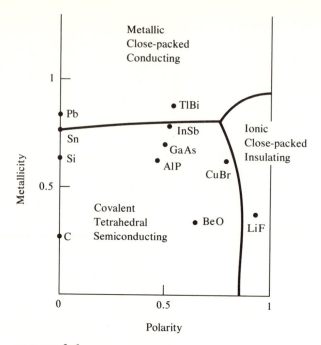

FIGURE 2-6

A schematic phase diagram indicating the three qualitatively different types of solids discussed in the book. The phase boundaries are topologically correct but details of shape are only schematic.

stable. In this case the electronic structure ordinarily approximates that of a free-electron gas and may be analyzed with methods appropriate to free-electron gases. Again, the crystal structure is the determining feature for the classification. When tin has a tetrahedral structure it is a covalent solid; when it has a close-packed white-tin structure, it is a metal. Even silicon and germanium, when melted, become close-packed and liquid metals.

To complete the "phase diagram," there must also be a line separating metallic and ionic systems. Materials near this line are called intermetallic compounds; they can lie on the metallic side (an example is Mg_2Pb) or on the ionic side (for example, CsAu). Consideration of intermetallic compounds takes the trends far beyond the isoelectronic series that we have been discussing.

The sharp distinction between ionic and covalent solids is maintained in a rearrangement of the periodic table of elements made by Pantelides and Harrison (1975). In this table, the alkali metals and some of their neighbors are transferred to the right (see Fig. 2-7). The elements of the carbon column (column 4) and compounds made from elements to either side of that column (such as GaAs or CdS) are covalent solids with tetrahedral structures. Compounds made from elements to either side of the helium column of rare gases (such as KCl or CaO) are ionic compounds with characteristic ionic structures. A few ionic and covalent compounds do not fit this correlation; notably, MgO, AgF, AgCl, and AgBr are

1 H	2 He	3 Li

Main group elements (columns 1–11)

1	2	3	4	5	6	7	8	9	10 / 11
4 Be	5 B	6 C	7 N	8 O	9 F	10 Ne	11 Na	12 Mg	21 Sc
12 Mg	13 Al	14 Si	15 P	16 S	17 Cl	18 Ar	19 K	20 Ca	39 Y
30 Zn	31 Ga	32 Ge	33 As	34 Se	35 Br	36 Kr	37 Rb	38 Sr	
48 Cd	49 In	50 Sn	51 Sb	52 Te	53 I	54 Xe	55 Cs	56 Ba	
80 Hg	81 Tl	82 Pb	83 Bi	84 Po	85 At	86 Rn	87 Fr	88 Ra	

Covalent solids |←— (columns 3–4) —→| Nonmetals |←— —→| Ionic solids

Transition metals

	D3	D4	D5	D6	D7	D8	D9	D10	D11
3d	21 Sc	22 Ti	23 V	24 Cr	25 Mn	26 Fe	27 Co	28 Ni	29 Cu
4d	39 Y	40 Zr	41 Nb	42 Mo	43 Tc	44 Ru	45 Rh	46 Pd	47 Ag
5d	71 Lu	72 Hf	73 Ta	74 W	75 Re	76 Os	77 Ir	78 Pt	79 Au

F-shell metals

		F4	F5	F6	F7	F8	F9	F10	F11	F12	F13	F14	F15	F16	F17
4f	57 La	58 Ce	59 Pr	60 Nd	61 Pm	62 Sm	63 Eu	64 Gd	65 Tb	66 Dy	67 Ho	68 Er	69 Tm	70 Yb	71 Lu
5f	89 Ac	90 Th	91 Pa	92 U	93 Np	94 Pu	95 Am	96 Cm	97 Bk	98 Cf	99 Es	100 Fm	101 Md	102 No	103 Lw

FIGURE 2-7

The Solid State Table of the elements, arranged to reflect the characteristic classes of solids. This table is shown schematically here. At the back of the book, a larger version of the table contains parameters which are used to calculate properties throughout the text.

ionic compounds, and MgS and MgSe can occur in either ionic or covalent structures. (Notice that Mg is found both in column 2 and column 10). The interesting isoelectronic series for ionic compounds will be those such as Ar, KCl, CaS, and ScP, obtained from argon by transferring protons between argon nuclei. In this case the ion *receiving* the proton is the metallic ion and the electronic structure is thought of as a slightly distorted rare gas structure. This model leads to a theory of ionic-compound bonding that is even simpler than the bonding theory for covalent solids. The Pantiledes-Harrison rearrangement of the periodic table is used as the format for the Solid State Table, where the parameters needed for the calculation of properties have been gathered.

2-D   Solid State Matrix Elements

Almost all of the discussion of covalent and ionic solids in this book is based upon descriptions of electron states as linear combinations of atomic orbitals. In order to obtain numerical estimates of properties we need numerical values for the matrix elements giving rise to the covalent and polar energies for the properties being considered. There is no *best* choice for these parameters since a trade-off must be made between simplicity (or universality) of the choice and accuracy of the predictions that result when they are used. Clearly if different values are used for each property of each material, exact values of the properties can be accommodated. We shall follow a procedure near the opposite extreme, by introducing four universal parameters in terms of which all interatomic matrix elements between s and p states for all systems can be estimated. We shall also use a single set of atomic s and p orbital energies throughout. These are the principal parameters needed for the entire range of properties, though the accuracy of the corresponding predictions is limited.

One might at first think that interatomic matrix elements could be calculated by using tabulated atomic wave functions and potentials estimated for the various solids. Such approaches have a long history of giving poor numerical results and have tended to discredit the LCAO method itself. However, the difficulty seems to be that though true *atomic* orbitals do not provide a good basis for describing electronic structure, there are *atomiclike* orbitals that can provide a very good description. One can therefore obtain a useful theory by using LCAO formalism but obtaining the necessary matrix elements by empirical or semiempirical methods.

One of the oldest and most familiar such approaches is the "Extended Hueckel Approximation" (Hoffman, 1963.) Let us take a moment to examine this approach, though later we shall choose an alternative scheme. Detailed rationalizations of the approach are given in Blyholder and Coulson (1968), and in Gilbert (1970, p. 244); a crude intuitive derivation will suffice for our purposes, as follows. We seek matrix elements of the Hamiltonian between atomic orbitals on adjacent atoms, $\langle \beta | H | \alpha \rangle$. If $|\alpha\rangle$ were an eigenstate of the Hamiltonian, we could replace $H|\alpha\rangle$ by $\varepsilon_\alpha |\alpha\rangle$, where ε_α is the eigenvalue. Then if the overlap $\langle \beta | \alpha \rangle$ is written

$S_{\beta\alpha}$, the matrix element becomes $\varepsilon_\alpha S_{\beta\alpha}$. This, however, treats the two orbitals differently, so we might use the average instead of ε_α. Finding that this does not give good values, we introduce a scale factor G, to be adjusted to fit the properties of heavy molecules; this leads to the extended Hueckel formula:

$$\langle \beta | H | \alpha \rangle = G S_{\beta\alpha}(\varepsilon_\beta + \varepsilon_\alpha)/2. \tag{2-12}$$

These matrix elements are substituted into the Hamiltonian matrix of Eq. (2-2) for a molecule, or a cluster of atoms, and the matrix is diagonalized. A value of $G = 1.75$ is usually taken; the difference from unity presumably arises from the peculiar manner in which nonorthogonality is incorporated.

The Extended Hueckel Approximation and a wide range of methods that may be considered as descendents of it (e.g., the CNDO method—Complete Neglect of Differential Overlap) have enjoyed considerable success in theoretical chemistry. Some machine calculation is required, first in determining the parameters S from tabulated wave functions or numerical approximations to them, and second in solving the resulting simultaneous equations, as at Eq. (2-2). This difficulty is exacerbated by the fact that S drops rather slowly with increasing distance between atoms, so a very large number of matrix elements are required. The computation required for any given system is very small, however, in comparison with what is required to obtain more accurate solutions. Once an Extended Hueckel Approximation has been made, direct machine computations of any property can be made and alternatives to the simplest approximations—e.g., Eq. (2-12)—can be made which improve agreement with the experimental values. Such improvements are described in detail by Pople and Beveridge (1970). Combining descriptions of electronic structure that are essentially correct, with the use of high-speed computers, and the results of a number of years of trial and error in correcting the simplest approximations, probably provide the most accurate predictions of the diverse properties of complex systems that are presently available. For isolated properties, such as the energy bands of solids, other computer methods are much more reliable and accurate.

The approach that will be used in this text is different, in that the description of electronic structures is greatly simplified to provide a more vivid understanding of the properties; numerical estimates of properties will be obtained with calculations that can be carried through by hand rather than machine. We shall concentrate on the "physics" of the problem. In this context a semiempirical determination of matrix elements is appropriate. The first attempt at this (Harrison, 1973c) followed Phillips (1970) in obtaining the principal matrix element V_2 from the measured dielectric constant. A second attempt (Harrison and Ciraci, 1974) used the principal peak in the optical reflectivity of the covalent solids, which we shall come to later, as the basis for the principal matrix element; this led to the remarkable finding that V_2 scaled from material to material quite accurately as the inverse square of the interatomic distance, the bond length d, between atoms. A subsequent study of the detailed form of valence bands (Pantelides and Harrison, 1975), combined with V_2 determined from the peak in optical reflectivity, gave a complete set of interatomic matrix elements for covalent solids with the finding that *all* of them varied approximately as d^{-2} from material to material.

The reason for this dependence recently became very clear in a study of the bands of covalent solids by Froyen and Harrison (1979). They took advantage of the similarity of the LCAO bands and free-electron bands, noted in Fig. 2-2. By equating selected energy differences obtained in the two limits, they derived formulae that had this dependence for all of the interatomic matrix elements. We may in fact see in detail how this occurs by considering Fig. 2-2. The lowest band, labelled s in Fig. 2-2,a, was given by Eq. (2-5). For **k** in an x-direction, it becomes $E(k) = \varepsilon_s - 4V_2 - 2V_2 \cos ka$, varying by $4V_2$ from Γ (where $k = 0$) to X (where $k = \pi/a$). The free-electron energy in Fig. 2-2,b varies by $(\hbar^2/2m)(\pi/a)^2$ over the same region of wave number space for the lowest band. Thus, if both limiting models are to be appropriate, and therefore consistent with each other, it must follow that $V_2 = \eta \hbar^2/(ma^2)$ with $\eta = \pi^2/8 = 1.23$. This predicts the dependence upon the inverse square of interatomic distance and a coefficient that depends only upon crystal structure. A similar comparison of the second band gives the same form with a different coefficient for the matrix element V_2' between p states. This simplest model is not so relevant, but it illustrates the point nicely. Before going to more relevant systems we must define more precisely the notation to be used for general interatomic matrix elements.

These matrix elements will be important throughout the text; they are specified here following the conventions used by Slater and Koster (1954) and used earlier while discussing the diatomic molecule. In general, for a matrix element $\langle \alpha | H | \beta \rangle$ between orbitals on different atoms we construct the vector **d**, from the nucleus of the atom of which $|\alpha\rangle$ is an orbital (the "left" atom) to that of the atom of which $|\beta\rangle$ is an orbital (the "right" atom). Then spherical coordinate systems are constructed with the z-axes parallel to **d**, and with origins at each atom; the angular form of the orbitals can be taken as $Y_l^m(\theta, \phi)$ for the left orbital and $Y_{l'}^{m'}(\theta', \phi)$ for the right orbital. The angular factors depending upon ϕ combine to $e^{i(m' - m)\phi}$. (Notice that the wave function $\langle \alpha |$ is the complex conjugate of $|\alpha\rangle$.) The integration over ϕ gives zero unless $m' = m$. Then all matrix elements $\langle \alpha | H | \beta \rangle$ vanish unless $m' = m$, and these are labelled by σ, π, or δ (in analogy with s, p, d) for $m = 0$, 1, and 2 respectively. Thus, for example, the matrix element $V_{sp\sigma}$ corresponds to $l = 0$, $l' = 1$, $m = 0$. Slater and Koster (1954) designated matrix elements by enclosing the indices within parentheses; thus, the element $V_{ll'm}$ used in this book and their $(ll'm)$ are the same.

We saw how formulae for the matrix elements can be obtained by equating band energies from LCAO theory and from free-electron theory in Fig. 2-2. Froyen and Harrison (1979) made the corresponding treatment of the tetrahedral solids, again including only matrix elements between nearest-neighbor atoms. The form of their results is just as found for the simple cubic case

$$V_{ll'm} = \eta_{ll'm} \hbar^2/(md^2). \qquad (2\text{-}13)$$

Notice that the subscript m is a quantum number but the m in the denominator

TABLE 2-1

Dimensionless coefficients in Eq. (2-13) determining approximate interatomic matrix elements.

Coefficient	Theoretical values		Adjusted value
	Simple cubic structure	Tetrahedral structure	
$\eta_{ss\sigma}$	$-\dfrac{\pi^2}{8} = -1.23$	$\dfrac{-9\pi^2}{64} = -1.39$	-1.40
$\eta_{sp\sigma}$	$\dfrac{\pi}{2}\sqrt{\dfrac{\pi^2}{4}} - 1 = 1.90$	$\dfrac{9\pi^2}{32}\sqrt{1 - \dfrac{16}{3\pi^2}} = 1.88$	1.84
$\eta_{pp\sigma}$	$\dfrac{3\pi^2}{8} = 3.70$	$\dfrac{21}{64}\pi^2 = 3.24$	3.24
$\eta_{pp\pi}$	$\dfrac{-\pi^2}{8} = -1.23$	$\dfrac{-3}{32}\pi^2 = -0.93$	-0.81

NOTE: Theoretical values (Froyen and Harrison, 1979) were obtained by equating band energies from LCAO and free-electron theory, as described in the text. Adjusted values (Harrison, 1976b) were obtained by fitting the energy bands of silicon and germanium; the adjusted values appear in the Solid State Table.

is the electron mass. The length d is the internuclear distance, equal to a in the simple cubic structure. If d is given in angstroms, this form is easily evaluated, using $\hbar^2/m = 7.62$ eV-Å^2. In Table 2-1 we give the values of the dimensionless coefficients obtained by Froyen and Harrison for both the simple cubic and tetrahedral structures. The calculation is closely related to that just carried through for the bands of Fig. 2-2, and in fact, the $V_{ss\sigma}$ matrix element for the simple cubic case is just the negative of the V_2 value evaluated there, leading to the $\eta_{ss\sigma} = -\pi^2/8$. We shall see in Section 18-A exactly how the other theoretical coefficients listed were obtained.

Notice that the coefficients obtained for the tetrahedral structure differ from those obtained for the simple cubic structure and indeed the coefficients for any one structure depend somewhat upon which band energies are used. However, the differences are not great and we shall neglect them. The coefficients we shall use are close to those given by Froyen and Harrison (1979) for the tetrahedral structure, but were obtained somewhat earlier by Harrison (1976b), who adjusted them to give the interatomic matrix elements found by Chadi and Cohen (1975) in fitting the known energy bands of silicon and germanium. The average of the coefficients so obtained for silicon and germanium is listed in Table 2-1 in the column headed "Adjusted," and these are the values listed in the Solid State Table and used throughout this text. Also listed in the Solid State Table are forms for predicting matrix elements involving atomic d states, formulae which will be developed in Chapter 20.

The coefficients in Table 2-1 have been obtained entirely in the context of nearest-neighbor coupling between states. They would have been different if a

TABLE 2-2

Atomic term values from Herman and Skillman (1963), or extrapolated from their values.

Element	$-\varepsilon_s$	$-\varepsilon_p$	$-\varepsilon_h = -(\varepsilon_s + 3\varepsilon_p)/4$	$V_1^{a,\,c} = (\varepsilon_p - \varepsilon_s)/4$
			Atomic term value (eV)	
			$2s, 2p$	
Li	5.48	—	—	—
Be	8.17	4.14	5.15	1.01
B	12.54	6.64	8.12	1.47
C	17.52	8.97	11.11	2.13
N	23.04	11.47	14.36	2.88
O	29.14	14.13	17.88	3.76
F	35.80	16.99	21.69	4.71
			$3s, 3p$	
Na	5.13	—	—	—
Mg	6.86	2.99	3.95	0.97
Al	10.11	4.86	6.18	1.33
Si	13.55	6.52	8.27	1.76
P	17.10	8.33	10.52	2.19
S	20.80	10.27	12.90	2.63
Cl	24.63	12.31	15.38	3.08
			$4s, 4p$	
K	4.19	—	—	—
Ca	5.41	—	—	—
Cu	6.92	1.83	3.10	1.27
Zn	8.40	3.38	4.64	1.26
Ga	11.37	4.90	6.52	1.62
Ge	14.38	6.36	8.37	2.01
As	17.33	7.91	10.27	2.36
Se	20.32	9.53	12.23	2.70
Br	23.35	11.20	14.27	3.02

second-neighbor LCAO fit had been used, for example, and it would not therefore be appropriate to use them if the description of the bands were to be extended to second-neighbor interactions.

It will ordinarily be more convenient in solids to use the forms for angular dependence, x/r, y/r, and z/r, as in Eq. (1-20), rather than the forms $Y_1^m(\theta, \varphi)$. Then in order to obtain matrix elements involving these orbitals, we need to expand the

	Atomic term value (eV)			
lement	$-\varepsilon_s$	$-\varepsilon_p$	$-\varepsilon_h = -(\varepsilon_s + 3\varepsilon_p)/4$	$V_1^{a,c} = (\varepsilon_p - \varepsilon_s)/4$
		5s, 5p		
Rb	3.94	—	—	—
Sr	5.00	—	—	—
Ag	6.41	2.05	3.14	1.09
Cd	7.70	3.38	4.46	1.08
In	10.12	4.69	6.05	1.36
Sn	12.50	5.94	7.58	1.64
Sb	14.80	7.24	9.13	1.89
Te	17.11	8.59	10.72	2.13
I	19.42	9.97	12.33	2.36
		6s, 6p		
Cs	3.56	—	—	—
Ba	4.45	—	—	—
Au	6.48	2.38	3.41	1.03
Hg	7.68	3.48	4.53	1.05
Tl	9.92	4.61	5.94	1.33
Pb	12.07	5.77	7.35	1.58
Bi	14.15	6.97	8.77	1.80
Po	16.21	8.19	10.20	2.01
At	18.24	9.44	11.64	2.20
		7s		
Fr	3.40	—	—	—
Ra	4.24	—	—	—

NOTE: These values appear also in the Solid State Table.

p orbital in question in terms of Y_1^m, which are defined with respect to the coordinate system discussed above. For *p* orbitals this is quite simple. For the simplest geometries it leads to the identification of matrix elements shown in the upper four diagrams of Fig. 2-8. For arbitrary geometries the result depends upon the direction cosines giving the vector **d** in the coordinate system of *x*, *y*, and *z*; this is illustrated at the bottom in Fig. 2-8. The corresponding transformations for *d*

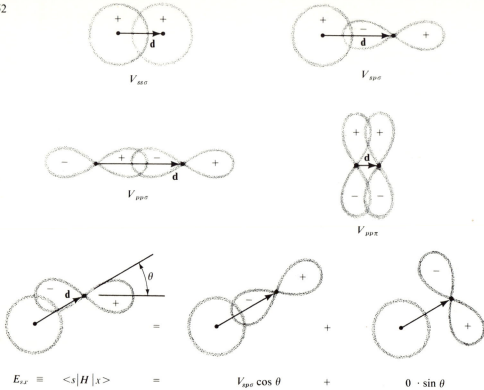

$$E_{sx} \equiv \quad <s|H|x> \quad = \quad V_{sp\sigma}\cos\theta \quad + \quad 0 \cdot \sin\theta$$

FIGURE 2-8

The four types of interatomic matrix elements entering the study of s- and p-bonded systems are chosen as for diatomic molecules as shown in Fig. 1-11. Approximate values for each are obtained from the bond length, or internuclear distance, d, by $V_{ij} = \eta_{ij}\hbar^2/md^2$, with η_{ij} taking values given in Table 2-1 and in the Solid State Table at the back of the book. When p orbitals are not oriented simply as shown in the upper diagrams, they may be decomposed geometrically as vectors in order to evaluate matrix elements as illustrated in the bottom diagrams. It can be seen that the interatomic matrix element at the bottom right consists of cancelling the contributions that lead to a vanishing matrix element.

orbitals as well as p orbitals will be given in detail in Table 20-1, but for s and p orbitals the simple vector transformations illustrated in Fig. 2-8 should be sufficient; the results can be checked with Table 20-1.

When we give the Froyen–Harrison analysis in Chapter 18-A, we shall see that the same procedure can give an estimate of the energy difference $\varepsilon_p - \varepsilon_s$. It is of the correct general magnitude but fails to describe the important trend in the energy bands among the covalent solids C, Si, Ge, and Sn. Furthermore, it does not provide a means of estimating term-value differences such as $\varepsilon_p^c - \varepsilon_p^a$ in polar solids. Thus, for these intra-atomic parameters we shall use calculated atomic term values, which are listed in Table 2-2. A comparison shows them to be roughly consistent with term values obtained in the fit to known bands done by

Chadi and Cohen (1975) for the polar semiconductors as well as for silicon and germanium.

This particular set of calculated values (by Herman and Skillman, 1963) was chosen since the approximations used in the calculation were very similar to those used in determining the energy bands that led to the parameters in Table 2-1. The values would not have differed greatly if they were taken from Hartree-Fock calculations (such values are tabulated in Appendix A). Values based on Hartree-Fock calculations have the advantage of giving good values for d states. Therefore, though the calculations in this book are based upon the Herman-Skillman values, for some applications the Hartree-Fock values may be better suited.

Notice that as absolute numbers the atomic energy values have only limited meaning in any case. Imagine, for example, that the value ε_p for oxygen correctly gives the energy required to remove an electron from an isolated oxygen atom in space. If this atom is brought close to the surface of a metal (or, almost equivalently, to the surface of a covalent solid with a large dielectric constant) but not close enough for any chemical bonding to take place, how much energy is now required to remove the electron from the oxygen? One way to calculate this is to move the neutral atom to infinity, with no work required, remove the electron requiring ε_p, and then return the oxygen ion to its initial position; as it returns it gains an energy $e^2/4d$ from the image field, where d is the final distance from the surface. The resultant correction of ε_p, with d equal to 2 Å, is 1.8 eV, far from negligible. The precise value is uncertain because of the dielectric approximation, the uncertainty in the d used, and other effects, but we may expect that significant corrections of the absolute energies are needed relative to the values in vacuum. The reason that the values are nevertheless useful as parameters is that in solids such corrections are similar for all atoms involved and the relative values are meaningful.

How do the values obtained from Tables 2-1 and 2-2 compare with the values obtained directly by fitting energy bands? This comparison is made in Table 2-3 for the covalent systems studied by Chadi and Cohen. Agreement is semiquantitative throughout and all trends are reproduced except the splitting of values for $V_{sp\sigma}$ in the compounds. The discrepancies are comparable to the differences between different fits (the most recent fits are used here), thus justifying the use of the simple forms in our studies. Significantly different values are obtained if one includes a greater number of matrix elements in the fit (Pandey, 1976) and would be appropriate if we were to include these matrix elements in the calculation of properties other than the bands themselves. Significantly different values have also been given by Levin (1974).

The coefficients from Table 2-1 and atomic term values from Table 2-2 will suffice for calculation of an extraordinarily wide range of properties of covalent and ionic solids using only a standard hand-held calculator. This is impressive testimony to the simplicity of the electronic structure and bonding in these systems. Indeed the same parameters gave a semiquantitative prediction of the one-electron energy levels of diatomic molecules in Table 1-1. However, that theory is intrinsically approximate and not always subject to successive correc-

TABLE 2-3

Matrix elements from the Solid State Table, compared with values (in parentheses) from fits to individual bands. All values are in eV.

Matrix element	C	Si	Ge	GaAs	ZnSe
$-V_{ss\sigma}$	4.50	1.93	1.79	1.79	1.79
	(5.55)	(2.03)	(1.70)	(1.70)	(1.54)
$V_{sp\sigma}$	5.91	2.54	2.36	2.36	2.36
	(5.91)	(2.55)	(2.30)	(2.4, 1.9)	(2.6, 1.4)
$V_{pp\sigma}$	10.41	4.47	4.15	4.15	4.15
	(7.78)	(4.55)	(4.07)	(3.44)	(3.20)
$-V_{pp\pi}$	2.60	1.12	1.04	1.04	1.04
	(2.50)	(1.09)	(1.05)	(0.89)	(0.92)
$(\varepsilon_p - \varepsilon_s)/4$	2.14	1.76	2.01	1.62, 2.36	1.26, 2.70
	(1.7)	(1.80)	(2.10)	(1.61, 2.41)	(1.47, 3.10)
$(\varepsilon_p^{(1)} - \varepsilon_p^{(2)})/2$	0	0	0	1.40	3.08
	(0)	(0)	(0)	(0.96)	(3.29)

SOURCES of data in parentheses: C from Chadi and Martin (1976); Si and Ge from Chadi and Cohen (1975); GaAs and ZnSe from Chadi and Martin (1976).

NOTE: Where two values of $V_{sp\sigma}$ are given for compounds, the first value is for an s state in the nonmetallic atom and p state in the metallic atom. States are reversed for the second value. Where two values of $(\varepsilon_p - \varepsilon_s)/4$ are listed, the first value is for the metallic atom, the second for the nonmetallic atom.

tions and improvements. In most cases our predictions of properties will be accurate on a scale reflected in Table 2-3, and though the introduction of further parameters allows a more accurate *fit* to the data, it may be that improvements at a more fundamental level are required for a more realistic treatment and that these improvements cannot be made without sacrificing the conceptual and computational simplicity of the picture that will be constructed in the course of this book.

Before proceeding to quantitative studies of the covalent solids it is appropriate to comment on the concept of "electronegativity," introduced by Pauling to denote the tendency of atoms to attract electrons to themselves (discussed recently, for example, by Phillips, 1973b, p. 32). It may be an unfortunate term since the positive terminal of a battery has greater electronegativity than the negative terminal. Furthermore, it was defined to be dimensionless rather than to have more natural values in electron volts. It would be tempting to take the hybrid energy values of Table 2-2 as the definition of electronegativity, but it will be seen that in some properties the energy ε_p is a more appropriate measure. Therefore it will be a wiser choice to use the term only qualitatively. Then from Table 2-2 (or from Fig. 1-8) we see that the principal trend is an increase in electronegativity with increasing atomic number proceeding horizontally from one inert gas to the next (e.g., from neon, Na, Mg, Al, Si, P, S, and Cl to argon). In addition, the elements between helium and neon have greater electronegativity than the heavier elements. It is useful to retain "electronegativity" to describe these two qualitative trends.

2-E   Calculation of Spectra

We have seen that in solids, bands of electron energies exist rather than the discrete levels of atoms or molecules. Similarly there are bands of vibration frequencies rather than discrete modes. Thus, to show electron eigenvalues, a curve was given in Fig. 2-2 rather than a table of values. However, a complete specification of the energies within the bands for a three-dimensional solid requires a three-dimensional plot and that cannot be made; even in two dimensions an attempt is of limited use. Instead, a convenient representation of electronic structure can be made by plotting the number of states, per unit energy, as a function of energy. This loses the information about, for example, electron velocity, since that requires a knowledge of energy as a function of wave number. However, it is all that is needed to sum the energies of the electrons for given atomic arrangements.

Calculation of such a spectrum might seem straightforward, but if done by sampling, it requires an inordinate amount of calculation. For example, to produce a plot we might divide the energy region of interest into one thousand intervals and then evaluate the energies (as we did in Section 2-A) over a closely spaced grid in the Brillouin Zone, keeping track of the number of eigenvalues obtained in each interval. A great increase in efficiency can be obtained by noting that the energy bands have the full symmetry of the Brillouin Zone—in the case of CsCl, a cube—so that the entire Brillouin Zone need not be sampled. One could sample one half the zone and multiply the results by two, one eighth and multiply by eight, or in fact, for a cube, one forty-eighth suffices. However, even in a sample of thousands of values, the resulting histogram shows large statistical fluctuations. Therefore an alternative approach is required.

The approach most commonly used, and used extensively in the curves in this book, is the Gilat-Raubenheimer scheme (Raubenheimer and Gilat, 1966). In this scheme, the idea is to replace the true bands by approximate bands, but then to calculate the density of levels for that spectrum accurately. This is done by dividing up the Brillouin Zone, or a forty-eighth of the zone for cubic symmetry, into cells; of the order of fifty may be appropriate; Raubenheimer and Gilat used cells in the shape of cubes. They then fit each band in each cell by a linear expression, $E_k = E_0 + A_1 k_x + A_2 k_y + A_3 k_z$, with $\mathbf{k}$ measured from the center of the cell. Then the energy region of interest for the system is divided into some 1000 energy intervals and the contribution to each of these intervals is accurately and analytically obtained from the linear values of the bands in each cell. This is illustrated for one dimension in Fig. 2-9. We see that the distribution of the approximate bands is obtained exactly. This turns out to eliminate most of the statistical error and to give very good results.

In the Gilat-Raubenheimer scheme it is inconvenient to obtain the necessary values of the gradient of the energy with respect to wave number in each cell, and the cubes do not fit the Brillouin Zone section exactly, so there are problems in calculating the energy at the surface of the section. For this reason Jepsen and Andersen (1971) and later, independently, Lehman and Taut (1972) replaced

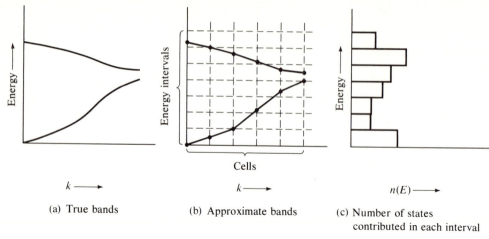

(a) True bands   (b) Approximate bands   (c) Number of states contributed in each interval

FIGURE 2-9

A schematic representation of the Gilat-Raubenheimer scheme for calculating densities of states. The energy bands (a) are replaced by linear bands (b) in each cell. The contribution by each cell to each of a set of small energy intervals (c) is then obtained analytically.

cubes by tetrahedra and wrote the distribution of energies in terms of the values at the four corners. A clear description of this much simpler approach is given by Rath and Freeman (1975), who include the necessary formulae. It is also helpful to see one manner in which the Brillouin Zone can be divided into cells. This is shown in Fig. 2-10. This procedure has been discussed also by Gilat and Bharatiya (1975). Another scheme, utilizing a more accurate approximation to the bands, has been considered recently by Chen (1976).

In some sense this is a computational detail, but the resulting curves are so essential to solid state properties that the detail is important. Once a program has been written for a given Brillouin Zone, any of the spectra for the corresponding structure can be efficiently and accurately obtained from the bands themselves.

PROBLEM 2-1   *Calculating one-dimensional energy bands*

Let us make an elementary calculation of energy bands, using the notation of LCAO theory. For many readers the procedure will be familiar. Consider a ring of N atoms, each with an s orbital. We seek an electronic state in the form of an LCAO,

$$|\psi\rangle = \sum_{\alpha=1}^{N} u_\alpha |\alpha\rangle,$$

where the integers α number the atoms. We can evaluate the expectation value of the energy, considering all atoms to be identical, so $\langle\alpha|H|\alpha\rangle = \varepsilon$ is the same for all α. We can also neglect all matrix elements $\langle\alpha|H|\beta\rangle$, except if α and β differ by one; we write that

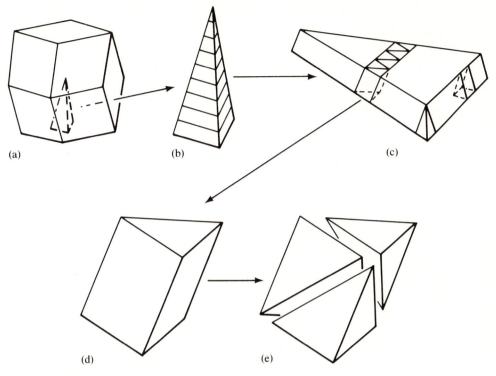

(a) (b) (c)

(d) (e)

FIGURE 2-10

(a) The body-centered-cubic Brillouin Zone is divided into 48 equivalent pyramidal segments. (Two such pyramids are required for face-centered cubic zones.) (b) The pyramid is cut by equally spaced planes parallel to the base. (c) Most of the slab may be subdivided into triangular prisms. An edge is left over on the right which can be divided into triangular prisms with one tetrahedron left over. Each triangular prism (d) may finally be divided into three tetrahedra, (e). This divides the Brillouin Zone entirely into tetrahedra of equal volume. The bands are taken to be linear in wave number within each tetrahedron.

matrix element $-V_2$. We obtain

$$E = \frac{\langle \psi | H | \psi \rangle}{\langle \psi | \psi \rangle} = \frac{\varepsilon \sum_\alpha u_\alpha^* u_\alpha - V_2 \sum_\alpha (u_{\alpha+1}^* u_\alpha + u_{\alpha-1}^* u_\alpha)}{\sum_\alpha u_\alpha^* u_\alpha}.$$

We shall treat the u_α^* as independent of u_α and minimize the expression with respect to u_α^*, giving a linear algebraic equation for each α.

(a) Show that for any integer n there is a solution for all of these equations of the form
$u_\alpha = A e^{2\pi i n \alpha / N}$.

(b) Give the energy as a function of n, and sketch it as a function of n/N for large N. Include positive and negative n.

(c) Obtain the value of A that normalizes the electron state.

(d) Show that for an n outside the range $-N/2 \leq n < N/2$, the electron state obtained is identical to that for some n within this range (within the Brillouin Zone). It suffices to prove that for given n the u_α are unchanged by the addition of N to n.

PROBLEM 2-2   *Electron dynamics*

Consider an electron in a one-dimensional energy band given by $E(k) = -V_2 \cos ka$ in a Brillouin Zone, $-\pi/a < k \leq \pi/a$. At time $t = 0$, with the electron having wave number $k = 0$, apply an electric field $\mathscr{E}$.

Obtain the energy, the speed, and the position of the electron as a function of time. The behavior will be oscillatory. It can be thought of as acceleration of the electron followed by gradual diffraction caused by the lattice.

How many lattice distances (each distance $a = 2$ Å) does the electron go if $V_2 = 2$ eV and the field is 100 volts per centimeter?

PART II

COVALENT SOLIDS

IN THIS PART of the book we shall consider the elements in column 4 of the Solid State Table shown in Fig. 2-7 and at the back of the book. These are said to be covalent solids when they form crystal structures in which the nearest neighbors to each atom are arranged as a regular tetrahedron. The compounds such as AlP or CuBr that straddle that column (except for the few that are not tetrahedrally coordinated) are also said to be covalent. Such compounds have polar bonding but they are nevertheless considered covalent rather than ionic; each bond contains two electrons. These simple tetrahedral solids are semiconductors and are therefore an important technological group. Covalent solids also include mixed tetrahedral solids, solids in which some but not all of the atoms have tetrahedrally coordinated neighbors; SiO_2 is an example. The bonding in covalent solids is simple and well understood, and is a good starting point for our discussion. The preceding chapter described the forest; we now begin an examination of the trees.

That chemical bonds are an aspect of electronic structure became evident with the discovery of the electron by J. J. Thomson in 1897 and with the observation by G. N. Lewis in 1916 that two electrons make up the covalent bond. With the advent of quantum mechanics it became clear that the electrons in covalent solids, as well as in metals, are not localized statically in bonds but are itinerent; each electron state belongs to the entire crystal. Although the existence of the corresponding Bloch electronic states and energy bands was known from about 1928, the actual form of the bands in semiconductors was not known till 1954 when F. Herman, using a combination of crude calculations and experimental information, deduced the essential features of the bands of germanium and silicon. Subsequent calculations and experiments by numerous workers have confirmed these features and now provide very precise knowledge of electronic structure.

As this precise knowledge has evolved, there have simultaneously developed simplified views of electronic structure that are of greater relevance to the chemistry of solids than the precise descriptions are. For example, our discussion of covalent solids will focus on LCAO descriptions of electronic structure, which were made first by G. G. Hall (1952) even before there was any basis on which to judge their reliability. The application of LCAO theory to explain bonding properties was taken up by groups led by Coulson (Coulson, Rèdei, and Stocker, 1962) and Friedel (Leman and Friedel, 1962), and more recently by Harrison (1973c), Harrison and Ciraci (1974), and Lannoo and Decarpigny (1973). An alternative view, based upon nearly free electrons, can be traced back to the description of the Jones Zone given several decades ago by N. F. Mott and H. Jones (Mott and Jones, 1936, 1958). This view has underlain the discussions of covalent solids by J. C. Phillips and J. A. Van Vechten (most completely described by Phillips, 1973) and the pseudopotential theory of covalent bonding given by Harrison (1976c). We shall return to a discussion of that view after formulating the pseudopotential theory for simple metals, and shall make some reference to the findings during discussions of LCAO theory.

The spirit of the LCAO approach, as it was developed by the series of authors listed above, was to make quantitative the concept of the two-electron bond by constructing bond orbitals in analogy with the molecular bonds discussed in Chapter 1. The most complete formulation of this type was the Bond Orbital Model (Harrison, 1973c, and Harrison and Ciraci, 1974), in which matrix elements coupling bond orbitals and antibonding orbitals were neglected so that the broadening of the bond states into valence bands did not change the energy; energies could then be rigorously calculated bond by bond. This procedure allowed an extraordinary simplification of the problem, and made possible the construction of a general theory of covalent bonding. Unfortunately, subsequent applications of the approach and tests of the specific Bond Orbital Approximation (the neglect of coupling between bonding and antibonding states) have indicated that it is frequently *not* a good approximation unless corrections are made; the many successes of the resulting theory were possible only because parameters were taken from experiment, which therefore incorporated corrections that had not been calculated explicitly. What will be undertaken in this book is a full LCAO theory of covalent solids from which a relatively simple bonding theory can be extracted.

The spirit of the approach will be essentially the same, in that bond states are separated from antibonding states, as was seen to be appropriate for covalent solids in Fig. 2-3. This allows the construction of a theory of covalent bonding, based upon an LCAO description; corrections can be included, or parameters properly chosen, to provide the benefits of the Bond Orbital Model, though without the conceptual simplicity afforded by that model. This development requires sufficient understanding of the energy-band structure to know which corrections and which approximations are appropriate for each property. We therefore begin with a study of the energy bands in the covalent solids.

CHAPTER 3

Electronic Structure
of Simple
Tetrahedral Solids

SUMMARY

Bond orbitals are constructed from sp^3 hybrids for the simple covalent tetrahedral structure; energies are written in terms of a covalent energy V_2 and a polar energy V_3. There are matrix elements between bond orbitals that broaden the electron levels into bands. In a preliminary study of the bands for perfect crystals, the energies for all bands at $\mathbf{k} = 0$ are written in terms of matrix elements from the Solid State Table. For calculation of other properties, a Bond Orbital Approximation eliminates the need to find the bands themselves and permits the description of bonds in imperfect and noncrystalline solids. Errors in the Bond Orbital Approximation can be corrected by using perturbation theory to construct extended bond orbitals. Two major trends in covalent bonds over the periodic table, polarity and metallicity, are both defined in terms of parameters from the Solid State Table. This representation of the electronic structure extends to covalent planar and filamentary structures.

In the discussion of the simple molecule ethylene, it was noted that atoms arrange themselves in order to achieve a favorable electronic structure, the electronic structure in turn stabilizing the atomic arrangement. Tetrahedral solids are the result of this kind of cooperation, and electronic structure in tetrahedral solids is determined in a fundamental way by the crystal structure. If we force the atoms into a different arrangement, the electronic structure can be made fundamentally different; for example, putting silicon in a closely packed structure (either by applying pressure or by melting) makes it a metal. As a first step in examining tetrahedral solids, let us specify the tetrahedral crystal structures, the geometry of which can be simplified by choosing notation and coordinate systems carefully.

3-A   Crystal Structures

We shall focus first upon the commonest crystal structure, the *zincblende structure*; the structure of diamond is a special case of this. The zincblende structure is based upon a *face-centered cubic* arrangement of zinc atoms, as shown in Fig. 3-1. This may be thought of as a simple repeating cubic lattice of zinc atoms, with a zinc atom also at the center of every resulting cube face. This makes a natural choice for a Cartesian coordinate system with axes along the cube edges. It is customary to indicate the direction of a vector in the crystal by writing its components along those axes within brackets. Thus a vector [100] lies along a cube edge; [110] lies along a diagonal across a face of the cube; and [111] lies along a diagonal through the cube. Although it is natural to think of the structure in terms of cubic units (the conventional cell or basic repetitive cube), it is important to note that the environment of *every* zinc atom within the arrangement is identical. Thus the crystal may also be thought of as made up of smaller unit cells (the *primitive cells*, each containing one zinc atom). The edges of the primitive cell of zincblende are of the type [110]$a/2$, rather than the type [100]a of the cube (a being the cube edge); vectors such as [011]$a/2$ are considered to be of the type [110]$a/2$.

In the zincblende structure there are an equal number of sulphur atoms, each in a position [111]$a/4$ relative to one of the zinc atoms (these displacements are represented by arrows in Fig. 3-1,a—the four shaded atoms represent sulphur atoms); other sulphur atoms are likewise displaced [111]$a/4$ from the other zinc atoms in the figure, and lie outside of the cube, so are not shown. The zinc atom at the front lower-left corner of the cube and those at the center of the bottom face, left face, and front face form the corners of a regular tetrahedron that has a sulphur atom at its center. Every other sulphur atom in zincblende is also tetrahedrally surrounded by zinc atoms in exactly the same way. Similarly, every

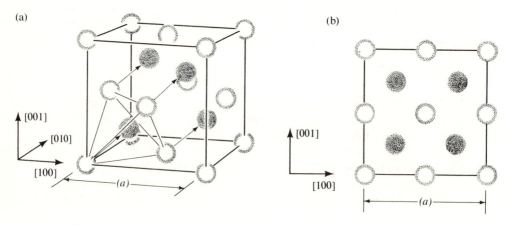

FIGURE 3-1

A unit cube of the zincblende structure (a) rotated and (b) viewed in the [010] direction.

zinc atom is surrounded tetrahedrally by sulphur atoms; these tetrahedra are inverted.

The cube in Fig. 3-1,a is shown in perspective, or slightly rotated, to illustrate atomic positions. In Fig. 3-1,b the cube is rotated back to provide a more convenient diagram for purposes of calculations, since in terms of it the position of each atom can very easily be written down.

The diamond structure is identical to the zincblende structure, the sulphur and zinc atoms being replaced by carbon atoms in the diamond structure: although every atom is of the same element, we may distinguish between two types of sites, depending upon the orientation of the tetrahadron formed by its nearest neighbors. In just this way we distinguished between the sites at the right and left ends of the horizontal bonds in the analogous two-dimensional structure of Fig. 2-3.

Another conception of the zincblende structure may be obtained by noting in Fig. 3-1,a that the zinc atoms lie in planes perpendicular to the [111] axis. If we view one of these planes, we see that within this plane the atoms form a hexagonal array, a few atoms of which are shown in Fig. 3-2 as circles labeled A. Another, parallel layer of zinc atoms beyond this contains atoms at the staggered positions marked B. Notice that were the first layer (the A layer) to be made up of hard spheres, tightly packed so that each touched six neighbors in the plane, the second plane would form a second close-packed layer of spheres, each touching three spheres in the preceding layer. A third layer contains atoms in positions marked C; each sphere C would be in contact with three spheres in the B layer. Next comes another A layer; and the series repeats in ABCABCABC fashion. The staggered positions in A, B, and C layers allow the densest possible lattice packing of hard spheres; thus the face-centered cubic structure is also called a *cubic close-packed structure*. In zincblende, a layer of sulphur atoms is located between each

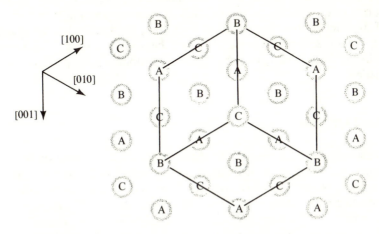

FIGURE 3-2

A zincblende lattice view along a [111] direction. The face-centered cubic lattice looks exactly the same viewed in this way.

layer of zinc atoms with each sulphur atom being directly above a zinc atom of the layer below, three quarters of the way to the next zinc layer; that is, a sulphur layer A' is positioned three quarters of the way from the A to the B zinc layers, etc. It is easy to see that in this configuration each atom is tetrahedrally surrounded by four atoms of the opposite species.

Understanding cubic close-packed structures makes it easier to understand another tetrahedral structure, the **wurtzite structure**. In constructing the close-packed structure we could alternatively have made a pattern ABABABAB, which is equally close-packed; this structure is called the **hexagonal close-packed structure**. Both the hexagonal and the cubic close-packed structures are important metallic structures. In hexagonal structures the spacing between planes is never exactly equal to the *ideal* value associated with hard spheres. This is because the ideal ratio does not lead to higher symmetry in hexagonal structures the way it does in the cubic close-packed structure. We may also construct a tetrahedral structure from the hexagonal close-packed structure by inserting a second atom species directly above each of the close-packed planes, just as we inserted sulphur atoms to make the zincblende structure. In wurtzite, as in zincblende, every atom is tetrahedrally surrounded by atoms of the opposite species. The wurtzite structure tends to occur in semiconductors of great polarity.

In connection with representation of tetrahedral structures as "layers" of atoms, notice that within the close-packed structure a **stacking fault** may exist—that is, an error in the stacking that does not disrupt the closeness of the packing. For example, the arrangement ABCABABCABC contains a stacking fault between the fifth and sixth layers. In such a structure, another kind of atom can be inserted where the stacking is at fault (for example, sulphur atoms can be added to a zinc structure) to make a tetrahedral structure, and every atom will still be tetrahedrally surrounded. Such faults in the stable structure increase the energy; however, it would be possible to have a metastable structure based upon quite random ordering of planes, ABACBCACB . . . , corresponding to a very simple form of disordered structure, without disrupting the tetrahedral arrangements. True amorphous semiconductors, which will be described later, have disorder in all three dimensions, which inevitably leads to disruption of the perfect tetrahedral coordination.

3-B   Bond Orbitals

A basic assumption made earlier in the discussion of diatomic molecules and crystalline CsCl is that electronic states can be written as linear combinations of atomic orbitals. We do not need to depart from that assumption now as we begin to describe electronic states in solids as linear combinations of bond orbitals, since bond orbitals can be written as linear combinations of atomic orbitals, and vice versa. Bond orbitals and atomic orbitals are equivalent representations, but thinking in terms of bond and antibonding orbitals, which can be made to correspond with occupied and empty states of the covalent solid (as was shown in Fig. 2-3), is essential in making approximations.

Let us use a process similar to finding solutions to ordinary simultaneous equations, in which variables are changed by taking linear combinations such that the new variables form sets that are not coupled to each other. Then, smaller sets of equations can be solved, and finally the equations can be reduced to coupled pairs that can be solved by using the formula for the roots of a quadratic equation. In dealing with bond orbitals the procedure leads toward an additional goal: the proper linear combinations will provide a context in which important approximations can be made. Our procedure will be to choose linear combinations first and then to consider matrix elements of the Hamiltonian between them, since these matrix elements, as in Eq. (1-26), are the coefficients that are needed in the calculation of bonding states and bonding energies. The series of steps is diagrammed in Fig. 3-3.

Construction of sp^3 Hybrids

A linear combination of two p orbitals is simply another p orbital with a different orientation (see Fig. 2-8). However, a linear combination of a p orbital and an s orbital (called an *sp hybrid*) corresponds to an unsymmetric probability distribution, a distribution of electron charge that "leans" in the direction of the axis of the p orbital (see Fig. 3-3). In tetrahedral solids we pick four sp hybrids, the wave functions of which are orthogonal to each other, and in which the electron charge density is greatest in the direction of the nearest neighboring atoms. This choice is completely natural and has been made in many chemical studies (see Coulson, 1970, pp. 288–437, for a recent review and historical background; natural choices for other configurations of atoms and for hybrid d orbitals are given in Coulson, 1970, p. 351). We write the normalized wave functions for p orbitals oriented along the three coordinate axes as $|p_x\rangle$, $|p_y\rangle$, and $|p_z\rangle$. In terms of these and wave functions for s orbitals, $|s\rangle$, the sp^3 hybrids (3 indicates there is three times as much probability of finding an electron in a p state as of finding it in an s state) and the directions in which the charge density is greatest are

$$\left.\begin{aligned}
|h_1\rangle &= \tfrac{1}{2}[|s\rangle + |p_x\rangle + |p_y\rangle + |p_z\rangle] \text{ with } [111] \text{ orientation}\\
|h_2\rangle &= \tfrac{1}{2}[|s\rangle + |p_x\rangle - |p_y\rangle - |p_z\rangle] \quad ,, \quad [1\bar{1}\bar{1}] \quad\quad ,,\\
|h_3\rangle &= \tfrac{1}{2}[|s\rangle - |p_x\rangle + |p_y\rangle - |p_z\rangle] \quad ,, \quad [\bar{1}1\bar{1}] \quad\quad ,,\\
|h_4\rangle &= \tfrac{1}{2}[|s\rangle - |p_x\rangle - |p_y\rangle + |p_z\rangle] \quad ,, \quad [\bar{1}\bar{1}1] \quad\quad ,,
\end{aligned}\right\} \tag{3-1}$$

(In crystallographic notation it is customary to place a bar over lattice index numbers to indicate negative direction relative to the origin of the axes.)

It is easy to show that hybrid wave functions are orthonormal, that is,

$$\langle h_i | h_j \rangle = \delta_{ij}, \tag{3-2}$$

by using the condition that nonhybrid atomic wave functions are orthonormal. Any rigid rotation of this set will give another orthonormal set, oriented in the direction of a rotated tetrahedron.

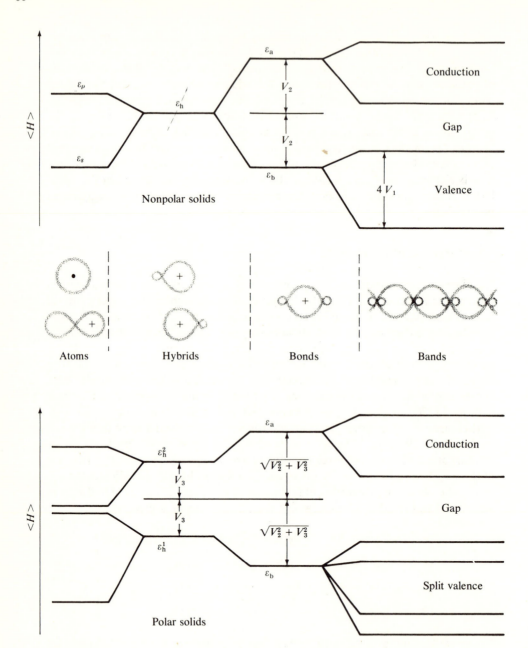

FIGURE 3-3

Successive transformations of linear combinations of atomic orbitals, beginning with atomic s and p orbitals and proceeding to sp^3 hybrids, to bond orbitals, and finally to band states. The band states represent exact solution of the LCAO problem.

These hybrids are not energy eigenstates; the expectation value of the energy (from Eq. 1-3) is an average energy called the **hybrid energy**,

$$\varepsilon_h = (\varepsilon_s + 3\varepsilon_p)/4, \tag{3-3}$$

where we have written $\varepsilon_s = \langle s|H|s \rangle$ and $\varepsilon_p = \langle p_x|H|p_x \rangle = \langle p_y|H|p_y \rangle = \langle p_z|H|p_z \rangle$; these values could differ somewhat in the solid from the corresponding values defined for the free atom but we will use free-atom values. The measurement of the energy of an electron described by such a hybrid wave function would give ε_s 25 percent of the time and ε_p 75 percent of the time.

For polar covalent solids, sp^3 hybrids of the form of Eq. (3-1) can be constructed on each of the atom types present and oriented in the directions of its nearest neighbors. The hybrid energies will be different; we call the lesser energy ε_h^1 and the greater energy ε_h^2 and, in direct analogy with Eq. (1-32), define a **hybrid polar energy** proportional to their difference:

$$V_3^h = (\varepsilon_h^2 - \varepsilon_h^1)/2. \tag{3-4}$$

For any compound, the hybrid polar energy can be estimated by using the term values given in Table 2-2 or the Solid State Table.

Inasmuch as the hybrids are not eigenstates, nonzero matrix elements of the Hamiltonian can exist between the hybrids; they may be evaluated by setting all matrix elements of the Hamiltonian between orthogonal *atomic* orbitals on the same atom equal to zero; in the applications considered in this book the symmetry will always be great enough to make the matrix elements between the atomic orbitals zero. The magnitude of the resulting matrix element is called V_1, the **metallic energy**. We obtain

$$\langle h_i|H|h_j \rangle = -1/4(\varepsilon_p - \varepsilon_s) = -V_1. \tag{3-5}$$

The metallic energy depends upon the sp splitting (that is, $\varepsilon_p - \varepsilon_s$) of the atom shared by the hybrids, and it will have different values for the two atom types for polar crystals.

A particularly important matrix element is that between hybrids pointed at each other, or into the bond, from two neighboring atoms. We call the magnitude of this matrix element the **hybrid covalent energy**,

$$V_2^h = -\langle h^1|H|h^2 \rangle$$
$$= (-V_{ss\sigma} + 2\sqrt{3}V_{sp\sigma} + 3V_{pp\sigma})/4. \tag{3-6}$$

Use of Table 2-1 gives a value of V_2^h of 4.37 $\hbar^2/md^2$.

We must now make a sharp distinction between the concepts we are using and the parameters of the theory. The V_2^h and V_3^h defined in the preceding paragraphs for the hybrid bond are the direct counterparts of covalent and polar energies introduced in discussing bonds in Chapter 1. Hybrid covalent and hybrid polar energies also give rise to bonding–antibonding splitting and to bond polarity. Because of the interactions between neighboring bonds, however, the energy difference that corresponds to bonding–antibonding splitting in the analysis of dielectric properties of solids is not $2(V_2^{h2} + V_3^{h2})^{1/2}$, but an expression depending principally upon the p orbitals. How this happens will become clear as we proceed with the LCAO theory. Although the concept of covalent energy remains valid and is extremely useful through this process, in the end, the parameters that are of greatest interest do not have the values given by Eqs. (3-4) and (3-6).

As a solution to this dilemma, let us retain the formulation of covalent and polar energy but drop the superscript h; then we can continue to associate V_2 and V_3 with the expressions defined by Eqs. (3-4) and (3-6) as we perform a series of rigorous transformations. Later in the book it will become apparent that some of the most important formulae can also be obtained directly by using a simplifying Bond Orbital Approximation in which the covalent and polar energies V_2 and V_3 are merely assigned different values, values that we can write in terms of the parameters from the Solid State Table, just as we have written expressions for V_2^h and V_3^h. For now, we may think of V_2 and V_3 as covalent and polar energies that take either the values given by Eqs. (3-4) and (3-6) or other values, depending on the property.

Construction of Bond Orbitals

At this point we switch from consideration of pairs of hybrid orbitals in each bond to bonding and antibonding combinations in each bond. We do this in the way discussed for a simple polar bond in diatomic molecules (Section 1-D) by writing a general linear combination of the two hybrids,

$$|\psi\rangle = u_1|h^1\rangle + u_2|h^2\rangle, \tag{3-7}$$

and minimizing the energy expectation value,

$$\frac{\langle\psi|H|\psi\rangle}{\langle\psi|\psi\rangle} = \frac{u_1^2\varepsilon_h^1 - 2u_1u_2V_2 + u_2^2\varepsilon_h^2}{u_1^2 + u_2^2}, \tag{3-8}$$

with respect to variations of u_1 and u_2.

Here, as in the discussion of simple molecules, we consider the two hybrids that make up the bond to be orthogonal, that is, $\langle h^1|h^2\rangle = 0$, so that there is no term in u_1u_2 in the denominator. We shall see in Appendix B that this is permissible since the two effects that nonorthogonality has can be absorbed in a redefinition of the covalent and polar energies and in a contribution to the overlap interaction between atoms, which is discussed in Chapter 7. Similarly, we have neglected the more complicated aspects of electron–electron interaction, treating it instead in terms of an average potential associated with each electron. The neglect of "electron correlation," which is what these aspects of electron–electron interaction are called, has been studied closely (Huang, Moriarty, Sher, and Breckenridge, 1975, and Huang, Moriarty, and Sher, 1976); the conclusion is that much of the effect of electron correlation can be absorbed within the parameters of the simple theory.

The minimization leads to two equations,

$$\varepsilon_h^1 u_1 - V_2 u_2 = E u_1;$$
$$-V_2 u_1 + \varepsilon_h^2 u_2 = E u_2. \tag{3-9}$$

It is convenient to rewrite these in terms of the average hybrid energy,

$$\bar{\varepsilon} = (\varepsilon_h^1 + \varepsilon_h^2)/2, \tag{3-10}$$

and the polar energy of Eq. (3-4). This gives

$$-V_3 u_1 - V_2 u_2 = (E - \bar{\varepsilon}) u_1,$$
$$-V_2 u_1 + V_3 u_2 = (E - \bar{\varepsilon}) u_2, \tag{3-11}$$

which may be solved immediately to obtain the bond energy,

$$\varepsilon_b = \bar{\varepsilon} - \sqrt{V_2^2 + V_3^2}, \tag{3-12}$$

and values of u_1 and u_2 of the bond orbital,

$$u_1^b = \left(\frac{1 + \alpha_p}{2}\right)^{1/2}, \qquad u_2^b = \left(\frac{1 - \alpha_p}{2}\right)^{1/2}. \tag{3-13}$$

These have been chosen to satisfy the normalization $u_1^2 + u_2^2 = 1$. The term α_p is the bond polarity, defined as in Eq. (1-37) as

$$\alpha_p = V_3/(V_2^2 + V_3^2)^{1/2}. \tag{3-14}$$

Similarly, the antibonding state, wave functions of which are orthogonal to those of the bonding state, has energy

$$\varepsilon_a = \bar{\varepsilon} + \sqrt{V_2^2 + V_3^2} \tag{3-15}$$

with parameters

$$u_1^a = \left(\frac{1 - \alpha_p}{2}\right)^{1/2}, \qquad u_2^a = -\left(\frac{1 + \alpha_p}{2}\right)^{1/2}. \tag{3-16}$$

These transformations have enabled us to construct bonding and antibonding orbitals from our starting s and p orbitals without any additional approximations.

The energy expectation values for these orbitals, and the orbitals themselves, depend only upon the covalent energy V_2 and the polar energy V_3, as indicated in Fig. 3-3. These orbitals do not represent eigenstates of the Hamiltonian, however; there are still matrix elements between the various bonding and antibonding orbitals. By solving the equations for eigenstates, we are in effect broadening the bonding and antibonding energy levels into bands, as indicated on the far right in Fig. 3-3 and explained in Chapter 2. In fact, if we were to solve the equations exactly, the transformations we have made would gain us nothing, for we would still need to solve just as many simultaneous equations as before. We will not solve the equations exactly, however, and a representation based upon bond orbitals will be an invaluable guide both for estimating the parameters that describe the electronic structure and for extracting physical information from that representation of the electronic structure.

Construction Viewed Mathematically

It will be helpful to have a purely mathematical description of the constructions we have made. The problem of finding electronic eigenstates $|k\rangle$ can be solved by approximating each eigenstate as a linear combination of atomic orbitals $|\alpha\rangle$:

$$|k\rangle = \sum_{\alpha} u_{k\alpha}|\alpha\rangle. \tag{3-17}$$

The solution, based on the variational method, leads to a set of simultaneous algebraic equations, like what was found in Eq. (1-26) for diatomic molecules:

$$\sum_{\alpha} \langle \beta | H | \alpha \rangle u_{k\alpha} - E_k u_{k\beta} = 0. \tag{3-18}$$

For a solid there would be as many equations as there are atomic orbitals, or for our simple LCAO description, four times as many equations as there are atoms in the LCAO description.

The solution of these equations proceeds by mathematical diagonalization of the Hamiltonian matrix $\langle \beta | H | \alpha \rangle$, by use of the *unitary transformation* $U^{-1}HU$, where the elements of the unitary matrix U are $u_{k\alpha}$. The transformations made in constructing bond orbitals correspond to a series of unitary transformations leading to a new form of the Hamiltonian matrix, again not a precisely diagonal form (in a strict diagonal form, all *off-diagonal matrix elements* would be zero), which has exactly the same eigenvalues as the starting matrix. The new Hamiltonian is huge, having some 10^{23} columns and rows. However, it can be shown schematically as in Fig. 3-4. Each row and each column corresponds to the bond orbital or the antibonding orbital at one bond site. In Fig. 3-4, bond orbitals are given first, in the rows at the top and the columns at the left. Most of the off-diagonal matrix elements are zero, since matrix elements between bonds separated by more than a few atoms are negligible. However, those that are nonvanishing

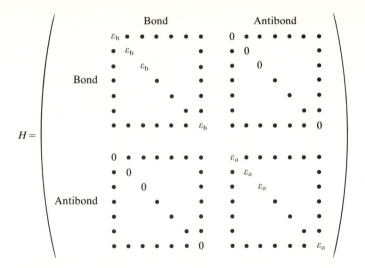

FIGURE 3-4

The form of the LCAO Hamiltonian matrix after
transformation to bond orbitals and antibonding orbitals. Each
diagonal element is either ε_b or ε_a. Solution of the variational
calculation in each bond has reduced the off-diagonal matrix
elements between bonding and antibonding orbitals in the *same*
bond site to zero, but matrix elements between neighboring
bonding and antibonding orbitals are nonzero.

may be classified as indicated in the figure. Our approximations will correspond
to various ways of treating the off-diagonal matrix elements.

Some of the discussion of bonding theory will concern distorted crystals or
crystals with defects; then description in terms of bond orbitals will be essential.
Description of electronic states is relatively simple for a *perfect* crystalline solid, as
was shown for CsCl in Chapter 2; for these, use of bond orbitals is not essential
and in fact, in the end, is an inconvenience. We shall nevertheless base the formulation of energy bands in crystalline solids on bond orbitals, because this formulation will be needed in other discussions; at the point where matrix elements must
be dealt with, we shall use the LCAO basis. The detailed discussion of bands in
Chapter 6 is done by returning to the bonding and antibonding basis.

3-C The LCAO Bands

Bloch Sums of Bond Orbitals

Energy bands are accurately known for most tetrahedral semiconductors;
Chapter 6 contains a discussion of the details of these bands as obtained by more
accurate methods than will be discussed here. In this section we shall continue the
construction of electron states in terms of the bond orbitals introduced in the

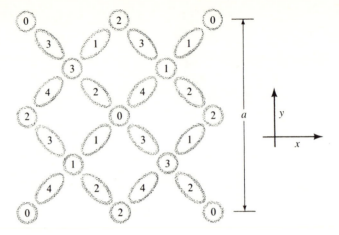

FIGURE 3-5

The numbers within the ovals represent the numbering of bond orbitals in a zincblende lattice. The small circles represent atoms; the numbers within them represent the distance below the plane of the figure in units of $a/4$.

previous section, following a procedure similar to that used for CsCl in Chapter 2. There we had a particularly simple system of identical orbitals (initially chlorine s orbitals) on a simple cubic lattice; each orbital was separated from every other by integral multiples of lattice vectors of length a in the directions of cube edges. In that case the correct linear combination of orbitals could be written immediately as $N_p^{-1/2}\sum_i e^{i\mathbf{k}\cdot\mathbf{r}_i}|s_i\rangle$, where $\mathbf{k}$ is the wave number characterizing the state and $|s_i\rangle$ is the atomic orbital at the site $\mathbf{r}_i$. Now we have four distinct types of bond orbitals, as indicated in Fig. 3-5 and in analogy with the three different types of p orbitals in CsCl. We also have four distinct types of antibonding orbitals. In constructing a band state as a linear combination of the eight different kinds of basis orbitals, it remains true that the coefficients for any two orbitals of the same type are related by a factor $e^{i\mathbf{k}\cdot(\mathbf{r}_i-\mathbf{r}_j)}$, where $\mathbf{k}$ is again characteristic of the state and $\mathbf{r}_i$ and $\mathbf{r}_j$ are the sites of the two orbitals. (This can be shown by using group theory and the translational symmetry of the lattice.) Symmetry alone does not determine relative values of coefficients for the eight different orbital types. However, knowledge of the translational symmetry can reduce the size of the Hamiltonian matrix shown schematically in Fig. 3-4 from 10^{23} by 10^{23} to 8 by 8, and that is the difference between an insoluble and a soluble problem.

A wave number must again be associated with every state; for each wave number we construct eight different **Bloch sums**, four for the bond orbital types,

$$|\chi_\alpha(\mathbf{k})\rangle = \sum_i e^{i\mathbf{k}\cdot\mathbf{r}_i}|b_\alpha(\mathbf{r}-\mathbf{r}_i)\rangle/\sqrt{N_p}, \qquad (3\text{-}19)$$

and four for the antibonding orbital types. The value N_p is the number of atom pairs, equal to the number of bonds of each type. The position $\mathbf{r}_i$ is taken to be the midpoint of the corresponding bond. Each eigenstate of wave number $\mathbf{k}$ can be written as a linear combination of the eight Bloch sums for that wave number:

$$|\psi_{\mathbf{k}}\rangle = \sum_{\alpha} u_{\alpha} |\chi_{\alpha}(\mathbf{k})\rangle. \tag{3-20}$$

In contrast, in CsCl we sufficiently simplified the matrix elements that we were able to use single Bloch sums, of the kind in Eq. (3-19), as eigenstates; we cannot do that here.

As in CsCl, the wave number is restricted so that it lies in a Brillouin Zone, but the shape of that zone is different for a zincblende or face-centered cubic crystal structure. Equivalent bond orbitals are separated by *primitive translations* of $[011]a/2$, $[101]a/2$, $[110]a/2$, or integral multiples of primitive translations. (This crystallographic notation was defined in Section 3-A.) Thus the addition of *primitive lattice wave numbers* $[111]2\pi/a$, $[1\bar{1}\bar{1}]2\pi/a$, $[\bar{1}\bar{1}1]2\pi/a$, or integral multiples of them, to the **k** appearing in Eq. (3-19), multiplies the entire sum by the same phase factor. (This is easily verified.) This is not an independent Bloch sum; the corresponding wave numbers are said to be *equivalent*, and we need not consider any wave number that has an equivalent wave number of smaller magnitude. The Brillouin Zone shown in Fig. 3-6 is that of a zincblende lattice (or that of a face-centered cubic lattice), and is the one we shall be using. A full treatment of the topology of wave number space for this system is not possible in a single paragraph. The reader who is not familiar with this topology will find fuller accounts in any solid state physics text, such as Kittel (1976).

For any wave number in the Brillouin Zone we seek the lowest energy solution by minimizing $\langle \psi_{\mathbf{k}} | H | \psi_{\mathbf{k}} \rangle / \langle \psi_{\mathbf{k}} | \psi_{\mathbf{k}} \rangle$ with respect to u_i, in direct analogy to the minimization of energy for the system based upon two basis states, as taken up in

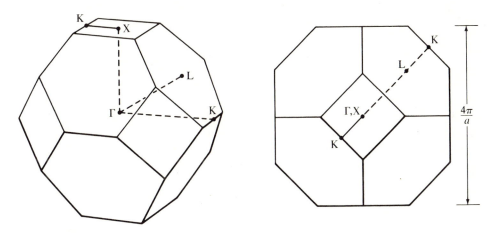

FIGURE 3-6

The Brillouin Zone and symmetry points for the zincblende (diamond, or face-centered cubic) lattice. The view at the right is along a [110] axis; that at the left has been tilted.

Chapter 1. The equations analogous to Eq. (1-10) are

$$\sum_{\beta} H_{\alpha\beta}(\mathbf{k})u_{\beta} = E_{\mathbf{k}}u_{\alpha}, \tag{3-21}$$

where the matrix elements and eigenvalues depend upon $\mathbf{k}$; that is,

$$H_{\alpha\beta} = \langle \chi_{\alpha}(\mathbf{k})|H|\chi_{\beta}(\mathbf{k})\rangle. \tag{3-22}$$

The eight solutions to Eqs. (3-21) are called the *eigenstates* of the matrix $H_{\alpha\beta}$ and are the electron states in the approximation that each eigenstate can be written in the form of Eq. (3-20).

A Return to Atomic Orbitals

At this stage we must evaluate the matrix elements and then solve Eqs. (3-21) to obtain eigenstates and energy bands. However, for our method of determining interbond matrix elements from the interatomic matrix elements of Table 2-1, the use of bonding orbitals and antibonding orbitals becomes inconvenient. For example, in Fig. 3-5, notice that we seek matrix elements between Bloch sums based upon bond orbitals of type 1 and type 2. We may focus on a particular bond orbital of type 2 and see that there are matrix elements between it and bond orbitals in adjacent bonds, and also matrix elements between it and bond orbitals in second-neighbor bonds arising from hybrids on the two atoms between, which are nearest-neighbor atoms. Thus we lose the simplicity provided by nearest-neighbor matrix elements because of our choice of basis. This will nevertheless be convenient in Chapter 7 when we seek approximate solutions. For now, we seek an exact solution, and for this it is much better to follow Chadi and Cohen (1975) in using atomic orbitals as the basis. This is obviously equivalent since the bond orbitals have been written as sums of atomic orbitals. The result of a transformation from the eight types of bond orbitals and antibonding orbitals to the six *p* orbitals and two *s* orbitals on the two atom types is a set of eight simultaneous equations of the form of Eq. (3-21), but with the Bloch sums of Eq. (3-19) replaced by sums over equivalent atoms (sums over cations or anions) at positions $\mathbf{r}_i$ of atomic *s* orbitals, or sums over one of the orientations of *p* orbitals on equivalent atoms. It will be more convenient to obtain matrix elements between such Bloch sums, and the diagonalization of the corresponding matrix will give exactly the same bands as the diagonalization of the equivalent matrix based upon bonding and antibonding orbitals. Making this change also bypasses the ambiguity we discussed concerning covalent and polar energies, since all matrix elements can now be written explicitly in terms of matrix elements between atomic orbitals.

We redraw some of the bond orbitals of Fig. 3-5 to show atomic orbitals, orienting them along the cube directions. The set of five atoms in the upper right corner of Fig. 3-5 will suffice; the corresponding orbitals are sketched in Fig. 3-7. Imagine the central atom to be metallic.

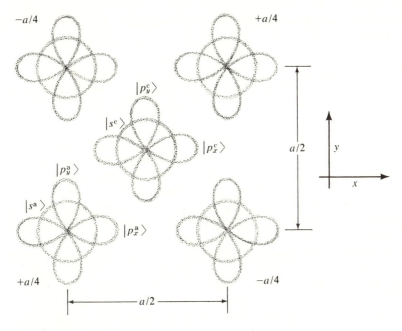

-a/4   +a/4

$|p_y^c\rangle$

$|s^c\rangle$

$|p_x^c\rangle$

$|p_y^a\rangle$

$|s^a\rangle$

$|p_x^a\rangle$

a/2   y

x

+a/4   -a/4

a/2

FIGURE 3-7

Numbering of atomic orbitals in a zincblende lattice. The central atom
is imagined to be a metallic atom, a cation; then the four neighbors are
anions. The central atom is imagined to be in the plane of the figure,
the plane $z = 0$. The z-coordinate of each anion is indicated as $\pm a/4$.
Orbitals $|p_z\rangle$ are not shown.

The Hamiltonian Matrix

Consider first a matrix element $H_{\alpha\alpha}(\mathbf{k})$, from Eq. (3-22), where both Bloch sums
are taken over a p orbital on the metallic atom, oriented in the x-direction. In
zincblende, since all nearest neighbors are nonmetallic atoms, no nearest-neighbor
matrix elements enter the calculation; the matrix element $H_{p_x^c p_x^c}(\mathbf{k})$ of Eq. (3-22)
becomes simply a sum over N_p p orbitals oriented in the x-direction, divided by
N_p, which is simply ε_p^c, with c denoting "cation." Similarly, every other diagonal
matrix element is simply the corresponding atomic term value. Furthermore, no
off-diagonal matrix elements $H_{\alpha\beta}(\mathbf{k})$ couple Bloch sums of cation states with those
of other cation states, or Bloch sums of anion states with those of other anion
states.

Let us now look at $H_{\alpha\beta}(\mathbf{k})$, from Eq. (3-22), where α corresponds to a cation s
orbital and β to an anion s orbital. Each orbital $\langle s^c|$ in the sum making up the
state $\langle \chi_\alpha(\mathbf{k})|$ will have matrix elements $V_{ss\sigma}$ between it and the four neighboring
orbitals $|s^a\rangle$, indicated in Fig. 3-7; each of these neighboring orbitals enters the
matrix element $V_{ss\sigma}$ with a phase factor differing from that of the $|s^c\rangle$ orbital by

$e^{i\mathbf{k}\cdot\mathbf{d}_i}$, with $\mathbf{d}_i$ being the vector distance to that neighbor. The two factors of $N_p^{-1/2}$ are cancelled by the number of terms in the χ_α sum, leading to a matrix element

$$H_{s^c s^a}(\mathbf{k}) = V_{ss\sigma} \sum_i e^{i\mathbf{k}\cdot\mathbf{d}_i}$$

$$= V_{ss\sigma}[e^{i(k_x+k_y+k_z)a/4} + e^{i(k_x-k_y-k_z)a/4} + e^{i(-k_x+k_y-k_z)a/4} + e^{i(-k_x-k_y+k_z)a/4}].$$

$$(3\text{-}23)$$

The matrix element $H_{s^a s^c}(\mathbf{k})$ is easily seen to be the complex conjugate of this, and in fact $H_{\alpha\beta}(\mathbf{k}) = H_{\beta\alpha}(\mathbf{k})^*$ in general; that is, the matrix is Hermitian.

Similarly, the matrix element $H_{p_x^c s^a}(\mathbf{k})$ can be seen to consist of four terms, and each of the four terms is obtained by decomposing the orbital p_x^c into σ and π components for each one of the four neighbors; between the π component and the neighboring s orbital the matrix element is zero but between the σ component and the neighboring s orbital the matrix element is $\pm V_{sp\sigma}/3^{1/2}$. The $3^{-1/2}$ is the coefficient in the decomposition; and the sign of the matrix element depends upon whether the s orbital lies in the direction of the positive or negative lobe of the p orbital. This is sufficient to show how the matrix elements are evaluated.

We now simply give the results obtained by Chadi and Cohen (1975) for each matrix element. The neighbors to the cation (designated with subscript numerals) and their vector distances are given as

$$\begin{aligned} \mathbf{d}_1 &= [111]a/4; \\ \mathbf{d}_2 &= [1\bar{1}\bar{1}]a/4; \\ \mathbf{d}_3 &= [\bar{1}1\bar{1}]a/4; \\ \mathbf{d}_4 &= [\bar{1}\bar{1}1]a/4. \end{aligned} \right\} \qquad (3\text{-}24)$$

Then four sums of phase factors are defined by

$$\begin{aligned} g_0(\mathbf{k}) &= e^{i\mathbf{k}\cdot\mathbf{d}_1} + e^{i\mathbf{k}\cdot\mathbf{d}_2} + e^{i\mathbf{k}\cdot\mathbf{d}_3} + e^{i\mathbf{k}\cdot\mathbf{d}_4}; \\ g_1(\mathbf{k}) &= e^{i\mathbf{k}\cdot\mathbf{d}_1} + e^{i\mathbf{k}\cdot\mathbf{d}_2} - e^{i\mathbf{k}\cdot\mathbf{d}_3} - e^{i\mathbf{k}\cdot\mathbf{d}_4}; \\ g_2(\mathbf{k}) &= e^{i\mathbf{k}\cdot\mathbf{d}_1} - e^{i\mathbf{k}\cdot\mathbf{d}_2} + e^{i\mathbf{k}\cdot\mathbf{d}_3} - e^{i\mathbf{k}\cdot\mathbf{d}_4}; \\ g_3(\mathbf{k}) &= e^{i\mathbf{k}\cdot\mathbf{d}_1} - e^{i\mathbf{k}\cdot\mathbf{d}_2} - e^{i\mathbf{k}\cdot\mathbf{d}_3} + e^{i\mathbf{k}\cdot\mathbf{d}_4}. \end{aligned} \right\} \qquad (3\text{-}25)$$

Finally, composite matrix elements are defined by

$$\begin{aligned} E_{ss} &= V_{ss\sigma}; \\ E_{sp} &= -V_{sp\sigma}/3^{1/2}; \\ E_{xx} &= 1/3V_{pp\sigma} + 2/3V_{pp\pi}; \\ E_{xy} &= 1/3V_{pp\sigma} - 1/3V_{pp\pi}. \end{aligned} \right\} \qquad (3\text{-}26)$$

TABLE 3-1

The LCAO Hamiltonian matrix for the zincblende structure. Parameters are defined in Eqs. (3-25) and (3-26).

	s^c	s^a	p_x^c	p_y^c	p_z^c	p_x^a	p_y^a	p_z^a
s^c	ε_s^c	$E_{ss}g_0$	0	0	0	$E_{sp}g_1$	$E_{sp}g_2$	$E_{sp}g_3$
s^a	$E_{ss}g_0^*$	ε_s^a	$-E_{sp}g_1^*$	$-E_{sp}g_2^*$	$-E_{sp}g_3^*$	0	0	0
p_x^c	0	$-E_{sp}g_1$	ε_p^c	0	0	$E_{xx}g_0$	$E_{xy}g_3$	$E_{xy}g_1$
p_y^c	0	$-E_{sp}g_2$	0	ε_p^c	0	$E_{xy}g_3$	$E_{xx}g_0$	$E_{xy}g_1$
p_z^c	0	$-E_{sp}g_3$	0	0	ε_p^c	$E_{xy}g_1$	$E_{xy}g_2$	$E_{xx}g_0$
p_x^a	$E_{sp}g_1^*$	0	$E_{xx}g_0^*$	$E_{xy}g_3^*$	$E_{xy}g_1^*$	ε_p^a	0	0
p_y^a	$E_{sp}g_2^*$	0	$E_{xy}g_3^*$	$E_{xx}g_0^*$	$E_{xy}g_2^*$	0	ε_p^a	0
p_z^a	$E_{sp}g_3^*$	0	$E_{xy}g_1^*$	$E_{xy}g_1^*$	$E_{xx}g_0^*$	0	0	ε_p^a

SOURCE: Obtained from form given by Chadi and Cohen (1975).

(In Eqs. 3-25 and 3-26 our definitions differ by a factor of four from those used by Chadi and Cohen, but the final matrix is equivalent, except that we do not distinguish two types of E_{sp}.) Now the Hamiltonian matrix can be written as in Table 3-1.

The Energy Bands

All parameters have now been specified in detail; energy bands can be obtained as a function of $\mathbf{k}$ by diagonalizing the matrix in Table 3-1 for each $\mathbf{k}$. For arbitrary wave numbers this would need to be accomplished numerically, but at special wave numbers or for wave numbers along symmetry lines in the Brillouin Zone it can be accomplished analytically. Let us diagonalize the matrix analytically for the point Γ at the center of the Brillouin Zone, $\mathbf{k} = 0$. At $\mathbf{k} = 0$, $g_1 = g_2 = g_3 = 0$ and $g_0 = 4$. Thus all off-diagonal matrix elements in Table 3-1 vanish except those coupling s^c with s^a, those coupling p_x^c and p_x^a, and so on. The solution for each coupled pair of Bloch sums is immediate, leading to energies of

$$
\left.\begin{aligned}
E &= \frac{\varepsilon_s^c + \varepsilon_s^a}{2} \pm \sqrt{\left(\frac{\varepsilon_s^c - \varepsilon_s^a}{2}\right)^2 + (4E_{ss})^2}\,; \\
E &= \frac{\varepsilon_p^c + \varepsilon_p^a}{2} \pm \sqrt{\left(\frac{\varepsilon_p^c - \varepsilon_p^a}{2}\right)^2 + (4E_{xx})^2}.
\end{aligned}\right\}
$$

(3-27)

The energy in the second equation is triply degenerate, giving one state for p_x, one for p_y, and one for p_z orbitals. The expressions in Eqs. (3-27) will prove extremely important in the discussions that follow.

Chadi and Cohen (1975) gave expressions for the values at other symmetry points and along some symmetry lines; they are analogous in form to those at Eqs. (3-27); some are listed in Chapter 6. Chadi and Cohen then adjusted the matrix elements E_{ss}, and so on, to fit the bands of diamond, silicon, germanium, and a few compounds. (In part of their study a second-neighbor interatomic matrix element was also included; in this book that matrix element is suppressed.) The fit of their bands for silicon and germanium led to the values used in this book to determine the parameters for Table 2-1 and the Solid State Table. Thus a plot of the bands for germanium, given in Fig. 3-8,a, is in some sense a fit to the bands of germanium. Comparison with the known bands, shown in Fig. 3-8,b, indicates that at least for the lower bands the fit is very good.

At the same time, calling this a *fit* to the bands is very much understating the accomplishment. The set of four parameters in Table 2-1 and the term values in Table 2-2 (all in the Solid State Table) allow calculation of energy bands for any of the homopolar semiconductors or any of the zincblende-structure compounds, as simply for one as for the other, without computers, with consistent accuracy, and without need for a previous accurate calculation for that compound. Only in first-row compounds is there indication of significant uncertainty in the results. Furthermore, as we noted in Table 2-1, the theoretical matrix elements are very nearly equal to the ones obtained by fitting bands; thus, if we had plotted bands in Fig. 3-8,a that were based upon purely theoretical parameters, the curves would have been hardly distinguishable.

For germanium, the lowest four bands contain electrons and are the valence bands; notice that they can be counted at the point K,U where they are all nondegenerate. They are separated in energy from the conduction bands at higher energies; the minimum conduction-band energy, L_{1c} in germanium, lies higher than the peak of the valence band at Γ. In the LCAO calculation, there are twice as many bands as electrons since twice as many orbitals as electrons were included. The gap between valence and conduction bands is small compared to the valence-band width, indicating that germanium is near the metallic end of the covalent range of Fig. 2-3.

Comparison of LCAO bands and true bands indicates that indeed the valence bands are very accurately represented by simple LCAO theory. The conduction bands are not nearly so accurately given, though they are qualitatively correct.

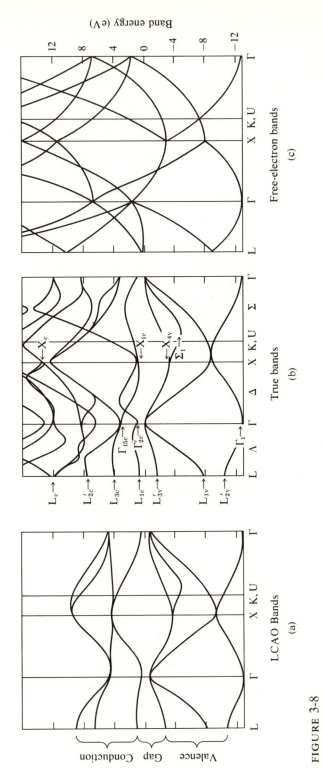

FIGURE 3-8

The energy bands of germanium. Part (a) gives the bands based upon s and p valence states and matrix elements taken from the Solid State Table. Second-neighbor interatomic matrix elements are neglected but otherwise the calculation is exact. Part (b) gives the bands obtained by Grobman, Eastman, and Freeouf (1975) by combining pseudopotential calculations with experimental optical studies; the points indicated with arrows were associated with experimental determinations. Part (c) gives the free-electron bands, $\hbar^2 k^2/(2m)$ with each state shifted into the Brillouin Zone by addition of the appropriate lattice wave number.

Since LCAO theory uses eight basis states, only eight bands can be produced; thus there is no possibility for a good description at higher energy, where many other bands are in evidence. Using additional matrix elements for more distant neighbors allows the bands to be matched at a greater number of points but does not remedy the insufficiency in the number of bands, and the refinements it produces in the valence bands seem unimportant.

Recall that though the Hamiltonian matrix constructed in this section was based on atomic states, its formulation in terms of bonding and antibonding states is entirely equivalent and must give the same results. When we proceed to a more approximate treatment of the bands in order to compute other properties, the bond orbital framework will be more convenient.

It is of interest also to compare the true bands with the free-electron bands, as we did in CsCl; free-electron bands are shown in Fig. 3-8,c. It is remarkable how close the relation is between free-electron bands and LCAO bands at low energies. The resemblance between the free-electron bands and the true bands is good at high energies, and clearly, if we wished to study highly excited states, the free-electron basis would provide a much better starting point for this than the LCAO bands.

It was of course just the similarity of the free-electron and LCAO bands that gave the theoretical matrix elements listed in Table 2-1. The first Eq. (3-27) gives the energy levels labelled Γ'_{2c} and Γ_1 in Fig. 3-8,b; for the homopolar case ($\varepsilon_s^c = \varepsilon_s^a$), that energy difference is $8E_{ss} = 8V_{ss\sigma}$. It is easy to identify the corresponding energy difference in the free-electron bands in Fig. 3-8,c; that energy difference is readily evaluated and is $9\pi^2\hbar^2/(8md^2)$. Equating the two and noting that $V_{ss\sigma} < 0$ gives precisely the theoretical tetrahedral $\eta_{ss\sigma}$ listed in Table 2-1, differing from the adjusted value by less than a percent. The corresponding evaluation of the other coefficients will be carried out in Section 18-A. In the present chapter, the band gap that separates the valence and conduction bands in the LCAO description, but not in the free-electron description, is one of the most central features.

3-D   The Bond Orbital Approximation and Extended Bond Orbitals

The energy bands represent the electronic structure of solids, whether we choose to discuss simple LCAO bands or to discuss more accurate bands obtained by other methods. However, two fundamental obstacles prevent the use of this electronic structure to understand bonding properties of solids. First, bonding properties are dependent on the total energy and, therefore, on calculation of an integral of the energy over the occupied bands. This can only be accomplished numerically since, as we indicated, even in the LCAO description, analytic forms for the bands exist only along certain symmetry directions. This is an inconvenience, though sufficiently accurate integration over the bands can be obtained by using a technique such as that of Gilat and Raubenheimer (1966), discussed in Section

2-E, or by the ingenious special point method (see, for example, Baldereschi, 1973, or Chadi and Cohen, 1973a,b) which will be discussed in Chapter 9. However, the second obstacle is more fundamental: band energies can be obtained only for perfect crystals, whereas many interesting bonding properties involve surfaces, defects, distorted lattices, or impurities. To some extent this difficulty can be surmounted by, for example, making a lattice of impurities in a crystal and calculating bands with a large primitive cell. This, however, can only be done at the cost of complex calculations, and with some uncertainty about how valid the results are for describing the systems that occur in nature.

Both of these obstacles can be overcome completely if we use a Bond Orbital Approximation. In the context of the Hamiltonian matrix of Fig. 3-4, *the Bond Orbital Approximation is the neglect of all matrix elements coupling bonding states to antibonding states.* (Use of the term in this sense was first made by Pantelides and Harrison, 1975. The approximation itself was apparently first made by Hall, 1952, to allow analytic forms to be obtained for the bands.) This reduces the Hamiltonian matrix to a valence-band matrix and a conduction-band matrix, which are not coupled and could in principle be diagonalized independently. However, it can be demonstrated that such a diagonalization would not change the sum of the diagonal elements, so the sum of all energies in the valence bands could be obtained without carrying out the diagonalization. Thus even if the size of the Hamiltonian is 10^{23} by 10^{23} the energy is given bond by bond, and the change in energy that accompanies a bond distortion is very simply obtained by considering only the bonds that are distorted. These consequences derive from the mathematical statement that a unitary transformation does not modify the trace of a Hermitian matrix. The point is so central to what follows that we should make it explicit before proceeding.

The kth valence band state is written as a linear combination of bond orbitals

$$|k\rangle = \sum_i u_{ki} |b_i\rangle. \tag{3-28}$$

Since the Hamiltonian matrix is Hermitian, the transformation matrix u_{ki} will be unitary; that is,

$$\sum_k u_{ki}^* u_{kj} = \delta_{ij}, \tag{3-29}$$

where the sum must be taken over all k. By substitution, we can write the expectation value of any one-electron operator, such as one-electron energy or one-electron density, in terms of the bond orbitals; thus,

$$\langle o \rangle_k = \langle k|o|k\rangle = \sum_{ij} u_{ki}^* u_{kj} \langle b_i|o|b_j\rangle. \tag{3-30}$$

Finally, we can take a sum over all states in the band, using the unitarity condition

(Eq. 3-29), to obtain

$$\sum_k \langle o \rangle_k = \sum_{ij} \delta_{ij} \langle b_i | o | b_j \rangle = \sum_i \langle b_i | o | b_i \rangle. \tag{3-31}$$

In the cases of interest to us, the operator o will stand for energy or electron density, and we see that these can be obtained in terms of the simple bond orbitals derived algebraically in Eqs. (3-7) through (3-12); thus a very wide range of properties can be treated in a very simple way.

The Bond Orbital Approximation not only greatly simplifies the bonding problem; it also greatly increases the number of problems that can be treated. This principal approximation of the Bond Orbital Model (Harrison, 1973c, and Harrison and Ciraci, 1974) provided elementary theories explaining a large number of properties of tetrahedral solids. Unfortunately, the neglected matrix elements in Fig. 3-4 are not zero, and subsequent studies have indicated that the errors in ignoring them can be significant.

The Bond Orbital Approximation is nevertheless essential to an understanding of the electronic structure of tetrahedral solids. The transformation from atomic states to bond orbitals has inserted a gap between occupied and empty energy levels, and approximate bands can be obtained in the Bond Orbital Approximation. There is no *qualitative* change in these bands if we reintroduce the matrix elements that the Bond Orbital Approximation neglects. In this sense, the Bond Orbital Approximation is *conceptually* correct and hence it is legitimate to discuss all properties of tetrahedral solids in terms of it. *Quantitatively* the approximation is not always adequate. Sometimes this drawback can be alleviated by suitable choice of parameters; sometimes it is necessary to compute corrections. The remainder of this section will deal with the choice of parameters and corrections for Bond Orbital Approximations, though these are quite inessential to a basic understanding of the concept.

Two Choices of Covalent and Polar Energy

It was stated that the Hamiltonian matrix given in Fig. 3-4 is equivalent to that of Table 3-1 and would therefore lead to the bands shown in Fig. 3-8,a, if solved exactly. On the other hand, if it were solved by using the simplest Bond Orbital Approximation (equivalent to the left panel in Fig. 6-3, with $V_1^c = V_1^a$ for germanium), identical valence and conduction bands would be obtained, with each conduction band having an energy $2V_2^h$ above the corresponding valence band. From Fig. 3-8, we see that such a description is at great variance with the true bands as well as with those obtained without the Bond Orbital Approximation. In spite of this, we can retain the important benefits of the Bond Orbital Approximation by choosing suitable parameters.

If we wish to calculate the total energy by summing (or integrating) the energy over the occupied bands, it is important only that the average energy be correct, and this is very nearly accomplished (except for small corrections that will be discussed soon) if we use the hybrid covalent energy V_2^h and, for polar semiconductors, the hybrid polar energy V_3^h. The energy $-(V_2^{h2} + V_3^{h2})^{1/2}$ is a good estimate of the average energy of the valence bands relative to the average hybrid energy.

On the other hand, we shall see in Chapter 4 that dielectric properties are dominated by p-like bands, bands which become degenerate at Γ, the energies of which were obtained in the second of Eqs. (3-27). Thus, to describe dielectric properties and other properties depending upon bond dipoles, in Eqs. (4-16) and (4-17) we will define a covalent and polar energy by using the second of Eqs. (3-27), which will describe the bonding–antibonding splitting appropriate to those properties. That covalent energy will turn out to be smaller, by a factor of about two, than the corresponding value defined in terms of the hybrids. Use of these values in computing dielectric properties corrects the largest errors that result from the use of the Bond Orbital Approximation to study those properties.

Corrections for Total Energy

Use of V_2^h and V_3^h does give correct ε_b values for the diagonal elements in Fig. 3-4, but error in the total energy arises when the Bond Orbital Approximation is used, because the approximation neglects matrix elements that couple bonding states with neighboring antibonding states. This error can be nearly eliminated by including the effects of these matrix elements in perturbation theory, which will lead to construction of extended bond orbitals for which the matrix elements we wish to neglect are indeed small. This will also give a measure of the error that is created by neglecting the bond–antibond matrix elements in the original matrix (Fig. 3-4). This analysis is somewhat intricate, but the gist of the argument can be gotten from Fig. 3-9 without following the details.

We define *extended bond orbitals* to be bond orbitals corrected in perturbation theory for those matrix elements that couple bonding and antibonding orbitals. This is done directly by using Eq. (1-16). Then corrections are added to every bond orbital $|b\rangle$; these arise from the interaction, V_{ab}, with neighboring antibond orbitals $|a\rangle$,

$$|B\rangle = |b\rangle + \sum_a |a\rangle V_{ab}/(\varepsilon_b - \varepsilon_a). \tag{3-32}$$

Similarly *extended antibonding orbitals* are defined by

$$|A\rangle = |a\rangle + \sum_b |b\rangle V_{ba}/(\varepsilon_a - \varepsilon_b). \tag{3-33}$$

Construction of extended orbitals results in a transformation to a new set of orbitals; the transformation is in fact unitary up to second order in $V_{ab}/(\varepsilon_b - \varepsilon_a)$, which is the ratio taken to be negligible in the Bond Orbital Approximation. (We are now adding corrections for its nonzero value.) The unitarity is seen by direct evaluation. Notice that (as in Fig. 3-4), V_{ab} is zero if the antibonding orbital $|a\rangle$ and the bonding orbital $|b\rangle$ are on the same site; then $\langle B'|B\rangle$ for $|b\rangle$ and $\langle b'|$ on different sites is seen to have contributions only from their common neighbors and these are proportional to $[V_{ab}/(\varepsilon_b - \varepsilon_a)^2]$. The same is true for $\langle A'|A\rangle$. On the other hand, $\langle A'|B\rangle$ has no contributions from common neighbors and the contributions from the sites of $\langle a'|$ and $|b\rangle$ are equal and opposite; $V_{ba} = V_{ab}$; and the energy denominators are opposite. The normalization is also maintained in first order, and a transformation from one orthonormal set to another is unitary.

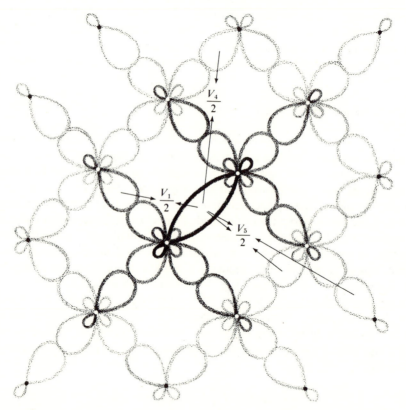

FIGURE 3-9

The bond orbital (center) and neighboring antibonding orbitals (figure 8's) that make up a single extended bond orbital. The matrix elements between nearest- and next-nearest-neighbor orbitals are specified for homopolar semiconductors; they are half the value of the matrix elements coupling the appropriate hybrids. For silicon, the total probability density on the nearest neighbors is 3 percent, that on the second neighbors is 1 percent; the remaining 96 percent is in the central bond orbital. For this reason the 8's representing second neighbors were drawn lighter. Also for this reason the corona of antibonding orbitals can frequently be neglected.

It may be helpful to sketch an extended bond orbital. For a given bond orbital $|b\rangle$, matrix elements couple nearest-neighbor antibonding orbitals and second-nearest-neighbor antibonding orbitals (as noted earlier about matrix elements between bonding orbitals), because matrix elements couple nearest-neighbor hybrids to the given orbital; we can neglect all matrix elements between hybrids more widely separated. Each of these matrix elements can be evaluated in terms of the interatomic matrix elements of Table 2-1 and the V_1 of Eq. (3-5) by decomposing the bonding or antibonding orbital into hybrids and the hybrids into atomic orbitals. Such a decomposition will be made in Chapter 6; here, it is sufficient simply to sketch, in Fig. 3-9, all of the orbitals making up an extended bond orbital. There are six nearest-neighbor antibonding orbitals and eighteen second-neighbor antibonding orbitals. (Two second-neighbor antibonding orbitals are superimposed, so only sixteen distinct second neighbors appear in Fig. 3-9.) The total probability density for electrons in antibonding orbitals (the sums of squares of coefficients of the antibonding orbitals) is quite small, indicating that the perturbation theory may be accurate; however, the probability density is not so small that the antibonding orbitals can be neglected. The intricate nature of the extended bond orbital might at first seem a very serious complication of the problem, but in fact we may continue to think in terms of a central bond orbital (the ellipse at the center of Fig. 3-9) and an outer "corona" that simply modifies the parameters that enter the theory.

The transformation to an extended-orbital basis modifies the Hamiltonian matrix of Fig. 3-4. We shall evaluate the new matrix elements here for a perfect crystal to illustrate the technique; bear in mind that it is very easy to carry out a similar analysis for imperfect crystals by recalculating the extended orbitals for such systems. We first obtain the diagonal element, $\langle B|H|B\rangle/\langle B|B\rangle$ (the denominator is required since $|B\rangle$ is not normalized to second order as required for the second-order diagonal terms). We do this by writing out the state

$$\langle B| = \langle b| + \sum_a V_{ba}\langle a|/(\varepsilon_b - \varepsilon_a), \tag{3-34}$$

using Eq. (3-32) for $|B\rangle$. Writing $\langle b|H|b\rangle = \varepsilon_b$, $\langle a|H|a\rangle = \varepsilon_a$, $\langle b|H|a\rangle = V_{ba}$, and $\langle a|H|b\rangle = V_{ab}$, we obtain

$$\frac{\langle B|H|B\rangle}{\langle B|B\rangle} = \frac{\varepsilon_b + \sum_a 2V_{ba}V_{ab}/(\varepsilon_b - \varepsilon_a) + \sum_a V_{ab}\varepsilon_a V_{ba}/(\varepsilon_b - \varepsilon_a)^2}{1 + \sum_a V_{ba}V_{ab}/(\varepsilon_b - \varepsilon_a)^2}$$

$$= \varepsilon_b + \sum_a V_{ba}V_{ab}/(\varepsilon_b - \varepsilon_a) + 0(V_{ba}^3), \tag{3-35}$$

which is exactly the energy expected by perturbation theory (Eq. 1-14). The matrix elements V_{ab} are just those that would have been neglected had we made a Bond Orbital Approximation at the outset. The corrections are not very large.

Taking the matrix elements from Fig. 3-9 for the homopolar material, and noticing that $\varepsilon_b = \bar{\varepsilon} - V_2$ (Eq. 3-12), we immediately obtain

$$\frac{\langle B|H|B\rangle}{\langle B|B\rangle} = \bar{\varepsilon} - V_2 - \frac{6(V_1/2)^2}{2V_2} - \frac{6(V_4/2)^2}{2V_2} - \frac{12(V_5/2)^2}{2V_2}$$

$$= \bar{\varepsilon} - V_2 \left[1 + \frac{3}{4}\left(\frac{V_1}{V_2}\right)^2 + 0.02\right]. \tag{3-36}$$

For the last step of this equation, the terms containing V_4 and V_5 were evaluated by using matrix elements that will be given in Chapter 6 and the fact that since all vary as d^{-2}, corrections can be written in terms of a dimensionless coefficient. The correction factor for V_2 is 1.04, 1.08, 1.12, and 1.13 for diamond, silicon, germanium, and tin, respectively. The corrections, though small, vary rapidly with distortion of the lattice and therefore are required if, for example, evaluation of the change in energy resulting from distortion (that is, evaluation of the elastic rigidity) is desired.

The matrix element $\langle A|H|A\rangle$ is obtained in just the same way, with an equation of the form of Eq. (3-35) in which the bonding and antibonding designations are interchanged. This simply changes the sign of the energy denominator, and changes ε_b to ε_a. The resulting antibonding energy is the same as Eq. (3-36) but with the sign preceding V_2 changed to $+$. The matrix element between an extended bond orbital and an extended antibonding orbital on the same site remains zero, as in Fig. 3-4, since $\langle a|H|b\rangle$ is zero in one site and for each cross-term with an $|a\rangle$ and a $|b\rangle$ in neighboring sites, one appears with a denominator $\varepsilon_b - \varepsilon_a$ and the other with $\varepsilon_a - \varepsilon_b$; these cancel term by term.

The matrix elements between extended bond orbitals on different sites will not be evaluated until Chapter 6; for now, examine the matrix elements between extended bonding and antibonding orbitals on neighboring sites to affirm that they are of second order in $\langle b|H|a\rangle$, rather than first order as in Fig. 3-4. As long as they are of second order, their neglect (implicit in the Bond Orbital Approximation for extended orbitals) is justified.

Consider then the matrix element $\langle B|H|A\rangle$, with the central site for the $\langle B|$ different from the central site for the $|A\rangle$. If the two sites are sufficiently far apart that the corresponding $\langle b|H|a\rangle$ is zero, then the corona of one extended orbital will not overlap the central site of the other; only corona–corona terms will enter the matrix element and they will be of order $V_{ab}^2/(\varepsilon_a - \varepsilon_b)$. If the sites are close enough that $\langle b|H|a\rangle = V_{ba}$ is nonzero, then the corona of each will overlap the central cell of the other, giving two terms, $V_{ba}\langle a|H|a\rangle/(\varepsilon_b - \varepsilon_a)$ and $\langle b|H|b\rangle V_{ba}/(\varepsilon_a - \varepsilon_b)$. The sum of these just cancels the direct V_{ba} contribution, leaving only contributions of order $V_{ab}^2/(\varepsilon_b - \varepsilon_a)$. This was in fact the point of using extended bond orbitals.

The net result is that a Hamiltonian matrix for the extended orbitals, written in the form of Fig. 3-4, gives modified diagonal elements ε_b and ε_a (neither of which, however, vary from one site to another in the perfect crystal), modified bond–

bond and antibond–antibond matrix elements, and bond–antibond matrix elements that vanish to first order in V_{ab}. Again thinking in terms of perturbation theory, we can see that these matrix elements of order V_{ab}^2 only modify the energy to order V_{ab}^4; it should be a very good approximation to neglect them.

The extended bond orbitals that we have defined are an approximation to what are called **Wannier functions** in solid state physics (see, for example, Weinreich, 1965, p. 127), and identifying them as Wannier functions will shed some light on the approach being used here. A Wannier function is defined in terms of the exact electronic eigenstates of the crystal $\psi_\mathbf{k}$ by taking the inverse of the unitary transformation that led to approximate eigenstates in the LCAO approach. This is simplest to do for an isolated energy band, for which an approximate LCAO state is a single Bloch sum, as in Eq. (3-19). Then a localized state can be written also in terms of a Bloch sum. For example, this can be done with the chlorine s band, which we discussed in Chapter 2. There we wrote an approximate state as

$$|\psi_\mathbf{k}\rangle = N^{-1/2} \sum_i e^{i\mathbf{k}\cdot\mathbf{r}_i} |s_i\rangle. \qquad (3\text{-}37)$$

We could write the atomic s orbital in terms of all of the Bloch states in the band by using the unitarity of the transformation,

$$|s_i\rangle = N^{-1/2} \sum_\mathbf{k} e^{-i\mathbf{k}\cdot\mathbf{r}_i} |\psi_\mathbf{k}\rangle. \qquad (3\text{-}38)$$

The interesting point is that if we know the exact eigenstates of the crystal, we can substitute them into Eq. (3-38) to obtain localized Wannier functions $|s_i\rangle$; if substituted into Eq. (3-37), the localized Wannier functions will give exact eigenstates of the crystal.

Although Wannier functions were conceived long ago (Wannier, 1937), they have not proved very useful except for formal analysis. They were first constructed for silicon by Callaway and Hughes (1967), but not in a form that has been helpful. Recently, Kane and Kane (1978) provided a set of Wannier functions for silicon that describes the valence bands very accurately and may provide a basis for accurate calculation of other properties. A similar approach has been made by Tejedor and Vergés (1978).

Notice that Wannier functions are orthogonal to each other, as we assumed our bond orbitals and extended bond orbitals to be, because the crystal states are orthogonal to each other. Notice also that there are no matrix elements of the Hamiltonian between Wannier functions associated with different energy bands (we were able to eliminate matrix elements to lowest order with the extended bond orbitals), because the different Wannier functions are linear combinations of a completely separate set of energy eigenstates. There are, however, matrix elements between Wannier functions from the same energy band, the counterpart of metallic energy, which broaden the levels into bands. Since the parameters of our theory are obtained by fitting energy bands rather than by using atomic wave functions explicitly, we could say that indeed the theory is based on Wannier functions rather than atomic orbitals. Wannier functions provide an exact formulation of energy bands.

There are, however, some very important differences between extended bond orbitals and Wannier functions. First, there is not a one-to-one correspondence between extended orbitals and bands; four orbitals collectively describe the four valence bands. This is a relatively minor generalization, made also by Kane and Kane (1978); it is nevertheless essential to an understanding of the tetrahedral system. More important, extended bond orbitals can be obtained just as easily for defective or noncrystalline systems, whereas Wannier functions have only been defined for perfect crystals. It is this feature that enables us to use the extended orbitals to study the entire range of properties of a system. It enables us to correct the bond orbitals for the new environment, and it now appears that these corrections are essential. A third important feature of extended orbitals is their identification with atomic orbitals, which allows the parameters to be related to atomic term values and to interatomic matrix elements dependent on bond length. Because of this, we begin the study of any property with the parameters needed for a theory of that property. The simplest approximation is the neglect of the corona and direct use of covalent and polar energies associated with bond orbitals, to obtain at least a partly quantitative theory. Then it is possible to improve the calculation by including the effects of the corona in perturbation theory.

3-E   Metallicity

The use of Wannier functions eliminates the bonding–antibonding matrix elements of Fig. 3-4, but once these have been eliminated, there remain two separate features to be dealt with: the diagonal terms ε_b, which can be called Wannier-function energy levels, and the bond–bond coupling terms, which describe the broadening of the levels into bands. An essential quality of covalent solids is the dominance of bonding–antibonding splitting over the broadening of energy levels into bands, but the smaller the remaining gap between conduction and valence bands, the closer we are to the metallic transition region of Fig. 2-3. We can make a quantitative measure of this "metallicity" in covalent solids by considering the magnitude of the gap between conduction and valence bands.

The maximum energy in the valence band is the lower degenerate level given in Eq. (3-27), and ordinarily, the minimum energy in the conduction band occurs at the upper nondegenerate level of Eq. (3-27). (This is discussed in Chapter 6.) We may subtract one from the other, and writing $\varepsilon_p - \varepsilon_s = 4V_1$ (the metallic energy) for each of the atom types, as in Eq. (3-5), we obtain a difference known as an optical threshold (discussed in Chapter 4). It is

$$
E_0 = -2(V_1^c + V_1^a) + \left[\left(\frac{\varepsilon_s^c - \varepsilon_s^a}{2} \right)^2 + (4E_{ss})^2 \right]^{1/2}
$$

$$
+ \left[\left(\frac{\varepsilon_p^c - \varepsilon_p^a}{2} \right)^2 + (4E_{xx})^2 \right]^{1/2} . \tag{3-39}
$$

If the term with the metallic energies, $2(V_1^c + V_1^a)$, becomes larger than the other terms, the gap disappears altogether, so that the ratio becomes a natural definition of the *metallicity*, α_m; then

$$\alpha_m = 2(V_1^c + V_1^a) \Bigg/ \left\{ \left[\left(\frac{\varepsilon_s^c - \varepsilon_s^a}{2} \right)^2 + (4E_{ss})^2 \right]^{1/2} \right.$$
$$\left. + \left[\left(\frac{\varepsilon_p^c - \varepsilon_p^a}{2} \right)^2 + (4E_{xx})^2 \right]^{1/2} \right\}. \tag{3-40}$$

This somewhat complicated form is in fact nearly the same as the simple form, $V_1/(V_2^2 + V_3^2)^{1/2}$, introduced earlier in terms of a much cruder theory (discussed in Harrison, 1973c). To see this, look first at homopolar semiconductors, for which Eq. (3-40) reduces to

$$\alpha_m = 4V_1/(4|E_{ss}| + 4|E_{xx}|). \tag{3-41}$$

The value $4E_{xx}$ is just the magnitude of the covalent energy V_2, which will be introduced in Chapter 5 (it is half the splitting between triply degenerate states at Γ). Both E_{xx} and E_{ss} are functions only of d, and from Eq. (3-26) and the Solid State Table, we see that E_{ss} is $2.59E_{xx}$. Thus Eq. (3-41) becomes $\alpha_m = 1.11V_1/V_2$, exactly the earlier definition scaled up by 11 percent. The relation for the polar semiconductors is almost as close. One may verify, for the series Ge, GaAs, ZnSe, for example, that $\overline{V_1} = (V_1^a + V_1^c)/2$ is almost independent of polarity ($\overline{V_1}$ differs slightly from the earlier definition of V_1), and that $\varepsilon_s^c - \varepsilon_s^a$ is approximately twice $\varepsilon_p^c - \varepsilon_p^a$, so that the two terms in the denominator scale nearly the same with polarity. (They would scale exactly the same if the ratio were 2.59 rather than 2.) Thus, to a very good approximation,

$$\alpha_m \approx 1.11\overline{V_1}/(V_2^2 + V_3^2)^{1/2}. \tag{3-42}$$

Using this approximate form in Eq. (3-39), we may write the optical threshold E_0 as

$$E_0 \approx 3.60(V_2^2 + V_3^2)^{1/2}(1 - \alpha_m). \tag{3-43}$$

At the critical metallicity of 1, the gap can be expected to vanish.

Values can be obtained from the exact definition for the homopolar semiconductors (Eq. 3-41), by using the metallic energies of Table 2-2 and the Solid State Table. The results are given in Table 3-2. The predicted critical value of unity is only approximate. We will see in Chapter 6 that the gap E_0 in grey tin is zero or slightly negative. In Chapter 6 the quantitative values for E_0 will also be discussed.

The metallicity defined here, as well as the simpler earlier form, corresponds roughly to the *metallization* introduced earlier by Mooser and Pearson (1959). It is also the parameter (based upon earlier values) that formed the ordinate in the schematic phase diagram shown in Fig. 2-6.

TABLE 3-2

Metallic energy and covalent energy (in eV), and metallicity for the homopolar semiconductors.

Element	Metallic energy V_1	Covalent energy V_2	Metallicity α_m
C	2.13	6.94	0.34
Si	1.76	2.98	0.66
Ge	2.01	2.76	0.81
Sn	1.64	2.10	0.87

3-F Planar and Filamentary Structures

Although our discussion in the next seven chapters will center on simple tetrahedral structures, in which all electrons form simple two-electron bonds, it is desirable to introduce two other types of structures; in these, *some* of the electrons form two-electron bonds (and are understandable in the same terms used for the tetrahedral solids) and other electrons are accommodated in pure *p* states, similar to the π states discussed in Chapter 1 for diatomic molecules. For a discussion of the stability of these structures, see Friedel (1978). The two-electron bonds are simpler geometrically in these systems than in the tetrahedral solids and will provide very good problems. (Many of the problems at the ends of the following chapters will explore the theory of the bonds for these simpler systems.)

Graphite

Graphite is a simple planar structure in which the carbon atoms are arranged as in Fig. 3-10. The nearest neighbors are separated by 1.42 Å (in diamond they are separated by 1.54 Å), but the distance between successive planes is much larger (3.4 Å), so that to a good approximation we may think of the planes as isolated sets of atoms when we discuss the electronic structure. Successive planes are stacked one above the other in the crystal. (The successive planes are also displaced laterally.)

The electronic structure in graphite may be understood in terms of sp^2 hybrids (see Problem 3-2) oriented in the direction of the bond and bond orbitals constructed from these hybrids. The shorter bond length and different composition of the hybrid lead to a covalent energy value, V_2, that is also different in graphite than it is in diamond.

The bond orbitals for graphite can accommodate three electrons per carbon atom. The remaining electrons go into *p* states oriented perpendicular to the plane in Fig. 3-10 and are analogous to the π states of diatomic molecules, as discussed in Chapter 1. The π states are coupled by small matrix elements and broaden into a rather narrow band. There are enough electrons to half-fill this band (because of the two spin states). By filling only the lower half of the band these electrons

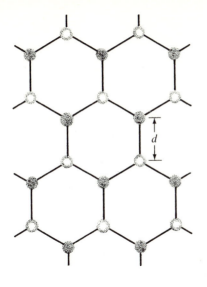

FIGURE 3-10

The arrangement of carbon atoms in one plane of the graphite structure. Boron nitride can also form this structure (hexagonal BN) with atoms of B and N alternating as indicated by shaded and open circles.

contribute to the cohesion weakly, so weakly, in fact, that they are often called "nonbonding" electrons. Bonding in which the lower half of a band is filled is more like the metallic bonding discussed in Chapter 15 than the covalent bonding discussed so far or the ionic bonding discussed in Chapter 13. It contributes to the reduced interatomic distance mentioned above.

The flatness of the graphite structure can be said to come ultimately from the *empty* part of the π-state bands rather than from the full part. As will be discussed in detail in connection with surfaces in Chapter 10, systems tend to deform to increase the hybrid energy of any empty states (since by the unitarity of the corresponding transformation of orbitals, the sum of the energies of occupied and empty states remains fixed). Making all the orbitals purely p-like optimizes this "dehybridization energy" and gives the flat structure. Traditionally, the tendency toward flat configurations has been associated with the *occupied* part of the energy bands by the concept that the double bond is rigid and favors a flat configuration. However, it is reasonable to envisage a quite different origin for the flatness of graphite than for the flatness of ethylene, discussed in Chapter 1; the latter comes from a cooperative shift of π electrons into similarly oriented p orbitals and a shift of the protons into the same plane. No dehybridization energy is involved in ethylene; a rigid-bond interpretation is reasonable.

It turns out that there is no band gap between the empty and full states in graphite but there are only a very tiny number of states near the top of the occupied region. Because of this, graphite has weak metallike conductivity and is called a *semimetal*. The corresponding band structure is studied in Problems 3-3 and 6-1.

Boron nitride forms the same structure that graphite does, as indicated in Fig. 3-10. The sp^2-hybrid energy-difference may be evaluated by using the Solid State Table and used to estimate the corresponding V_3^h. It is possible to treat the counterparts of all the properties of polar tetrahedral semiconductors also for the layered system. See, for example, Problem 3-4. In boron nitride the π electrons lie predominantly on the nitrogen atoms, because the p-state energy is lower there, though there will be some mixture with boron π electrons (see Problem 5-2). The presence of the matrix element V_3 opens up a gap between the occupied and unoccupied states in hexagonal BN, so it is semiconducting rather than semimetallic, and the π states are analogous to the states of ionic solids that will be studied in Chapter 13.

Chalcogenides

Sulphur, selenium, and tellurium, called the chalcogenides, have six valence electrons per atom. Placing them in the diamond structure, with four neighboring atoms, would require putting two electrons per atom in antibonding states, which is energetically unfavorable. Putting them in a graphite structure, with three neighbors, means that one of the extra electrons is in a nonbonding π state and the other is in an antibonding state, a structure also unfavorable. The atoms avoid occupying antibonding states if they have only two neighbors, leading to a filamentary structure, as shown in Fig. 3-11, where we have allowed the chain to pucker. This is an easy system to study in terms of LCAO electronic structure and is very instructive. Recent studies of this kind have been made by Joannopoulos (1974) and by Shevchik (1976). In such a structure it is easy to see that the energy tends to be minimized when the bond orbitals form a puckering angle of 90°, as drawn in Fig. 3-11. (A similar kind of puckering occurs on the surfaces of tetrahedral semiconductors and in the structure of mixed tetrahedral solids.) Then the low-energy s states on each atom are doubly occupied and bonding combinations of p states in each bond are doubly occupied. This leaves the nonbonding π state on each atom also doubly occupied and leaves a gap between occupied and empty states.

We may also imagine a system with p-state energy that is different for alternate atoms in the zig-zag structure, in order to have a polar as well as a covalent energy. This becomes the simplest structure to have essentially the same character as the tetrahedrally bonded structure, and it will be very useful for illustrating the theory of covalent solids; Problems 8-1 and 8-2 are based on this system.

Notice that the electronic structure in the chalcogenides favors 90° angles at each atom, but does not favor or disfavor a coiling of the structure. The observed structure *is* coiled and can be understood in terms of Figure 3-12. There, a coiling of the structure is matched with cube edges, giving an average orientation along the [111] direction. If other coils are added to occupy all of the other cube sites, and the structure is distorted slightly, the observed selenium structure is obtained. The distortion brings atoms in the same chain closer together and separates the chains, so it makes sense to describe the covalent bonding within the chains rather than to give attention to coupling between chains. However, the coupling is not negligible, as has been shown by Joannopoulos, Schlüter, and Cohen (1974) for the electronic structure of chalcogenides, and by Martin, Lucovsky, and Helliwell

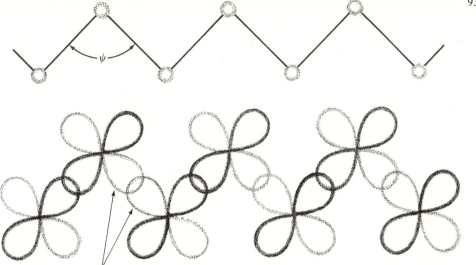

Two atomic p orbitals making up one σ-bond orbital

FIGURE 3-11

A zig-zag chain of atoms approximating the structure of a filament of selenium or tellurium (upper part). The p orbitals that make up corresponding bond orbitals are shown in lower part.

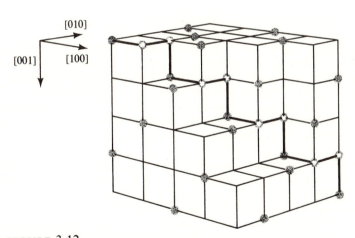

FIGURE 3-12

A schematic sketch of the selenium structure showing how the zig-zag chain can be coiled up along a [111] direction in a simple cubic lattice. Every third atom along the coil is translationally equivalent; one such set is shaded. Other chains may be added such that the isolated shaded atoms shown are translationally equivalent to the shaded atoms in the chain and there is an atom site at every cube corner. The structure can then be distorted to reduce intrachain atom distances, increase interchain distances, and increase the bond angles to 105°. This is then the observed structure of S, Se, and Te.

(1976) for the vibrational properties of chalcogenides. It is interesting to notice that the coiling of the chains makes the matrix elements between π orbitals on adjacent atoms small (they would be zero with the 90° coiling of Fig. 3-12), and therefore narrows the corresponding nonbonding π bands. If these were half full, as in graphite, this would cost significant energy and favor a flat configuration. However, since these bands are full in the chalcogenides, the change is of little consequence.

Planar and filamentary structures are of interest in their own right, but understanding their electronic structure and properties seems to depend upon a combination of concepts applicable also to other, more general systems rather than upon a unique set of concepts. We will not undertake a special study of them, therefore, but will carry them along as illustrative examples at the end of each chapter.

PROBLEM 3-1   *Tetrahedral crystal structure*

Sketch the diamond structure as viewed along a [110] direction, analogous to Fig. 3-1,b. Shade atoms not in the plane of the figure and draw lines connecting nearest neighbors.

If spheres are packed in a face-centered cubic structure, what fraction of the space lies within a sphere? This is called the "packing fraction."

What is the packing fraction for the diamond structure?

PROBLEM 3-2   *Hybrid and bond orbitals in graphite*

The graphite structure was discussed in Section 3-F and illustrated in Fig. 3-10. If we take the p_z orbitals to be oriented perpendicular to the plane of the graphite (these become the π energy levels), we may proceed to construct the σ bands from the s, p_x, and p_y orbitals in close analogy to the construction of the bands in the diamond structure.

(a) Construct sp^2 hybrid states in analogy with the diamond-structure states of Eq. (3-1), choosing coefficients such that the hybrid wave functions are orthogonal to each other and normalized, and oriented in the direction of each of the three neighbors in the plane.

(b) Obtain the hybrid energy in terms of atomic s and p energies. Obtain also $-V_1$, the matrix element of the Hamiltonian between hybrids on the same atom, in analogy with Eq. (3-5). Values may all be obtained from the Solid State Table.

(c) The matrix elements between two hybrids in the same bond, V_2^h, can also be obtained from the Solid State Table in terms of the $V_{ss\sigma}$, the $V_{sp\sigma}$, and the $V_{pp\sigma}$ for $d = 1.42$ Å. Sketch the energy levels as far as the formation of bands, as on the upper left of Fig. 3-3. Think of the π-band energy as being the same as the p-orbital energy, neglecting the broadening into bands.

PROBLEM 3-3   *Band energies in graphite*

The LCAO Bloch states at Γ ($\mathbf{k} = 0$) in graphite, as in diamond, can be written as pure sums of each orbital type; for example, there is one state given by $(2N_p)^{-1/2}$ times a sum of p states oriented in the x-direction chosen in the plane of graphite; another is given by a

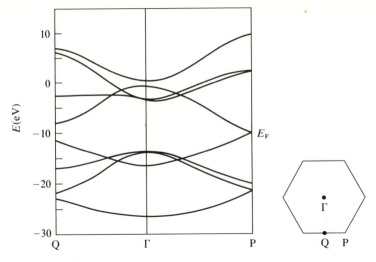

FIGURE 3-13

The energy bands of graphite given by Painter and Ellis (1970), corresponding to a separated graphite plane. The zero of energy has been selected to correspond roughly to an ε_p taken from the Solid State Table. Symmetry points in the Brillouin Zone are shown to the right.

similar sum but with each p state from the second atom in each primitive cell multiplied by -1. (You will find this a bonding, low-energy state.) Find the energies of the eight bands of graphite at Γ, in analogy with Eq. (3-27). Label them σ valence, s valence, π conduction, and so on.

The bands obtained by Painter and Ellis (1970) are shown in Fig. 3-13 for comparison. We have chosen the zero of energy to make comparison easy. The comparison would suggest that we may have overestimated $V_{pp\sigma}$ by a factor of two for this case, a difficulty that may be associated with carbon's being a first-row element.

PROBLEM 3-4   *Hybrid polarity in hexagonal* BN

Consider hexagonal boron nitride, shown in Fig. 3-10. Define a hybrid covalent and hybrid polar energy for this structure, in analogy with the corresponding definitions for tetrahedral solids in Eqs. (3-4) and (3-6). Compare the polarity obtained from these values with that of tetrahedral BN, listed in Table 7-2. Take $d = 1.42$ Å as in graphite. Notice that many values are modified by having sp^2 hybrids rather than sp^3 hybrids.

CHAPTER 4

Optical Spectra

SUMMARY

Optical studies of solids probe the details of the electronic structure. The central property is the response of the solid to electromagnetic fields of sufficiently high frequency that motion of the atoms themselves can be neglected. The response is measured in terms of a dielectric susceptibility that has a real part and an imaginary part, related by the Kramers-Kronig relations. The imaginary (absorptive) part at each energy (Planck's constant times frequency) depends on the density of occupied and empty electron-states differing by that energy and on the oscillator strength associated with the coupling of those states. Both are obtainable once the electronic structure has been determined. The principal features of the absorption spectrum are directly related to the parameters listed in the Solid State Table. A scale parameter γ for oscillator strengths is needed, however, to obtain absolute values for the absorption or for the real part of the dielectric susceptibility. In these, as in other dielectric properties, the parameters serving to define the covalent and polar energies, V_2 and V_3, are based on p orbitals rather than sp^3 orbitals.

Most properties of solids represent the response of the solid to some external electromagnetic field or mechanical force. The response of the electronic states to electromagnetic fields is so directly related to electronic structure that it is appropriate to discuss it first. We are interested in absorption and reflection of light as well as in the ordinary dielectric and diamagnetic properties of solids. When static or slowly varying electric fields act on a polar semiconductor, the two types of atoms in the semiconductor will move with respect to each other, giving a lattice contribution, in addition to the electronic contribution, to the dielectric constant; this is a consideration best put aside until mechanical properties of solids are treated in Chapter 9. Thus, for the present, let us restrict discussion to frequencies that are high in comparison to the natural vibrational frequencies of the lattice, which are in the infrared region of the spectrum. Then the response of the lattice itself can be neglected. We shall discuss properties as a function of light frequency: the low-frequency limits will correspond to frequencies small compared to characteristic *electronic* excitation energies, but still well above the *vibrational* frequencies. In covalent solids, the electronic responses are sufficiently large that the results are also approximately applicable when the electric fields are static.

4-A   Dielectric Susceptibility

Let us initially consider electric fields that are spatially uniform. This still allows us to consider light since the wavelengths of light will in all cases contain many interatomic distances and the fields may be considered uniform. The assumed uniformity does imply, however, that the net field within the material is also uniform, though certainly a field varies over the atomic cell even if the applied field is uniform; the field at any one bond is affected by the polarization of a neighboring bond. Such variations are called *local field corrections* and have been treated by a number of authors, among them, Decarpigny and Lannoo (1976), Lannoo (1977), Louie, Chelikowsky, and Cohen (1975), and Hanke and Sham (1974). Local field corrections may reduce the static dielectric constant by the order of 10 percent; despite their influence, they will not be treated here in detail. Ignoring local field effects is like assuming that electric polarization arises by passing charge from bond to bond through the crystal without distorting the charge distribution locally, and this is essentially how we shall picture the system.

   The response of the system to a uniform electric field $\mathscr{E}$ within the material is designated by an electric polarization density, **P**; initially we shall examine a polarization density that is linear in the electric field (appropriate for small fields where terms in $\mathscr{E}^2$, $\mathscr{E}^3$, ..., and so on, are negligible). The proportionality constant is the susceptibility, χ, such that

$$\mathbf{P} = \chi(\omega)\mathscr{E}, \tag{4-1}$$

where χ is dependent upon angular frequency ω for a field varying periodically with time. Since we seek only the linear response, we could make a Fourier transform of any time dependence and obtain the linear response component by component, by using Eq. (4-1). Thus, use of a single frequency is not a limitation on the generality. Because of the linearity, we may also write the field $e^{-i\omega t}$ (recognizing that only the real part should be taken as meaningful). Then if the phase of the polarization is different from that of the field, it can be described by a complex susceptibility; a phase factor $e^{i\varphi}$ in the susceptibility leads to a polarization containing a factor $\exp[-i(\omega t - \varphi)]$ (again since only the real part is meaningful); therefore it is out of phase with the field by φ. The complex susceptibility can be written as

$$\chi(\omega) = \chi_1(\omega) + i\chi_2(\omega). \tag{4-2}$$

The complex susceptibility can also be obtained from the electronic structure by treating the electric field as a perturbation. The procedure is quite simple and it gives all the expressions needed. However, it is necessary to choose the path to be followed carefully, in order to obtain the relevant limits, because the long-range nature of the Coulomb potential can give rise to different answers depending upon the order in which the wavelength and the crystal dimensions are taken to infinity. There is an additional problem in that the time dependence of the field requires a generalization of the perturbation formula that we have been using, Eq. (1-14). A way through this is to obtain the perturbing term in the Hamiltonian arising from the electric field, and then simply to write down the expression for χ_2 appropriate for optical absorption. This form is quite plausible and understandable. Then we can obtain χ_1 from χ_2 and see that in the limit of vanishing frequency, χ_1 gives the result expected in terms of the static perturbation formula, Eq. (1-14).

The most convenient way to introduce the electric field is in terms of the vector potential $\mathbf{A}$, by adding $+e\mathbf{A}/c$ to the momentum operator in the Hamiltonian, with $\mathbf{A}$ chosen to give the field $\mathcal{E} = -(1/c) \partial\mathbf{A}/\partial t = (i\omega/c)\mathbf{A}$. We shall use such an approach also when we compute the static magnetic susceptibility later, in Section 4-E. The perturbation becomes $(e/2mc)2\mathbf{p} \cdot \mathbf{A} = -(eh/m\omega)\mathcal{E} \cdot \mathbf{V}$. The polarization density, or alternatively, the current density, can then be calculated by using perturbation theory (see, for example, Harrison, 1970, p. 317ff), and leads to an out-of-phase polarization corresponding to a χ_2 of

$$\chi_2(\omega) = \frac{\pi}{\Omega} \sum_{\substack{k \text{ (full)} \\ k' \text{ (empty)}}} \left| \left\langle \psi_{k'} \left| \frac{eh}{m\omega} \frac{\partial}{\partial x} \right| \psi_k \right\rangle \right|^2 \delta(\varepsilon_{k'} - \varepsilon_k - \hbar\omega). \tag{4-3}$$

The electric field has now been taken in the x-direction, and the polarization is assumed to be parallel to the field. For cubic crystals the susceptibility is independent of field direction. The k values are to be summed over all valence states and the k' values are to be summed over all conduction-band states.

The interpretation of Eq. (4-3) is direct and the formula could almost be written down immediately in terms of Fermi's familiar " Golden Rule " of quantum theory (see, for example, Schiff, 1968, p. 314). The absorption of light arises from the coupling, caused by the electric field, of occupied states with unoccupied states. However the absorption can only occur if the two states differ in energy by $\hbar\omega$, so that energy can be conserved if the transition is made and one photon is absorbed. It should also be noted that in a perfect crystal the matrix elements will vanish unless the wave numbers of the coupled states in the Brillouin Zone are the same. That is, the transitions must be between states that are on the same vertical line in a diagram such as Fig. 4-7. We shall return to a more complete description of absorption later.

The real and the imaginary parts of the susceptibility are related by the Kramers-Kronig relations. (A derivation of these equations is given in Kittel, 1976, p. 324; Kittel used analytic continuation of the susceptibility into the complex plane.) One relation is

$$\chi_1(\omega) = \frac{2}{\pi} P \int_0^\infty \frac{\omega' \chi_2(\omega') \, d\omega'}{\omega'^2 - \omega^2} . \tag{4-4}$$

The P indicates that principal parts are to be taken in the integral. Of particular interest here is the real part at vanishing frequency. By setting $\omega = 0$ in Eq. (4-4) and substituting Eq. (4-3), we may use the delta function to evaluate the integral over ω', to obtain

$$\chi_1(0) = \frac{2\hbar^4 e^2}{m^2 \Omega} \sum_{\substack{k \text{ (full)} \\ k' \text{ (empty)}}} \frac{\left| \langle \psi_{k'} | \frac{\partial}{\partial x} | \psi_k \rangle \right|^2}{(\varepsilon_{k'} - \varepsilon_k)^3} , \tag{4-5}$$

which is probably the most useful form for the susceptibility. It can be rewritten in a form more closely related to perturbation theory, Eq. (1-14), by using a quantum-mechanical relation (see Schiff, 1968, p. 404), which is applicable if the electron states are localized so that the matrix elements of x do not diverge; that is,

$$\langle a | x | b \rangle = \{\hbar^2 / [m(\varepsilon_b - \varepsilon_a)]\} \langle a | \partial/\partial x | b \rangle. \tag{4-6}$$

Then Eq. (4-5) becomes

$$\chi_1(0) = \frac{2e^2}{\Omega} \sum_{\substack{k \text{ (full)} \\ k' \text{ (empty)}}} \frac{|\langle \psi_{k'} | x | \psi_k \rangle|^2}{\varepsilon_{k'} - \varepsilon_k} . \tag{4-7}$$

We may complete the identification of the susceptibility by noting that the applied electric field $\mathscr{E}$ corresponds to a perturbing potential $e\mathscr{E}x$, so that the shift in energy of the state k can be written $e\mathscr{E} \sum_{k'} \langle \psi_{k'} | x | \psi_k \rangle^2 / (\varepsilon_k - \varepsilon_{k'})$, and thus the right side of Eq. (4-7) will be minus twice the shift in electron energy density, divided by $\mathscr{E}^2$. However, the shift in electron energy density can be written directly in terms of the susceptibility. It is the work done by the field on the dielectric, that is, $\int \mathscr{E} dP = \chi \mathscr{E}^2 / 2$, plus the resulting electrostatic energy $-\mathscr{E} \cdot \mathbf{P} = -\chi \mathscr{E}^2$, giving finally $-\chi \mathscr{E}^2 / 2$. Thus, the left side of Eq. (4-7) is also minus twice the electron energy density divided by $\mathscr{E}^2$, and the various expressions for the susceptibility given in the preceding discussion are consistent with ordinary perturbation theory, Eq. (1-14).

One other form for the susceptibility should be written because it occurs frequently in the physics literature and because it provides a very convenient description. A *dimensionless oscillator strength* $f_{k'k}$ can be defined by

$$f_{k'k} = 2\hbar^2 |\langle \psi_{k'}| \partial/\partial x |\psi_k\rangle|^2/m(\varepsilon_{k'} - \varepsilon_k).$$ (4-8)

In terms of the dimensionless oscillator strength, Eqs. (4-3) and (4-5) may be rewritten as

$$\chi_2(\omega) = \frac{\pi \hbar e^2}{2m\omega\Omega} \sum_{\substack{k \text{ (full)} \\ k' \text{ (empty)}}} f_{k'k} \delta(\varepsilon_{k'} - \varepsilon_k - \hbar\omega)$$ (4-9)

and

$$\chi_1(0) = \frac{\hbar^2 e^2}{m\Omega} \sum_{\substack{k \text{ (full)} \\ k' \text{ (empty)}}} \frac{f_{k'k}}{(\varepsilon_{k'} - \varepsilon_k)^2}.$$ (4-10)

Full calculation of the susceptibility requires a knowledge of oscillator strengths as well as of the energies of the electronic states we have been discussing. We can learn what approximations for the oscillator strengths are appropriate from a consideration of the optical absorption in perfect crystals and then proceed to use these approximations to consider other properties of covalent systems.

4-B   Optical Properties and Oscillator Strengths

The absorptive term in the susceptibility, arising from electron transitions between states in the solid, is one of the most direct probes of the electronic structure of a solid. In atoms and molecules, absorption reflects very directly the differences in energies between different, sharply defined levels, and the resulting absorption lines can be used directly to deduce "term values," which may be thought of as one-electron energy values for the atoms or molecules. In solids, the sharp spectral absorption lines are replaced by bands, and though it is not possible to work back to determine the bands from the spectrum, it is possible to extract specific energy differences between bands; this was done to obtain the arrows in the germanium band-structure shown in Fig. 3-8,b. Any theoretical model of the band structure can be tested in terms of the absorptive term, though the absorption $\chi_2(\omega)$ depends upon the oscillator strengths as well as the bands themselves.

In principle, it should be possible to make a full energy-band calculation that would give wave functions and energies from which all of the above expressions might be evaluated. This was nearly accomplished in empirical calculations with nonlocal pseudopotentials by Chelikowsky and Cohen (1976b) for eleven solids with diamond and zincblende structures. Among the approximations used in their approach was the neglect of the difference between matrix elements coupling pseudo wave functions and matrix elements coupling true wave functions (see Harrison, 1970, p. 322, where such corrections were as large as a factor of two

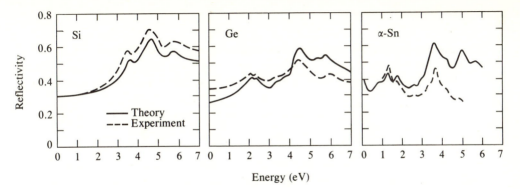

FIGURE 4-1

Theoretical and experimental optical reflectivity based upon empirical calculations of non-local pseudopotentials by Chelikowsky and Cohen (1976b). Experimental curves are taken from Philipp and Ehrenreich (1963) for Si and Ge, and from Cardona et al. (1966) for Sn.

in the χ_2 for simple metals, larger, most likely, than for semiconductors). These calculations produced susceptibilities and then reflectivity as a function of frequency, in extraordinary agreement with experiment, as indicated in Fig. 4-1, where calculated reflectivity is compared with experiment for Si, Ge, and Sn; the agreement was comparable for the compounds treated. It is fair to say that reflectivity is now understood in terms of the known electronic structure. We should like to extract the physics from that understanding, hoping that it will enable us to treat the many properties which cannot be directly obtained by computer calculation. With that hope in mind, let us look first at the oscillator strengths and then at the relevant features of the energy bands.

Two separate details must be dealt with concerning oscillator strengths. First is the variation of some average oscillator strength from material to material. There has been dispute about this subject (fully discussed by Harrison and Pantelides, 1976), and there may be several valid ways to formulate and discuss the variation. In this section, only those features will be given that are relevant to the discussion of general properties made subsequently in the book. Second is the variation of oscillator strength over the bands in a given material.

Average Oscillator Strength in One Material

Let us examine first the average, over the bands, of the oscillator strength. The average that is most conveniently considered is

$$\sum_{\substack{k \text{ (full)} \\ k' \text{ (empty)}}} |\langle \psi_{k'} | \partial/\partial x | \psi_k \rangle|^2 = - \sum_{\substack{k \text{ (full)} \\ k' \text{ (empty)}}} \langle \psi_k | \partial/\partial x | \psi_{k'} \rangle \langle \psi_{k'} | \partial/\partial x | \psi_k \rangle. \quad (4\text{-}11)$$

The minus sign comes from the identity, obtainable by partial integration,

$$\langle \psi_{k'} | \partial/\partial x | \psi_k \rangle = - \langle \psi_k | \partial/\partial x | \psi_{k'} \rangle^*. \quad (4\text{-}12)$$

The average is convenient because the unitarity argument of Eq. (3-31) can be applied first to k and then to k' to show that in the Bond Orbital Approximation, or when $|b\rangle$ and $|a\rangle$ are regarded as Wannier functions, this average is equal to

$$\sum_{\substack{k\,(\text{full}) \\ k'\,(\text{empty})}} |\langle\psi_{k'}|\partial/\partial x|\psi_k\rangle|^2 = \sum_{b,\,a} \langle a|\partial/\partial x|b\rangle^2. \tag{4-13}$$

To obtain an idea of what variation from material to material is expected, we may evaluate such matrix elements by ignoring the corona of the extended bond orbitals and in fact by taking only the matrix element for a bond orbital and an antibonding orbital at the same bond site. Using Eqs. (3-13) and (3-16),

$$\langle a|\partial/\partial x|b\rangle = \left[\left(\frac{1-\alpha_p}{2}\right)^{1/2}\langle h_1| - \left(\frac{1+\alpha_p}{2}\right)^{1/2}\langle h_2|\right]\frac{\partial}{\partial x}$$

$$\times \left[\left(\frac{1+\alpha_p}{2}\right)^{1/2}|h_1\rangle + \left(\frac{1-\alpha_p}{2}\right)^{1/2}|h_2\rangle\right]$$

$$= \frac{1-\alpha_p}{2}\langle h_1|\frac{\partial}{\partial x}|h_2\rangle - \frac{1+\alpha_p}{2}\langle h_2|\frac{\partial}{\partial x}|h_1\rangle = \langle h_1|\frac{\partial}{\partial x}|h_2\rangle. \tag{4-14}$$

Eq. (4-12), as applied to hybrid states, is used in the last step; it shows also that $\langle h_i|\partial/\partial x|h_i\rangle = 0$. The remarkable result is that the matrix elements are not expected to vary with polarity in an isoelectronic series. This feature also applies to the intra-atomic terms in the matrix element between bonding and antibonding orbitals on adjacent sites. A similar treatment of the matrix elements of x indicates that they vary inversely with covalency $\alpha_c = V_2/(V_2^2 + V_3^2)^{1/2}$ in an isoelectronic series; indeed, this follows from Eq. (4-13) and the relation shown in Eq. (4-6), since the bonding–antibonding splitting varies as α_c^{-1} in an isoelectronic series. Similarly, the dimensionless oscillator strength of Eq. (4-8) varies in proportion to the covalency in such a series. The last point is of some interest since a rigorous sum rule,

$$\sum_{\text{all }k'} f_{k'k} = 1, \tag{4-15}$$

with the sum taken over occupied as well as unoccupied k' (see, for example, Harrison and Pantelides, 1976), has led some workers to consider $f_{k'k}$ constant in isoelectronic series (Phillips, 1975). This ignores important differences in the extent to which the sum is "exhausted" in the range of energy of interest for different systems (Ehrenreich, 1967; Cardona et al., 1970).

We conclude here that the matrix elements of $\partial/\partial x$ are constant. This conclusion will follow also from the pseudopotential theory of covalent bonding in Chapter 18, and was found to be true of the matrix elements in the nonlocal pseudopotential calculations of Chelikowsky and Cohen (quoted by Phillips,

1975). It was also found in experimental studies by Lawaetz (1971). We shall use this result in treating $\chi_1(0)$.

We might also expect the matrix elements of $\partial/\partial x$ to scale inversely with d among the homopolar semiconductors (and correspondingly, for the matrix elements of x to scale with d) and this is in fact predicted by the pseudopotential theory of Chapter 18. However, that does not describe the trends in $\chi_1(0)$ well, and in Section 4-C, we shall allow the proportionality constant to vary from row to row in the Periodic Table.

Variation of Oscillator Strength over the Bands

The variation of matrix elements over the band is more subtle. The values of $\chi_1(0)$ are very large in covalent solids as compared to atoms, a fact that will be discussed in the next section, and their size would suggest that the interatomic matrix elements, such as those in Eq. (4-14), are dominant. The simplest approximation is the neglect of any dependence of the matrix elements upon initial and final states; one should notice that this will give different answers depending upon whether one assumes that matrix elements of $\partial/\partial x$ are constant or that the $f_{k'k}$ are constant. As usual, the assumption of equal probability on the basis of lack of information is not unique. A common approximation is that the matrix elements of x (or $1/\omega$ times the matrix elements of $\partial/\partial x$) are constant, which from Eq. (4-9) gives a $\chi_2(\omega)$ value directly proportional to the joint density of states:

$$\chi_2(\omega) \approx C \sum_{k,\, k'} \delta(\varepsilon_{k'} - \varepsilon_k - \hbar\omega). \qquad (4\text{-}16)$$

This may be written as an integral and is illustrated in Fig. 4-2. Notice that the joint density of states is smoother and less angular in shape than the separate densities of states. In particular, it can be seen that the joint density of states rises proportional to $(\hbar\omega - E_0)^2$ just above the threshold, E_0. A more realistic assumption of transitions allowed only between states of the same wave number leads to a much more abrupt rise proportional to $(\hbar\omega - E_0)^{1/2}$ (see, for example, Phillips, 1973a, p. 158). Thus the joint-density-of-states model gives poor detail for $\chi_2(\omega)$ even though it describes some of the gross features. In particular, the large peak in the joint density of states (which we associate with the energy E_2 in the real spectrum) is seen to arise from transitions from p-like states in the upper two valence bands to the corresponding p-like states in the conduction bands. This important feature will be discussed in Section 4-C. The large quantitative discrepancy between the peak position in Fig. 4-2 and that in real germanium is attributable to the inadequacy of the density of states used and is not significant.

The most serious error encountered when using joint density of states as a model for χ_2 is its neglect of the fact that for a perfect crystal, the matrix elements $\langle \psi_{k'} | \partial/\partial x | \psi_k \rangle$ are zero unless both states correspond to the same wave number in the Brillouin Zone; that is, only *vertical transitions* in a band diagram (such as is

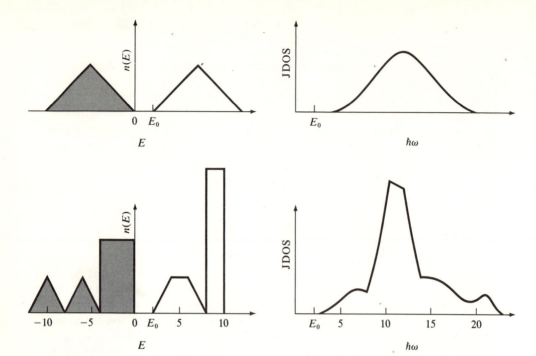

FIGURE 4-2
A simple approximation to the optical absorption spectrum is the joint density of states, corresponding to the assumption that equal matrix elements of x couple occupied states of density $n_v(E)$ with empty states of density $n_c(E)$. In the upper diagram, the joint density of states, $\mathrm{JDOS} = \int dE' n_v(E') n_c(E' + \hbar\omega)$, is shown on the right for the triangular $n_v(E)$ (shaded) and triangular $n_c(E)$ (unshaded), shown on the left. Below, the densities of states shown on the left—an approximation to those given for Ge by Chadi and Cohen (1975)—leads to the joint density of states shown on the right.

shown in Fig. 3-8) are allowed. The use of this restriction, combined again with a simple assumption about the magnitudes of the matrix elements, can be used with the scheme outlined in Section 2-E to give a much more realistic prediction. An assumption of equal matrix elements of $\partial/\partial x$, rather than of x, would also be much better.

This still leaves an important error in the matrix elements. It can be shown by symmetry that matrix elements between certain wave functions vanish; e.g., the matrix element of $\partial/\partial x$ between two band states of Γ_1 symmetry is zero just as it is for $\partial/\partial x$ between two s states in the free atom. Such dependence on band and wave number could also be readily incorporated by using the decomposition of each state into hybrids or atomic orbitals, again, if one desired, with a single-parameter description of the remaining matrix elements. It would be interesting to see how adequate a description of the spectrum, made entirely in terms of the parameters of the Solid State Table, this would produce. Certainly the main features of the spectrum would be predicted, and a more complete calculation, such as that which is illustrated in Fig. 4-1, can very accurately produce the real spectrum.

The discussion in this section has entirely concerned electron transitions between states within the solid. If the final state is of sufficient energy, the electron may emerge from the crystal; this is known as the ***photoelectric effect***. By measuring the energy of the emerging electron, one learns not only the energy difference obtained in optical absorption measurements but the absolute energy of the initial and final states, a method for studying electronic structure that has been used extensively and fruitfully. (For a recent reference, see Chye et al., 1977.) Even further information is obtained by measuring the angle of exit of the electron, the ***angle-resolved photoelectric emission***; it gives the transverse component of the wave number of the states involved. The variation of emission intensity with angle can also give information about the type of initial state involved. When the final state energy is fairly high (photon energies of some 20 eV) one may approximate the final state by a plane wave, as proposed by Gadzuk (1974). Then the matrix elements appearing, for example, in Eq. (4-3), are seen to contain a factor $\langle e^{i\mathbf{k}'\cdot\mathbf{r}}|\partial/\partial x|\psi_{\mathbf{k}}\rangle$, where the field is in the x-direction. Using Eq. (4-12), this becomes $ik'_x\langle e^{i\mathbf{k}'\cdot\mathbf{r}}|\psi_{\mathbf{k}}\rangle$. For a given LCAO composition of the initial state, one may predict the variation of the intensity with orientation of final-state wave number $\mathbf{k}'$, thus identifying the initial state symmetry. This is beautifully illustrated with a recent treatment of angle-resolved photoemission of chlorine chemisorbed on silicon by Larsen et al. (1978). It is not clear, however, how generally this simple approximation will apply.

4-C   Features of the Absorption Spectrum

Let us now turn to the relation between energy bands and the spectrum. For this, see Fig. 4-3, which shows the absorption spectrum for InAs; the corresponding energy bands are shown in Fig. 4-4. InAs is convenient for identifying features of the spectrum.

The Threshold, E_0

Notice first that the minimum gap between the valence and conduction bands in Fig. 4-4 occurs at Γ; this is the minimum gap for the entire Brillouin Zone. The corresponding energy difference, in this case less than 1 eV, is called E_0 and, as indicated in Fig. 4-3, is the minimum energy at which absorption occurs. It is the same E_0 evaluated in Eq. (3-39) and discussed there. The situation has a complication in the homopolar semiconductors in that the minimum energy in the conduction band does not occur at Γ where the valence-band maximum is. The difference is called an ***indirect gap*** and absorption cannot occur at that energy in the absence of other perturbations such as thermal vibrations. For this reason we chose InAs as better suited for discussion than silicon.

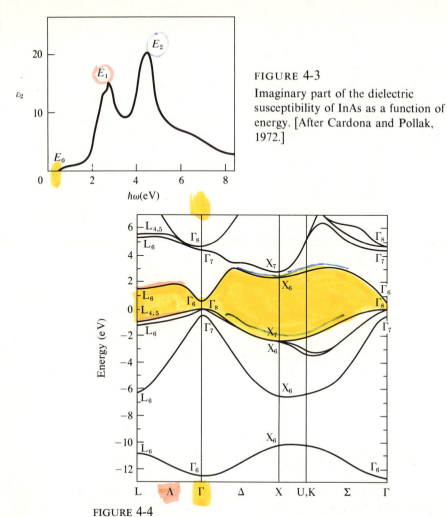

FIGURE 4-3

Imaginary part of the dielectric susceptibility of InAs as a function of energy. [After Cardona and Pollak, 1972.]

FIGURE 4-4

Energy bands of InAs. [After Chelikowsky and Cohen, 1976b.]

The E_1 Peak

The first peak in the absorption, labelled E_1 in Fig. 4-3, occurs at an energy slightly above 2 eV and can be identified with the parallel bands along Λ at the left of Fig. 4-4. The occurrence of parallel bands is very common in semiconductors and is easily understandable in terms of the pseudopotential approach (Harrison, 1966b), which will be discussed in Chapter 18. Because the bands are nearly parallel, many pairs of states have the same energy difference and contribute absorption to a single peak. Careful studies of the energy contributed from the entire Brillouin Zone show that large amounts come from large regions of the zone and not just along symmetry lines (Kane, 1966), but certainly the region along Λ makes an important contribution to the E_1 peak. Ren (1979) has suggested that this peak be associated with the average splitting between conduction s and valence p bands.

The E_2 Peak

The largest peak in Fig. 4-3, labelled E_2, can be easily identified with the fairly parallel bands separated by between 4 and 6 eV over most of the region shown in Fig. 4-4. (A careful study of this was made recently by Kondo and Moritani, 1977.) These bands arise from the Jones Zone, which will be discussed in detail in the treatment of tetrahedral semiconductors with pseudopotentials in Chapter 18. The energy at which this peak occurs was used earlier as a basis for obtaining experimental values for the covalent energy V_2 (Harrison and Ciraci, 1974).

The separation of this particular set of parallel bands can be estimated by observing the separation of the Γ_8 levels (or the Γ_7 levels) at the center of the Brillouin Zone. (Notice, incidentally, that though both sets of levels are purely p-like, they are nevertheless coupled by light owing to the interatomic matrix elements of the perturbation.) The slight splitting between the Γ_7 and Γ_8 levels arises from spin-orbit coupling, which will be discussed in Chapter 6. If that splitting is eliminated, the levels coalesce into the triply degenerate levels for which we have LCAO energy eigenvalues from the second equation of the pair at Eqs. (3-27). The separation between the two triply degenerate sets, which we now identify as the energy of the E_2 peak, is

$$E_2 \approx 2 \left[\left(\frac{\varepsilon_p^c - \varepsilon_p^a}{2} \right)^2 + \left(\frac{4}{3} V_{pp\sigma} - \frac{8}{3} V_{pp\pi} \right)^2 \right]^{1/2}. \tag{4-17}$$

We may substitute values for InAs from the Solid State Table to obtain 5.8 eV directly, in rough agreement with the observed peak position of 4.5 eV. We shall see soon that the ratio of estimated to observed peak energy is very much the same in tetrahedral solids other than InAs. The discrepancy comes principally from the error in our estimates of the energies of conduction bands. It does not come principally from differences in the splitting at Γ (from accurate band structures) and the observed peak energy. The error in scale will be absorbed in the parameters that will be introduced in the discussion of χ_1 in Section 4-D.

The picture that emerges is remarkable and applies to all tetrahedral semiconductors. The principal peak in the optical absorption comes at an energy determined by matrix elements between p orbitals, rather than between hybrids. To be sure, the s orbitals are necessary for any reasonable description of the energy bands or for calculating the full absorption spectrum, but the strongest features in $\chi_2(\omega)$ and, by Eq. (4-4), in $\chi_1(\omega)$, and perhaps in all dielectric properties, are dominated by p orbitals.

Covalent and Polar Energies

Let us make the formulation specific. The principal optical peak E_2 is the relevant measurement of bonding–antibonding splitting for dielectric properties.

For homopolar semiconductors, we see from Eq. (4-17) that it is twice the *covalent energy*—in contrast to the hybrid covalent energy of Eq. (3-6)—defined by

$$V_2 = \frac{4}{3}V_{pp\sigma} - \frac{8}{3}V_{pp\pi} = 2.16\hbar^2/(md^2). \tag{4-18}$$

The last form is obtained by using the expressions for the interatomic matrix elements from the Solid State Table. For polar semiconductors the splitting becomes $2(V_2^2 + V_3^2)^{1/2}$, with the *polar energy*—in contrast to the hybrid polar energy of Eq. (3-4)—defined by

$$V_3 = (\varepsilon_p^c - \varepsilon_p^a)/2. \tag{4-19}$$

By fitting the parameters of the Bond Orbital Model to the observed E_2 peaks, Harrison and Ciraci (1974) obtained experimental values for the covalent and polar energies. Predicted values will be used here instead, obtained from Eq. (4-17) and the parameters in the Solid State Table. Some discrepancy will be found between these values and the experimental E_2 values of Harrison and Circaci. A comparison is made in Fig. 4-5 for the systems for which experimental values of E_2 were found. The line drawn there has slope 0.84; this 16 percent

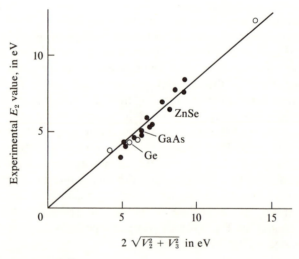

FIGURE 4-5

The observed energies of the E_2 peak, as compiled by Van Vechten (1969b), plotted against the prediction based upon atomic matrix elements and atomic term values from the Solid State Table. The empty circles are the homopolar systems Sn, Ge, Si, and C, in order of increasing gap.

discrepancy is not at all inconsistent with variations in the separation of the bands reflected in Fig. 4-4; notice, for example, the difference between the splitting at Γ and the splitting at X. The scatter in the points around this line is typical of the scatter found in computing other properties in terms of the LCAO basis. Adjusting the energies to fit E_2 exactly will not eliminate the scatter found for other properties; the choice in this book to use the Chadi-Cohen values for Si and Ge, the d^{-2} scaling for V_2, and the Herman-Skillman term values for V_3 seems to give an adequate set of values. Notice in particular that the labelled points give both the trend with metallicity and the trend with polarity correctly. Scaling all the values so that the slope for the points in Fig. 4-5 would be unity does not significantly improve the picture. When we discuss other dielectric properties, we shall find that they depend upon oscillator strengths and that a scale factor γ must be introduced. A scaling of the covalent and polar energies by 15 percent would simply change the value of γ by 15 percent and leave the calculated susceptibilities the same.

Although we have chosen to define covalent and polar energies for the solid by using formulae for the predicted position of the E_2 peak, we should not forget that ultimately the theory of the properties is based upon the full LCAO theory and that ignoring all matrix elements except V_2 and V_3, as in the Bond Orbital Approximation (Harrison, 1973; Harrison and Ciraci, 1974), can lead to significant errors. If we are willing to tolerate those errors, then this choice of parameters is a good one, and better than the hybrid matrix elements given in Eqs. (3-4) and (3-6). The polarity $\alpha_p = V_3/(V_2^2 + V_3^2)^{1/2}$ is a good measure of the polarity of the bond. In discussing dielectric properties, we shall also find it convenient to use the complementary quantity, covalency, which was introduced for molecular bonds in Eq. (1-38). The covalency is given by

$$\alpha_c = V_2/(V_2^2 + V_3^2)^{1/2}. \tag{4-20}$$

Values of the covalent and polar energies defined here are included in Table 4-1 in the following section.

Effects of Pressure

All of the discussion in this section has concerned systems at the equilibrium value of bond length d, though we have discussed variations from material to material that depend upon variations in bond length from material to material. We may also discuss variations in the parameters for a given material under pressure; there is no obvious reason why these two variations should be the same. However, the universal coefficients, $\eta_{ll'm}$, derived by matching the LCAO bands and free-electron bands would suggest the same dependence might be expected. (This remarkable result suggests that one could estimate the pressure dependence of the optical absorption peak for silicon by interpolating, as a function of bond length, between equilibrium values for silicon and diamond.) Indeed, the pressure dependence of E_2 has been measured by Zallen and Paul (1967), who found values of -2.0 and -3.2 for $\partial \ln E_2 / \partial \ln d$ for silicon and germanium, respectively. A value of -2 would be predicted for all semiconductors by the d^{-2} dependence, and this is in semiquantitative agreement. Our estimate of V_3, on the other hand, came from atomic term values and we might expect it to be independent of d. It would follow immediately that $\partial \ln E_2 / \partial \ln d$ for the polar semiconductors would be $-2(1 - \alpha_p^2)$, or about -1.5 for semiconducting compounds consisting of one element from column 3 and the other from column 5 in the Solid State Table. Pseudopotential theory, on the other hand, would again suggest a universal value of -2. Measurements by Zallen and Paul (1967) gave values of -2.1 and -1.7 for GaSb and InSb, respectively. The measurements do not seem extensive enough or certain enough to make either of the two suggested dependences a better choice than the other.

4-D χ_1 and the Dielectric Constant

Although $\chi_2(\omega)$ has a rather complete and direct relation to electronic structure, and though the low-frequency (in comparison to $E_0 / \hbar$) susceptibility is only a single number, it is a number of very great significance. First, it directly expresses the electronic contribution to the ordinary static dielectric constant. This may be seen by imagining a static electric field $\mathscr{E}_0$ applied perpendicular to a slab of a covalent solid, as illustrated in Fig. 4-6. The resulting field inside the material, $\mathscr{E}$, will be different because of surface charges. These are calculated by writing the polarization density, from Eq. (4-1), as $\mathbf{P} = \chi \mathscr{E}$, giving a surface charge density of the same magnitude and, by Gauss's law, a discontinuity in the field of 4π times the same magnitude, or

$$\mathscr{E}_0 - \mathscr{E} = 4\pi \mathbf{P} = 4\pi \chi \mathscr{E}. \tag{4-21}$$

The dielectric constant ε_1 is defined to be the ratio of the applied field $\mathscr{E}_0$ to the field inside the material, $\mathscr{E}$; and from Eq. (4-21), the dielectric constant is given by

$$\varepsilon_1 = 1 + 4\pi \chi_1. \tag{4-22}$$

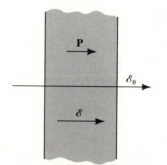

FIGURE 4-6

A field $\mathscr{E}_0$ applied to a dielectric gives rise to a smaller field $\mathscr{E}$ within the material. The difference is calculated by noting that the polarization inside, $\mathbf{P} = \chi \mathscr{E}$, causes a surface charge density P and a resulting opposing field $\mathscr{E}_0 - \mathscr{E} = 4\pi \mathbf{P}$.

It is also a familiar result of electromagnetic theory that the *refractive index*, n (the ratio of the speed of light in vacuum to the speed of light in the material), is given by

$$n = \varepsilon_1^{1/2},$$

if the magnetic susceptibility is very small; in Chapter 5 it will be seen that this is so. The refractive index should be determined at frequencies sufficiently low that χ_2 is zero and sufficiently high that any displacements of the atoms themselves can be neglected.

Repeating the formula for susceptibility written at Eq. (4-5),

$$\chi_1(0) = \frac{2\hbar^4 e^2}{m^2 \Omega} \sum_{\substack{k \text{ (full)} \\ k' \text{ (empty)}}} \frac{|\langle \psi_{k'} | \frac{\partial}{\partial x} | \psi_k \rangle|^2}{(\varepsilon_{k'} - \varepsilon_k)^3}, \tag{4-23}$$

observe that the expression to be summed is directly proportional to $\chi_2[(\varepsilon_{k'} - \varepsilon_k)/\hbar]$, as may be seen from Eqs. (4-8), (4-9), and (4-10), and is strongly peaked at $\varepsilon_{k'} - \varepsilon_k$ equal to the energy E_2 (Fig. 4-3). We have estimated that energy as $2(V_2^2 + V_3^2)^{1/2}$ and may therefore replace the denominator in Eq. (4-23) by the cube of this value and take it out of the sum. This further enables us to use unitarity, as in Eq. (3-31), to write the expression for the susceptibility in terms of bonding and antibonding orbitals:

$$\chi_1(0) = \frac{\hbar^4 N e^2}{4m^2 (V_2^2 + V_3^2)^{3/2}} \sum_a \langle a | \frac{\partial}{\partial x} | b \rangle^2. \tag{4-24}$$

Here also we have performed the sum over bonding orbitals to write the result in terms of the electron density N.

It would have been tempting at some stage in this argument to use the sum rule, Eq. (4-14), to eliminate the matrix elements. This cannot be done without making corrections (Phillips, 1970), because the sum is over all states k', not just the empty ones. In addition, the fact that the energy difference $\varepsilon_{k'} - \varepsilon_k$ enters the sum rule and the susceptibility with a different power leads to a serious error in this case. We noted following Eq. (4-13) that the dimensionless oscillator strength varies in proportion to the covalency in an isoelectronic series, giving an extra factor of $V_2/(V_2^2 + V_3^2)^{1/2}$ in the susceptibility. The existence of this factor is confirmed by a number of experimental and theoretical tests (Harrison and Pantelides, 1976; Phillips, 1975; and Harrison, 1975). The physical origin of this effect is a transfer of probability density for the bonding states to the anion and for the antibonding states to the cation, with a corresponding decrease in dimensionless oscillator strength. The sum rule tells us that oscillator strength cannot disappear, but it is shifted to higher energies where it does not contribute importantly to the dielectric susceptibility.

We have written the susceptibility in terms of matrix elements of $\partial/\partial x$, but we can use Eq. (4-6) to write it immediately in terms of matrix elements of x,

$$\chi_1(0) = \frac{Ne^2}{(V_2^2 + V_3^2)^{1/2}} \sum_a \langle a|x|b\rangle^2, \tag{4-25}$$

or in terms of the dimensionless oscillator strengths,

$$\chi_1(0) = \frac{Ne^2}{4m(V_2^2 + V_3^2)} \sum_a f_{ab}. \tag{4-26}$$

Making these transformations assumes that contributions to the susceptibility come from a sufficiently narrow range of energies that only a single energy denominator need be considered. The form in terms of the matrix elements of x is particularly appealing because it makes it rather easy to estimate the value. We proceed to that next.

Estimate of Values

For each bond orbital $|b\rangle$ there remains a sum over antibonding orbitals $\langle a|$, with the largest term coming, presumably, from the antibonding orbital in the same site. Let us consider that term first. The matrix elements can be written in terms of matrix elements of the hybrids on the two atoms by using the coefficients defined in Eqs. (3-13) and (3-16) for the bonding and antibonding states:

$$\begin{aligned}
\langle b|x|a\rangle &= u_1^b u_1^a \langle h_1|x|h_1\rangle + u_2^b u_2^a \langle h_2|x|h_2\rangle + (u_1^b u_2^a + u_1^a u_2^b)\langle h_1|x|h_2\rangle \\
&= \tfrac{1}{2}(1 - \alpha_p^2)^{1/2}(\langle h_1|x|h_1\rangle - \langle h_2|x|h_2\rangle) - \alpha_p\langle h_1|x|h_2\rangle.
\end{aligned} \tag{4-27}$$

The calculation is illustrated in Fig. 4-7. If we take our origin of coordinates midway between atoms, and if $|h_1\rangle$ and $|h_2\rangle$ are symmetrically disposed, we see that $\langle h_1|x|h_2\rangle = 0$; since $|h_1\rangle$ and $|h_2\rangle$ are taken to be orthogonal to each other, adding a constant to x (changing the origin of coordinates) does not change the result. Now $\langle h_1|x|h_1\rangle - \langle h_2|x|h_2\rangle$ is just the separation of the center of

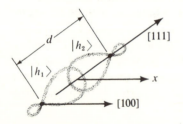

FIGURE 4-7

A schematic diagram of the hybrids making up a bond, and the calculation of the bonding–antibonding matrix element for a field in a [100] direction. For this choice of field-direction, all bonds make the same angle (with cosine $= 3^{-1/2}$) with the field.

gravity in the x-direction of the two hybrids; it would be $d/3^{1/2}$ if the hybrids were localized at the nuclei.

In keeping with other procedures in this text, we should take this value for the matrix elements and proceed to the calculation of the susceptibility with no parameters other than the V_2, V_3, and d, which are obtainable from the Solid State Table. This is the way susceptibility is treated for all other systems. However, the errors become sizable (up to a factor of 3.4) when the susceptibility becomes very large, so we introduce a scale parameter γ that multiplies the $d/3^{1/2}$ value. Deviations of γ from unity may be taken as a measure of the error in the simplest concept of the susceptibility of covalent solids. Proceeding this way will also enable us to systematize the corrections to this simplest concept. It will also enable us to see, for example, that the simplest treatment *does* give the correct dependence on polarity, though it fails to give the correct dependence on metallicity. Thus we take $\langle b|x|a\rangle = (\gamma d/2\sqrt{3})(1 - \alpha_p^2)^{1/2}$. The susceptibility then becomes

$$\chi_1 = Ne^2 d^2 \gamma^2 (1 - \alpha_p^2)/[12(V_2^2 + V_3^2)^{1/2}]$$
$$= Ne^2 \gamma^2 d^2 \alpha_c^3/(12V_2). \tag{4-28}$$

This calculation has included only the matrix element with the antibond lying in the same site with the bond state. There are also matrix elements of x with the antibonds on the nearest neighbor sites. For those arising from hybrids on atoms shared by the bond and antibond, the same combination of coefficients $u_1^a u_1^b$ and $u_2^a u_2^b$ enters into the calculation, so these simply make an additional contribution to γ. On the other hand, contributions arising from nearest-neighbor hybrids enter with a different combination of coefficients; these contributions would have a different dependence upon polarity, but it can be hoped the contributions are smaller. Similarly, if contributions from the corona of the extended bond orbital (Fig. 3-9) are made, some will add directly to γ and others will enter in different form. It was indicated earlier that a different choice of scale for the covalent and polar energies will also shift the γ value. Even the local field corrections discussed at the beginning of the chapter may be thought of as corrections to γ. All in all, many different corrections are absorbed in γ and even without those corrections we can only estimate that it is of order one. We shall fit it to the homopolar susceptibility $\chi_1(0)$ and then use the same value for all other dielectric properties and for the polar compounds. For solids other than the tetrahedral semiconductors we will drop the factor of γ.

The formula at Eq. (4-28) applies to the wurtzite structure if nearest neighbors form a regular tetrahedron (the ideal axial ratio). We see this by observing that the same answer would have been obtained in the zincblende structure with a field along a [111] direction, and that one can go from the zincblende to the wurzite structure with bond rotations all made around that [111] axis.

Equation (4-28) has given a form for $\chi_1(0)$ with all parameters known except γ. Experimental values of $\chi_1(0)$ then give experimental values for γ for the semiconductors, but a consideration of those values indicates that the only important

TABLE 4-1

Parameters for tetrahedral semiconductors, and predicted and experimental dielectric constants. A compilation for other classes of solids has been recently made by Wemple (1977).

Material	$d(\text{Å})$	V_2 (eV)	V_3 (eV)	α_c	α_m from Eq. (3-42)	γ	ε_1 from Eqs. (4-22), (4-28)	ε_1 (experimental)
C	1.54	6.94	0	1	0.34	1.13	5.7†	5.7
BN	1.57	6.68	2.42	0.94	0.34	1.13	4.97	4.5
BeO*	1.65	6.05	3.75	0.85	0.37	1.13	4.08	3.0
Si	2.35	2.98	0	1	0.66	1.40	12.0†	12.0
AlP	2.36	2.96	1.74	0.86	0.57	1.40	8.06	8.0
Ge	2.44	2.76	0	1	0.81	1.60	16.0†	16.0
GaAs	2.45	2.74	1.51	0.88	0.71	1.60	11.03	10.9
ZnSe	2.45	2.74	3.08	0.66	0.53	1.60	5.39	5.9
CuBr	2.49	2.65	4.69	0.49	0.44	1.60	2.81	4.4
Sn	2.80	2.10	0	1	0.87	1.84	24.00†	24.0
InSb	2.81	2.08	1.28	0.85	0.74	1.84	15.00	15.7
CdTe	2.81	2.08	2.61	0.62	0.49	1.84	6.48	7.2
AgI	2.80	2.10	3.96	0.47	0.43	1.84	3.32	4.9
SiC	1.88	4.66	1.23	0.97	0.45	1.26	7.42	6.7
BP	1.97	4.24	0.85	0.98	0.47	1.26	8.02	—
AlN*	1.89	4.61	3.31	0.81	0.41	1.26	4.83	4.8
BeS	2.10	3.73	3.04	0.78	0.42	1.26	4.70	—
BAs	2.07	3.84	0.64	0.99	0.55	1.34	9.49	—
GaN*	1.94	4.37	3.29	0.80	0.46	1.34	5.23	—
BeSe	2.20	3.40	2.67	0.79	0.47	1.34	5.58	—
ZnO*	1.98	4.20	5.38	0.62	0.41	1.34	2.91	4.0
CuF	1.84	4.86	7.58	0.54	0.37	1.34	2.24	—

trend is with metallicity. We therefore fit γ for C, Si, Ge, and Sn (they are indeed of order one), take the geometric mean of the values in each of the two rows for *skew compounds* (those composed of two elements, each from a different row in the Solid State Table), and obtain predictions for all of the others. The parameters that enter the calculation, the predicted dielectric constants, and experimental values are listed in Table 4-1. As a set of predictions, what results is not so impressive, because the variation from material to material is systematic, as is the theory, so that with the few parameters being used, one can expect success. The parameters can also be used to discuss a wide variety of other properties, however.

Wemple (1977) has made a far-ranging study of oscillator strengths and found a correlation corresponding to a dependence $\gamma^2 \propto d$ with a proportionality constant

Material	d(Å)	V_2 (eV)	V_3 (eV)	α_c	α_m from Eq. (3-42)	γ	ε_1 from Eqs. (4-22), (4-28)	ε_1 (experimental)
InN*	2.15	3.56	3.39	0.72	0.48	1.44	5.03	—
BeTe	2.40	2.86	2.20	0.79	0.48	1.44	6.90	—
AlAs	2.43	2.79	1.53	0.88	0.64	1.50	9.78	9.1
GaP	2.36	2.92	1.72	0.86	0.62	1.50	9.02	9.1
ZnS	2.34	3.01	3.45	0.66	0.47	1.50	4.56	5.2
CuCl	2.34	3.01	5.24	0.50	0.40	1.50	2.55	5.7
AlSb	2.66	2.33	1.19	0.89	0.68	1.60	12.45	10.2
InP	2.54	2.55	1.82	0.81	0.63	1.60	9.35	9.6
MgTe*	2.76	2.16	2.80	0.61	0.49	1.60	4.83	—
CdS	2.53	2.57	3.45	0.60	0.48	1.60	4.29	5.2
GaSb	2.65	2.34	1.17	0.89	0.74	1.72	14.36	14.4
InAs	2.61	2.42	1.61	0.83	0.71	1.72	11.61	12.3
ZnTe	2.64	2.36	2.61	0.67	0.53	1.72	6.61	7.3
CdSe*	2.63	2.38	3.08	0.61	0.54	1.72	5.24	5.8
CuI	2.62	2.40	4.07	0.51	0.43	1.72	3.42	5.5

SOURCES of experimental values of ε_1: C, Si, Ge from Goryuneva (1965); BN from Gielisse et al. (1967); BeO, CuBr, AgI, ZnO, CuCl, CuI from Weast (1965); AlP from Tsang et al. (1975); GaAs, InSb, SiC, AlSb, InP, GaSb, InAs from Burstein et al. (1967), and from Hass and Henvis (1962); ZnSe, CdTe, ZnTe from Marple (1964); Sn from Lindquist and Ewald (1964); AlN from Collins et al. (1967); AlAs, GaP from Barker (1968); ZnS from Czyzak et al. (1957); CdS from Bieniewski and Czyzak (1963); and CdSe from Manabe et al. (1967).

* Wurtzite structure.

† Values of γ have been adjusted to bring these into agreement with experiment.

depending on the class of solid considered. The values in Table 4-1 illustrate this for the tetrahedral solids.

Pressure Dependence

Before formulating the susceptibility in more detail in Chapter 5, we should examine the dependence of the susceptibility upon pressure. Noting that N is proportional to d^{-3} and V_2 is proportional to d^{-2}, we see from Eq. (4-9) that the susceptibility is expected to vary linearly with bond length for an elemental semi-conductor. A recent measurement of $\partial \ln \chi / \partial \ln d$ for silicon by Biegelsen (1975) gave 0.71, in reasonable agreement with the prediction of unity. Measurements compiled by Kastner (1972)—$\partial \ln \chi / \partial \ln d$ is twice his parameter n—give 2.2 for diamond, 2 for silicon, and 4.6 for germanium, at variance with theory and with Biegelsen's experiment. Effects of anisotropic strains have also been studied extensively, by Cardona (1972), Yu et al. (1971), Higginbotham et al. (1969), Shileika et al. (1969), Gavini and Cardona (1969), and Sanchez and Cardona (1971), but this work will not be discussed here.

The observation made earlier, that the E_2 values in compounds vary similarly with variation in bond length would in general suggest a value of $\partial \ln \chi / \partial \ln d$ of order unity for compounds as well as homopolar semiconductors (Kastner, 1972, lists 1.8 and 3.6 for GaP and GaAs, respectively).

The increase in band gap and decrease in oscillator strength tend to outweigh the increase attributable simply to an increase in the density of polarizable bonds. The exception is in materials with strong directional bonding, such as graphite. When directional bonding is strong the application of pressure is expected simply to move the "molecular" units closer together without greatly modifying the bond lengths or band gaps. We may then expect an increase in χ with pressure corresponding roughly to $\partial \ln \chi / \partial \ln d = -3$. Kastner (1972) has obtained values between -5 and -8 for several amorphous chalcogenide semiconductors that fit this description.

This perturbation-theoretic formulation of the dielectric susceptibility is most appropriate and will be used in the treatment of dielectric properties based upon pseudopotentials in Chapter 18. However, for the calculation of higher-order susceptibilities, a more direct approach in terms of bond dipoles is more convenient. Because the two derivations are equivalent, the bond-dipole approach will also enable us to establish parameters for the bond dipoles and effective charges in terms of the parameters listed in Table 4-1.

Recent Developments

In a very recent unpublished study of birefringence made by the author and by M. Ewbank and E. Kraut of the Rockwell International Science Center, it was noticed that the matrix element of $\partial / \partial x$ between two atomic states on adjacent atoms can be evaluated by isolating those states and constructing the corresponding "molecular" bonding and antibonding states, using the term value for each and the interatomic matrix element $V_{ll'm}$, all from the Solid State Table. The matrix element of x between these bonding and antibonding states is then found to be $\alpha_c d_x / 2$, where d_x is the x-component of $\mathbf{d}$; this assumed only that the matrix element of x between the two atomic states is zero, as was assumed following Eq. (4-27). Eq. (4-6) can then be used, since the bonding and antibonding states are localized, to write this matrix element of x in terms of the matrix element of $\partial / \partial x$. This gives the remarkably simple result that the matrix element of the gradient between the two atomic states is simply $-\mathbf{d}/d^2$ times the $\eta_{ll'm}$ from Table 2-1 for this pair of states. This elegant result was not available at the time the analysis in this chapter was made. Its use with p states to obtain the matrix elements in Eq. (4-23) leads to a susceptibility of the form given in Eq. (4-28), with a value of γ close to that for germanium; however, it fails to give a variation of γ from row to row in the Solid State Table.

PROBLEM 4-1 *Features in the optical spectra*

Estimate E_0 (see Eq. 3-39) and E_2 (see Eq. 4-17) from the Solid State Table, for Ge, GaAs, and ZnSe. Experimental values obtained from Phillips (1973a, p. 169) for E_0 are 0.89, 1.55, and 2.68 eV, respectively. For E_2 they are 4.3, 4.85, and 6.4 eV, respectively.

PROBLEM 4-2 *Polarity in* BN

Rederive an expression for the covalent energy (see Eq. 4-18) for the graphite structure, and evaluate the corresponding polarity for hexagonal BN, assuming that it has the graphite bond length (see Problem 3-1).

PROBLEM 4-3 *The dielectric susceptibility*

Rederive the contribution to the susceptibility (see Eq. 4-28) for sp^2 hybrids in the graphite structure and fields in the plane of the graphite. The electron density that enters is $\frac{3}{4}$ of the total density, since we exclude the π electrons. To calculate that density, you need the spacing between layers, 3.4 Å. Evaluate the result by taking the same γ as for diamond.

CHAPTER 5

Other Dielectric Properties

SUMMARY

An alternative formulation of the dielectric susceptibility can be made in terms of bond dipoles, by using the γ value from Chapter 4 to set the scale of the dipole. This allows elementary prediction of the higher-order susceptibilities, though the accuracy of the prediction decreases with each order. The analysis of charge distribution within the bonds, from Chapter 4, provides a definition of effective atomic charges in the covalent solids and covalent alloys, and a theory of the screening of impurities. For calculation of the magnetic susceptibility we return to a perturbative approach analogous to that used in Chapter 4, and calculate both diamagnetic and paramagnetic contributions.

The discussion in Chapter 4 was directed at the study of the linear response of covalent systems to light. We turn next to a series of related properties that can be studied, using the parameters developed in Chapter 4 and presented in Table 4-1. The first is the response of a higher-than-linear order in the applied field. To study this property we shall reformulate the response problem in a form that will be very useful for understanding other properties as well.

5-A   Bond Dipoles and Higher-Order Susceptibilities

Polarizations of higher order in the applied field have become important recently because of their use in generating harmonic signals with lasers and because lasers have made available fields sufficiently large that higher-order corrections become significant. A treatment of the third-order susceptibility based on a model similar to the one that will be used here has been made by Flytzanis (1970) and a rather direct application to second-order susceptibility was made by Choy, Ciraci, and Byer (1975). For both of these susceptibilities a number of corrections were incorporated in accounting for observed values. Rather than examine them, it will be of more interest here to approach second-order and third-order susceptibilities directly without additional parameters or corrections to see what can be learned. For a complete description of the theory and a detailing of its complexities, see Flytzanis (1975).

The procedure here will be to compute the dipole to be associated with a bond, proportional to $\langle b|x|b \rangle$, multiply it by $e\mathscr{E}$ to obtain a correction to the bond energy, and again minimize the energy to obtain the equilibrium bond dipole in the presence of a field. This will give an alternative derivation of Eq. (4-28), in fact, the way the formula was first derived (Harrison, 1973). It will then also directly give terms of higher order in the electric field. Note that here the field is independent of time, so that $\chi = \chi_1(0)$ is real; we can drop the subscript and parentheses and write χ.

First-Order Susceptibilities

Writing the bond orbital $|\psi\rangle = u_1|h^1\rangle + u_2|h^2\rangle$, as in Eq. (3-7), we seek

$$\langle \psi|x|\psi \rangle = u_1^2\langle h^1|x|h^1\rangle + u_2^2\langle h^2|x|h^2\rangle. \qquad (5\text{-}1)$$

(We have again taken $\langle h^1|x|h^2\rangle = 0$.) The value of $\langle \psi|x|\psi \rangle$ clearly depends upon the origin of coordinates, but we may measure it relative to the midpoint of the bond by subtracting $(u_1^2 + u_2^2)(\langle h^1|x|h^1\rangle + \langle h^2|x|h^2\rangle)/2$ from Eq. (4-27). This immediately gives

$$\langle \psi|x|\psi \rangle = (u_2^2 - u_1^2)(\langle h^2|x|h^2\rangle - \langle h^1|x|h^1\rangle)/2. \qquad (5\text{-}2)$$

We have given the value $\gamma d/\sqrt{3}$ to $\langle h^2|x|h^2\rangle - \langle h^1|x|h^1\rangle$, following Eq. (4-27), so that the moment becomes

$$\langle \psi|x|\psi \rangle = (u_2^2 - u_1^2)\gamma d/(2\sqrt{3}), \qquad (5\text{-}3)$$

It is desirable here to change to vector notation, writing the vector distance from the nucleus of atom 1 to that of atom 2 as $\mathbf{d}$, so that this result becomes

$$\langle \psi|\mathbf{r}|\psi \rangle = (u_2^2 - u_1^2)\gamma\mathbf{d}/2. \qquad (5\text{-}4)$$

This gives the center of gravity of the bond electron density relative to the bond center. The corresponding shift in electron energy due to an electric field $\mathscr{E}$ is simply $-(-e)\mathscr{E} \cdot \langle \psi | \mathbf{r} | \psi \rangle$, and for this formula the field need not be considered small. This shift may be added to the expectation value of the energy given in Eq. (3-8); it is seen to have exactly the effect of adding $e\gamma\mathbf{d} \cdot \mathscr{E}/2$ to ε_h^2 and subtracting the same amount from ε_h^1. This is equivalent to adding $e\gamma\mathbf{d} \cdot \mathscr{E}/2$ to V_3 so that when the energy is minimized to obtain u_1 and u_2, the expression for $u_2^2 - u_1^2$, which was $V_3/(V_2^2 + V_3^2)^{1/2}$—see Eqs. (3-13) and (3-14)—becomes

$$u_2^2 - u_1^2 = (V_3 + e\gamma\mathbf{d} \cdot \mathscr{E}/2)/[V_2^2 + (V_3 + e\gamma\mathbf{d} \cdot \mathscr{E}/2)^2]^{1/2}. \qquad (5\text{-}5)$$

This equilibrium value for $u_2^2 - u_1^2$ may be substituted into Eq. (5-4) and multiplied by $-e$ to obtain the electric dipole due to the corresponding electron, and multiplying further by two gives the dipole associated with the bond:

$$\mathbf{p} = e\gamma\mathbf{d}(V_3 + e\gamma\mathbf{d} \cdot \mathscr{E}/2)/[V_2^2 + (V_3 + e\gamma\mathbf{d} \cdot \mathscr{E}/2)^2]^{1/2}. \qquad (5\text{-}6)$$

Since the field is not assumed to be small, the result may be summed over bonds and used to calculate higher-order susceptibilities. Note that if the field is taken equal to zero, the dipole $e\gamma\mathbf{d}\alpha_p$ summed over the four bonds surrounding each atom is zero. The value to first order in the field is

$$\mathbf{p}^{(1)} = \frac{e^2\gamma^2(1 - \alpha_p^2)\mathbf{d} \cdot \mathscr{E}\mathbf{d}}{2(V_2^2 + V_3^2)^{1/2}}. \qquad (5\text{-}7)$$

To obtain the linear susceptibility, we must average over bond angles, which gives a factor of 1/3, and multiply by the bond density $N/2$ to obtain directly the result obtained earlier, Eq. (4-28).

Second-Order Susceptibilities

The higher-order susceptibilities are defined by an expansion of the polarization density in powers of the electric field, in the form

$$P_i = \sum_j \chi_{ij}^{(1)}\mathscr{E}_j + \sum_{jk} \chi_{ijk}^{(2)}\mathscr{E}_j\mathscr{E}_k + \sum_{jkl} \chi_{ijkl}^{(3)}\mathscr{E}_j\mathscr{E}_k\mathscr{E}_l + \cdots. \qquad (5\text{-}8)$$

The term $\chi_{ij}^{(1)}$ is the linear susceptibility we have just discussed; in the zincblende structure it was isotropic, $\chi_{11} = \chi_{22} = \chi_{33} = \chi$, and all other components were zero. All elements of the second-order susceptibility for a crystal having the zincblende symmetry can be shown to be zero unless all of the three indices are different (that is, when the coordinate axes are taken along the cube directions). The single element, usually called $\chi_{14}^{(2)}$ (the reasons for this notation will be more apparent when we discuss elastic constants in Chapter 8), can be evaluated by introducing a field $\mathscr{E}_1$, which we write $\mathscr{E}_x$, and a field $\mathscr{E}_3$, which we write $\mathscr{E}_z$, and computing the polarization in the *y*-direction. The terms χ_{213} and χ_{231} add, so the second-order terms in Eq. (5-8) become

$$P_y^{(2)} = 2\chi_{14}^{(2)} \mathcal{E}_x \mathcal{E}_z. \tag{5-9}$$

The evaluation is made by expanding Eq. (5-6) to second order in the field rather than only first as we did in Eq. (5-7). The second-order polarization for a particular bond is

$$\mathbf{p}^{(2)} = -(3/2)\gamma ed\alpha_p (1 - \alpha_p^2)^2 (\gamma e\mathbf{d} \cdot \mathcal{E}/2V_2)^2. \tag{5-10}$$

We again view a zincblende crystal along a cubic direction and construct the four bond vectors $\mathbf{d}_i$ around a nonmetallic atom as shown in Fig. 5-1.

We may immediately evaluate the geometrical quantities that enter Eq. (5-10), to obtain

Internuclear vector	$\mathbf{d} \cdot \mathcal{E}$	$\mathbf{d}(\mathbf{d} \cdot \mathcal{E})^2/(d/\sqrt{3})^3$	
$\mathbf{d}_1 = [111]d/\sqrt{3}$	$d(\mathcal{E}_x + \mathcal{E}_z)/\sqrt{3}$	$(\mathcal{E}_x^2 + 2\mathcal{E}_x\mathcal{E}_z + \mathcal{E}_z^2)[111]$	
$\mathbf{d}_2 = [1\bar{1}\bar{1}]d/\sqrt{3}$	$d(\mathcal{E}_x - \mathcal{E}_z)/\sqrt{3}$	$(\mathcal{E}_x^2 - 2\mathcal{E}_x\mathcal{E}_z - \mathcal{E}_z^2)[1\bar{1}\bar{1}]$	(5-11)
$\mathbf{d}_3 = [\bar{1}1\bar{1}]d/\sqrt{3}$	$d(-\mathcal{E}_x - \mathcal{E}_z)/\sqrt{3}$	$(\mathcal{E}_x^2 + 2\mathcal{E}_x\mathcal{E}_z + \mathcal{E}_z^2)[\bar{1}1\bar{1}]$	
$\mathbf{d}_4 = [\bar{1}\bar{1}1]d/\sqrt{3}$	$d(-\mathcal{E}_x + \mathcal{E}_z)/\sqrt{3}$	$(\mathcal{E}_x^2 - 2\mathcal{E}_x\mathcal{E}_z + \mathcal{E}_z^2)[\bar{1}\bar{1}1]$	

By summing over the four bonds we see that only the *y*-component, along [010], sums to a nonzero value; the contribution per bond is substituted into Eq. (5-10) to give

$$p_y^{(2)} = -3^{-1/2}(\gamma ed)^3 \alpha_p (1 - \alpha_p^2)^2 \mathcal{E}_x \mathcal{E}_z/(2V_2)^2. \tag{5-12}$$

We multiply by the bond density, $N/2$, to get the polarization density, which corresponds to a second-order susceptibility of magnitude

$$\chi_{14}^{(2)} = \frac{\sqrt{3}\,N}{48} \frac{(\gamma ed)^3}{V_2^2} \alpha_p (1 - \alpha_p^2)^2. \tag{5-13}$$

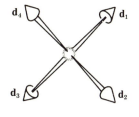

FIGURE 5-1

Bond orientations for a nonmetallic atom in a zincblende structure.

TABLE 5-1

Comparison between theoretical and experimental values of $\chi_{14}^{(2)}$.

| Compound | $\chi_{14}^{(2)}$ (10^{-7} electrostatic units) | |
	Theoretical	Experimental
GaP	4.4	5.2
GaAs	6.2	6.4
GaSb	10.6	20.0
InP	6.4	8.0
InAs	9.2	17.4
InSb	15.8	24.8
ZnTe	5.5	7.3
ZnSe	3.1	3.7
CdTe	6.7	8.0
CuCl	0.8	0.2
CuBr	1.2	0.4
CuI	2.1	0.4

SOURCES of data in right-hand column: GaP from Wynne and Bloembergen (1969); GaAs from McFee et al. (1970); GaSb, InAs, InSb from Chang et al. (1965); InP from Braunstein and Ockman (1964); ZnTe, CdTe from Soref and Moos (1964); ZnSe from Patel (1966); and CuCl, CuBr, CuI from Chemla et al. (1971).

Choy et al. (1975) noted that this peaks at a polarity of $5^{-1/2} = 0.45$, near that for compounds of groups 3–5 in the Solid State Table, and that it varies with bond length approximately as d^4; thus a high atomic number is indicated. One may conclude that InSb would be the best material of this category, and this turns out to be true.

For a more detailed comparison with experiment, we may write N in terms of d^3, substitute for e, and simplify Eq. (5-13) to read

$$\chi_{14}^{(2)} = (40 \times 10^{-7})\gamma^3 \alpha_p (1 - \alpha_p^2)^2 / V_2^2. \qquad (5\text{-}14)$$

The result is in electrostatic units if V_2 is in electron volts. We may substitute values from Table 4-1; the results are given in Table 5-1 along with experimental values.

The general scale of the results is correct, as is almost guaranteed by dimensional arguments (the charges that enter the calculations are electron charges, the distances are interatomic distances, and the energies are of the order of electron volts). However, the numerical agreement in most cases is quite remarkable in view of the fact that none of the experimental input has depended on higher-order susceptibilities.

There are sizable discrepancies for GaSb, InAs, and InSb—the three materials with very small gaps between the conduction and valence bands. It is not surprising that we would underestimate these; if these quantities were calculated fully by using perturbation theory, as in Eq. (4-7), for the linear susceptibility, we would find three powers of the matrix element of x and two powers of the energy denominator. This is in fact apparent in Eq. (5-13). To some degree the Bond Orbital Approximation replaces all bonding–antibonding gaps with an average and makes up the difference for $\chi^{(1)}$ by adjusting γ. A full calculation of $\chi^{(2)}$, with varying gaps entering the calculation squared in the denominator, will give a larger value than the approximation in which the average is used:

$$\left\langle \frac{1}{\Delta E^2} \right\rangle > \left\langle \frac{1}{\Delta E} \right\rangle^2. \tag{5-15}$$

We may expect this difficulty to be worse in susceptibilities of still higher order for any of the small-gap materials.

There are also significant errors in the noble-metal halides; certainly these are owing to influence by the noble-metal d bands. These contribute differently to the first-order and second-order susceptibility and as with measurements of many properties in the noble-metal halides, the discrepancies are larger than for other materials.

Third-Order Susceptibilities

We may expect all discrepancies found in second-order susceptibilities to become still worse in higher order, and this is certainly the case. The calculation of third-order susceptibilities is a straightforward extension of the calculation of second-order susceptibilities. Symmetry analysis of the zincblende structure shows that there are two independent components: those in which all indexes are the same (χ_{1111} is an example) and those in which the indexes are in pairs (χ_{1122} is an example). We obtain the susceptibility by obtaining the polarization per bond to third order in the field, in analogy with Eq. (5-10):

$$\mathbf{p}^{(3)} = \tfrac{1}{2}\gamma e\mathbf{d}(1 - \alpha_p^2)^{5/2}(5\alpha_p^2 - 1)(\gamma e\mathbf{d} \cdot \mathscr{E}/2V_2)^3. \tag{5-16}$$

If we introduce an electric field in the x-direction, each $\mathbf{d} \cdot \mathscr{E}$ becomes $d\mathscr{E}/\sqrt{3}$ and, multiplying by the bond density, we obtain the susceptibility

$$\chi^{(3)}_{1111} = \frac{N}{288} \frac{(\gamma ed)^4}{V_2^3} \alpha_c^5(5\alpha_p^2 - 1). \tag{5-17}$$

By introducing components along two axes, one sees that the corresponding susceptibility is the same for this model; that is, $\chi^{(3)}_{1122} = \chi^{(3)}_{1111}$, though experimental values for zincblende-type semiconductors give $\chi_{1122} \approx 0.5 \chi_{1111}$ (see Wynne, 1969, Wang and Ressler, 1970, and Yablonovitch et al., 1972).

Eq. (5-17) gives negative values for polarities less than 0.45, whereas experimental values are positive for homopolar as well as for polar semiconductors. Substitution of values into Eq. (5-17) leads to a χ_{1111} value for GaAs of 7×10^{-14} electrostatic units. The experimental value of Yablonovitch et al. (1972) is 970×10^{-14}. This large discrepancy reflects the extreme sensitivity of these higher-order susceptibilities to the details of the calculation, reflected by the high powers with which parameters such as the band gap enter the calculations. Indeed, because of these, the property tends to be dominated by states near the band edges and associated with the E_0 of Fig. 4-3, a fact noted earlier by Van Vechten et al. (1970).

In none of the discussions of dielectric response have we allowed a change in the angles between hybrids (such a change would correspond to a physical bending of the bond away from the line joining the atoms). Such changes are expected to produce small discrepancies, though the observation of χ_{1122}/χ_{1111} values near 0.5 suggests that they are far from negligible (Flytzanis, 1975, Section III,B,4b). In the context of bond orbitals, we might retain the orthogonality of the four hybrids around each atom but rotate them by changing the linear combination of s and p orbitals that enter each, in order to rotate the hybrid toward the direction of the field. However, we lose as much in one hybrid as we gain in the other. This follows from the unitarity of the transformation; the expectation value of the dipole associated with each hybrid, $\langle h_i | x | h_i \rangle$, summed over hybrids, is independent of the choice of orthogonal hybrids. The inclusion of bond-bending requires the addition of corrections to the wave function outside the context of those bonding and antibonding states that provide a description of the largest effects.

An important message has been gained from this study. First, the direct use of the bond-dipole formula, Eq. (5-6), in more complicated situations appears meaningful, though uncertainties grow very rapidly at higher order and in materials with very small band gaps. A consideration of the perturbation-theoretic formulation of other problems may be helpful in indicating the kinds of discrepancies that should be expected.

Choy et al. (1975) have considered also the second-order susceptibilities of several wurtzite crystals. The situation appears much the same as in the wide-gap zincblende structures.

5-B   Effective Atomic Charge

There is a danger in defining a charge associated with an atom in a solid. As plausible as it seems at the outset to assign a definite charge, we must realize that even if we knew precisely the charge distribution at all points in a crystalline structure, there is no unique way to divide it up and associate different parts with different atoms. Thus there is no unique atomic charge. This would be quibbling if a range of sensible ways of doing it gave nearly the same answer, but this is not the case in tetrahedrally coordinated solids. We will see, for example, that the two most direct definitions of an effective charge in terms of experiment give $+2.16$

and -0.47 for GaAs; they are not even of the same sign, and this ambiguity is typical. Nevertheless, there is a rather natural definition that will be an aid to thinking about polar covalent solids and it will be the starting point from which we can consider the various experimental charges and the very large corrections which lead to the ambiguity mentioned above. Furthermore, these effective charges give us a way of guessing the magnitudes and signs of local electric fields and field gradients that occur in solids. Such effects can be measured in nuclear magnetic resonance or in nuclear acoustic resonance. Sundfors (1974) has in fact been successful in predicting the tensor that relates electric-field gradients to elastic strains in terms of this effective charge and comparing these predictions with his nuclear acoustic resonance experiments. They are also relevant to the shift in core-level energies in compounds, but surface effects (to be discussed) may be equally important.

Whenever an ambiguity arises about the meaning of a quantity such as the effective charge, it can be removed by discussing only quantities that can be experimentally determined, at least in principle. Thus, though there is not a unique amount of charge to be associated with an atom, in an undistorted crystal there is a unique dipole moment induced when an atom is displaced, and that dipole moment, divided by the magnitude of the displacement, defines a charge that can be measured experimentally. This is in fact the transverse charge e_T^*, which we shall examine in Chapter 9. A careful way of formulating the problem is to introduce simultaneously two perturbations to the system, the displacement $\mathbf{u}$ of a single atom, and a small electric field $\mathscr{E}$. The change in energy can be calculated, and the term bilinear in $\mathbf{u}$ and $\mathscr{E}$ can be divided by $(-e)\,\mathscr{E}\cdot\mathbf{u}$ to obtain the magnitude of the effective charge. This avoids a discussion of the details of the charge distribution, which enters only through the bond dipole given in Eq. (5-4), and the assumption that γ remains constant as the bond deforms. It will be seen in Chapter 9 that a good agreement with experimental values is obtained. It will also be seen that there are two contributions to the dipole, one associated with the atom being moved, and a second arising from the transfer of charge among neighboring atoms. Thus there arises a natural definition of static charge: it is that charge which moves with the atom. The static charge contributes to any experimental definition of charge, though the transfer contributions will be seen to differ, for example, in the piezoelectric effect, from those associated with the transverse charge. The effective atomic charge that will be introduced here turns out to be equal to the static part of e_T^*.

Calculation of Effective Charge

We constructed bond orbitals as linear combinations of two hybrids, $u_1\,|h^1\rangle + u_2\,|h^2\rangle$, in Eq. (3-7), and we associated a fraction u_1^2 of the corresponding electron with atom 1 and a fraction u_2^2 with atom 2. We then found that if $|h^1\rangle$ has the lower hybrid energy (we shall call atom 1 the nonmetallic atom, and atom 2 the metallic one), $u_1^2 = (1 + \alpha_p)/2$ and $u_2^2 = (1 - \alpha_p)/2$; in the context of Eq. (3-7) these were hybrid polarities, but for effective charges, as for other dielectric properties, we use polarities based upon p states. There are two electrons per bond, so each bond contributes $1 - \alpha_p$ electrons to the metallic atom and $1 + \alpha_p$ electrons to the nonmetallic atom, consistent with the bond dipole, Eq. (5-4), with a bond length **d** scaled by γ. The total charge Z^* associated with an atom in a column Z of the periodic table is given by a number Z of protons in the nucleus minus the number of electrons contributed from the four neighboring bonds. This is called the *effective atomic charge* and is given by

$$Z^* = Z - 4 + 4\alpha_p \qquad \text{for metallic atoms,}$$
$$Z^* = Z - 4 - 4\alpha_p \qquad \text{for nonmetallic atoms,} \tag{5-18}$$

in units of the magnitude of the electronic charge. It is ordinarily positive for metallic atoms and negative for nonmetallic atoms. We should note again that though this is a natural choice of atomic charge, we should be cautious about drawing physical conclusions from it.

We may immediately evaluate the effective charges from Table 4-1; typically (though there is considerable variation), magnitudes of about one are found for compounds with elements from groups 3 and 4 or groups 2 and 6 in the Solid State Table, and magnitudes of about one-half are found for compounds with constituents from groups 1 and 7.

At first glance this result might seem surprising. It suggests that *the more polar the compound, the more neutral its constituents*. However, recall that polarity, and the related valence difference between constituents, is a measure of the *electronic* dipole to be associated with a bond. Even if the copper and bromine atoms in CuBr were completely neutral, the presence of 7 of 8 electrons on the bromine would constitute a strong electronic dipole in each bond, or a large polarity. In terms of potentials, a very strongly asymmetric potential is required to keep the large number of electrons on the nonmetallic atom; the potential is in fact sufficiently asymmetric to put more than seven electrons on the bromine atom, so the nonmetallic atom is negative. However, the common intuitive feeling, based largely on the idea of electronegativity, that the effective net atomic charges should be larger in highly polar materials, is probably misguided; there is no direct experimental reason to support this view, and the effective charge in Eq. (5-18), which decreases with polarity, is consistent, as we shall see, with the related experiments.

We should remark that of the compounds listed in Table 4-1, BP and BAs have negative metallic atoms ($Z^* < 0$), and BeO is near zero. Similarly, the electronegativity tables of Pauling or of Phillips (these can be compared in Phillips, 1973a, p. 54) suggest that Be and B may be negative in some compounds.

Another way to calculate the same effective atomic charge may be useful in other contexts. It uses the derivative of the total energy with respect to V_3 and the fact that in equilibrium the derivative with respect to charge transfer must vanish. One can then show that the difference in the number of electrons on the two atoms (with a total charge of eight electrons per pair) is $8 \, d\varepsilon_b/dV_3$, or, by using Eq. (3-12), $8\alpha_p$.

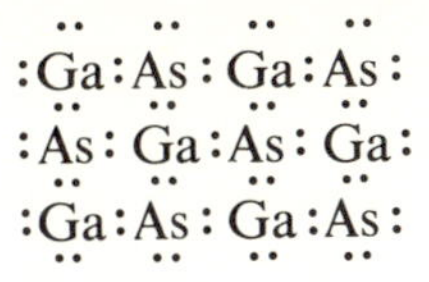

$$
\begin{array}{l}
\ddot{\;}\;\;\;\ddot{\;}\;\;\;\ddot{\;}\;\;\;\ddot{\;} \\
:Ga:As:Ga:As: \\
\ddot{\;}\;\;\;\ddot{\;}\;\;\;\ddot{\;}\;\;\;\ddot{\;} \\
:As:Ga:As:Ga: \\
\ddot{\;}\;\;\;\ddot{\;}\;\;\;\ddot{\;}\;\;\;\ddot{\;} \\
:Ga:As:Ga:As: \\
\ddot{\;}\;\;\;\ddot{\;}\;\;\;\ddot{\;}\;\;\;\ddot{\;}
\end{array}
$$

FIGURE 5-2

Lewis structure for GaAs.

Lewis Structures

It is clear from the foregoing discussion that one should be careful not to attach undue meaning to the effective atomic charge. In many ways intuition is better served by the diagrams called *Lewis structures* (after the chemist Gilbert N. Lewis; Lewis structures are explained in standard chemistry texts—see, for example, Andrews and Kokes, 1965). Such a diagram is shown in Fig. 5-2. Each dot represents an electron; three are contributed by each gallium atom and five by each arsenic atom. The two dots between any two neighbors represent the two electrons in the bonding state, both of which are shared by the two atoms, so that the environment of each atom is very much as if that atom had a rare-gas configuration (for GaAs, a krypton configuration). It is important that this rare-gas configuration consists of shared electrons. We shall see that it is convenient to think of ionic solids also as rare-gas configurations, but the rare-gas configurations for ionic solids do not consist of shared electrons. The traditional chemical terms for a system such as that in Fig. 5-2 are *saturated bonds* or *single bonds*. In this context there is limited meaning in saying quantitatively what fraction of the bond belongs to each atom; we can only make truly meaningful definitions of effective charge when we destroy the translational periodicity of the lattice. There are many different ways that this can be done. One way is by the introduction of impurities or defects, which will be discussed in the next section.

5-C Dielectric Screening

In computing the susceptibility for uniform electric fields, we neglected local field effects and thought of the charge as simply being passed through the crystal without local-charge distortions. This view is consistent with the calculation in terms of bond dipoles in Section 5-A, since the charge at the end of one bond dipole could be cancelled by the charge from the neighboring bond dipoles. There is, however, no such cancellation at the surface of the solid and it is the resulting surface-charge density that we illustrated in Fig. 4-6, where we related the dielectric constant to the susceptibility.

We now wish to turn to electric fields that vary with position within the system, which therefore give rise to charge accumulation within the material. If variation over an atomic distance is small we may still divide the applied field by the dielectric constant, Eq. (4-22), to obtain the net field. In particular, a localized charge Q will give an "applied field" of $Q\mathbf{r}/r^3$ far from the charge, and therefore a net field $Q\mathbf{r}/(\varepsilon_1 r^3)$. The *net* charge localized is then Q/ε_1, which corresponds to *dielectric screening* of the charge Q. Such screening becomes important when the crystal is disrupted by crystalline defects.

The simplest defect in a periodic lattice is an impurity such as a phosphorus atom substituted for a silicon atom in an otherwise perfect silicon crystal. Such a substitution is in fact simply the addition of a proton to one of the silicon nuclei and the addition of one electron to the system. Neglecting the electron for the moment, we write the screened potential due to the proton as $e/r\varepsilon_1$, where the dielectric constant for silicon is 12. The field we estimate is slowly varying except very close to the impurity atom, and our predicted field should be reliable in the distant region.

The extra electron that has been added to the crystal will be attracted to the impurity nucleus just as an electron is attracted to a proton in free space, but the attraction is greatly reduced by the dielectric constant. In addition, the electron will behave as if its mass is very small, as will be seen in Chapter 6. Both effects expand the radius of the bound orbital so that it lies mainly in the region where the dielectric screening approximation is good, and so that the binding may be only of the order of 0.01 electron volts; at normal temperature these *impurity states* will be empty and the corresponding electrons will move freely through the crystal. In Chapter 6 we shall examine such states.

It is of some interest to look more closely at the electronic structure near the impurity, as illustrated in Fig. 5-3: consider the bonds adjacent to the phosphorus atom. Taking the bond length to be the same as in silicon, we may associate a covalent energy with this bond, equal to that in silicon, of 2.98 eV. The polar energy for the bond can be determined from the difference in p-state energies, $V_3 = (8.33 - 6.52)/2 = 0.91$ eV, by using the Solid State Table or Table 2-2, just as the polar energies were computed in the compounds. The corresponding polarity of 0.29 for the four bonds transfers 1.16 electrons to the phosphorus, giving a net *negative* charge of 0.16 electrons on the phosphorus. This is consistent with the greater electronegativity of phosphorus in comparison to silicon, but not with the result, $+1/\varepsilon_1$, suggested by the dielectric constant. This, however, has left a deficit of 0.29 electrons on each neighboring silicon so that the *net* local charge is still $+1$ and the potential at large distances is still $+e/(r\varepsilon_1)$. In a similar way a carbon atom substituted for a silicon would be expected to take on a negative charge of

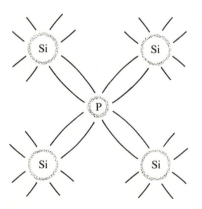

FIGURE 5-3

The bonds adjacent to a phosphorus impurity atom in silicon become polar and transfer charge to the phosphorus atom, compensating the extra charge on the nucleus.

1.52 electrons, leaving positive nearest neighbors of charge 0.38 and no long-range potential.

In this context we should say something about ionic conductivity and diffusion in covalent solids. This will be discussed for ionic crystals in Chapter 14. In most cases impurities appear substitutionally, as we have described. The principal manner in which such impurities can move through the lattice is by a vacancy mechanism. There will always be some vacancies present in the lattice in thermal equilibrium; in polar semiconductors the ratio of metallic and nonmetallic vacancies in equilibrium can be varied by varying the stoichiometry. A vacancy moves when a neighboring atom jumps into the vacant site. As a vacancy moves through a region it causes atoms in its path to be displaced to neighboring sites and this is the principal mechanism for diffusion. Frank and Turnbull (1956) noted that in the case of copper impurities in germanium a tiny fraction of the impurities will instead be in interstitial positions—that is, in the empty spaces between the germanium atoms. Although the number is small, interstitial atoms move so much more easily that they may dominate the diffusion. Such substitutional–interstitial diffusion has since been observed in the diffusion of other metal atom impurities in polar, as well as homopolar, solids.

The application of an electric field can cause diffusion along the field, which is called *ionic conductivity*. One may ask what effective charge is appropriate in that context. We should consider the geometry of interest. The correct answer is then obtained by asking how much the energy changes when, in the presence of a field, an impurity moves from one site to the next. If we think of the specimen as a long wire parallel to the field, this clearly depends upon the total charge localized at the impurity or at the surfaces of the specimen near the impurity, and the displacement of charge in the neighboring bonds does not affect this; that is, the appropriate charge to use is the charge ΔZ that determined the long-range fields in the discussion earlier in this section, and not Z^*. Although a carbon atom in silicon acquires negative effective charge Z^*, this will not make a corresponding contribution to ionic conductivity, and a zinc atom substituted for gallium in gallium arsenide will be caused to drift in a field, as if it had a unit negative charge. We will be thinking about it correctly if we consider the Lewis diagram of Fig. 5-2; in substituting a zinc for a gallium we are simply removing a proton from a gallium nucleus; any slight displacement of the neighboring dots is of no consequence.

There are, however, other contributions to the effective charge entering the ionic conductivity. If the empty state corresponding to the electron removed from the system during the substitution becomes bound to the zinc (it would be in an impurity state), the conglomerate would be neutral and would no longer contribute to the ionic conductivity.

5-D   Ternary Compounds

The generalization of the theory of dielectric properties to compounds containing more than two atom types follows very directly from the foregoing discussion of impurities and has the same uncertainties. Compounds like GaAs, but with

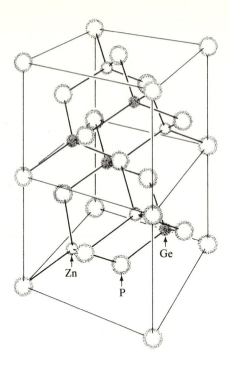

FIGURE 5-4

Atomic arrangement in ZnGeP$_2$.
[Based on de Alvarez et al., 1974.]

indium substituted at some sites for gallium, are analogous to the C impurity in Si, since the substituted atom is from the same column in the periodic table as the atom replaced. We can proceed with some confidence; we can predict that properties such as the dielectric constant will take on values that are the appropriate weighted average of the values for GaAs and InAs, and we may expect these predictions to be reliable.

More uncertain and somewhat more interesting are compounds such as ZnGeP$_2$, which is like GaP but with gallium replaced alternately by zinc and germanium so that the *average* column number is the same. There are a variety of such compounds, called chalcopyrites; in ZnGeP$_2$ (discussed by de Alvarez et al., 1974), there is an ordered arrangement of the zinc and germanium atoms as shown in Fig. 5-4. The lattice is compressed slightly along the vertical axis so that the boxes shown are not quite cubes and there is a slight displacement of the phosphorus atoms from the positions they would have in the zincblende lattice. Both distortions alter the lattice spacings and angles by only a few percent. There are now two kinds of bonds, ZnP bonds and GeP bonds, neither of which occurred in the binary tetrahedral structures we considered before. We may nonetheless proceed by using the *p*-state energies from the Solid State Table and the covalent energy (2.92 eV) for GaP. We find the polarity of the ZnP bond to be 0.65 and that of the GeP bond to be 0.32, leading to effective charges on the Zn, Ge, and P ions of 0.60, 1.28 and -0.94, respectively. The bonding–antibonding splitting for the two bonds suggests two E_2 peaks at 6.2 eV and 7.7 eV, but the observed peak is broad and there is no apparent splitting. Perhaps it is not surprising that the splitting is washed out in the experiment.

Following the procedure used for binary compounds, we may compute the susceptibility by simply adding the effects of the two kinds of bonds. Neglecting the effects of distortion, we obtain a susceptibility within 1 percent of that for GaP as expected. The computed second-order susceptibility shows a larger deviation, a decrease of about 18 percent, which should be observable if it is not dominated by other effects. A quite complete study of the susceptibility of ternary compounds has been made along these lines by Lines and Waszczak (1977).

In general, all of the properties of ternary compounds can be treated just as we will treat them for binary compounds. In each case there will be uncertainties, as there are in the treatment of susceptibilities, about the parameters used and whether the dominant effects have been included. For a complete and current review of ternary semiconductors, see Kaufmann and Schneider (1974, p. 229).

5-E   Magnetic Susceptibility

The response of covalent crystals to magnetic fields is very weak and of less interest than the dielectric response. It may nevertheless prove useful as a probe of the electronic structure. An early discussion of the problem, with references to still earlier work, was given by Krumhansl (1959). Two recent treatments in terms of LCAO theory have been given by Sukhatme and Wolff (1975) and by Chadi, White, and Harrison (1975). We shall follow the latter treatment but give a more complete formulation than given there.

The Field as a Perturbation

The response may be found by treating the magnetic field in perturbation theory, as we treated the electric response in Chapter 4; however, there are additional complications. To begin the calculation, we add the effect of a magnetic field to the Hamiltonian by adding a term $+e\mathbf{A}/c$ to the momentum operator in the Hamiltonian; $\mathbf{A}$ is a vector potential given by $-\mathbf{r} \times \mathscr{H}/2$ for a uniform magnetic field $\mathscr{H}$; again, we have written the electronic charge as $-e$. Thus, in the Hamiltonian, the simple kinetic energy, $p^2/2m$, is replaced by

$$\frac{1}{2m}\left(\mathbf{p} + \frac{e\mathbf{A}}{c}\right)^2 = \frac{p^2}{2m} + \frac{e}{mc}\mathbf{p} \cdot \mathbf{A} + \frac{e^2 A^2}{2mc^2}. \tag{5-19}$$

Notice that $\mathbf{p} \cdot \mathbf{A} = \mathbf{A} \cdot \mathbf{p}$, since $\nabla \cdot \mathbf{A} = 0$. The terms in $\mathbf{A}$ are to be treated as a perturbation, but we immediately see a difficulty in that $\mathbf{A}$ is proportional to r and becomes very large in a large system. That difficulty was removed for a tight-binding expansion of the wave function by White (1974); we use a similar but more direct approach here: instead of writing the electron states in the crystal as a simple linear combination of bond orbitals, we write

$$|k\rangle = \sum_i u_{ki} e^{-ie\mathbf{A}(\mathbf{r}_i)\cdot\mathbf{r}/\hbar c} |b_i\rangle, \tag{5-20}$$

which reduces to the simpler form of Eq. (3-17) when the magnetic field goes to zero. The expression $\mathbf{A}(\mathbf{r}_i)$ is the vector potential evaluated at the center of the ith bond. The purpose of using this form is to eliminate the large terms; notice that

$$\left(\mathbf{p} + e\frac{\mathbf{A}}{c}\right) e^{-ie\mathbf{A}(\mathbf{r}_i)\cdot\mathbf{r}/\hbar c} |b_i\rangle = e^{-ie\mathbf{A}(\mathbf{r}_i)\cdot\mathbf{r}/\hbar c} \left(\mathbf{p} + e\frac{\mathbf{A} - \mathbf{A}(\mathbf{r}_i)}{c}\right) |b_i\rangle, \tag{5-21}$$

so that in evaluating matrix elements it will always be small differences in position, $\mathbf{r} - \mathbf{r}_i$, that enter the expression. The choice of bond center is appropriate for weakly coupled bonds. The corresponding expansion is made also for the anti-bonding states.

We write down the perturbation-theoretic expression for the electron energy, Eq. (1-16), to second order in $\mathbf{A}$:

$$E_k = \langle k| \frac{1}{2m}\left(\mathbf{p} + \frac{e\mathbf{A}}{c}\right)^2 + V |k\rangle + \frac{e^2}{m^2 c^2} \sum_{k'} \frac{\langle k'|\mathbf{p}\cdot\mathbf{A}|k\rangle^2}{\varepsilon_k - \varepsilon_{k'}}. \tag{5-22}$$

We again sum over k and use the unitarity of the transformation, as in Eq. (3-31), to replace the sum over k by a sum over bonds and the sum over k' by a sum over antibonding states; notice that in this way we neglect variations in the $\varepsilon_k - \varepsilon_{k'}$ in the sum, as we did in the dielectric susceptibility.

The first term in Eq. (5-22) may then be evaluated by using Eq. (5-21). The term independent of field gives the zero-field bond energy, ε_b. The term linear in field vanishes, and the quadratic term may be written explicitly by using the form of the vector potential. The final term in Eq. (5-22) is explicitly second-order in field, so the phase factors may be dropped. This leads to a field-dependent bond energy given by

$$E_b(\mathcal{H}) = \varepsilon_b + \frac{e^2}{8mc^2} \langle b_i| [(\mathbf{r} - \mathbf{r}_i) \times \mathcal{H}]^2 |b_i\rangle$$

$$+ \frac{e^2}{4m^2 c^2} \sum_j \frac{\langle a_j|\mathbf{p}\cdot(\mathbf{r} - \mathbf{r}_i) \times \mathcal{H}|b_i\rangle^2}{\varepsilon_b - \varepsilon_a}. \tag{5-23}$$

The first term is seen to be positive, corresponding to a diamagnetic contribution to the susceptibility, χ_L, called the **Langevin term**. The second is seen to be nega-tive, corresponding to a paramagnetic contribution χ_p, called the **Van Vleck term**. We may obtain these terms directly by multiplying Eq. (5-23) by the electron density N and equating them to the corresponding change in energy, written in terms of the susceptibility, $-\chi\mathscr{H}^2/2$. We may note that in a cubic system the expectation value of $[(\mathbf{r} - \mathbf{r}_i) \times \mathscr{H}]^2$ is twice the expectation value of $(z - z_i)^2\mathscr{H}^2$, where we picked the z-direction to be specific. Then the Langevin term becomes

$$\chi_L = \frac{-Ne^2}{2mc^2} \langle b_i | (z - z_i)^2 | b_i \rangle. \tag{5-24}$$

Notice that $\mathbf{p} \cdot (\mathbf{r} - \mathbf{r}_i) \times \mathscr{H} = \mathscr{H} \cdot \mathbf{p} \times (\mathbf{r} - \mathbf{r}_i) = -\mathscr{H} \cdot \mathbf{l}_i$, where $\mathbf{l}_i$ is the angular momentum operator measured from the ith bond site. Taking the field in the z-direction, we write the Van Vleck term

$$\chi_p = \frac{Ne^2}{2m^2c^2(\varepsilon_a - \varepsilon_b)} \sum_j \langle a_j | l_{zi} | b_i \rangle^2. \tag{5-25}$$

There is also a contribution, χ_c, of form analogous to Eq. (5-24), due to the response of the core electrons, so that the total diamagnetic susceptibility is given by

$$\chi = \chi_c + \chi_L + \chi_p. \tag{5-26}$$

The Diamagnetic Contribution

The matrix element in the Langevin term, Eq. (5-24), will clearly be of the order of the value $(d/2\sqrt{3})^2$ it would have if the charge were concentrated at the bond extremities. Therefore it is convenient to introduce a magnetic γ_m, analogous to the γ introduced for the electric susceptibility. Then $\langle b_i | (z - z_i)^2 | b_i \rangle$ can be written $\gamma_m d^2/12$. (D. J. Chadi and R. M. White have been able to write an approximate relation for γ_m in terms of γ by using a sum rule. However, their analysis is intricate, and their corrections to γ are only of the order of 5 percent, so we will not reproduce it here.) The Langevin term becomes simply

$$\chi_L = -\frac{\gamma_m^2 Ne^2 d^2}{24mc^2}. \tag{5-27}$$

This may be equated to the experimental estimate of this term (Hudgens, Kastner, and Fritzsche, 1974, and Hudgens, 1973) for silicon, to give a value $\gamma_m = 1.12$. Following Chadi, White, and Harrison (1975), let us use this same value for all materials in estimating the susceptibility. Notice that the order of magnitude of magnetic susceptibility is e^2/d divided by mc^2. Thus, in contrast to the dielectric susceptibility (which is of order unity), it is very small.

Paramagnetic Contribution

The evaluation of the matrix elements in the paramagnetic term of Eq. (5-25) is somewhat tricky. The matrix elements of l_{zi} vanish unless the states a_j and b_i are on different sites. (Since both a_j and b_i are cylindrically symmetric, the component of $\mathbf{l}$ along the axis is zero, and the transverse components vanish by reflection symmetry.) Thus restriction of matrix elements to only intrasite terms, as described at the beginning of Section 4-D and as might be suggested by the Bond Orbital Approximation, would eliminate all but the Langevin term. However, in the magnetic case, the coupling between bond states and neighboring antibonding states is seen to give terms of different form than the Langevin terms, so they cannot be absorbed in a modified γ_m. These must be obtained explicitly. We will sketch enough of the derivation to see the form of the terms that enter the calculation, but not in enough detail to obtain the numerical factors.

In Fig. 5-5 is shown a bond orbital and a neighboring antibonding orbital. We include only such nearest-neighbor pairs and will, in fact, include only the contribution of the two hybrids that share the intervening atom. The value l_{zi} is an angular momentum measured from the bond center, but can be related to the value l_z measured from an origin at the intervening atom; thus, $\mathbf{l}_i = (\mathbf{r} - \mathbf{d}/2) \times \mathbf{p} = \mathbf{l} - \mathbf{d} \times \mathbf{p}/2$. The evaluation of the matrix elements is straightforward but tedious. The bonding and antibonding orbitals are written in terms of hybrids, by using Eqs. (3-13) and (3-16), so that

$$\langle a_j | l_{zi} | b_i \rangle = \pm \alpha_c \langle h_j | l_{zi} | h_i \rangle /2. \tag{5-28}$$

The hybrids are then written in terms of atomic states, as in Eq. (3-1), and familiar properties of the angular momentum operators are used to evaluate the intra-atomic matrix elements. In terms of our notation in Eq. (3-1), $\langle p_y | l_z | p_x \rangle = i\hbar$, but all the other intra-atomic matrix elements of l_z vanish. Thus the matrix element $\langle a_j | l_z | b_i \rangle$ becomes a numerical constant times $\alpha_c \hbar$. (Since we shall not have occasion to use this aspect of quantum theory elsewhere, we can omit the details. A good account is given by Messiah, 1962, p. 507ff.)

Matrix elements of $\mathbf{p}$ also enter the paramagnetic term. These are conveniently written in terms of the dimensionless oscillator strength f_{ps}, defined in Eq. (4-8) but with the bonding state replaced here by an atomic s state and the antibonding state replaced by a p state oriented parallel to the momentum operator. These terms in $\mathbf{p}$ have the same proportionality to α_c, and are found to have the effect of

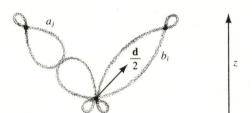

FIGURE 5-5

A bond state, and an antibonding state on a neighboring site.

simply scaling the matrix elements of l_z by a factor

$$\lambda_m = 1 - [m(\varepsilon_p - \varepsilon_s)\, d^2 f_{ps}/24\hbar^2]^{1/2}. \tag{5-29}$$

The final result for the Van Vleck term in the susceptibility obtained by Chadi et al. (1975) may be written

$$\chi_p = \frac{Ne^2\hbar^2\lambda_m^2\alpha_c^3}{16m^2c^2V_2}. \tag{5-30}$$

Substituting suitable values into Eq. (5-29) for λ_m gives values varying from 0.52

TABLE 5-2
Magnetic susceptibilities (multiplied by 10^6) from Eqs. (5-27) ($\gamma_m = 1.12$) and (5-30) ($\lambda_m = 1.31$), and from experiment.

Material	χ_c	χ_L	χ_L (Experimental)	χ_p	χ_p (Experimental)	χ	χ (Experimental)
C	−0.03	−2.47	−3.56	2.38	1.91	−0.12	−1.70
BN	−0.04	−2.43	—	1.94	—	−0.53	—
BeO	−0.05	−2.32	—	1.36	—	−1.01	−1.43
Si	−0.19	−1.63	−1.63	1.56	1.56	−0.26	−0.26
Ge	−0.51	−1.57	−1.84	1.50	1.87	−0.58	−0.58
GaAs	−0.51	−1.56	−1.73	1.02	1.12	−1.05	−1.22
ZnSe	−0.55	−1.56	—	0.43	—	−1.68	−1.70
CuBr	−0.49	−1.54	—	0.17	—	−1.86	−1.96
Sn	−0.78	−1.37	—	1.31	—	− .84	−1.55
InSb	−0.80	−1.36	—	0.80	—	−1.37	−1.60
CdTe	−0.82	−1.36	—	0.31	—	−1.87	−2.04
AgI	−0.83	−1.37	—	0.14	—	−2.06	−2.22
GaP	−0.37	−1.62	−1.90	0.99	1.15	−1.00	−1.23
ZnS	−0.46	−1.64	—	0.45	—	−1.64	−1.66
CuCl	−0.50	−1.64	—	0.20	—	−1.94	−1.93
InP	−0.66	−1.51	—	0.77	—	−1.40	−1.50
CdS	−0.76	−1.51	—	0.31	—	−1.96	−1.77
GaSb	−0.64	−1.45	—	0.98	—	−1.11	−1.11
InAs	−0.76	−1.47	—	0.80	—	−1.43	−1.67
CdSe	−0.80	−1.46	—	0.32	—	−1.94	−1.89

SOURCES: χ_c from Selwood (1956); χ_L experimental and χ_p experimental data as well as χ experimental data for the corresponding materials, from Hudgens et al. (1974) and Hudgens (1973); χ experimental value for BeO from Bailly and Manca (1972); and the remaining χ experimental data from Landolt-Börnstein (1966).

for tin to 0.71 for diamond. Comparison of Eq. (5-30) with the experimental values of the paramagnetic term for silicon estimated by Hudgens et al. (1974) gives a value 1.31. Chadi et al. (1975) took the silicon experimental value of λ_m as universal, just as they took the experimental value of γ_m from silicon for the Langevin term as universal, and predicted values of the various terms in the susceptibility for all semiconductors for which these values had been measured. The results, recalculated and presented in Table 5-2, are impressive indeed. Note that the covalent energy based on p states was used in the calculation. Use of the hybrid values would change λ_m but lead to approximately the same susceptibilities.

It is interesting to notice that if we substitute the expression $V_2 = 2.16\hbar^2/md^2$ into Eq. (5-30), we may combine the Van Vleck and Langevin terms to obtain

$$\chi - \chi_c = -\frac{Ne^2d^2\gamma_m^2}{24mc^2}(1 - 0.96\alpha_c^3). \tag{5-31}$$

This form separates the metallicity dependence (both terms vary inversely with the bond length) and the dependence on covalency ($\chi - \chi_c$ varies as $1 - 0.96\alpha_c^3$). From Table 5-2 it is seen that the experiments clearly show both trends and both are very well described by the theory.

PROBLEM 5-1   *Local theory of susceptibility*

Notice that Eq. (5-6) is valid for a σ bond in graphite, just as it is for a σ bond in diamond, if appropriate parameters are used. Confirm that this leads to the same *linear* susceptibility obtained in Problem 4-3.

PROBLEM 5-2   *Effective atomic charge*

Evaluate the effective charge Z^* for a hexagonal BN structure. You may first obtain the contributions from the σ bonds by using the polar and covalent energies of Problem 4-2. The fraction of the π-electron density on the nitrogen can be estimated by using the states at Γ only, writing a boron state $\psi_B = N_p^{-1/2}\Sigma_{i(boron)}\psi_\pi(r - r_i)$ with diagonal energy ε_p (boron) and a similar nitrogen state $\psi_N = N_p^{-1/2}\Sigma_{i(nitrogen)}\psi_\pi(r - r_i)$ with energy ε_p(nitrogen). The matrix element between these is simply $3V_{pp\pi}$ since each atom is coupled with three neighbors.

CHAPTER 6

The Energy Bands

SUMMARY

Accurate energy bands obtained from first principles by computer calculation are available for most covalent solids. A display of the bands obtained by the Empirical Pseudopotential Method for Si, Ge, and Sn and for the compounds of groups 3–5 and 2–6 that are isoelectronic with Ge and Sn shows the principal trends with metallicity and polarity. The interpretation of trends is refined and extended on the basis of the LCAO fitting of the bands, which provides bands of almost equal accuracy in the form of analytic formulae. This fitting is the basis of the parameters of the Solid State Table, and a plot of the values provides the test of the d^{-2} dependence of interatomic matrix elements.

The conduction-band minima and valence-band maxima are studied in terms of the $\mathbf{k} \cdot \mathbf{p}$ method, which relates the effective masses to the oscillator strengths discussed in Chapter 4. Wannier excitons and impurity states are also understandable in this context.

In Chapter 3 we gave a preliminary discussion of the energy bands in terms of the simple LCAO theory, and illustrated, in Fig. 3-7, the form of more accurately determined energy bands. For most of the studies made in this text, that description will be sufficient. However, the bands are of some interest in their own right and are important to the understanding of the electronic properties of semiconductors, and a consideration of them increases one's understanding of the electronic structure of covalent solids. In this chapter, therefore, we shall look at a more extensive set of accurate bands and at their interpretation in terms of the concepts of the LCAO representation of the electronic structure. We shall look also at the effective masses that are essential to an understanding of the transport properties of semiconductors.

6-A   Accurate Band Structures

Energy bands can be calculated from first principles, without any experimental input. The main approximation required is the one-electron approximation (see Appendix A), which we use throughout this text. Then the two remaining questions are: what does one use for the potential?; and what representation does one use to describe the wave function? At present the same essential view of the potential is taken by almost all workers, based upon free-electron exchange and little, if any, modification for correlation. (This is discussed in Appendixes A and C.) The principal differences in different calculations are in the accuracy with which the potentials are determined self-consistently with the charge densities from the states that are being calculated. It may well be that the principal remaining inaccuracies are in use of the one-electron approximation itself, and that little is to be gained from further improvements within that context.

The OPW Method

Perhaps the most successful representation of the wave functions for band calculations for semiconductors has been the OPW method (orthogonalized plane-wave method), developed by Herring (1940). The success of the method has been due to the ease of obtaining and using realistic potentials in the calculation, in contrast to methods that utilize the "muffin-tin" approximation to the potential (discussed in Chapter 20). Only recently have difficulties with the application of muffin-tin potentials to semiconductors been overcome. (For discussion and references see Johnson, Norman, and Connolly, 1973.) For any given potential, any of the accurate methods should give the same bands if the necessary effort is applied.

In developing the OPW method, Herring recognized that the wave function could be written as accurately as one wished by expanding in plane waves, but that a vast number would be required to describe the structure of each wave function near the nucleus. He further recognized that the final wave function would be orthogonal to the wave functions of the core states. (This follows from the discussion after Eq. (1-5).) By making the individual plane waves orthogonal to the cores states (*orthogonalized plane waves*), one can produce the necessary structure near the nuclei and greatly reduce the number of waves needed in the expansion. In terms of this small set, one can proceed to construct a Hamiltonian matrix, just as was done using LCAO's in Chapter 3, and diagonalize it to obtain the bands. The principal proponents of the OPW method have been F. Herman and co-workers (see Bibliography for citations). Even with these most accurate methods, the validity of the bands can be improved by making adjustments to the potentials to fit some experimentally determined values.

Calculating Bands from Pseudopotentials

In the pseudopotential approach, which we discuss in Chapter 18, the off-diagonal matrix elements of the Hamiltonian between OPW's are regarded as matrix elements of a pseudopotential between plane waves. No approximation is made in taking this view, though it suggests a range of approximations that one might make, all of which are then called pseudopotential methods. In the *Empirical Pseudopotential Method* of M. L. Cohen, the pseudopotential is written in terms of a very small number of parameters (frequently three), which are adjusted to fit the calculated band structure to the observed optical spectra. This was the essential method used to obtain the "true bands" for germanium shown in Fig. 3-7. It was also used recently by Chelikowsky and Cohen (1976b) to calculate the band structures of eleven diamond and zincblende semiconductors. Such a set of bands, all obtained with the same approximations, provides the ideal basis for a discussion of the bands. In Fig. 6-1 we show those bands for silicon, germanium, and tin, and for the polar semiconductors isoelectronic with Ge and Sn. We can see among them the trends noted in the discussion of optical properties.

To see trends with metallicity and with polarity, notice the levels Γ'_{25} and Γ_{15} in the silicon bands. These levels in the other systems have been split by spin-orbit coupling, which will be discussed later, but it is not difficult to locate the corresponding levels and see approximately where Γ'_{25} and Γ_{15} would have occurred in the absence of spin-orbit coupling. The separation between these two sets of levels was associated with the optical peak at E_2 in Section 4-C and was written as $2(V_2 + V_3^2)^{1/2} = 2V_2(1 - \alpha_p^2)^{-1/2}$. For the homopolar materials, with $V_3 = 0$, this gap is expected to decrease with metallicity (Si to Ge to Sn) and this is seen by comparing the bands. With increasing polarity (for example, Ge to GaAs to ZnSe) this gap is expected to increase and that is also clearly seen.

In the "normal" band structure, at Γ the nondegenerate level (Γ_7 in Ge) lies below the triply degenerate level (Γ_6 and Γ_8 in Ge) in the conduction band, just as it does in the valence band. We see that silicon is not normal in this sense. In Section 3-E we identified the energy difference between the nondegenerate level in the conduction band and the degenerate level in the valence band as the optical threshold, E_0. (That identification only applies to normal band structures.) We found E_0 to be given approximately by $3.6V_2(\alpha_m - 1)/(1 - \alpha_p^2)^{1/2}$. Both the decrease with increasing metallicity and the increase with polarity are reflected in the bands of Fig. 6-1, and in fact, there is fair agreement between this equation and the bands of Fig. 6-1. The total valence band width (at Γ) does not show strong trends because more than half of it arises from the metallic energy, the contribution $4V_1$, which also shows no important trends.

Another important trend is the drop seen in the nondegenerate level at Γ in the

140

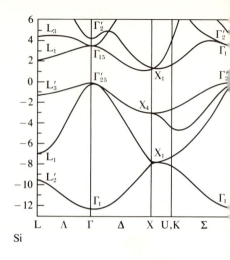

Si

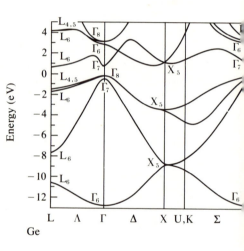

FIGURE 6-1
The energy bands of Si, Ge, and Sn and the energy bands of the polar semiconductors isoelectronic with Ge and Sn, as calculated by Chelikowsky and Cohen (1976b) by means of the Empirical Pseudopotential Method. Metallicity increases, material by material, downward; polarity increases from left to right.

Ge

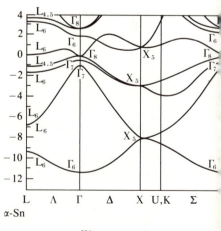

α-Sn

Wave vector **k**

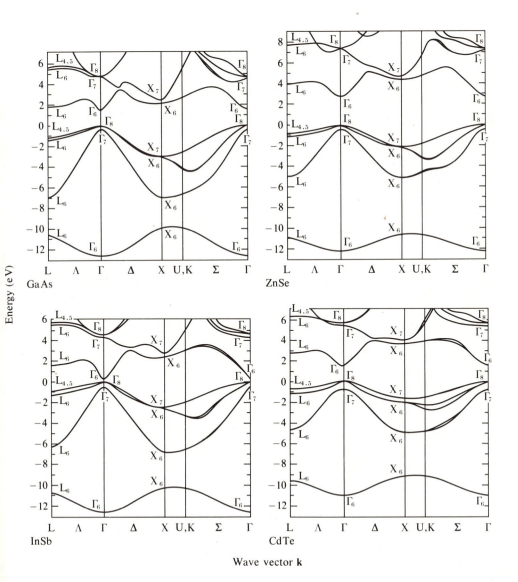

Energy (eV)

GaAs

ZnSe

InSb

CdTe

Wave vector **k**

conduction band relative to the conduction-band levels at X and L, with increasing metallicity and with increasing polarity. Thus the minimum gap between the valence and conduction bands tends to be between levels at Γ, a vertical transition allowed for optical absorption; these are called *direct-gap semiconductors*. At low metallicity and polarity this need not be true, Si and Ge being the important *indirect-gap semiconductors*.

A final important trend is the separation of the lowest valence band from the other three as the polarity increases, visible particularly at X. In order to understand trends within the valence bands we must carry the LCAO description somewhat further than we did in Chapter 3.

6-B   LCAO Interpretation of the Bands

The energy bands of tetrahedral solids have been studied in terms of LCAO's for many years; the first study was that of Hall (1952), who used a Bond Orbital Approximation, keeping only nearest-neighbor interbond matrix elements in order to obtain analytic expressions for the bands over the entire Brillouin Zone. The recent study by Chadi and Cohen (1975), which did not use either of Hall's approximations, is the source of the interatomic matrix elements between s and p orbitals, which appear in the Solid State Table. Pantelides and Harrison (1975) used the Bond Orbital Approximation but not the nearest-neighbor approximation and found that accurate valence bands could be obtained by adjusting a few matrix elements; at the same time very clear interpretations of many features of the bands were achieved. The main features of the Pantiledes-Harrison interpretation will be presented here.

Even if one were to solve for the bands exactly (as Chadi and Cohen did), using the parameters from the Solid State Table, agreement with the true bands shown in Section 6-A would not be very impressive. Those parameters formed a systematic set and were chosen by using d^{-2}-scaling and atomic term values, rather than by using all of the matrix elements as free parameters to be redetermined for each material. In short, the choice that is suitable for a general study of properties does not produce such accurate bands. The differences can be seen from the upper two entries for each matrix element in Table 6-1.

An additional complication arises if the Bond Orbital Approximation is used, since then the matrix elements should be determined for extended bond orbitals. Sokel (1976) made these determinations, using the Chadi-Cohen parameters, and Sokel's results should approximate the more accurate bands; the corrections are sizable and, in most cases, bring the matrix elements into better agreement with the Pantelides-Harrison values (which were fitted, using the Bond Orbital Approximation, to obtain accurate bands); again, see Table 6-1. The comparison in Table 6-1 is ambiguous since, to some extent, a modified B_1^s can be compensated for by modifying B_5; hence, the matrix elements are not uniquely determined. For the purpose of interpreting the bands it matters little which of the sets of values in Table 6-1 one thinks of.

TABLE 6-1   143

Comparison of interbond matrix elements (eV) obtained in different ways. The first entry in each list is based upon the parameters of the Solid State Table. The second entry is the matrix element based upon the Chadi-Cohen parameters (1975). The third entry is the value for extended bond orbitals obtained by Sokel (1976) by using perturbation theory and the Chadi-Cohen parameters. The fourth entry is the value obtained by Pantelides and Harrison by fitting the observed bands.

Matrix element	Si	Ge	GaAs	ZnSe
B_1^s	0.93	1.05	1.12	1.15
	1.01	1.12	1.24	1.55
	1.16	1.34	1.46	1.77
	1.55	1.62	1.59	1.57
B_1^a	0.0	0.0	0.66	1.10
	0.0	0.0	0.34	0.89
	0.0	0.0	0.39	0.95
	0.0	0.0	0.55	0.87
B_4	-0.31	-0.29	-0.25	-0.19
	-0.29	-0.30	-0.24	-0.20
	-0.28	-0.31	-0.27	-0.24
	-0.31	-0.35	-0.30	-0.24
B_5	0.26	0.23	0.20	0.15
	0.26	0.23	0.19	0.14
	0.19	0.11	0.08	0.03
	0.01	-0.01	-0.01	-0.01

It will be very convenient at some stages to make reference also to the results of the full diagonalization of the eight-by-eight matrix, so we shall begin by listing the formulae resulting from that diagonalization. Expressions by Chadi and Cohen (1975) for the energies at Γ are repeated (these were given in Chapter 3), and their values at X and one set at L are also given, all for the zincblende structure:

$$E(\Gamma_1) = \frac{\varepsilon_s^c + \varepsilon_s^a}{2} \pm \sqrt{\left(\frac{\varepsilon_s^c - \varepsilon_s^a}{2}\right)^2 + (4E_{ss})^2}; \qquad (6\text{-}1)$$

$$E(\Gamma_{15}) = \frac{\varepsilon_p^c + \varepsilon_p^a}{2} \pm \sqrt{\left(\frac{\varepsilon_p^c - \varepsilon_p^a}{2}\right)^2 + (4E_{xx})^2}; \qquad (6\text{-}2)$$

$$E(X_1) = \frac{\varepsilon_s^c + \varepsilon_p^a}{2} \pm \sqrt{\left(\frac{\varepsilon_p^a - \varepsilon_s^c}{2}\right)^2 + (4E_{sp})^2}; \qquad (6\text{-}3)$$

$$E(X_3) = \frac{\varepsilon_p^c + \varepsilon_s^a}{2} \pm \sqrt{\left(\frac{\varepsilon_p^c - \varepsilon_s^a}{2}\right)^2 + (4E_{sp})^2};$$ (6-4)

$$E(X_5) = \frac{\varepsilon_p^c + \varepsilon_p^a}{2} \pm \sqrt{\left(\frac{\varepsilon_p^c - \varepsilon_p^a}{2}\right)^2 + (4E_{xy})^2};$$ (6-5)

$$E(L_3) = \frac{\varepsilon_p^c + \varepsilon_p^a}{2} \pm \sqrt{\left(\frac{\varepsilon_p^c - \varepsilon_p^a}{2}\right)^2 + (2E_{xx} + 2E_{xy})^2}.$$ (6-6)

Bond Orbitals as a Basis

We return now to a basis of bond orbitals and antibond orbitals as illustrated in Fig. 3-4. We make the Bond Orbital Approximation in order that we may consider only the four bond orbitals and a four-by-four matrix. We also initially include only matrix elements between nearest-neighbor bonds; subsequently, matrix elements between second-neighbor bonds will be added. The resulting bands for the homopolar solids are exactly the same as the bands given by Hall (1952). They were extended to heteropolar solids by Coulson et al. (1962); with the use of parameters from Harrison (1973c), they become the bands of the Bond Orbital Model. It is a popular, widely used model. There are conspicuous discrepancies between this model and accurate bands; nevertheless, it can be informative to study the model before adding other matrix elements.

The rows and columns of the four-by-four Hamiltonian matrix correspond to Bloch sums of equivalent bond orbitals of the form

$$|\chi_\alpha(\mathbf{k})\rangle = \sum_i e^{i\mathbf{k} \cdot \mathbf{r}_i} |b_\alpha(\mathbf{r} - \mathbf{r}_i)\rangle / \sqrt{N_p}.$$ (6-7)

The four bond types were indicated in Fig. 3-5, a portion of which is redrawn in Fig. 6-2. To illustrate the calculation, we choose $\mathbf{k}$ to lie in a [100] direction. We may then immediately construct the matrix elements.

Consider first $H_{43} = \langle \chi_4 | H | \chi_3 \rangle$. Let one of the bond orbitals from the Bloch sum for the wave function on the right be the orbital labelled 3 in the center of Fig. 6-2. There are then matrix elements with nearest-neighbor bonds of type 4 lying above and below the central orbital 3 in the figure, and both of these have phase factors $e^{i\mathbf{k} \cdot \mathbf{r}_i}$, the same as the phase factor for the central orbital (that is, for this choice of wave number). One is coupled through an atom of type A and one through an atom of type C; we call these interbond matrix elements B^A and B^C, respectively. Thus if we retain only nearest-neighbor matrix elements, we obtain immediately

$$H_{43} = H_{21} = B^A + B^C.$$ (6-8)

Here we have performed the sum over the N_p orbitals of type 3, and have also noted that the matrix element H_{21} has the same value.

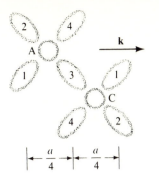

FIGURE 6-2

A set of neighboring bond orbitals from Fig. 3-5, showing the numbering of the four orientations of orbitals. The two atom types are indicated by A and C. A wave number in the [100] direction is also indicated.

For the matrix element H_{13}, on the other hand, the phase of the orbital of type 1 to the right of 3 differs from the central orbital by a factor $e^{ika/4}$ and that to the left of 3 by $e^{-ika/4}$, so matrix element H_{13}, and matrix elements equivalent to it, are all equal to

$$H_{13} = H_{23} = H_{14} = H_{24} = B^A e^{ika/4} + B^C e^{-ika/4}. \tag{6-9}$$

Using also the fact that the matrix is Hermitian, we can write the Hamiltonian matrix in detail for wave numbers in the [100] direction:

$$H = \begin{pmatrix} \varepsilon_b & H_{12} & H_{13} & H_{13} \\ H_{12} & \varepsilon_b & H_{13} & H_{13} \\ H_{13}^* & H_{13}^* & \varepsilon_b & H_{12} \\ H_{13}^* & H_{13}^* & H_{12} & \varepsilon_b \end{pmatrix}. \tag{6-10}$$

The eigenvectors (u_1, u_2, u_3, u_4) and eigenvalues can be guessed, and can be confirmed by matrix multiplication. Two of the eigenvectors, $(1, -1, 1, -1)/2$ and $(1, -1, -1, 1)/2$, have the same eigenvalue $\varepsilon_b - H_{12}$. The other two can be obtained by substituting the form (u_1, u_1, u_3, u_3) and solving the resulting quadratic equation to obtain $\varepsilon_b + H_{12} \pm 2|H_{13}|$. Substituting Eqs. (6-8) and (6-9) gives the forms

$$\Delta_5 = \varepsilon_b - B^A - B^C, \tag{6-11}$$

and

$$\Delta_1 = \varepsilon_b + B^A + B^C \pm 2|B^A e^{ika/4} + B^C e^{-ika/4}|. \tag{6-12}$$

Here we have used the conventional notation of representing the energies in the bands by the symmetry of the wave function. We are not concerned with the symmetry here, but specific labels will help. The designation Δ indicates $\mathbf{k}$ in a [100] direction. Bands indicated by Δ_5 are doubly degenerate and in this case are the upper bands which are made up of p-like states. The Δ_1 bands are nondegenerate and, here, are the lower bands with mixed s and p symmetry.

Inclusion of Only V_1

To establish values for the parameters B^A and B^C, we must decompose the bond orbitals into hybrids, as will be illustrated in Fig. 6-4, taking the coefficients of the two hybrids from Eq. (3-13) and using the polarities obtainable from Table 4-1. Interatomic matrix elements will also be included in Fig. 6-4. However, we look first at the contributions arising only from the matrix elements V_1^a and V_1^c between hybrids on the same atoms. We call them V_1-*only bands*. In terms of these,

$$B^A = -\tfrac{1}{2}(1 + \alpha_p)V_1^a;$$
$$B^C = -\tfrac{1}{2}(1 - \alpha_p)V_1^c.$$

$$(6\text{-}13)$$

Values for V_1^a and V_1^c are given in Table 2-2, or the same values can be obtained from the Solid State Table. These parameters are included in the calculation of the first panel of Fig. 6-3.

These values give a total band width of $2(1 + \alpha_p)V_1^a + 2(1 - \alpha_p)V_1^c$. Notice that for the homopolar semiconductors, it becomes $4V_1 = \varepsilon_p - \varepsilon_s$; the band is

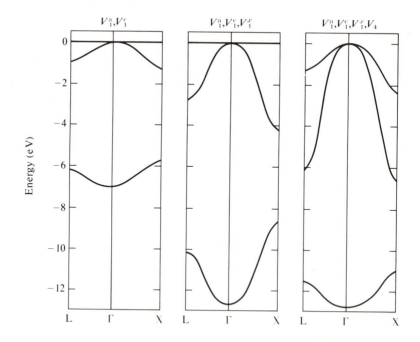

FIGURE 6-3

Illustration of the effect of the important matrix elements on the valence bands. The various matrix elements included in the calculation are indicated at the top of each panel. The material is GaAs. [The figure is based on one in Pantelides and Harrison, 1975.]

broadened out to a width corresponding to the initial *sp* splitting. This contribu- - - -
tion, however, accounts for only a little over a half of the observed band width.
Another interesting feature of the V_1-only bands is the gap between the lowest two
bands at X, which is given by

$$\Delta_X = 2(1 + \alpha_p)V_1^a - 2(1 - \alpha_p)V_1^c. \tag{6-14}$$

Notice that the gap vanishes for a homopolar semiconductor, which is true also
for the exact bands, and if V_1^a were equal to V_1^c, it would simply be equal to α_p
times the predicted band width, $\alpha_p 4V_1$. Thus, qualitatively, the gap is in very
simple correspondence with the polarity of the system. The observed splittings are
from 30 percent to 45 percent lower than those predicted in the V_1-only theory by
Eq. (6-12). The value is not modified as we add additional matrix elements within
the Bond Orbital Approximation (Pantelides and Harrison, 1975). From
Eqs. (6-3) and (6-4) we see that the situation is greatly complicated if the Bond
Orbital Approximation is not used (that is, bonding–antibonding matrix
elements are added), though of course the predicted gaps *do* go to zero as the
polarity goes to zero in any case.

The lowest gap at L (which requires a calculation similar to that given above)
does not go to zero for the homopolar semiconductors; for the homopolar V_1-only
bands it is given by $2V_1$; that is, the gap is half the predicted band width. For
heteropolar semiconductors, the lowest gap at L is given by an expression that is
complicated even with V_1-only bands, but an expression that is in almost as good
accord with the exact bands (Harrison and Ciraci, 1974) as the gap at X was. This
is somewhat of a surprise since, as can be seen from the second and third panels in
Fig. 6-3, the matrix elements we add individually give large shifts in this gap, yet
the net shift is small. This particular gap will become of special interest when we
develop pseudopotential theory, because in the simplest pseudopotential model it
is found to be equal to the gap between the doubly degenerate valence and
conduction bands at X, which, however, is the gap between parallel bands used in
the definition of covalent energy. Thus in LCAO terms, the simplest pseudopotential model corresponds to an assumption of unit metallicity. The point is rather
complicated and perhaps is not an important one since there are sizable
discrepancies between the true bands and both the V_1-only bands and the simplest
pseudopotential bands.

Effect of V_1^x

Notice from Fig. 6-4 that there are other contributions, designated by $-V_1^x$, to
the matrix elements between nearest-neighbor bonds. Two of these contribute for
each pair of bonds, each weighted by the coefficient for the anion hybrid and the
cation hybrid (Eq. 3-13). The two V_1^x values are not required to be equal by

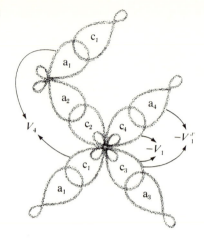

FIGURE 6-4

A set of neighboring bond orbitals made up of hybrids c on the metallic atom (cation) and hybrids a on the nonmetallic atom (anion). The subscripts denote the four bond orientations in the primitive cell, and the important interhybrid matrix elements are indicated. With the choice of signs shown, V_1, V_1^x, and V_4 are all positive. There is one other nearest-neighbor interhybrid matrix element, called V_5 by Pantelides and Harrison (1975); it connects, for example, c_4 and the a_1 above, and is small; hence it can be neglected altogether.

symmetry, though in Pantelides and Harrison (1975) they are equal. These contributions add $-\alpha_c V_1^x$ to each of the B^A and B^C of Eq. (6-13), which we rewrite

$$B_1^C = (B_1^s - B_1^a)/2 = B^C - \alpha_c V_1^x;$$
$$B_1^A = (B_1^s + B_1^a)/2 = B^A - \alpha_c V_1^x. \tag{6-15}$$

Notice that we have related B_1^C and B_1^A also to the matrix elements B_1^s and B_1^a (symmetric and antisymmetric) given in Table 6-1. These have the effect of broadening the band as seen in the second panel of Fig. 6-3, and Pantelides and Harrison adjusted the value to give the correct total band width. It does not, however, modify the splitting at X, given by Eq. (6-14) and roughly in accord with experiment already. Adding V_1^x simply modifies the parameters entering the V_1-only bands.

The π-Bonding Matrix Element

The matrix element $-V_1^x$, like $-V_2$, couples hybrids on nearest-neighbor atoms. Another such matrix element, V_4, indicated in Fig. 6-4, couples bonds of the same type that are second-neighbor bonds. Adding these goes beyond the Bond Orbital Model but does not greatly complicate the calculation. These matrix elements contribute to the diagonal elements, $H_{\alpha\alpha}$, of Eq. (6-10). They can be calculated just as the off-diagonal elements were. By returning to Fig. 3-5, we see that there are six contributions, weighted by the product of the coefficient (Eq. 3-13) for the cation hybrid and the coefficient for the anion hybrid. Each is written $B_4 = \alpha_c V_4/2$. For wave numbers in a [100] direction, four of these six occur with phase factors differing from one. It is not difficult to see that these have the effect of modifying the diagonal elements;

$$\varepsilon_b \to \varepsilon_b - \alpha_c V_4(1 - 2\cos ka/2) \quad \text{(k in [100] direction)} \tag{6-16}$$

The effect, shown in the third panel of Fig. 6-3, accounts for the curvature of the upper set of valence bands (and adds curvature to the others); this the V_1-only bands lacked. It does this without modifying the total band width nor the band gap at X. The matrix element V_4 was chosen by Pantelides and Harrison to give the observed energy of the upper level at X relative to that at Γ. There is one remaining nearest-neighbor interhybrid matrix element V_5, described in the caption of Fig. 6-4. Pantelides and Harrison found it to be very small (see $B_5 = \alpha_c V_5/2$, in Table 6-1) and its effect unimportant, so we will not discuss it. In fact, they chose V_1^x, V_4 and V_5, to fit the homopolar semiconductors, and used polarities based upon the optical absorption-peak energy to predict bands for all of the polar semiconductors, giving a fit very nearly on the scale to which the bands are known. Refinements can be obtained by including more-distant neighbors. Perhaps the most extensive study including additional neighbors is that by Dresselhaus and Dresselhaus (1967). A more recent study has been made by Levin (1974).

Pantelides and Harrison (1975) noted that if one took V_5 to be identically zero, and wrote both V_4 and V_5 in terms of atomic matrix elements, V_4 was found to be exactly $-V_{pp\pi}$. This remarkable result suggests that the curvature of the upper valence bands is to be associated with π bonding, with the antibonding states at Γ and bonding states at X. Consideration of the states themselves confirms this, and indeed, subtraction of Eq. (6-5) from Eq. (6-2) for homopolar semiconductors confirms that the difference in energy between these levels is $4V_{pp\pi}$, even when the Bond Orbital Approximation is not made. Thus V_4 may be called the **π-bonding energy**. Since both bonding and antibonding states are occupied (addition of V_4 does not change the energy average over the Brillouin Zone) the π-bonding energy does not have any effect on the bonding properties themselves.

The d^{-2}-Scaling

It was the finding that V_1^x and V_4 as well as E_2 scale as d^{-2} which first suggested a universal d^{-2}-scaling. One may in fact write each of V_1^x, V_2, V_4, and V_5 in terms of the four interatomic matrix elements ($V_{ss\sigma}$, and so on), and if one set scales, so must the other. It was only with the matching of free-electron and LCAO bands by Froyen and Harrison (1979) that the full significance of the d^{-2} dependence became clear. To indicate the extent of the validity of that scaling, the values obtained by Pantelides and Harrison are plotted against bond length in Fig. 6-5. We have also listed those values in Table 6-2, along with band energies predicted by Pantiledes and Harrison. The formulae for the energy values along symmetry lines obtained in the Bond Orbital Approximation are not repeated here; though a rather good fit was obtained for symmetry points, general points were not as well given.

We may summarize the LCAO interpretation of the energy bands. Accurate bands were displayed initially in Fig. 6-1. The energy difference between the upper valence bands and the conduction bands that run parallel to them was associated with twice the covalent energy for homopolar semiconductors, or twice the bonding energy $2(V_2^2 + V_3^2)^{1/2}$ in heteropolar semiconductors. The broadening of those two upper valence bands arises from the π-bonding energy, V_4. A major portion of the total valence-band width comes from the metallic energy. That portion is $4V_1$;

TABLE 6-2

Interhybrid matrix elements and predictions of band energies.

Material	α_p	Matrix element			Band energy						
		V_1^x	V_4	V_5	$-\Gamma_1$	$-X_5$	$-X_3$	$-X_1$	$-L_3$	$-L_1$	$-L_2$
C	0.00	1.77	+1.09	0.41	24.2	6.0	14.8	14.8	3.0	12.6	18.7
BN	0.41	1.77	+1.09	0.41	23.6	5.5	11.8	16.8	2.7	11.2	18.9
BeO	0.64	1.77	+1.09	0.41	23.0	4.6	9.2	18.0	2.3	9.1	19.3
Si	0.00	0.85	+0.61	0.01	12.5	2.5	8.6	8.6	1.2	6.8	10.5
AlP	0.47	0.85	+0.61	0.01	12.3	2.2	6.3	10.3	1.1	5.7	10.9
MgS	0.63	0.85	+0.61	0.01	12.6	1.9	5.1	11.3	1.0	4.7	11.7
Ge	0.00	0.82	+0.70	−0.02	12.8	2.7	9.3	9.3	1.3	7.4	11.1
GaAs	0.50	0.82	+0.70	−0.02	12.5	2.3	6.6	11.0	1.2	6.0	11.5
ZnSe	0.72	0.82	+0.70	−0.02	12.5	1.9	4.8	11.7	0.9	4.5	11.9
CuBr	0.79	0.82	+0.70	−0.02	12.5	1.7	4.1	12.0	0.8	3.9	12.1
α-Sn	0.00	0.68	+0.58	−0.03	10.4	2.2	7.7	7.7	1.1	6.1	9.1
InSb	0.51	0.68	+0.58	−0.03	10.1	1.9	5.4	8.9	0.9	4.9	9.3
CdTe	0.76	0.68	+0.58	−0.03	10.1	1.4	3.7	9.7	0.7	3.5	9.8
AgI	0.83	0.68	+0.58	−0.03	9.8	1.2	3.1	9.5	0.6	2.9	9.5
SiC	0.39	1.23	+0.82	0.21	17.0	3.8	9.2	12.2	1.9	8.4	13.8
BP	0.00	1.23	+0.82	0.21	17.7	4.1	10.7	11.8	2.1	9.3	14.1
AlN	0.59	1.23	+0.82	0.21	16.9	3.3	7.4	13.5	1.7	7.1	14.4
BeS	0.21	1.23	+0.82	0.21	17.9	4.0	9.4	13.3	2.0	8.8	14.8
BAs	0.00	1.21	+0.87	0.19	17.8	4.3	10.9	12.3	2.1	9.6	14.4
GaN	0.62	1.21	+0.87	0.19	16.6	3.3	7.4	13.5	1.7	7.1	14.4
BeSe	0.32	1.21	+0.87	0.19	18.1	4.0	9.2	14.0	2.0	8.7	15.2
ZnO	0.70	1.21	+0.87	0.19	17.4	3.0	6.3	15.0	1.5	6.2	15.6
CuF	0.83	1.21	+0.87	0.19	17.4	2.4	4.7	15.7	1.2	4.7	16.2
InN	0.64	1.10	+0.80	0.19	15.4	3.0	6.5	12.7	1.5	6.3	13.4
BeTe	0.00	1.10	+0.80	0.19	16.2	3.9	9.7	11.5	2.0	8.7	13.2
AlAs	0.44	0.84	+0.65	−0.01	12.7	2.3	6.6	10.9	1.2	6.0	11.4
GaP	0.52	0.84	+0.65	−0.01	12.1	2.2	6.3	10.3	1.1	5.7	10.9
MgSe	0.68	0.84	+0.65	−0.01	12.7	1.9	4.9	11.7	1.0	4.6	12.0
ZnS	0.73	0.84	+0.65	−0.01	12.1	1.8	4.6	11.2	0.9	4.3	11.4
CuCl	0.75	0.84	+0.65	−0.01	12.5	1.7	4.3	11.7	0.9	4.1	11.9
AlSb	0.54	0.76	+0.60	−0.01	10.9	2.0	5.6	9.4	1.0	5.1	9.9
InP	0.58	0.76	+0.60	−0.01	11.1	1.9	5.3	9.7	1.0	4.9	10.1
MgTe	0.67	0.76	+0.60	−0.01	11.1	1.7	4.6	10.1	0.9	4.3	10.4
CdS	0.77	0.76	+0.60	−0.01	11.1	1.5	3.8	10.4	0.7	3.6	10.6
GaSb	0.44	0.75	+0.64	−0.03	11.1	2.2	6.5	9.5	1.1	5.8	10.0
InAs	0.53	0.75	+0.64	−0.03	11.5	2.1	5.7	10.4	1.0	5.2	10.7
ZnTe	0.72	0.75	+0.64	−0.03	10.8	1.7	4.5	10.0	0.8	4.2	10.2
CdSe	0.77	0.75	+0.64	−0.03	11.5	1.6	3.9	11.0	0.8	3.7	11.1
CuI	0.78	0.75	+0.64	−0.03	10.8	1.5	3.9	10.2	0.8	3.7	10.3

SOURCE: Pantiledes and Harrison (1975).

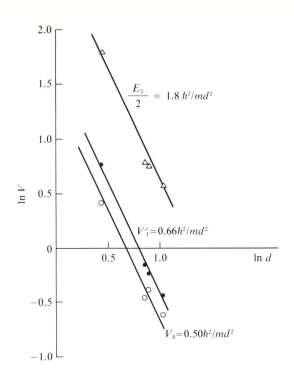

FIGURE 6-5

Variation of interhybrid matrix elements with bond length. The energy of the optical absorption peak is designated as E_2, approximately $2V_2$. The four points on each line, in order of increasing d, are C, Si, Ge, and Sn.

the remainder comes from the interatomic matrix element V_1^x. A band gap appears at L, given approximately by $2V_1$, and thus the ratio of that gap to the $2(V_2^2 + V_3^2)^{1/2}$ gap is a measure of metallicity. There is a lower gap at X only in polar semiconductors and it is roughly in proportion to polarity. In highly polar materials the gaps grow, separating the valence bands into three narrow p-like bands and one narrow s-like band at much lower energy. Pantelides and Harrison (1975) have given a more complete study of the features of the valence bands, which will not be repeated here.

The conduction bands are less completely and accurately described by the simplest LCAO description, as we saw in Chapter 3. Nonetheless a number of things can be learned from such a study. We turn to that next.

6-C The Conduction Bands

In discussing the bonding properties of solids, the electronic energy levels that are occupied—the valence or bonding energy bands—have been the only ones of interest. The antibonding bands, ordinarily called *conduction bands*, are also of interest in a number of circumstances. In a pure material, electrons can be transferred to these bands by light absorption, as we discussed in Chapter 4. The transfer of an electron from the valence bands to the conduction bands allows the corresponding wave packet to move through the crystal and to contribute to the conductivity; similarly, a packet centered on a valence-band state which has been emptied can also migrate through the crystal and contribute to the conductivity just as a bubble rising in a liquid causes a corresponding, downward flow of the liquid. The empty state is called a *hole* and the valence bands are sometimes called *hole bands*. A material in the ground state with full valence bands and empty conduction bands is said to be an insulator; the term *semiconductor* describes a material that can be made conductive by the introduction of *carriers*—either holes or electrons. Such carriers may be introduced by absorption of light (giving *photoconductivity*); they may be thermally excited to these bands (this is called *intrinsic conduction*); or they may arise from impurities (the semiconductor is then said to be *doped*). The last category includes impurities that only slightly modify the energy bands but provide an extra electron that can only be accommodated in the conduction band. (Arsenic as an impurity in germanium is an example.) Such an impurity is called a *donor* since it donates an electron to the conduction band. Gallium as an impurity in germanium, on the other hand, provides one less electron to the system and therefore leaves one state in the valence band, ordinarily near the top, unoccupied. Such an impurity is called an *acceptor*.

However such carriers are introduced, when they come to equilibrium with the lattice, they can be expected to occupy states within the thermal energy kT (25 millivolts at room temperature) of the band edges—near the bottom of the conduction band or the top of the valence band. Thus, as we discuss the conduction band, it will be desirable to focus on the states of lowest energy and it will be convenient, at the same time, to consider the states near the top of the valence band.

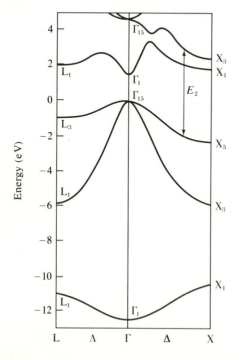

FIGURE 6-6

The energy bands of gallium arsenide. One of the transitions that contributes to the optical absorption peak, E_2, is shown. Energies are given in eV. [After Herman et al., 1968.]

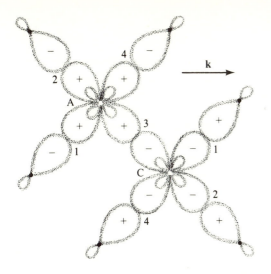

FIGURE 6-7

In constructing antibonding bands, we must be consistent in choosing the signs of the coefficients of the hybrids. We choose to have the signs positive on the anions and negative on the cations: the signs of all matrix elements used in the text will be consistent with this definition. This corresponds to taking u_1^a positive and u_2^a negative in accordance with the notation of Chapter 3, where subscripts 1 refer to anions.

We saw the general form of the valence bands in Fig. 6-1 for a number of semiconductors, and discussed the general features there. In Fig. 6-6 we show another version of the bands for GaAs, which will be useful for reference as we construct the conduction bands from LCAO theory.

Antibonding Orbitals as a Basis for Interpreting Band Features

The formulation proceeds exactly as it did for bond orbitals in Section 6-B, but the matrix elements B^A and B^C now refer to matrix elements between antibond orbitals. We take the convention of letting the coefficient of the hybrid on the anion be positive. (See Fig. 6-7.) Then, when we write these matrix elements in terms of interhybrid matrix elements, we use the coefficients from Eq. (3-16) rather than (3-13), so that Eq. (6-13) is modified for antibonds by changing the sign before α_p in each expression:

$$
\begin{aligned}
B^A \text{ (antibond)} &= -\tfrac{1}{2}(1 - \alpha_p)V_1^a; \\
B^C \text{ (antibond)} &= -\tfrac{1}{2}(1 + \alpha_p)V_1^c.
\end{aligned}
\qquad (6\text{-}17)
$$

Furthermore, the contribution of the interatomic matrix element V_1^x to the nearest-neighbor antibond matrix element is changed in sign, as can be seen in Fig. 6-7. Then Eq. (6-15) becomes

$$
\begin{aligned}
B_1^C \text{ (antibond)} &= B^C \text{ (antibond)} + \alpha_c V_1^x; \\
B_1^A \text{ (antibond)} &= B^A \text{ (antibond)} + \alpha_c V_1^x.
\end{aligned}
\qquad (6\text{-}18)
$$

The most important feature of this result, in comparison to the corresponding Eq. (6-15) for bonding states, is that the interatomic term V_1^x now cancels against the negative intra-atomic term, $-V_1$, rather than adding to it, and the splitting between the nondegenerate and triply degenerate levels at Γ, given by $2B_1^A + 2B_1^C$ (analogous to the total band width in the valence band), is predicted to be very small. This is confirmed by the correct bands given in Fig. 6-1 and, in fact, in silicon B_1 is positive though B is negative. For higher metallicity, V_1^x is reduced and the other systems correspond to negative B_1. We will see in Section 6-D that the quantitative agreement between this simple form and the accurate bands is reasonably good. The squeezing down of the conduction bands, as well as the admixture of higher bands, makes the shapes much less systematic within the conduction bands than within the valence bands. The matrix element V_4 gives curvature to the conduction bands that is opposite to the curvature of the valence bands. (This effect is seen in Fig. 3-7 or Fig. 6-1 by focusing on the degenerate conduction bands.) Thus V_4 gives curvature to the otherwise flat bands; it is *not* responsible for the tendency for conduction and valence bands to be parallel. In the LCAO context additional basis states or more distant neighbors are required to reproduce that tendency, a tendency that is more easily understood in the context of the pseudopotential theory.

The E_0 Gap

A final interesting feature of the conduction bands, which we can discuss in the context of the V_1-only bands, is the gap E_0 between the nondegenerate conduction-band level at Γ and the valence-band maximum at Γ. A formula for this gap was obtained from the full LCAO theory in Eq. (3-43), which is repeated here:

$$E_0 = 3.6(V_2^2 + V_3^2)^{1/2}(1 - \alpha_m). \tag{6-19}$$

The estimate obtained from the bands based only upon antibonding orbitals is complicated, as was the full expression given in Eq. (3-39). It is

$$E_0 = \varepsilon_a - \varepsilon_b - |B^A + B^C - 2\alpha_c V_1^x| - 3|B^A \text{ (antibond)} + B^C \text{ (antibond)}$$
$$+ 2\alpha_c V_1^x|. \tag{6-20}$$

For homopolar semiconductors it can be seen to reduce to the form

$$E_0 = 2V_2^h + 4V_1^x - 4V_1 = 5.27V_2 - 4V_1. \tag{6-21}$$

In the last step, we used Eq. (3-6) to write the hybrid covalent energy as $4.37\hbar^2/(md^2)$ and Fig. 6-5 to write V_1^x as $0.66\hbar^2/(md^2)$, and then used Eq. (4-16) to write $\hbar^2/(md^2)$ as $V_2/2.16$. This result can be compared with that resulting from Eq. (6-19), which is written in the form $E_0 = 3.60V_2 - 4.44V_1$, for the homopolar semiconductors. Neither is very accurate, but both correctly reflect a bonding–antibonding splitting from the covalent energy, reduced by the band-broadening

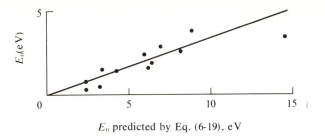

E_0 predicted by Eq. (6-19), eV

FIGURE 6-8

A plot comparing values for the optical gap, E_0, as observed (vertical scale) and as predicted from the LCAO bands by Eq. (6-19). The line corresponds to observed values 1/3 as great as predicted values. Data are taken from Table 6-3.

effect of the metallic energy. A plot of the observed gap, E_0, as a function of the prediction, Eq. (6-19), is shown in Fig. 6-8. Agreement with experiment would correspond to a slope of 1 rather than the slope of 1/3 that is shown in Fig. 6-8. The predictions are qualitatively correct, but not quantitatively so, partly because they are small differences in relatively large quantities. The main effects are increase in gap with decreasing metallicity and with increasing polarity.

The levels of particular interest in the conduction bands are those which lie near the band minimum since they are the only levels ever occupied in thermal equilibrium. We consider first the normal bands in which the conduction-band minimum occurs at Γ.

6-D Effective Masses

Band energy will vary in proportion to the square of k near the conduction-band minimum and, for the zincblende structure, must by symmetry be independent of the direction of $\mathbf{k}$. Thus, near the conduction-band minimum, we can write

$$E_k = \Gamma_{1c} + \frac{\hbar^2 k^2}{2m_c},\qquad (6\text{-}22)$$

where m_c is called the *effective mass* of an electron. The dimensionless parameter m_c/m is adjusted to give the correct energy dependence for each material. It varies greatly from material to material and can be very small. This is an appropriate term since, as we saw in Section 2-B, the dynamics of electron wave packets are determined by the energy bands, with $\mathbf{k}$ replaced by the canonical momentum divided by $\hbar$. Thus, Eq. (6-22) tells us that the wave packets for low-energy electrons in the conduction band may be described by a Hamiltonian, $H = p^2/2m_c$, and their dynamics is that of a free particle with an effective mass m_c. Transport properties can be studied by assigning an effective mass to the electron and forgetting altogether about the crystalline lattice that gave rise to the effective mass.

Effective Masses for Holes

In the same way, the energy varies quadratically with k near the maximum in the valence band. In terms of Eq. (6-22) this would correspond to a negative effective mass, but as is well known (e.g., Harrison, 1970, p. 148), a suitable change of variables allows us to describe empty states at the top of the band (that is, holes) as positively charged particles with positive mass. The magnitudes remain the same but the signs are changed. However, the region at the top of the valence band presents other complications, which are apparent in Fig. 6-6. There are three bands. The one with the greatest curvature, the band $L_1 \Gamma_{15} X_3$, is called a *light-hole* band because the magnitude of an effective-mass fit to the curvature is small. There are also two *heavy-hole* bands, $L_3 \Gamma_{15} X_5$. It can be seen from Fig. 6-6 that they have the same curvature for **k** in either a [100] (Γ to X) or [111] (Γ to L) direction, though the masses for the two directions are different. In other directions, the two bands have different masses. This is complicated indeed if we wish to describe it in detail, though we will see that the main features can be given rather simply.

We begin by expanding the LCAO bands for small k; we can, for example, expand the expressions in Eqs. (6-11) and (6-12). Expansion of Eq. (6-11), with ε_b taken from (6-16), gives the heavy-hole bands,

$$\Delta_5 = \Gamma_{15} - \alpha_c V_4 k^2 a^2 / 4. \tag{6-23}$$

This corresponds to a heavy-hole mass, m_h, given by

$$\frac{m}{m_h} = \frac{8\alpha_c}{3} \frac{md^2 V_4}{\hbar^2}. \tag{6-24}$$

Or, we may combine this result with the expression for V_4 given in Fig. 6-5, to obtain

$$m_h/m = 0.75/\alpha_c. \tag{6-25}$$

Again, for different directions of **k**, different values of the mass would be obtained.

What we should like to have is a suitable average over the two bands and all directions, for this will give the correct density of heavy-hole states per unit energy. Since the heavy holes dominate the total density of states, the corresponding mass will be appropriate for a description of properties that depend directly on the density of states. Such values have been estimated by Lawaetz (1971); they are given in Table 6-3 and are compared with the values of m_h obtained from

TABLE 6-3

Band parameters for direct-gap semiconductors.

Material	E_0 (Experimental)	m_h/m (Eq. 6-24)	m_h/m (Lawaetz)	m_c/m (Eq. 6-33)	m_c/m (Experimental)
GaAs	1.52	0.85	0.62	0.131	0.066
GaSb	0.81	0.84	0.49	0.075	0.045
InP	1.42	0.93	0.85	0.131	0.077
InAs	0.42	0.90	0.60	0.039	0.024
InSb	0.237	0.88	0.47	0.023	0.0137
ZnO	3.40	1.21	—	0.239	0.19
ZnS (hex)	3.80	1.14	1.76	0.281	0.28
ZnSe	2.82	1.14	1.44	0.218	0.17
ZnTe	2.39	1.12	1.27	0.192	0.16
CdS	2.56	1.25	—	0.213	0.20
CdSe	1.84	1.23	—	0.154	0.13
CdTe	1.60	1.21	1.38	0.136	0.096

SOURCES: Experimental E_0 were compiled by Lawaetz (1971), based on the following: GaAs values from Sturge (1962); GaSb, Reine, Aggarwal, and Lax (1970); InP, Turner et al. (1964); InAs, Pidgeon et al. (1967); InSb, Johnson (1967); ZnO, CdS, and CdSe, Langer et al. (1970); ZnS, Bir et al. (1970); ZnSe, Hite et al. (1967); ZnTe, Nabory and Fan (1966); and CdTe, Segall and Marple (1967).

Experimental m_h/m: Lawaetz (1971).

Experimental m_c/m: GaAs, Kaplan et al. (1969) and Stillman et al. (1969); GaSb, Adachi (1969); InP, Palik and Wallis (1961); InAs, Palik and Wallis (1961) and Pidgeon et al. (1967); InSb, Dickey et al. (1967) and Summers et al. (1968); ZnO, Reynolds and Collins (1969); ZnS (hexagonal), Miklosz and Wheeler (1967); ZnSe, Marple (1964); ZnTe, Riccius and Turner (1968); CdS, Hopfield and Thomas (1961); CdSe, Wheeler and Dimmock (1962); and CdTe, Kanazawa and Brown (1964).

Eq. (6-24). We see that they are in reasonably good agreement, though the values given by Lawaetz are preferred. Using them, it is appropriate in many circumstances to treat these two heavy-hole bands as isotropic and identical.

In a similar way we could expand the expression of Eq. (6-12) by using Eq. (6-13) or Eq. (6-15) for the light-hole band or the corresponding expression for the conduction-band minimum. The masses obtained are far too large in comparison to more accurate determinations. The origin of this error is largely the Bond Orbital Approximation, which has led to the simple sinusoidal bands of Eq. (6-12), which do not reproduce the rather sharp curvature of the true bands. This can be rectified, even within the Bond Orbital Approximation, if interactions with more-distant neighbors are included since, as was indicated at the end of Chapter 3, Wannier functions for the valence bands alone can produce exact valence bands. There is no fundamental flaw in Eq. (6-12), only a defect in our approximations. These sharp curvatures in the bands arise whenever two bands approach each other closely, as is illustrated at many points in the true bands of Fig. 3-8. We should treat this feature in some detail since it will give us also a good description of the conduction-band minima and valence-band maxima, the features of the energy bands that are central to an analysis of transport.

The $k \cdot p$ Method

We do this by using the **$k \cdot p$ method**, (called k-dot-p), which is based upon the perturbation theory of Eq. (1-14). In this method, energy is calculated near a band maximum or minimum by considering the wave number (measured from the extremum) as a perturbation. (The method is described in many solid state texts, such as Kittel, 1963, p. 186, or Harrison, 1970, p. 140.) The method was used for a study of effective masses by Cardona (1963, 1965). It was also used in the more extensive study by Lawaetz (1971) referred to in the discussion of heavy-hole bands. We shall discuss here only the conduction band and the light-hole band where the effects of interaction are great.

At small k, the conduction band is isotropic, as we indicated in writing Eq. (6-22), so we may consider wave numbers in a [100] direction without loss of generality. The **$k \cdot p$** expression for the energy takes the perturbation theoretic form of Eq. (1-14), which can be written schematically as

$$\Delta_1 = \Gamma_1 + \frac{\hbar^2 k^2}{2m} + \frac{\hbar^2 k^2}{m^2} \sum_n \frac{|\langle \Gamma_n | p_x | \Gamma_1 \rangle|^2}{\Gamma_1 - \Gamma_n}. \qquad (6\text{-}26)$$

Here the matrix element is that of the momentum operator p_x in the x-direction, since we have chosen **k** to lie in the x-direction. The matrix element is taken between the state Γ_1 of the conduction-band minimum and any other state Γ_n at Γ; the denominator is the energy difference between the two states. We drop all terms in this sum except those with the valence-band maximum, for which the energy denominator is the smallest and the contribution the largest. It can be shown by symmetry that the matrix element vanishes for the two heavy-hole bands (they correspond to p orbitals with an orientation perpendicular to the x-axis), so only the matrix element between wave functions for the conduction band and the light-hole band remains. The denominator is the band gap E_0, so we may extract a conduction-band mass from Eq. (6-26). This mass is given by

$$\frac{m}{m_c} = 1 + \frac{2}{mE_0} |\langle v | p_x | c \rangle|^2. \qquad (6\text{-}27)$$

We have written the state at the conduction-band minimum as $|c\rangle$ and that of the light hole as $|v\rangle$. In just the same way we can compute the effective mass of the light hole:

$$\frac{m}{m_l} = -1 + \frac{2}{mE_0} |\langle v | p_x | c \rangle|^2. \qquad (6\text{-}28)$$

In both equations the second term can be large, giving a small mass. In contrast, there are no such matrix elements between wave functions for the conduction band and the heavy-hole states, so their masses remain large.

The most important consequence of these two equations has been known for a long time: the effective mass varies with the smallest band-gap and goes to zero as that gap goes to zero.

We note now that the momentum operator is $p_x = (\hbar/i)\,\partial/\partial_x$, so *the matrix elements that enter the k · p method are exactly the same matrix elements that entered the calculations of optical absorption.* This remarkable fact enables us to obtain parameters from the LCAO theory given in Chapter 4.

The states $|v\rangle$ and $|c\rangle$ are obtained from solution of the Hamiltonian matrix, Eq. (6-10), for $\mathbf{k} = 0$, in which case all off-diagonal matrix elements take the same value. It can be easily verified that the two eigenvectors (other than the doubly degenerate ones corresponding to heavy holes) are $(1, 1, 1, 1)/2$ and $(1, 1, -1, -1)/2$. The Bond Orbital Approximation turns out to be exact for these states at Γ. The second of these eigenvectors corresponds to the light-hole band (the first corresponds to the bottom of the valence band). Thus the valence-band state at Γ that enters the calculation can be written as a sum of bond orbitals, as in Eq. (3-20), with $\mathbf{k} = 0$:

$$|v\rangle = \tfrac{1}{2}(|\chi_1\rangle + |\chi_2\rangle - |\chi_3\rangle - |\chi_4\rangle). \qquad (6\text{-}29)$$

The conduction-band state Γ_1 can be written as a sum of antibonding orbitals (chosen such that the coefficients of all anion hybrids are positive, as in Fig. 6-7); we write

$$|c\rangle = \tfrac{1}{2}(|\chi_1^a\rangle + |\chi_2^a\rangle + |\chi_3^a\rangle + |\chi_4^a\rangle). \qquad (6\text{-}30)$$

Keeping only matrix elements between bonding and antibonding orbitals on the same bond site, as we have done before, we obtain

$$
\begin{aligned}
\langle v|p_x|c\rangle &= 1/4(\langle b_1|p_x|a_1\rangle + \langle b_2|p_x|a_2\rangle \\
&\quad - \langle b_3|p_x|a_3\rangle - \langle b_4|p_x|a_4\rangle) \\
&= \langle b_1|p_x|a_1\rangle.
\end{aligned}
\qquad (6\text{-}31)
$$

In the first step in Eq. (6-31) we summed over all N_p bonds of each type appearing in the Bloch sums, giving N_p equal contributions for each bond type. In the second step in Eq. (6-31), we noted (see Fig. 6-7) that the magnitude of each matrix element is the same for fields in the [100] direction but that matrix elements for bonds 3 and 4 have a sign opposite those of 1 and 2.

The dielectric susceptibility was written in Eq. (4-5) directly in terms of the magnitude of the matrix elements of $\partial/\partial x$, which are matrix elements of $p_x/\hbar$, so by equating the forms for the susceptibility given in Eqs. (4-5) and (4-26), we may solve for the square of the magnitude of the matrix element. The result may be written

$$\langle b_1|p_x|a_1\rangle^2 = \gamma^2 m^2 d^2 V_2^2/(3\hbar^2). \qquad (6\text{-}32)$$

This can be directly substituted into Eq. (6-27) to obtain the effective mass:

$$\frac{m}{m_c} = 1 + \frac{2\gamma^2 md^2 V_2^2}{3E_0 h^2}.$$

(6-33)

We have retained E_0 explicitly, since that is exactly the energy denominator that enters the calculation. We may substitute for the covalent energy, $V_2 = 2.16h^2/(md^2)$, and obtain the values for m_c/m given in Table 6-3. The agreement is remarkably good, particularly for the 2–6 compounds, those with constituents from columns 2 and 6 in the Solid State Table. This is principally an achievement of the $\mathbf{k} \cdot \mathbf{p}$ method itself, since we have used experimental E_0 values, but it is gratifying that our matrix elements did so well. Dividing the final term in Eq. (6-33) by $2\alpha_p^2$ would improve the agreement but would be a purely empirical correction.

Light Holes and Spin-Orbit Splitting

The same matrix elements may be used in Eq. (6-28) to obtain the light-hole mass. This predicts effective masses slightly lower in magnitude than the conduction-band masses. However, the values given by Lawaetz are slightly higher than the conduction-band masses, owing to interaction with more-distant bands, which we have ignored. At the level of our approximations it is reasonable to neglect the difference and to think of the simplified bands of Fig. 6-9 as having

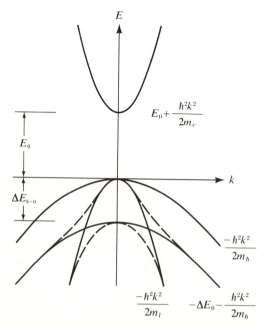

FIGURE 6-9

Spin-orbit coupling splits off one of the otherwise degenerate heavy-hole bands, as shown by the lowest solid line. Equations are given for each, with parameters given in Tables 6-3 and 6-4 (and with m_l taken equal to m_c). The heavy-hole band that is split away then becomes mixed with the light-hole band to give final energy bands, as shown by the dashed lines, corresponding to the bands for the heavier materials shown in Fig. 6-1. The bands for tin are complicated by a crossing of the kind illustrated in Fig. 6-10,b).

TABLE 6-4   161

Spin-orbit splittings of the valence-band maximum, in eV.

Material	ΔE_{s-o}	Eq. (6-34)
C	0.0	—
Si	0.044	—
Ge	0.29	—
Sn	0.64	—
Pb	~2.0	—
GaAs	0.34	0.29
InSb	0.82	0.64
GaSb	0.80	0.54
ZnTe	0.93	0.59
CdTe	0.91	0.64
HgS	0.32	0.29
HgSe	0.58	0.48
HgTe	1.05	0.80

SOURCE for ΔE_{s-o}: Si, Ge, GaAs, InSb, GaSb, ZnTe, and CdTe from Cardona et al. (1967). The rest are theoretical estimates by Herman et al. (1968).

$m_l = m_c$. We have included there a small splitting to be associated with spin-orbit coupling, which we shall not discuss in detail. That splitting is larger for heavier materials, in contrast to the interatomic matrix elements. Experimental values for the splitting have been collected by Herman et al. (1968) and are included in Table 6-4. The splittings in the homopolar semiconductors clearly show the trend of larger splitting for heavier materials. We might expect the splittings in the compounds to be given by a weighted average of the splittings for the two components, and assign to each component the value for the column 4 element in its row; thus,

$$\Delta E_{s-o} = (1 + \alpha_p)\, \Delta E_{s-o}^a/2 + (1 - \alpha_p)\, \Delta E_{s-o}^c/2. \tag{6-34}$$

These values are compared with experiment for a few compounds in Table 6-4. Clearly, there is an increase with polarity that is not described by Eq. (6-34). For further discussion and references, see Herman et al. (1968) and Chadi (1977).

Indirect-Gap Semiconductors

We turn finally to materials that are not direct-gap semiconductors. The conduction bands of semiconductors have an analogy in the conjugation of verbs: those encountered oftenest have the exceptional forms. Let us first consider the four homopolar semiconductors: diamond, Si, Ge, and Sn. In the text following

Eq. (6-18), we gave the energy difference between the conduction band levels at Γ as $2B_1^A + 2B_1^C$. For homopolar semiconductors it takes the simple form

$$\Gamma_{15c} - \Gamma_{1c} = 4(V_1 - 2V_1^x). \tag{6-35}$$

The value V_1 does not vary greatly with row in the periodic table, but V_1^x, an interatomic matrix element, does decrease with increasing lattice spacing. Evaluation of Eq. (6-35) shows this trend for diamond, Si, Ge, and Sn (-8.5, -0.2, 1.3, and 1.4 eV, respectively), while values obtained from band calculations are -5.8, -0.5, 2.5, and 2.7, respectively—the value for diamond is from Herman et al., 1967, and the others are from Herman et al., 1967, 1968). Thus in diamond and silicon, another band has dropped below the simple conduction band shown in Fig. 6-9. As we noted in Section 6-C, the reordering of levels at Γ is a special feature of materials of low metallicity and low polarity.

A second, more important, feature of this compression of the conduction-band levels is the dropping of the conduction bands at some other point of the Brillouin Zone below that at Γ. Materials for which this occurs are called **indirect-gap semiconductors**. We may see from Fig. 6-6 that in GaAs, the conduction bands at the points X and L nearly make that material an indirect-gap semiconductor. The conduction-band minimum appears to occur at X in diamond, Si, SiC, BP, GaP, and AlSb; it occurs at L in Ge (see, for example, Phillips, 1973a, p. 169). Some of the corresponding indirect gaps will be listed in Table 10-1. In such systems we can again expand energy as a function of wave number around each of the equivalent minima (that is, at each of the equivalent points in the Brillouin Zone). However, in this case symmetry no longer requires that the curvature be the same in each direction. It is customary to define a **longitudinal effective mass**, m_1, describing the variation of energy for wave numbers measured from the minimum and in the direction of Γ (m_1 is typically of the order of the true electron mass), and a **transverse effective mass**, m_t, for wave number variations perpendicular to this (m_t is typically of the order of a tenth of the true electron mass). Depending upon the property, it may be adequate to replace these by an isotropic density-of-states mass (given by $m_d = (m_1 m_t^2)^{1/3}$). It is not appropriate to explore this complicated situation further here.

Zero-Gap Semiconductors

Eq. (3-43) and Eq. (6-19) give the expression for the band gap E_0, which becomes smaller at higher metallicity. This formula gives the energy of the Γ_1^c state minus that of the Γ_{15}^v state (see Fig. 6-6); as was noted, it is possible, at sufficient metallicity, for the difference to be negative, corresponding to the top state of the valence band being nondegenerate and the bottom state of the conduction band being triply degenerate, as illustrated in Fig. 6-10. (We have ignored spin-orbit splitting in constructing that figure, though this crossing of bands only occurs in materials of atomic number sufficiently high that the splitting is important.) Of the

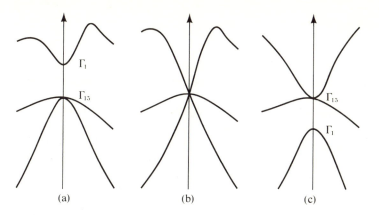

FIGURE 6-10

The lower conduction band and the upper valence band change as the gap goes through zero and as metallicity increases, from (a) to (c). Spin-orbit splitting is not included. Diagram (b) shows the crossing of the bands that is seen in the compound $Cd_{1-x}Hg_x$ Te. Then (a) corresponds to $x < 0.84$, (b) to $x = 0.84$, and (c) to $x > 0.84$.

homopolar materials, we may see from Fig. 6-1 that germanium corresponds to the ordering shown in part (a), tin approximately corresponds to the ordering shown in part (b), and if lead could be put in the diamond structure, it would be of type (c). In part (c), no longer does a gap separate the lowest four bands from the upper bands, and the system cannot be a semiconductor. Indeed, it would be a metal (or a *zero-gap semiconductor*) and would lie just to the left of the metal–covalent boundary shown in Fig. 2-3.

In fact, lead is not stable in the tetrahedral structure, so that associating Fig. 6-10,c with lead is academic. However, some stable compounds, such as HgTe, have an electronic structure corresponding to part (c). Indeed, one may substitute Hg for some of the Cd in a compound with a formula $Cd_{1-x}Hg_x$ Te; as x increases, the bands change from type (a) in the figure to type (b), and with a mercury concentration greater than $x = 0.84$, the compound becomes a gapless semiconductor (Vérié, 1972, p. 513). It is interesting that these compounds retain their stability beyond the point of the vanishing gap.

6-E   Impurity States and Excitons

It was noted in Section 5-C that when a donor atom gives up an electron to the conduction band (for example, when a phosphorus atom is an impurity in silicon), the donor becomes positively charged and can bind an electron, as a proton does in hydrogen. The binding energy in hydrogen was given in Eq. (1-22) as $me^4/2\hbar^2$ (the ground state). In a dielectric solid, each factor of e^2 should be reduced by the dielectric constant, and the electron mass should be the effective mass of the electron. In Table 6-3 we saw that the effective masses are typically 0.1, so if a dielectric constant is 10, the binding energy of hydrogen, 13.6 eV, is reduced to 0.01 eV; most donor-impurity states are empty—the donor is ionized and the electron is free.

In the semiconductors of greater polarity, the dielectric constants are smaller and the effective masses larger, and the same evaluation leads to 0.07 eV in zinc selenide, for example; many of the impurity states can be occupied at room temperature. As the energy of the impurity states becomes deeper, the effective Bohr radius becomes smaller and the use of the effective mass approximation becomes suspect; the error leads to an underestimation of the binding energy. Thus, in semiconductors of greatest polarity—and certainly in ionic crystals— impurity states can become very important and are then best understood in atomic terms. We will return to this topic in Chapter 14, in the discussion of ionic crystals.

An analogous condition can arise from the interaction between electrons and holes when both are present. A bound electron–hole pair is called an *exciton*; it is in many ways analogous to an electron–positron pair in a hydrogenlike state, which is known as *positronium*. Either the exciton or positronium pair can annihilate itself and give off energy, but both pairs may have a significant lifetime before that happens. Since the heavy-hole mass is ordinarily much larger than the effective mass of the electron, the reduced mass is determined by the effective electron mass and the binding-energy estimates described earlier remain appropriate. Such weakly bound excitons are called ***Wannier excitons***. (The electron bound to a light hole has even weaker binding.) Thus in homopolar materials, excitonic effects tend not to be important, though they have received considerable attention lately in circumstances when the exciton density is so large that the excitons condense into "electron–hole droplets" (see, for example, Pokrovsky et al. (1970), and Benoit à la Guillaume et al. (1970).

Again, the binding becomes larger in the more polar crystals, and we shall see that excitonic effects become extremely important in studying the optical properties of ionic solids: strong optical absorption from the ground state to the exciton state can occur, so that the optical absorption edge measures the band-gap *minus* a large exciton-binding energy. Since the exciton is neutral and cannot carry current, such absorption does not produce conductivity. Thus if an absorption edge (the energy above which absorption occurs) has lower energy than the ***photoconductivity edge***, this may be an indication of excitonic effects. Such effects appear not to be important in understanding optical properties of even the highly polar tetrahedral solids, though they are important in studying mixed tetrahedral solids such as SiO_2.

PROBLEM 6-1 *Calculation of the π bands in graphite*

First evaluate the nearest-neighbor matrix elements $V_{pp\pi}$, using the bond length of 1.42 Å (see Fig. 3-10.) We neglect second-neighbor matrix elements and matrix elements between successive graphite planes.

There are two atoms per primitive cell, so two Bloch sums of the form of Eq. (6-7) are needed (with a bond orbital replaced by p orbitals oriented perpendicular to the graphite plane). The expectation value of the Hamiltonian with respect to either of the Bloch sums taken individually is simply ε_p. The matrix element between the two is of a form analogous to Eq. (6-9), but with only a single type of nearest-neighbor matrix element and with three rather than two terms. They may be written for arbitrary $\mathbf{k}$ in terms of the three nearest-neighbor vectors $\mathbf{d}_1$, $\mathbf{d}_2$, and $\mathbf{d}_3$.

Obtain the bands explicitly and plot them for $\mathbf{k}$ along a nearest-neighbor vector, $\mathbf{d}_1$. This vector is in the direction of a zone edge of the hexagonal Brillouin Zone, reaching the edge at $k = 2\pi/(3d)$. The results may be compared with the π bonds of Painter and Ellis (1970), shown in Problem 3-3.

Also plot the bands along a direction perpendicular to this (towards a zone corner). Plot them only to the corner $[k = 4\pi/(3\sqrt{3}\,d)]$. (The points beyond can be seen to lie along a zone edge in the reduced zone.)

PROBLEM 6-2 π Bands in BN

Recalculate the π band for boron nitride, using the same structure and bond length as in Problem 6-1, but do not plot the results; just obtain the modified energies at the zone center and zone corner. The splitting at the corner makes BN insulating rather than semimetallic.

PROBLEM 6-3 Calculation of σ bands in the Bond Orbital Approximation

Extend the band calculation for graphite, Problem 6-1, to the valence σ-bands, using the Bond Orbital Approximation as described below, and plot the bands for wave numbers along the line from $k = 0$ to the zone edge Q. At the zone edge, $k = 2\pi/(3d)$; the direction is shown in Figure 6-11.

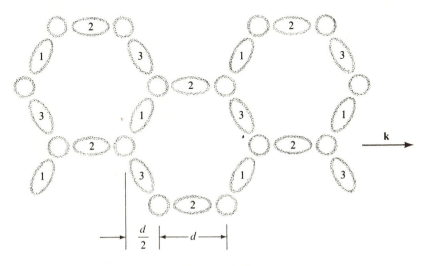

FIGURE 6-11

The hybrid energy and V_2^h were obtained in Problem 3-2, where a value of the bond energy ε_b was found. A value for the matrix element between two hybrids on the same atom, $-V_1$, was also obtained, and only its contribution, $-V_1/2$, to the interbond matrix element need be retained.

The state may be written as a linear combination of the three Bloch sums $|\chi_\alpha(\mathbf{k})\rangle$ from Eq. (6-7) for $\alpha = 1$, 2, and 3. The three bond types are illustrated below. Construct matrix elements between Bloch sums to obtain the three-by-three Hamiltonian matrix.

Symmetry could be used to deduce that one solution is of the form $(|\chi_1\rangle - |\chi_3\rangle)/\sqrt{2}$ and that the others are of the form

$$u|\chi_2\rangle + v(|\chi_1\rangle + |\chi_3\rangle)/\sqrt{2},$$

reducing the problem to the solution of a quadratic equation.

Recall that we found in Problem 3-3 that we had considerably over-estimated V_2. If you compare your results with Painter and Ellis (Problem 3-3), you should shift the zero of energy to fit theirs. Even then the reduction to a single matrix element has made our bands rather inaccurate.

PROBLEM 6-4 *Effective masses*

What effective masses should be associated with each valence band at Γ as obtained in Problem 6-3? (Expand the bands for small k.)

CHAPTER 7

The Total Energy

SUMMARY

The cohesive energy of covalent solids is obtained by calculating three component energies: first, a promotion energy to prepare the isolated atoms; second, an overlap interaction energy between atoms as they are brought together without bonding; and third, an energy gained in bond formation. The overlap interaction is obtained by means of an approximation in which atomic electron densities are superimposed, and potential, kinetic, and exchange energies are then obtained approximately in terms of the resulting distribution. The total energy of a solid is obtained next, as a function of bond length. This function can be used to predict equilibrium bond length, cohesive energy, and bulk modulus for the homopolar semiconductors. Trends are correctly predicted, but significant errors result from use of this approach for covalent solids. Empirically, the cohesive energy in an isoelectronic series of polar semiconductors is found to be given by the value for the homopolar semiconductor times the covalency.

Our discussion to this point has been of electronic energies, and one might assume that the total energy of a solid is simply the sum of individual electronic energies. It is immediately clear that this cannot be the whole story: if only the bonding energy changed as atoms were displaced, the total energy would decrease monotonically with decreasing lattice distance and the crystal would collapse. Terms that contribute to the bonding energy, such as the Coulomb interaction between nuclei, have not been discussed and, as will be seen, a very important contribution to the kinetic energy of the electrons needs to be added. For the properties we have discussed, omitting these contributions has not caused errors, but we now wish to understand why this is so, and we wish also to see how the additional terms should be calculated for the properties in which they become important.

7-A   The Overlap Interaction

We can at the outset list contributions to the energy of a crystal. There is the kinetic energy of the electrons, some of which is included in the sum of electronic energies which we have discussed. The kinetic energy of the nuclei arises in connection with lattice vibrations, but is unimportant in most aspects of the cohesion of the lattice. Then there are a series of coulombic contributions to the energy. One is the electron–nuclear interaction, which is included in the electronic energies. In fact, the core electrons are so tightly bound to the nuclei that we can consider them together as *atom cores* (with charge $+4$ for silicon, $+5$ for phosphorus, and so on) and deal only with valence-electron contributions to the energy; in these terms, the electron–core interaction is included in the individual valence-electron energies. Another coulombic type of energy is the electrostatic energy of interaction between the cores, which is not included in the electron energies we have discussed. We will need to add it now. Finally, there is the Coulomb interaction between the electrons, and this interaction is difficult to calculate.

The Coulomb interactions between electrons have been included conceptually in an average way: we have included, as part of the potential associated with each electron, the Coulomb potential arising from the electron distribution itself. Notice first that this counts the interaction between each pair of electrons *twice* when we sum the electron energies, once when we add the energy of the first electron, and once when we add the energy of the second. The doubly counted energy must be subtracted in obtaining the total energy. This property of the one-electron approximation seems strange the first time it is encountered. Notice, however, that if we consider the total energy of two neutral atoms that are well separated, the individual electron energies change only very slightly with additional separation, and the Coulomb interaction between the cores is just cancelled by the subtraction of the interaction between the valence-electron distributions; the resulting, negligible interaction is appropriate to the neutral atoms. When the atoms are close, as in the solid, the two terms must be considered separately. We do this quite automatically when we evaluate the total Coulomb energy of interaction and need worry about it only when we make explicit use of the one-electron energies. There are, in addition, a set of errors in our prescription for the electron–electron interaction energy owing to our use of a one-electron approximation— that is, our treating each electron as an independent particle moving in a fixed potential. The corrections of this error are called *exchange and correlation energy*. They are discussed in Appendix C, and we will not probe their origin further here.

Calculation of Component Energies

We can proceed much as we did in Chapter 2 by conceptually putting together the crystal from isolated atoms. To be specific, think of silicon, which, as an isolated atom, has two electrons in atomic s states and two electrons in atomic p states. The first step is to construct sp^3 hybrids states, costing an energy per atom equal to four times the hybrid energy, $4(\varepsilon_s + 3\varepsilon_p)/4$, minus the electron energy of the free atom, $2(\varepsilon_s + \varepsilon_p)$. This is called **promotion energy** and is given by $\varepsilon_p - \varepsilon_s$, or $4V_1$ per atom, where V_1 is the metallic energy introduced in Chapter 3. In putting together a polar crystal, there will be different numbers of electrons on the two atom types, and the promotion energy will be different. The promotion energy is easily evaluated, and is given by

$$E_{\mathrm{pro}} = \left(1 + \frac{\Delta Z}{4}\right) V_1^{\mathrm{c}} + \left(1 - \frac{\Delta Z}{4}\right) V_1^{\mathrm{a}} \qquad \text{per bond,} \qquad (7\text{-}1)$$

for $\Delta Z = 0$ (as in Ge), 1 (as in GaAs), and 2 (as in ZnSe). For $\Delta Z = 3$ (as in CuBr), this value should be reduced by V_1^{c}. Again, V_1^{c} is a matrix element for the metallic atom (Ga, Zn, or Cu) and V_1^{a} a matrix element for the nonmetallic atom (As, Se, or Br). It will be convenient to give all energies in eV per bond; notice that there are two bonds per atom in the crystal.

After promotion, one quarter of the charge density arises from s orbitals and three quarters from p orbitals, just as in a rare gas. However, the number of electrons on each atom is smaller than for a rare gas—four for silicon or germanium, and still less for the metallic atom in a polar system. For either the rare gas or the covalent solid, every atom is neutral and there is no change in electrostatic energy as the atoms are brought together until the electron wave functions begin to overlap. We imagine this overlap occuring without distortion of the atoms (as would be appropriate, for example, if all electrons were of the same spin, so that silicon would have a full shell for that spin); we can therefore calculate the overlap interaction just as in the statistical theory of closed-shell systems. Such a calculation was carried out by Sokel (Harrison and Sokel, 1976) for silicon; values form the graph shown in Fig. 7-1 as "overlap." The kinetic-energy change and the sum of Coulomb and exchange interactions are also plotted in Fig. 7-1. As indicated before, the overlap interaction provides a simple interatomic repulsion. (There is actually also a very shallow minimum at large d, which will not be of interest.)

Let us imagine bringing the atoms to their observed spacing in this undistorted state; then bonds can be formed as described in Chapter 3. This lowers the energy of each electron by the difference between the bond energy ε_b and the hybrid energy ε_h; for a nonpolar system this energy change per bond is $2(\varepsilon_h - \varepsilon_b) = 2V_2^{\mathrm{h}}$. There will have been some change in the hybrid energies as the atoms are brought together, owing to the influence of the overlapping potentials, but this was included in the electrostatic contribution to the overlap energy, so the energy difference $2V_2^{\mathrm{h}}$ remains appropriate. A small double-counting error is made here, proportional to the *square* of the change in charge distribution in the formation of

170

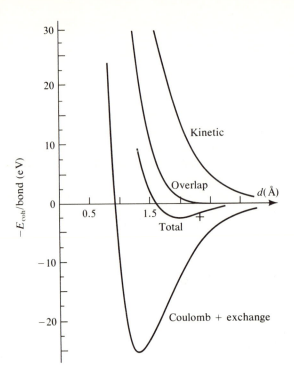

FIGURE 7-1

Various contributions to the total energy per bond as a silicon crystal is assembled by reducing the interatomic spacing d. The "overlap" contribution is $V_0(d)$. The contributions to "overlap" from "kinetic" energy and from "Coulomb and exchange" energies are shown separately; there is a small additional contribution from correlation. The effects of promotion and bonding are added to "overlap" to obtain "total." The cross indicates the observed spacing (internuclear distance) and cohesion. [After Harrison and Sokel, 1976.]

the bonds. In polar semiconductors, the gain in energy per bond is readily identified in the energy level diagram (refer to Fig. 3-3); notice that $1 + \Delta Z/4$ electrons per bond were initially in the lower hybrid and $1 - \Delta Z/4$ were initially in the upper hybrid. The magnitude of the energy gained in forming bonds, called *bond-formation energy*, is seen to be

$$E_{bond} = \left(1 + \frac{\Delta Z}{4}\right)\left(\sqrt{V_2^{h2} + V_3^{h2}} - V_3^h\right)$$

$$+ \left(1 - \frac{\Delta Z}{4}\right)\left(\sqrt{V_2^{h2} + V_3^{h2}} + V_3^h\right)$$

$$= 2\sqrt{V_2^{h2} + V_3^{h2}} - \Delta Z V_3^h/2 \qquad \text{per bond.} \qquad (7-2)$$

We can combine all of these contributions to obtain the energy required to separate the solid into isolated atoms; this is called *cohesive energy*. Notice in particular that each nearest-neighbor bond contributes $V_0(d)$ to the total energy, with V_0 the overlap interaction; the V_0 for next-nearest neighbors is negligible. Thus we are led to a cohesive energy of magnitude

$$E_{coh} = -E_{pro} - V_0(d) + E_{bond} \qquad (7-3)$$

per bond. The last two terms depend upon bond length and therefore may be used to predict the equilibrium bond length and bulk modulus. We turn to that evaluation next.

7-B   Bond Length, Cohesive Energy, and the Bulk Modulus

Bond Length

We may predict the bond length in a solid by minimizing the total energy, or maximizing the E_{coh} of Eq. (7-3), with respect to d. Thus, to the energy per atom labelled "overlap" in Fig. 7-1, from Eqs. (7-1) and (7-2) we add E_{pro}, which does not depend upon d, and $-E_{bond}$, which varies as d^{-2}, at least near the equilibrium spacing. Harrison and Sokel (1976) actually used values closer to V_2 and V_3 than to V_2^h and V_3^h. This gave the interaction energy labelled "total" in Fig. 7-1. The position of the minimum total energy is the bond length predicted by Harrison and Sokel. The observed bond length, and interaction energy, as deduced from the cohesive energy, are shown by the cross in the figure; the discrepancy in bond length is 15 percent. Table 7-1 lists the predicted and observed bond lengths obtained by Sokel for diamond and germanium, also. The discrepancies are similar but are larger than those encountered in the treatment of ionic crystals by Gordon and Kim. They would be even larger if the hybrid covalent energy, about twice V_2, had been used.

A consideration of Fig. 7-1 indicates that the discrepancies must arise from an error in the overlap interaction in which the kinetic term and the term from Coulomb and exchange interactions almost exactly cancel at the observed spacing. No minor modification of the bond energy could bring the minimum to the observed position, and a major modification would disrupt the agreement with

TABLE 7-1
Theoretical and experimental parameters for the homopolar semiconductors.

Parameter	C	Si	Ge
BOND LENGTH (Å)			
Theoretical	1.2	2.0	2.4
Experimental	1.54	2.35	2.45
COHESIVE ENERGY (eV/bond)			
Theoretical	9.18	2.45	0.40
Experimental	3.68	2.32	1.94
BULK MODULUS (10^{12} erg/cm^3)			
Theoretical	12.0	2.0	1.0
Experimental	5.45	0.988	0.772

SOURCE: Theoretical values are from Harrison and Sokel (1976). Experimental bond lengths are from Wyckoff (1963); other experimental values are as given by Kittel (1971).

experiment shown for the other properties that will be discussed. Harrison and Sokel suggested that the error in the overlap interaction in covalent solids comes from the violation of a condition basic to the justification of the method, namely, the condition that the electrons be in the ground state, as will be outlined in Appendix C. Here we have brought together atoms having electrons promoted to hybrid states, which may lead to errors in both the kinetic and the exchange energy calculations. It will also be noted in Appendix C that to go beyond the approximations used in the overlap interaction is a major undertaking, one which it is inappropriate to study here.

Cohesive Energy and Bulk Modulus

We may proceed to two other predictions that can be made on the basis of Fig. 7-1, the cohesive energy (the value of E_{coh} at the minimum) and the bulk modulus (proportional to the curvature, $\partial^2 E_{coh}/\partial d^2$, at the minimum). The results for these, obtained by Sokel for diamond, silicon, and germanium are given in Table 7-1, along with experimental values. We see that the results are only qualitatively in agreement with experiment; they describe all trends correctly and no experimental parameters have entered the calculation. We take this as evidence that we have found the correct physical origin of the bond-length-dependent energy, though the approximations have only limited accuracy in covalent solids. As fundamental as evaluations of bond length, cohesive energy, and bulk modulus are, discrepancies in them will cause us little problem elsewhere. We shall calculate all other properties at the spacing observed for each material, as we have been doing; we use experimental bulk moduli in Chapters 8 and 9 to obtain radial forces when we discuss elasticity and vibrational spectra; and the magnitude of the cohesive energy is not essential in other discussions. It is only for a surprisingly small number of properties that the radial dependence of the interatomic interaction needs to be understood in detail.

There is one very important correlation in the values for the bulk modulus that is not so striking here but that will recur in our discussions, and thus it takes on some significance. The bulk modulus is found to vary among the three homopolar semiconductors approximately as d^{-5}. Use of this form and of the observed value for silicon leads to predicted bulk moduli of 8.2 and 0.8 × 10^{12} erg/cm^3 for diamond and germanium, respectively, in rough agreement with experiment; the fit is even slightly better with experiment than with the theoretical values. The result is plausible in the sense that the variation of the covalent energy V_2^h with d^{-2} was derivable: energies tend to vary as d^{-2}, and since the bulk modulus is an energy per unit volume, it should vary as d^{-5}. Here we note that the important part of the overlap interaction is kinetic energy, and this fits particularly well, though in terms of the atomic calculations and all of the aspects entering the calculation, we might not expect a good correlation. Indeed, the argument could be made just as strongly for the ionic crystals, yet the bulk modulus for the alkali halides varies as d^{-3}. The d^{-5} trend for the bulk modulus will show up in the study of simple metals, and in terms of the pseudopotentials that will be used in the study of simple metals, d^{-5}-dependence takes on a particularly fundamental role. In Problem 15-3, the simple metal theory is used to give a good account of the bulk modulus in C, Si, and Ge. It should be noted also that the simple metal theory does not give a good account of cohesive energy itself; there is much cancellation between terms for that property, and there are important contributions (for example, E_{pro}) that do not vary as d^{-2}.

There is another important message contained in this finding. The cohesive energy per bond—originally called the "bond energy"—had a fundamental place in the early development of chemical theory. Aside from its intuitive importance, it was the parameter of obvious relevance, one that could be directly measured in the chemical laboratory. (The bond length was another such parameter.) Therefore, it served the function in early theory that has been served in our discussions by the bond energy $(V_2^2 + V_3^2)^{1/2}$. We have seen here that the cohesive energy contains terms from the overlap interaction that do not affect dielectric properties, the energy bands, nor even—as we shall see—most mechanical properties. Thus, if we wish to discuss this extensive range of properties, it is important to use a measure *other than the cohesive energy* as the experimental basis. This step was taken by Phillips (1973a), who used the dielectric constant to set an ionicity scale; following that example, optical and dielectric properties were used initially to determine the parameters of the Bond Orbital Model. Now, however, parameters can be taken more directly from electronic band structures.

7-C Cohesion in Polar Covalent Solids

Let us now discuss briefly the total energy in polar semiconductors. We have seen that the bond length is set principally by the overlap interaction. The overlap interaction increases so rapidly with decreasing d that a great change in bonding energy would be required to make only an appreciable change in bond length. This is also reflected by the fact that the magnitude of the contribution to the bulk modulus made by E_{bond} is very much smaller than the observed value. (The contribution to the bulk modulus is readily calculated; for homopolar semiconductors it is given by $-3^{1/2}V_2^h/d^3$.) Thus the fact that the bond length remains very nearly the same in isoelectronic series (and so, to some extent the bulk modulus remains nearly the same, as will be seen in Chapter 8) suggests that *the overlap interaction is relatively insensitive to polarity*. This is not surprising and certainly is an appropriate approximation on the scale of the accuracy of the overlap interaction found for homopolar semiconductors. This assertion describes the insensitivity of d and the bulk modulus to polarity, and only a discussion of the dependence of cohesion remains to be made.

Quantitative Estimates

To make this quantitative, we must return to the hybrid covalent energy and polarity, which were introduced in Chapter 3 and which related to average energies over the valence bands. These are listed in Table 7-2. The hybrid covalent energy is about twice the V_2 defined in Chapter 4, and the polarities defined in terms of the hybrid energies are seen in the table to be much smaller than those given earlier. An evaluation of the bond energy for silicon, using Table 7-2, gives $E_{\text{bond}} = 12.06$ eV, to be reduced by a promotion energy $E_{\text{pro}} = 3.52$ eV. This estimate of E_{coh} exceeds the observed cohesive energy of 2.32 eV by 6.22 eV, suggesting again a very large overlap interaction; this value is, in fact, of the order of the kinetic energy contribution alone (see Fig. 7-1). We can use the discrepancy as an experimental estimate of $V_0(d) = 6.22$ eV for silicon, again assume that it is constant in an isoelectronic series, and estimate the cohesive energy of aluminum phosphate to be 2.11 eV, in good agreement with the experimental 2.13 eV. Similarly, we can estimate the cohesive energy for all compounds isoelectronic with C, Si, Ge, and Sn; the results are displayed in Table 7-3.

We see that the predictions are very good, except for the noble-metal halides, which are much more strongly bound than is predicted; we associate the discrepancy with the noble-metal d bands. We also see that the predictions are very crude in the diamond row, as has happened with other properties.

Variation with Polarity

The main experimental trend of decreasing cohesive energy with increasing polarity is well described and arises from the decrease in the bond-formation energy, E_{bond}, the energy gained by forming the bonds *from neutral atoms*. (Notice in particular the term $-\Delta Z V_3^{\text{h}}/2$ in Eq. (7-2).) The bond-formation energy E_{bond} should be distinguished from the bond energy ε_{b} of Eq. (3-12), which gives the energy gained by forming a bond from electrons in hybrid states on the two atoms as $2(V_2^{\text{h}2} + V_3^{\text{h}2})^{1/2}$. That bond energy increases with polarity but is only part of the cohesion. This important distinction in covalent solids has frequently been overlooked and has led to confusion. The cohesive energy per bond in the homopolar semiconductors is comparable to the covalent energy V_2, showing the same trend with metallicity. If, however, one confuses the bond energy ε_{b} with the cohesive energy per bond E_{bond}, one guesses the wrong trend in cohesion with polarity. It is a similar confusion that incorrectly suggests increasing effective charges with increasing polarity. Concerning Table 7-3, notice finally that about half of the trend in cohesive energy owing to changing bond-formation energy is cancelled by a promotion energy decreasing with polarity.

The cohesive energy per bond is a quantity of sufficient fundamental importance that we have included in Table 7-3 also the experimental values for the tetrahedral solids other than those isoelectronic with C, Si, Ge, and Sn. We should

Hybrid covalent and hybrid polar energies, in eV. The hybrid polarity α_p^h can be compared with the polarity α_p based upon p states and used to describe dielectric properties.

Material	d (Å)	V_2^h	V_3^h	α_p^h	α_p
C	1.54	14.04	0.00	0.00	0.00
BN	1.57	13.51	3.12	0.23	0.34
BeO*	1.65	12.23	6.37	0.46	0.53
Si	2.35	6.03	0.00	0.00	0.00
AlP	2.36	5.98	2.17	0.34	0.51
Ge	2.44	5.59	0.00	0.00	0.00
GaAs	2.45	5.55	1.88	0.32	0.47
ZnSe	2.45	5.55	3.80	0.56	0.75
CuBr	2.49	5.37	5.59	0.72	0.87
Sn	2.80	4.25	0.00	0.00	0.00
InSb	2.81	4.22	1.54	0.34	0.53
CdTe	2.81	4.22	3.13	0.60	0.78
AgI	2.80	4.25	4.60	0.73	0.88
SiC	1.88	9.42	1.42	0.15	0.26
BP	1.97	8.58	1.20	0.14	0.20
AlN*	1.89	9.32	4.09	0.40	0.58
BeS	2.10	7.55	2.56	0.32	0.63
BAs	2.07	7.77	1.08	0.14	0.16
GaN*	1.94	8.85	3.92	0.40	0.60
BeSe	2.20	6.88	3.54	0.46	0.62
ZnO*	1.98	8.94	6.62	0.60	0.78
CuF	1.84	9.84	9.30	0.69	0.84
InN*	2.15	7.20	4.16	0.50	0.69
BeTe	2.40	5.78	2.79	0.43	0.61
AlAs	2.43	5.64	2.05	0.34	0.48
GaP	2.36	5.98	2.00	0.32	0.51
ZnS	2.34	6.08	4.13	0.56	0.75
CuCl	2.34	6.08	6.14	0.71	0.87
AlSb	2.66	4.71	1.48	0.30	0.45
InP	2.54	5.16	2.24	0.40	0.58
MgTe*	2.76	4.37	3.39	0.61	0.79
CdS	2.53	5.20	4.22	0.63	0.80
GaSb	2.65	4.74	1.31	0.27	0.45
InAs	2.61	4.89	2.11	0.40	0.55
ZnTe	2.64	4.78	3.04	0.54	0.74
CdSe*	2.63	4.81	3.89	0.63	0.79
CuI	2.62	4.85	4.62	0.69	0.86

* Wurtzite structures.

TABLE 7-3

Cohesive energy per bond, in eV (multiplying by 92.2 gives the value in kilocalories per mole for the polar semiconductors).

Material	E_{pro}	$V_o(d)$	E_{bond}	E_{coh} Theoretical	E_{coh} Experimental
C	4.26	20.14	28.08	3.68 =	3.68
BN	4.00	20.14	26.17	2.03	3.34
BeO	3.37	20.14	21.21	2.30	3.06
Si	3.52	6.22	12.06	2.32 =	2.32
AlP	3.31	6.22	11.64	2.11	2.13
Ge	4.02	5.22	11.18	1.94 =	1.94
GaAs	3.80	5.22	10.78	1.76	1.63
ZnSe	3.24	5.22	9.65	1.19	1.29
CuBr	1.71	5.22	7.12	0.19	1.45
Sn	3.28	3.66	8.50	1.56 =	1.56
InSb	3.12	3.66	8.21	1.43	1.40
CdTe	2.69	3.66	7.38	1.03	1.03
AgI	1.41	3.66	5.63	0.56	1.18
SiC	—	—	—	—	3.17
BP	—	—	—	—	2.52
AlN	—	—	—	—	2.88
GaN	—	—	—	—	2.24
ZnO	—	—	—	—	1.89
InN	—	—	—	—	1.93
AlAs	—	—	—	—	1.89
GaP	—	—	—	—	1.78
ZnS	—	—	—	—	1.59
CuCl	—	—	—	—	1.58
InP	—	—	—	—	1.74
MgTe	—	—	—	—	1.43
CdS	—	—	—	—	1.42
GaSb	—	—	—	—	1.48
InAs	—	—	—	—	1.55
ZnTe	—	—	—	—	1.14
CuI	—	—	—	—	1.33

SOURCES of data: Values of $V_o(d)$ were chosen to bring the total cohesive energy into agreement with experiment for the homopolar solid, and held fixed in the compounds isoelectronic with them. The theoretical E_{coh} is predicted by using Eq. (7-3) and the values from the first three columns. Experimental values were obtained by adding the heat of formation (the energy required to separate the compound into elements in the standard state), from Wagman et al. (1968), to the heat of atomization of the elements, from Kittel (1967, p. 98). A correction of about 0.01 eV/bond should be made to compensate for the different temperatures at which the heat was measured.

also discuss some empirical correlations which, from the point of view of the more complete LCAO description, seem somewhat accidental.

The first correlation is the decrease of E_{coh} with polarity, which was noted by Phillips and Van Vechten (1970), who found that it decreased linearly with their ionicity parameter. We noted in Section 2-C that the ionicity parameter is essentially proportional to α_p^2. In fact, a plot of cohesive energy against α_c (proportional to $1 - \alpha_p^2/2$ at small α_p) shown in Fig. 7-2 is quite linear; it is remarkable that these lines go through the origin, since *any* straight line would have been consistent with a decrease in ionicity that is linear at small ionicity. The simple result that cohesive energies tend to be proportional to α_c in an isoelectronic series is an empirical rule rather than a derived result. The same rule is followed to some extent in the skew compounds (though SiC is an exception). The noble-metal halides again do not satisfy this rule at all, as can be seen in Fig. 7-2, because of the

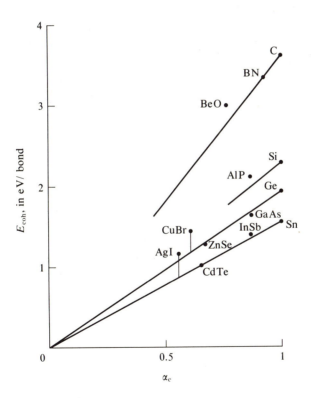

FIGURE 7-2

The cohesive energy per bond (commonly called bond energy) for the homopolar semiconductors and compounds isoelectronic with them, plotted against covalency. The lines represent an extremely simple empirical rule.

additional bonding arising from the noble-metal d electrons, as pointed out by Phillips and Van Vechten (1970). We shall discuss this contribution when we consider transition metals.

Variation with Metallicity

Another correlation can also be seen clearly in Fig. 7-2—the decrease in cohesive energy with increasing metallicity, indicated by the lower slopes for heavier compounds. This important qualitative trend is not described quantitatively by a proportionality either to α_m^{-1} or to V_2 itself.

We may remark in conclusion that separating out the overlap interaction energy and considering it to be a contribution to the total energy of covalent solids that is additive and independent of polarity seems a reasonably reliable course. The other two contributions are the promotion energy and the bond-formation energy, which are obtainable from the parameters of the Solid State Table. However, calculating the overlap interaction by using the statistical theory (which suffices for closed-shell systems) is not quantitative in covalent solids. The calculated overlap interaction is sufficiently accurate to give an understanding of the origin of the interaction and the trends, but it is best to use empirical values for bond lengths, radial force constants, and equilibrium overlap interactions in the homopolar semiconductors.

PROBLEM 7-1   *Cohesion in graphite*

What is the promotion energy per bond (three bonds for each two atoms) in graphite, with one electron per atom remaining in a p orbital and the others forming sp^2 hybrids? What is the bond-formation energy of the σ bonds in terms of d^2?

PROBLEM 7-2   *Equilibrium spacing in graphite*

We shall see in the next chapter that, experimentally, the change in energy in diamond, owing to a change in bond length d (differing from the equilibrium value $d_0 = 1.54$ Å), is $35.2(d - d_0)^2/d_0^2$ eV per bond (from Eq. (8-1) and Table 8-4). Assume that the overlap interaction is the same in diamond as in graphite and think of the difference between $\partial E_{\text{bond}}/\partial d$ found from Problem 7-1 and that obtained for diamond as an applied force on this springlike bond. Add a compression from the π states, estimated to contribute $V_{pp\pi}$ per bond, and estimate the bond length in graphite. (The estimate of $V_{pp\pi}$ per bond is an estimate of one quarter the π-band width of $6V_{pp\pi}$ divided by the $\frac{3}{2}$ bonds per π electron.)

PROBLEM 7-3   *Overlap interaction*

Is the sum of the charge densities of the promoted electrons in graphite the same as the sum for diamond, leading therefore to the same overlap interaction?

PROBLEM 7-4   *Cohesion in* BN

What is the promotion energy and the bond-formation energy in hexagonal boron nitride (see Fig. 3-10)? (Ignore the π electrons, as if $V_{pp\pi}$ were zero.)

CHAPTER 8

Elasticity

SUMMARY

Accurate LCAO calculations of the change in energy of covalent solids under uniform distortion have been made and accord well with experiment. These provide a basis for extracting the essential features and allowing more elementary calculation of lattice rigidity. The first such elementary calculation is of the change in bond energy for angular distortions of a type that do not permit reorientation of the hybrids making up the bonds. The corresponding force constant is proportional to the hybrid covalent energy and, in the Bond Orbital Approximation, to the covalency of polar semiconductors. Corrections to the Bond Orbital Approximation lead to an approximate proportionality to covalency cubed, in better accord with experiment. The angular constant is reduced for other distortions that permit hybrid reorientation. Such force constants, constituting the " valence force field," are fitted to the elastic constants to treat internal displacements and will be used in the study of lattice vibrations in Chapter 9. However, they do not include long-range forces that are present in the LCAO calculations.

In Chapter 7 we discussed changes in energy associated with small uniform changes in the volume of a system; the bulk modulus is an elastic constant that describes the rigidity of the system against such compressions. We now extend the discussion to uniform distortions that lower the symmetry of the system; the rigidity of the system against these distortions is described by shear constants. Having all of these elastic constants will be useful; they can also tell us something about the stability of the tetrahedral structure itself.

The restriction of the discussion to *uniform* distortions means we can retain the translational periodicity of the crystal in our treatment, and therefore, also, well-defined energy bands (though they are also distorted), and it is possible to devise rather accurate and complete theories. We discuss such theories briefly in Section 8-A, but the principal goal there is to clarify the nature of elasticity sufficiently to allow the introduction of approximate models that can be used when the distortions are not uniform.

8-A   Total Energy Calculations

The more general distortions that will be discussed here will often include changes in bond length as well as angular distortions. We can envision the complete calculation, as we did in Chapter 7, by thinking first of atoms, then of promoting the electrons to hybrids, then of bringing the atoms to their final positions without bond formation, of forming bonds, and finally of the energy-band states for the crystal. Bringing the atoms together without bond formation gives a change in energy that is simply a sum of overlap interactions between the various atoms. Since we are interested only in atomic arrangements very close to the equilibrium arrangement, it is appropriate to expand the total overlap interaction, sketched in Fig. 7-1, in terms of the deviation, $d - d_0$, of the bond length from its equilibrium value, d_0, keeping terms only to second order. In doing this we can use the interaction obtained from experiment, rather than the predicted one. Thus the interaction is simply a parabola fitted to the observed bond length, the cohesive energy, and the bulk modulus. The energy per bond is written as

$$E = -E_{coh}^0 + \tfrac{1}{2}C_0 \frac{(d - d_0)^2}{d_0^2}. \tag{8-1}$$

The term E_{coh}^0 is the observed cohesive energy; it will not interest us here. The constant C_0 is called the ***radial force constant***. Its value is about forty to fifty electron volts for most semiconductors and will be tabulated later. In this chapter, we focus on angular distortions and add this empirical radial interaction where it is needed.

To the radial interaction energy, then, must be added the change in energy that accompanies the formation of the final electronic states. Again, since only energy differences are needed, we can simply use the sum of electron energies to calculate those differences as we did in Chapter 7. We shall focus initially on distortions in which there is no change in bond length. Thus our task is simply to understand the changes in the sum of the one-electron energies as the lattice is distorted.

The Special Points Method

It is possible, by using the LCAO approach or a pseudopotential approach, to make a calculation of energy bands for the distorted lattice. There are still two atoms per primitive cell, so no serious difficulty is encountered. The sum of the energies of the occupied states can then be calculated with any desired precision, by using the Gilat-Raubenheimer scheme described in Section 2-E, so the accuracy is limited only by the accuracy of the energy-band calculation. The *special points method* (Baldareschi, 1973; Chadi and Cohen, 1973) provides a very simple alternative for summing the energies *over full bands*. It also can be made as precise as desired; that method was used by Chadi and Martin (1976) to study the elasticity of covalent solids. The method is sufficiently general that we should outline it before continuing with the discussion of elasticity theory.

Any energy band can be written as a Fourier series in the Brillouin Zone in the form

$$E(\mathbf{k}) = \sum_{\mathbf{T}} a_{\mathbf{T}} e^{i\mathbf{k} \cdot \mathbf{T}}, \tag{8-2}$$

where the $\mathbf{T}$ are lattice translations, that is, integral combinations of primitive lattice translations. (See Section 3-C and the references to band theory given there.) The value of $a_{\mathbf{T}}$ will be the same for all $\mathbf{T}$ that are related by a symmetry rotation or reflection of the crystal; for example, it will be the same for all 12 nearest-neighbor distances in the face-centered cubic lattice. Furthermore, we can expect it to decrease with increasing magnitude of $\mathbf{T}$ if the bands are reasonably smooth. (In fact, the bands have cusps at "points of contact" between bands, but these seem not to have caused any problem for Chadi and Martin.) The energy values at any wave number depend upon all of the $a_{\mathbf{T}}$, but the average clearly depends only upon a_0. The idea of the special points method is to select wave numbers as samples for which the other components in the Fourier series cancel out.

Notice how this is done, first, for a one-dimensional case for which the translations $\mathbf{T}$ are integral multiples of the atomic spacing a. If the special point $k^* = \pi/(2a)$ is chosen, the two terms $a_a e^{ik^*a} + a_{-a} e^{-ik^*a}$ sum to zero (note $a_a = a_{-a}$), so if the terms for larger $\mathbf{T}$ are negligible, the average of $E(k)$ is just $E(k^*)$. In two dimensions, we can do better. For a square lattice, the four terms $a_a(e^{ik_x a} + e^{-ik_x a} + e^{ik_y a} + e^{-ik_y a})$ sum to zero along a set of lines in k-space and we may choose k^* to make the contributions from the four translations of the type $[110]a$ also vanish simultaneously; the corresponding special point is $k^* = [110]\pi/(2a)$. In three dimensions, we can set the contributions of the first three sets of Fourier coefficients equal to zero (or set the first two equal to zero and minimize the third if there is no solution; this turns out to be the case for the face-centered cubic lattice). Baldareschi (1973) has obtained the corresponding special points for several lattices. That for the face-centered cubic lattice, and therefore the diamond and zincblende lattices, which have the same translational periodicity, is

$$k^* = [0.6223, 0.2953, 0]2\pi/a. \tag{8-3}$$

There are 24 such points in the Brillouin Zone, but only one need be used since the band energies are the same at the others. The same special point can be used to evaluate any other property (such as the charge density variation over the real lattice) that requires a sum over the Brillouin Zone. Chadi and Cohen (1973a) proposed a scheme for obtaining a larger set of points by requiring a weighted sum of terms in the Fourier expansion to be minimum. This may seem somewhat arbitrary, but it has the advantage of a larger sampling. For the face-centered cubic structure, they obtained two special points,

$$
\begin{aligned}
k_1^* &= [3/4,\ 1/4,\ 1/4]2\pi/a, \\
k_2^* &= [1/4,\ 1/4,\ 1/4]2\pi/a,
\end{aligned}
\right\}
\tag{8-4}
$$

with $E(k_1^*)$ to be weighted three times as strongly as $E(k_2^*)$. This was the scheme used by Chadi and Martin (1976). An indication of the accuracy of the calculation of two special points may be taken from their estimate of an energy per atom for silicon of -26.395 eV (relative to the silicon hybrid energy). Use of ten special points (Chadi and Cohen, 1973b) gave almost the same result, -26.399 eV. The result obtained from the special points method can be regarded as the exact sum over the band, which we will need in discussing the accuracy not only of the LCAO description of electronic structure but of other more approximate calculations as well.

Chadi and Martin (1976) used essentially the same LCAO parameters that are given in the Solid State Table to obtain the energies at the two special points in the Brillouin Zone. They then redetermined the wave numbers of the special points for the distorted crystal and recalculated the energy. The elastic distortion which they used is a shear strain, in which there is no change in bond length to first order in the strain; thus the radial force constant of Eq. (8-1) does not enter the calculation. That strain can be written as

$$
\begin{aligned}
\varepsilon_2 = \varepsilon_{yy} &= \frac{\partial v}{\partial y} = \varepsilon, \\
\varepsilon_3 = \varepsilon_{zz} &= \frac{\partial w}{\partial z} = -\varepsilon,
\end{aligned}
\right\}
\tag{8-5}
$$

where the notation is that of Kittel (1967, p. 11ff). This distortion is illustrated for the zincblende structure in Fig. 8-1. The fact that the relative displacements of nearest neighbors occur in directions perpendicular to the axes connecting those nearest neighbors can be seen by examining the central atom and its nearest neighbors.

The change in energy density can be written in terms of the elastic constants c_{11} and c_{12} (again, in notation used by Kittel, 1967):

$$
\delta E = (c_{11} - c_{12})\varepsilon^2.
\tag{8-6}
$$

Thus the change in electron energy obtained with the special points method can be equated to the expression in Eq. (8-6) to predict this combination of elastic constants. The results are compared with experiment in Table 8-1. Also listed is the calculated and experimental vibrational frequency for a vibrational mode (to be discussed in Chapter 9). The latter calculation required doubling the size of the primitive cell and determining new special points.

Considering that all theoretical parameters used to obtain Table 8-1 came from the energy bands and none came from measurement of elastic properties, the agreement with experiment is remarkably good. Although the agreement is semiquantitative, the theoretical values reflect correctly the decrease in rigidity with metallicity and with polarity. We conclude that this description of electronic structure does contain the physical origin of elastic rigidity and even predicts it rather well.

The Need for a Simpler Method

The special points method depends upon retention of the translational periodicity of a lattice, which is lost if we consider defects, surfaces, or lattice vibrations. (Even the special vibrational mode with frequency listed in Table 8-1 entailed a halving of the translational symmetry.) It is therefore extremely desirable to seek an approximate description in terms of bond orbitals, so that the energy can be summed bond by bond as discussed in Chapter 3. We proceed to that now.

Imagine first displacing a single atom in a zincblende crystal in a [100] direction. Then it should be possible to recalculate the electron states as in Chapter 3, but now for the distorted structure. We are confronted immediately with an uncertainty that was not present before. In the undistorted crystal, it was clearly appropriate to construct the four orthogonal hybrids at each atom so that their

TABLE 8-1

An elastic constant predicted by using LCAO bands and the special points method, compared with experiment; also, calculated and experimental values of a zone-boundary transverse acoustic mode frequency, $\omega_{TA}(X)$, to be discussed in Chapter 9.

Material	$c_{11} - c_{12}$ (in 10^{11} erg/cm^3)		$\omega_{TA}(X)$ (in 10^{12} Hz)	
	Theoretical	Experimental	Theoretical	Experimental
Si	10.80	10.20	6.52	4.49
Ge	7.84	8.02	3.15	2.39
GaAs	6.56	6.50	3.06	2.36
ZnSe	3.14	3.22	2.28	2.10

SOURCE: Martin and Chadi (1976).

maximum wave functions were in the direction of the tetrahedrally oriented near-est neighbors. Now, at the atom that has been moved, we might construct sp^3 hybrids oriented in the new directions of the neighbors nearest the central atom, but these would not be orthogonal to each other, and we have in all cases required orthogonal basis states. The correct approach, in principle, is to redo the varia-tional calculation in the distorted structure, optimizing the composition of the hybrids as we optimized the composition of the bond orbitals in Chapter 3. In doing this, we would make use of the atomic matrix elements from the Solid State Table. This task was undertaken by Sokel (1976, 1978). After optimizing the bond orbitals, he made a Bond Orbital Approximation, finally adding corrections to the Bond Orbital Approximation corresponding to the use of the extended bond orbitals described in Chapter 3. The final results were in very good agreement with the more accurate calculations of Chadi and Martin. In fact, the results obtained even with the uncorrected Bond Orbital Approximation were reasonably good for the homopolar semiconductors, but corrections to that approximation became very important for the polar semiconductors. This kind of approximate analysis sheds much light on the physics of covalent bonding; therefore we shall carry out the main parts of the analysis.

8-B   Rigid Hybrids

Let us return to the simple shear distortion that was illustrated in Fig. 8-1. Focus-ing on the central atom and the four nearest neighbors tetrahedrally bonded to it, we see that the distortion twists the tetrahedron as illustrated in Fig. 8-2, where

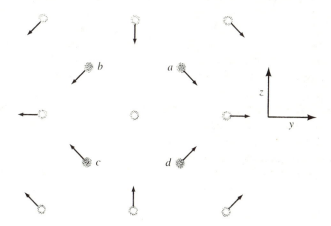

FIGURE 8-1

A distortion in the zincblende lattice, corresponding to
$\varepsilon_3 = -\varepsilon_2$; all other strains are zero. Atoms a and c lie in the positive x-direction from the origin by $a/4$; b and d lie in the negative x-direction by the same amount.

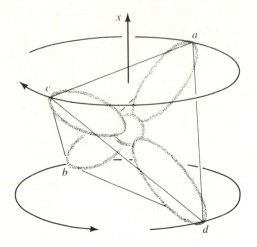

FIGURE 8-2

The shear distortion of Fig. 8-1 exerts a "twist" on each set of tetrahedral bonds. Rehybridization cannot accommodate to this so the hybrids may be imagined rigidly fixed.

one set of bonds is redrawn in perspective. The distortion can also be imagined as tending to draw all bonds toward the same plane. However, a unitary transformation (unitarity is required to keep the hybrids orthogonal, so that $\langle h_i | h_j \rangle = 0$) does not change the total charge distribution and cannot make hybrids that are more favorable than the starting tetrahedral hybrids. Correspondingly, the optimization of the hybrids in the distorted structure leaves them unchanged. The hybrids are "rigid" under this distortion.

The rigid-hybrid concept was introduced by Harrison and Phillips (1974) for the treatment of the $(c_{11} - c_{12})/2$ elastic constant. Harrison and Phillips needed a scale factor in that theory, partly because they used the covalent energy V_2 rather than the much larger hybrid covalent energy V_2^h, but with that scale factor, a very good fit was obtained, and the physics of the rigidity they envisaged is the same as that given here. In contrast, a treatment by Bullett (1975) discarded the orthogonality requirement and rotated the hybrids toward the nearest-neighbor nuclei. This gave a rigidity proportional to V_1 (how such a term arises may be seen from Appendix E) and the rigidity appeared to agree with experiment for diamond and silicon. However, it is difficult to see how Bullett obtained his results, and the conceptual and mathematical bases of the approach seem rather questionable. The proportionality to V_1 found by Bullett is not in accord with trends with metallicity and with polarity that have long been known (Keyes, 1962), trends that are reflected by the values in Table 8-1 and that will follow from the calculation here.

Calculation of the Change in Bond Energy

The calculation of the change in bond energy is rather straightforward. The distortion misaligns the two hybrids making up a bond, reducing the hybrid covalent energy (we seek here the average energy of the valence band, not the valence-band maximum) and thereby raising the energy of each bonding electron. To calculate the angle of misalignment of the two hybrids in each bond, take the dot product of the nearest-neighbor vector distance in the distorted and undistorted crystal and divide by the distance before and after distortion; that is, divide by $3^{1/2}a/4$ and $(3 + 2\varepsilon^2)^{1/2}a/4$, respectively:

$$\cos \theta = [1, 1, 1] \cdot [1, 1 + \varepsilon, 1 - \varepsilon]/[3^{1/2}(3 + 2\varepsilon^2)^{1/2}]. \tag{8-7}$$

The meaning of θ, the angle of misalignment, is seen in Fig. 8-3. Expanding Eq. (8-7) for small θ and small ε, we obtain $\theta^2 = 2\varepsilon^2/3$; the same result applies for every bond. We write the change in hybrid covalent energy, quadratic in θ, in the form

$$\delta V_2^h = -\lambda V_2^h \theta^2, \tag{8-8}$$

where λ is a dimensionless parameter that can be obtained in terms of interatomic matrix elements. The two misaligned hybrids shown in Fig. 8-3 can each be decomposed into s-orbitals and p-orbitals oriented parallel and perpendicular to the interatomic separation:

$$|h\rangle = \tfrac{1}{2}(|s\rangle + \sqrt{3} \cos \theta |p_\sigma\rangle + \sqrt{3} \sin \theta |p_\pi\rangle). \tag{8-9}$$

The matrix element can be directly evaluated; notice that $|p_\pi\rangle$ has nonvanishing matrix elements only with the $|p_\pi\rangle$ on the other hybrid and that the two are antiparallel. The result may be expanded for small θ and written in the form $V_2^h(1 - \lambda\theta^2)$, as in Eq. (8-8). One immediately obtains

$$\lambda = \frac{-\sqrt{3} V_{sp\sigma} - 3V_{pp\sigma} + 3V_{pp\pi}}{V_{ss\sigma} - 2\sqrt{3} V_{sp\sigma} - 3V_{pp\pi}} = 0.88 \tag{8-10}$$

The energy of each electron, in the Bond Orbital Approximation, is obtained by using hybrid parameters in Eq. (3-12):

$$\varepsilon_b = (\varepsilon_h^1 + \varepsilon_h^2)/2 - \sqrt{V_2^{h2} + V_3^{h2}}. \tag{8-11}$$

FIGURE 8-3

Definition of the angle of misalignment of the two hybrids making up a bond.

The other contributions to the energy, discussed in Chapter 7, contribute only to the radial energy of Eq. (8-1), so we may combine Eqs. (8-8) and (8-11), and the relation between ε and θ, to obtain the total change in energy per electron as

$$\delta\varepsilon_b = 2\lambda V_2^h \alpha_c^h \varepsilon^2/3, \qquad (8\text{-}12)$$

where $\alpha_c^h = V_2^h/(V_2^{h2} + V_3^{h2})^{1/2}$ is the hybrid covalency. Inasmuch as there are 32 electrons per unit cube of edge a, the same change in energy may be written in terms of the elastic constants (Kittel, 1967), as

$$\delta\varepsilon_b = \frac{a^3}{32}(c_{11} - c_{12})\varepsilon^2. \qquad (8\text{-}13)$$

These two equations can be combined to give an expression for $c_{11} - c_{12}$ in the Bond Orbital Approximation:

$$c_{11} - c_{12} = \sqrt{3}\,\lambda\alpha_c^h V_2^h/d^3 \qquad \text{in the Bond Orbital Approximation.} \qquad (8\text{-}14)$$

where $d = a3^{1/2}/4$. Using $\lambda = 0.88$ from Eq. (8-10) and other parameters from Table 7-2, we can immediately evaluate this expression for any material. The results for an interesting sample are given in Table 8-2.

Comparison with Experiment

Values of the elastic constant from Eq. (8-14) agree extremely well with experiment for C, Si, and Ge and, in fact, they agree almost as well as the values obtained by the more accurate calculations given in Table 8-1. The Bond Orbital Approxi-

TABLE 8-2

The elastic shear constant, $c_{11} - c_{12}$, as obtained in the Bond Orbital Approximation of Eq. (8-14), as obtained by Sokel (1976) and including the bonding–antibonding matrix elements from perturbation theory, and experimental values; all are in units of 10^{11} erg/cm^3.

Material	Eq. (8-14)	Sokel	Experiment
C	93.88	82.38	95.1
Si	11.35	11.48	10.20
Ge	9.39	8.83	8.02
GaAs	8.73	6.86	6.50
ZnSe	6.11	3.65	3.22

mation has worked very well here. Equation (8-14) predicts a variation among the homopolar semiconductors of exactly d^{-5} and, as can be seen from Table 8-2, and as was indicated in Chapter 7, this is very well followed by the experimental values. The coefficient is also very well given, though a similar expression given by Harrison and Phillips (1973) required a scale factor λ equal to 4.3. Part of the reason was their use of a V_2 roughly half the value of V_2^h given here. The other was the inclusion of a term in the nonorthogonality ($S = 0.5$ in Appendix B) of the hybrids. We have noted here that such effects are properly included in the overlap interaction and therefore do not contribute to $c_{11} - c_{12}$. The Harrison-Phillips model led to a factor $\alpha_c - S$ in Eq. (8-14). It is gratifying that a scaling of the theory for homopolar solids is not needed here, and the plausible physical effects of misaligned hybrids of Fig. 8-3 give a quantitative description of the rigidity.

The agreement in the polar series Ge, GaAs, and ZnSe is not so good. The decrease in angular rigidity with increasing polarity is qualitatively described by Eq. (8-14), but the observed trend is very much stronger. The error clearly comes from the use of the Bond Orbital Approximation, since the more accurate calculations of Table 8-1 quantitatively reflect this trend. Indeed, Sokel (1976) did include the matrix elements neglected in the Bond Orbital Approximation, just as we included them for the extended bond orbitals in Chapter 3, and consequently Sokel restored agreement with experiment, as indicated in Table 8-2.

The same effect can be seen in the zig-zag chain of Fig. 3-11. It is remarkable that we can compute the angular force constant in that model exactly, as well as in the Bond Orbital Approximation (see Problems 8-1 and 8-2). The results turn out to be identical for the homopolar semiconductors, but for polar semiconductors, the exact solution has α_c replaced by α_c^3. Sokel has shown that the result is not so simple for the tetrahedral solid, but turns out quantitatively to be very close to an α_c^3 dependence. We will also find an α_c^3 dependence when we treat tetrahedral solids in terms of the chemical grip in Section 19-F. This suggests the approximation to the full calculation,

$$c_{11} - c_{12} \approx \sqrt{3}\,\lambda\alpha_c^{h3}\,V_2^h/d^3, \qquad (8\text{-}15)$$

which gives the same values as Eq. (8-14) for the homopolar semiconductors, but reduces the values in GaAs and ZnSe down to 6.80 and 2.67×10^{11} erg/cm^3, in good agreement with the experimental 6.50 and 3.22×10^{11}. The comparison with experiment for a number of materials given in Fig. 8-4 indicates that Eq. (8-15) gives a very good account of all tetrahedral semiconductors, though the noble-metal halide CuCl would seem to be anomalously soft, an effect initially attributed to the d states of the noble metal. S. Froyen (private communication) has made a study of this and noted that the CuCl point in Fig. 8-3 is close to a line drawn through the isoelectronic series GeSi, GaP, ZnS (GeSi is taken halfway between Ge and Si), so perhaps there is not an anomaly to be explained. He nevertheless explored the role of the d shell in determining the angular rigidity.

The leading effect of the full d shells should come from the coupling with the empty states nearest in energy, the valence s states of the noble metal. Such

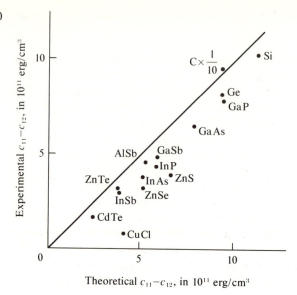

FIGURE 8-4

A plot of the experimental shear constants compiled by Martin (1970), against the predictions by Eq. (8-15). No experimental parameters (except bond length d) enter. Note both the experimental and theoretical values for diamond were scaled by 1/10 to bring that point into the figure.

coupling can arise from the nonspherical part of the potential near the noble-metal atom. Such coupling can only *lower* the energy of the d states; this follows from second-order perturbation theory, Eq. (1-14). However, it can readily be verified that for tetrahedral symmetry all such matrix elements vanish. Nonzero matrix elements from *distortion* will therefore lower the energy and reduce $c_{11} - c_{12}$ for CuCl, as suggested by Fig. 8-4. Froyen estimated these matrix elements and found their effect very small compared to the difference between the CuCl point and the line in Fig. 8-4. He therefore included an indirect coupling through the Cl p states in perturbation theory. This effect is fourth order in the interatomic matrix elements, as is the chemical grip discussed in Section 19-F, and is analogous to it. The effect in CuCl, however, arises from intra-atomic coupling and may be thought of as **ion distortion** rather than as a chemical grip. This term is much larger than that arising from nonspherical potentials but still gives an effect small compared to the separation between the CuCl point and the line in Fig. 8-4. The result, then, is consistent with the empirical suggestion given above that there is no anomalous effect of the d states on the angular rigidity of CuCl.

We may note in passing that the dependence upon covalency from the Harrison-Phillips model, $\alpha_c - 0.5$, does not look very similar to the present α_c but does give a reasonable account of experiment. In fact, both are approximately linear in α_p^2 over the range of values of α_p that occur. As we have noted about ionicity theories, a finding of linear plots is not compelling evidence for the validity of a model, particularly if a scale parameter is available.

The same analysis that has been made here for the zincblende semiconductors can of course be carried out for the wurtzite structure. For comparison with experiment, another approach is simpler; that is to compare the formulae obtained here with "effective" cubic elastic constants, which were obtained by Martin (1972b). These were estimates of what the constants of wurtzite compounds would be in the zincblende structure, based upon the measured constants of the wurtzite crystals.

The important accomplishment in this section has been the reduction of the problem to the point where the energy can be computed bond by bond, either approximately, by the Bond Orbital Approximation, or more accurately, by including the bonding–antibonding matrix elements in perturbation theory. It is not so important for the understanding of the uniform shears, but will be important in other properties. Before we can examine these, however, we must consider another mode of shear in the lattice.

8-C Rehybridization

For a crystal having the symmetry of diamond or zincblende (thus having cubic elasticity), there are three independent elastic constants, c_{11}, c_{12}, and c_{44}. The bulk modulus that was discussed in Chapter 7 is $B = (c_{11} + 2c_{12})/3$. We can discuss the bulk modulus, and the combination $c_{11} - c_{12}$, entirely in terms of rigid hybrids, and therefore the two elastic constants c_{11} and c_{12} do not require deviations from this simple picture. This will not be true for the strain ε_4, which is relevant to c_{44}, and this is a complication of some importance.

Complications in ε_4

The strain ε_4 is given by

$$\varepsilon_4 = \varepsilon_{yz} = \frac{\partial v}{\partial z} + \frac{\partial w}{\partial y} = \varepsilon, \tag{8-16}$$

where v is displacement in the y-direction and w is displacement in the z-direction; this is illustrated in Fig. 8-5. This strain may look even simpler than that illustrated in Fig. 8-1, but we will find at least four complicating features.

Let us consider the displacements of the four neighbors nearest to the central atom. These have components that rotate the entire tetrahedron, the first complication. These complicate the geometry slightly but do not affect the energy. They may be subtracted out, leaving the displacements shown in Fig. 8-6.

We see immediately that these residual displacements have components along the bonds so that, in contrast with the $c_{11} - c_{12}$ shear, the radial interaction contributes to the shear constant. This is the second complication.

The third is the possibility of rehybridization. Note that the nonradial components of that displacement tend to pull the bonds to atoms a and c apart and tend to pull those to b and d together, as illustrated in Fig. 8-6. In order to avoid the misalignment of hybrids of the type shown in Fig. 8-3, it becomes favorable to decrease the content of orbitals $|p_x\rangle$ (the x-direction is indicated in Fig. 8-6) in

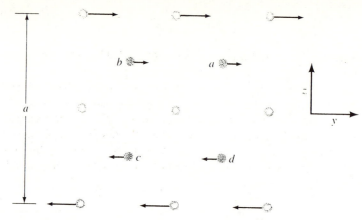

FIGURE 8-5

Atoms a and c lie in the positive x-direction from the origin atom by $a/4$; atoms b and d lie in the negative x-direction by the same amount. For zincblende the shaded atoms might be Zn and the empty atoms S.

hybrids to atoms c and a and increase the content of orbitals $|p_x\rangle$ in the other two. Orthogonality is maintained by modifying the content of s orbitals in each hybrid.

The fourth complication is the tendency toward *internal displacements*, displacements of the two atoms in the primitive cell with respect to each other. We see in particular that radial interactions tend to pull the central atom upward (in the x-direction) under this distortion. Similarly, the tendency to maintain tetrahedral angles will push the central atom in the opposite direction. Radial forces always win. These displacements are the origin of piezoelectricity and we shall return to them in Section 8-E.

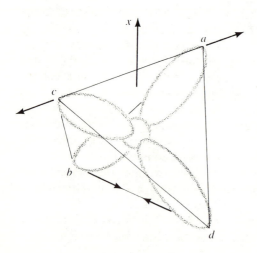

FIGURE 8-6

Tetrahedral distortions corresponding to the lattice shear distortion of Fig. 8-5. (There is also a rotation of the tetrahedron.) Note that there are radial components to the displacement though the elastic deformation is pure shear. In addition, the hybrids can accommodate to this distortion without disrupting the orthogonality by modifying the content of $|p_x\rangle$ and $|s\rangle$ orbitals. Notice finally that such a motion of the atoms a, b, c, and d tends to cause displacement of the central atom relative to them in the x-direction.

Predictions of c_{44} from different approximations,
all in 10^{11} erg/cm^3.

Source of data	C	Si	Ge
Sokel (1976)	43.46	7.17	5.66
Chadi-Martin (1976)	45.25	9.43	7.83
Experiment	57.7	7.96	6.71

SOURCES of experimental data: C, from McSkimin and Bond
(1957); Si and Ge, from Huntington (1958).

Calculation of c_{44}

Sokel carried out the variational calculations necessary to calculate c_{44} for a
few materials. In a similar way Chadi and Martin used the special points method
to calculate c_{44} with the artificial restriction that all nearest-neighbor distances
remain fixed. They then did a valence force field calculation, such as will be
discussed in the next section, to correct for internal displacements. The results of
these calculations are listed in Table 8-3. In spite of the extra complications, the
theory seems to be about as accurate as it was for the calculation of $c_{11} - c_{12}$.

If we had proceeded to the calculation of c_{44} by assuming rigid hybrids, we
would have overestimated c_{44}. To see this, imagine performing the full calculation
in steps: the crystal is distorted and the energy increase is calculated with rigid
hybrids. The elastic constant is proportional to that increase. We then allow
rehybridization, which lowers the energy, reducing the constant. Sokel found the
reduction due to rehybridization to be about 30 percent in silicon.

The final physical picture of the angular rigidity is quite appealing. There is
radial rigidity determining the constant $(c_{11} + 2c_{12})/3$, and an angular rigidity
from hybrid-misalignment determining $c_{11} - c_{12}$. Finally, the lattice is softened to
those distortions which can be accommodated by the hybrids; that softening is
reflected in the lower value of c_{44}. The three independent elastic constants reflect
the three important physical effects.

8-D The Valence Force Field

The three independent elastic constants describe the rigidity of the system under
uniform distortions but not under arbitrary nonuniform distortions. The elastic
energy of the system, E_{tot}, is a function of the positions of all of the atoms in the
crystal, or in particular, of the components u_i of the displacements of each of the
atoms from equilibrium; there are $6N_p$ such components for the N_p atom pairs.
For sufficiently small distortions we can expand that energy for small u_i:

$$E_{tot} = E_{tot}(0) + \sum_i \frac{\partial E_{tot}}{\partial u_i} u_i + \tfrac{1}{2} \sum_{ij} \frac{\partial^2 E_{tot}}{\partial u_i \partial u_j} u_i u_j. \tag{8-17}$$

All $\partial E_{tot}/\partial u_i$ vanish identically, since they are evaluated at the equilibrium positions. The $\partial^2 E_{tot}/\partial u_i \partial u_j$ are constants of the crystal and function as spring constants or *force constants*. Many are related by symmetry, but there are nevertheless a number (of the order of the number of atoms in the crystal) that are independent and they must be known if one wishes to calculate all the normal modes of vibration of the crystal.

It is possible in principle to calculate all of these modes from the theory of the electronic structure, which is equivalent to the calculation of all the force constants. Indeed we will see that this is possible in practice for the simple metals by using pseudopotential theory. In covalent solids, even within the Bond Orbital Approximation, this proves extremely difficult because of the need to rotate and to optimize the hybrids, and it has not been attempted. The other alternative is to make a model of the interactions, which reduces the number of parameters. The most direct approach of this kind is to reduce the force constants to as few as possible by symmetry, and then to include only interactions with as many sets of neighbors as one has data to fit—for example, interactions with nearest and next-nearest neighbors. This is the Born-von Karman expansion, and it has somewhat surprisingly proved to be very poorly convergent. This simply means that in all systems there are rather long-ranged forces.

A second possibility is to model the interactions on the basis of theoretical or intuitive ideas in order to reduce the number of independent parameters. Perhaps the simplest such model for the diamond or zincblende structure is to assume the "bond-stretching" force constant C_0 introduced in Eq. (8-1) and "bond-bending" interactions, which are written

$$\delta E_1 = \tfrac{1}{2}C_1(\delta\theta_{ij})^2, \tag{8-18}$$

where the $\delta\theta_{ij}$ are the deviations of the angle between adjacent bonds (as defined by the internuclear vector distances) from the tetrahedral angle; these must be summed over all adjacent-bond angles. This model is one of many force-constant models, sometimes called collectively the "valence force field approach" (Musgrave and Pople, 1962). A more recent and often quoted model by Keating (1966) used parameters α and β, analogous to the C_0 and C_1 used here, but since the angular energy was proportional to the dot product of adjacent bond vectors, it included some radial energy as well as angular energy. We shall use a more complete separation, as did Musgrave and Pople, in this book. Martin (1972b) used an extension of Keating's model in computing the "effective" cubic force constants for the wurtzite structure. He argued that Keating's two-parameter model gave a better fit to the elastic properties than the two-parameter model we are using; this seems to be true, but we nevertheless seek the conceptual simplicity of the model based on C_0 and C_1.

In all such models, the more constants one introduces, the more experimental constants one can fit and, it is always hoped, the more accurately also can one predict other quantities, though the process has not proved rapidly convergent. The particular model that will be used here fits nicely with the spirit of the approach used in this book: it contains the minimum number of parameters allowing even a qualitative description of the elastic properties. The model makes it quite simple to estimate the parameter C_1 in terms of tables of elastic constants such as we have given, and it will allow us to make very elementary but reasonably quantitative discussions of a number of properties. The appropriate first step is the determination of the constants C_0 and C_1 from the experimental elastic constants.

Fitting the Parameters

In the zincblende structure, with three independent elastic constants, the use of a model with only two parameters will allow a test and also allows alternate ways of obtaining the parameters. We obtain the radial force constant from a uniform compression, $\varepsilon_1 = \varepsilon_2 = \varepsilon_3 = \varepsilon$, from which we obtain a change in energy per bond of

$$\tfrac{1}{2}C_0\varepsilon^2 = [\tfrac{1}{2}c_{11}(3\varepsilon^2) + \tfrac{1}{2}c_{12}(6\varepsilon^2)]a^3/16. \tag{8-19}$$

We can use this to obtain C_0 in terms of c_{11} and c_{12}:

$$C_0 = (3a^3/16)(c_{11} + 2c_{12}). \tag{8-20}$$

We use the distortion given in Fig. 8-1 and an analysis similar to that leading to Eq. (8-7) to obtain C_1 in terms of c_{11} and c_{12}. We see that of the six bond angles with apex at a given atom, two $\delta\theta_{ij}$ are $2^{1/2}\varepsilon$, two are zero, and two are $-2^{1/2}\varepsilon$; the change in energy per atom is then $\tfrac{1}{2}C_1(2 + 0 + 2)(2\varepsilon^2)$ and the change in energy per electron is a quarter of this. Equating the result to the form in Eq. (8-13), we obtain

$$C_1 = a^3(c_{11} - c_{12})/32. \tag{8-21}$$

Eqs. (8-20) and (8-21) are in some sense backwards since the parameters C_0 and C_1 are the microscopic parameters of the model, which give rise to the macroscopic elastic constants. However, by solving for C_0 and C_1, we may obtain "experimental" values for these parameters. Such parameters are given in Table 8-4, based upon the elastic constants compiled by Martin (1970). Later we shall give a different set based upon two important vibrational frequencies.

We note first that the radial force constants are quite large and only qualitatively follow the d^{-2} trend that would correspond to the d^{-5} bulk modulus variation discussed in Chapter 7. There is a weak tendency for C_0 to decrease both with metallicity and with polarity.

The angular forces are considerably weaker. However, they are strong compared to angular forces in other types of solids and they represent the rigidity

TABLE 8-4

Elastic constants (in 10^{11} erg/cm^3) and force constants (in eV) for tetrahedral semiconductors.

Material	Experiment			Theory			
	c_{11}	c_{12}	c_{44}	c_{44}	C_0	C_1	ζ
C	107.6	12.50	57.7	48.6	70.4	8.4	0.267
Si	16.57	6.39	7.96	6.99	55.0	3.2	0.524
Ge	12.89	4.83	6.71	5.40	47.7	2.8	0.521
AlSb	8.94	4.43	4.16	3.33	48.1	2.0	0.626
GaP	14.12	6.25	7.05	5.67	50.4	2.5	0.574
GaAs	11.81	5.32	5.92	4.71	47.4	2.3	0.581
GaSb	8.84	4.03	4.32	3.43	45.2	2.1	0.594
InP	10.22	5.76	4.60	3.57	51.4	1.8	0.672
InAs	8.33	4.53	3.96	2.90	44.7	1.6	0.666
InSb	6.67	3.65	3.02	2.34	44.4	1.6	0.664
ZnS	10.40	6.50	4.62	3.18	43.3	1.2	0.728
ZnSe	8.10	4.88	4.41	2.52	38.0	1.1	0.718
ZnTe	7.13	4.07	3.12	2.35	40.3	1.3	0.692
CdTe	5.35	3.68	1.99	1.43	40.4	0.89	0.775
CuCl	2.72	1.87	1.57	0.72	11.9	0.26	0.777

SOURCES of experimental data: C, from McSkimin and Bond (1957); Si and Ge, Huntington (1958); AlSb, GaAs, InAs, InSb, and CdTe, Landolt and Börnstein (1966); GaP, Weil and Groves (1968); GaSb, McSkimin et al. (1968); InP, Hickernell and Gayton (1966); ZnS, ZnSe, and ZnTe, Berlincourt et al. (1963); and CuCl, Inoguchi et al. (1969).

that stabilizes the open tetrahedral structure. We can write our own expression for C_1 by combining Eqs. (8-15) and (8-21). This leads to

$$C_1 = 2\lambda V_2^h \alpha_c^{h3}/3. \tag{8-22}$$

Figure 8-4 may be regarded as a direct test of this relation. We see from Eq. (8-22) that the rigidity tends to decrease with the square of bond length, as does V_2, and drops with decreasing covalency.

It is clear that the valence force field model based upon C_0 and C_1 is less general than even the simplest LCAO representation that we introduced. This is apparent first in that it contains only two independent constants while clearly the three elastic constants depended on different parameters in the LCAO model. A more subtle, and probably more essential, distinction is that the valence force field gives strictly short-range interactions. If only a single atom is displaced, forces arise on its nearest and next-nearest neighbors but no further. One might at first think that this was true also in the nearest-neighbor LCAO theory, but it is easy to see that electronic readjustment leads to long-range forces. (This was pointed out to the author by R. M. Martin; see also Chadi and Martin, 1976.) The displacement of a single atom tends to rotate the neighboring tetrahedra of hybrids, which in turn

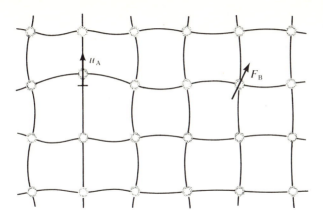

FIGURE 8-7

A schematic diagram showing how the displacement of a single atom by u_A can cause a disruption of the bonds and a force F_B on a distant atom even though the intervening atoms are not displaced. This long-range interaction is omitted in the valence force field model.

rotate their neighbors, and so on, propagating a disturbance to large distances even though the intervening nuclei remain in their initial positions. This is illustrated schematically in Fig. 8-7, where the bonds are represented by lines which must enter the atoms at 90°. Misaligned hybrids then are represented by curved bond lines; it is seen how the disruption propagates. The forces drop exponentially with distance but may still be of long-enough range to be important. We shall return to examine the effects of these forces in more detail in Chapter 9. For the present, the simple valence force field will suffice.

8-E Internal Displacements, and Prediction of c_{44}

We saw in Section 8-C that the strain ε_4 introduced four complications in the problem that were not present in the strain ε_1. The valence force field bypasses three of these, but leaves us with internal displacements. These are of interest in their own right and must be included if we wish to predict c_{44} in terms of the valence force field and parameters obtained from c_{11} and c_{12}. That will be an interesting prediction since it gives some measure of the validity of the valence force field model, so let us proceed with it.

The physical origin of the internal displacements is more easily seen in a two-dimensional analogue of the three-dimensional lattice illustrated in Fig. 8-8 and treated in Problem 8-3. If we imagine the bonds as freely rotating springs (setting the constant C_1 equal to zero), we see that the elastic distortion tends to pull the shaded atoms upward with respect to the others. These are conventionally thought of as positive internal displacements. Note that if in contrast there were

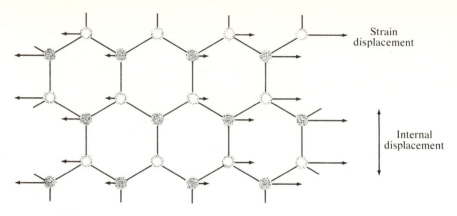

Strain displacement

Internal displacement

FIGURE 8-8

A strain field ε_1 in a BN lattice will tend to cause internal displacement of the B atoms (shaded) upward with respect to the N atoms.

only angular rigidity (setting the constant C_0 equal to zero), the internal displacements would be negative. This same competition occurs in the three-dimensional crystal but since, as we have seen, the angular forces are always weak in comparison to the radial forces, the latter win out and the internal displacements will always be positive.

We can also notice from Fig. 8-8, and it remains true in three dimensions, that the internal displacements occur whether or not the atoms marked plus and minus are different from each other; that is, whether or not the midpoint between the atoms is a center of inversion symmetry for the crystal. However, if the atoms are identical, there will be no electric polarization arising from the internal displacements and the piezoelectric constant will vanish. This is associated with a well-known proof (see, for example, Nye, 1957) that the piezoelectric constant vanishes if there exists a center of symmetry in the crystal. A related theorem, that the internal displacement of an atom will vanish *if that atomic site is a center of inversion*, could easily be proved.

Let us proceed to the evaluation of the elastic energy under the strain:

$$\varepsilon_4 = \varepsilon_{yz} = \frac{\partial v}{\partial z} + \frac{\partial w}{\partial y} = \varepsilon, \tag{8-23}$$

which was illustrated in Fig. 8-5; let there be y displacements only. We focus on the central atom at the origin and write the magnitude of the y-displacements of its four neighbors as $v = \varepsilon a/4$. If we imagine the bonds to these four neighbors as springs, as in the discussion of Fig. 8-8, we see that each bond tends to push the central atom in the positive x-direction, but that all other forces cancel out. We may therefore allow an x-displacement of the central atom (and all other atoms drawn as open circles) of $u = \zeta \varepsilon a/4$ with respect to the atoms drawn as solid circles. For given ε the energy is to be minimized with respect to the internal displacement. The value indicated by ζ is called **Kleinman's internal displacement** *parameter* (Kleinman, 1962). Notice that it would vanish if there were no internal displacement; it would equal 1 if there were no change in bond length linear in ε; and it would equal $-\frac{1}{2}$ if there were no change in angle linear in ε. The change in the bond length from the origin-atom to the atom a is easily obtained:

$$\delta d = v/\sqrt{3} - u/\sqrt{3} = (\varepsilon a/4\sqrt{3})(1 - \zeta) = \varepsilon(1 - \zeta)d/3. \tag{8-24}$$

The magnitude of the changes in lengths of the other bonds are the same, though some are of opposite sign. Thus the change in radial energy per atom pair can be obtained by summing four bonds:

$$\delta E_0 = 4 \cdot \frac{1}{2} C_0 \left[(1 - \zeta)\frac{\varepsilon}{3}\right]^2 = \frac{2C_0}{9} \varepsilon^2 (1 - \zeta)^2. \tag{8-25}$$

Of the bond angles with vertices at the origin-atom, only the angles $a{-}o{-}c$ and $b{-}o{-}d$ change to first order in ε and therefore contribute to the angular distortion energy. The change in angle of the first of these is readily obtained just as was Eq. (8-7) and is found to be

$$\delta\theta = \frac{2\sqrt{2}}{3}\left(\frac{1}{2} + \zeta\right)\varepsilon. \tag{8-26}$$

The change in the second is equal and opposite. There are equal changes in two of the angles centered at each atom of the other type, and in our model we have used the same angular force constant for each so that the total angular energy change per atom pair is

$$\delta E_1 = 4 \cdot \tfrac{1}{2} C_1 \left[\frac{2\sqrt{2}}{3}\left(\frac{1}{2} + \zeta\right)\varepsilon\right]^2 = \frac{16}{9} C_1 \left(\frac{1}{2} + \zeta\right)^2 \varepsilon^2. \tag{8-27}$$

We may add Eq. (8-27) to Eq. (8-25) and minimize with respect to ζ to obtain

$$\zeta = \frac{C_0 - 4C_1}{C_0 + 8C_1}. \tag{8-28}$$

This is the most interesting result of the calculation; the values obtained using C_0 and C_1 from Table 8-4 are also listed in Table 8-4. These values are measurable in nonpolar solids by X-ray techniques, which give $\zeta = 0.63 \pm 0.04$ for both silicon and germanium (Segmuller and Neyer, 1965). Values found by Martin (1970) are a few percent higher than ours and are probably more reliable. We will therefore use Martin's values, or values estimated in terms of his, in our discussion of piezoelectricity. We may also substitute Eq. (8-28) back into Eqs. (8-25) and (8-27) to obtain the total energy per atom pair:

$$\delta E_0 + \delta E_1 = \frac{4C_0 C_1}{C_0 + 8C_1} \varepsilon^2. \tag{8-29}$$

This can be set equal to the same energy written in terms of the elastic constant, $\frac{1}{2}c_{44}\,\varepsilon^2(a^3/4)$. Solving for c_{44}, we obtain

$$c_{44} = \frac{32}{a^3} \frac{C_0 C_1}{C_0 + 8C_1}, \qquad (8\text{-}30)$$

which has also been evaluated from the C_0 and C_1 of Table 8-4. They are listed there along with the experimental values compiled by Martin (1970). The estimates of c_{44} based upon the model typically are too small by from 15 percent to 30 percent; these can of course be corrected by including additional interactions.

The wurtzite structure has lower symmetry and six, rather than three, independent elastic constants. These could be estimated by obtaining C_0 from the measured bulk modulus and C_1 from Eq. (8-22), if one wished them.

PROBLEM 8-1   *Angular rigidity in the Bond Orbital Approximation*

Consider the zig-zag coplanar structure shown in Fig. 8-9 with alternate p states of energy $\pm V_3$. (The p orbitals of only a few of the atoms are shown.) Let the matrix elements for nearest neighbors be $V_{pp\sigma} = V_2$ and $V_{pp\pi} = 0$. (Our leaving s orbitals out of the problem is like taking $V_{ss\sigma} = V_{sp\sigma} = 0$.) The system can be deformed by changing the angle ψ uniformly through the system while holding d fixed. Notice that these are analogous to rigid hybrids (the p orbitals must remain perpendicular to each other if they are to remain orthogonal), which simplifies the energy calculation.

Find the total energy $(2\varepsilon_b$, as in Eq. (8-11)) as a function of bond angle ψ in the Bond Orbital Approximation. Verify that it is minimum at $\psi = \pi/2$.

From an expansion of this energy for small deviations from $\pi/2$ (or from the second derivative of the energy with respect to ψ), obtain the angular force constant C_1, defined in analogy with Eq. (8-18), such that the total energy change for the N_a atoms is equal to

$$2N_a\varepsilon_b = \tfrac{1}{2}C_1 \sum_i \delta\psi_i^2.$$

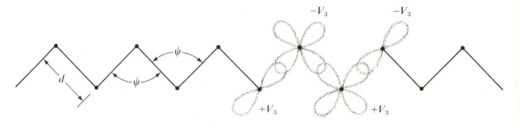

FIGURE 8-9

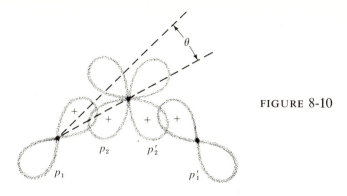

FIGURE 8-10

PROBLEM 8-2   *Corrections to the Bond Orbital Approximation*

The Bond Orbital Approximation neglects matrix elements between a bond orbital and the antibonding orbitals in the neighboring sites. It is not difficult to see that for the atomic matrix elements retained in Problem 8-1, these bonding–antibonding matrix elements all vanish at $\psi = \pi/2$, so the approximation is exact at that angle. However, when ψ differs from $\pi/2$, each bond orbital acquires matrix elements with its neighboring antibonding orbitals. These may be evaluated by using Fig. 8-10. The angle of misalignment of the p orbitals with the nearest-neighbor vector is $\theta = \frac{1}{2}(\psi - \frac{1}{2}\pi)$. We may construct a bond orbital in analogy with Eq. (3-7) for the bond on the left, $|b\rangle = u_1^b |p_1\rangle + u_2^b |p_2\rangle$, and similarly construct an antibonding orbital on the right, $|a'\rangle = u_2^a |p_2'\rangle + u_1^a |p_1'\rangle$, where the primed orbitals are perpendicular to the unprimed ones. Then, keeping only the intera-tomic matrix element $V_{pp\sigma}$ and performing the decomposition of each p orbital into σ and π parts, we obtain

$$\langle a' | H | b \rangle = -(u_b^1 u_a^2 + u_b^2 u_a^1) \sin\theta \, \cos\theta \, V_{pp\sigma}.$$

Again writing $V_{pp\sigma} = V_2$, the algebra becomes identical to that following Eq. (3-7), and we may write the coefficients u_1^b, and so on, using the equations there.

Use these matrix elements to obtain the shift in bond energy to second order in θ. Use this to correct the value of C_1 obtained in Problem 8-1 for errors arising from the Bond Orbital Approximation. (Notice that $\delta\psi_i = 2\theta$.) The finding that there are only corrections for polar semiconductors carries over to the tetrahedral case.

You should notice that there are also nearest-neighbor atomic matrix elements between a bond orbital and the next-nearest-neighbor antibonding orbital; however, they are are of order θ^2, and only contribute to the energy to order θ^4; thus they do not affect C_1. We have, in fact, included all terms of order θ^2, so the calculation of C_1 is exact.

PROBLEM 8-3   *Calculation of internal displacements using force constants.*

Consider a layer of BN, or graphite, uniformly stretched in the x-direction with displace-ment $u_x = \varepsilon x_i$, where ε is constant and x_i is the x-coordinate of the atom in the undistorted lattice. There may also be a movement of all shaded atoms, in the plane of Fig. 8-11 but perpendicular to the x-direction, by v compared to all open circles.

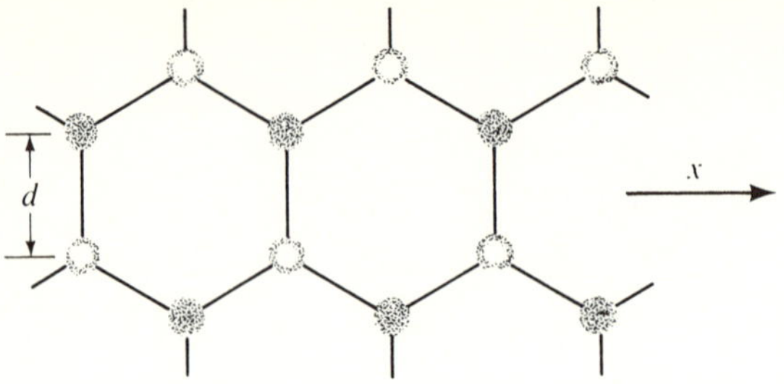

FIGURE 8-11

Let there be an energy change $\frac{1}{2}C_0(\delta d/d)^2$ for every bond-length change δd (3 bonds per atom pair, some δd's are different) and $\frac{1}{2}C_1(\delta\psi)^2$ for every bond-angle change $\delta\psi$ (six angles per atom pair, some at ⬭ and some at ⬤).

Calculate the ratio of v/d to ε (at small ε, of course). It should change sign depending on the relative size of C_0 and C_1.

. *Note*: the vertical bonds remain vertical.

CHAPTER 9

Lattice
Vibrations

SUMMARY

The calculation of vibration spectra in terms of force constants is similar to the calculation of energy bands in terms of interatomic matrix elements. Force constants based upon elasticity lead to optical modes, as well as acoustical modes, in reasonable accord with experiment, the principal error being in transverse acoustical modes. The depression of these frequencies can be understood in terms of long-range electronic forces, which were omitted in calculations using the valence force field. The calculation of specific heat in terms of the vibration spectrum can be greatly simplified by making a natural Einstein approximation.

By using the LCAO description, the charge redistribution arising from atomic displacements is calculated to give the transverse charge, which represents the coupling between vibrations and electromagnetic fields. Both this charge and the very much smaller piezoelectric charge are well described by theory. Thus, the corresponding contributions to the electron–phonon interaction are also well described; however, the deformation-potential contribution, also present in homopolar semiconductors, appears not to be so well given.

In our discussion of elastic constants we have imagined uniform distortions of the crystal. An elastic medium can sustain vibrations, which at any instant consist of nonuniform distortions. The normal modes of an elastic *continuum* are sound waves (longitudinal and transverse) propagating in the medium, and these normal modes will also exist in the crystal. Indeed, viewing the crystal as an elastic continuum will be appropriate for wavelengths that are long compared to the interatomic distances, but the view must be refined for a discussion of modes corresponding to shorter wavelengths. The calculation of vibration spectra in detail is a somewhat specialized task; here we can make an illustrative calculation, note one or two salient features of the results, and discuss the appropriate approximations for using these spectra in the calculation of other properties. A rather complete account of the theory of vibration spectra, with many useful tables, is given by Born and Huang (1954). A more recent account is given by Wallis (1965), and there are many articles in the current literature on this general topic of lattice dynamics.

9-A   The Vibration Spectrum

A lattice vibration can be specified by describing, for each atom in the crystal, the displacement of the atom from its equilibrium position, as a function of time. By numbering the atoms i, such a set of displacements can be designated as $\mathbf{u}_i(t)$. In a *normal mode* of vibration, however, the time variation for every atom is sinusoidal and has the same frequency. It is therefore possible to specify these displacements as the real part of $\mathbf{u}_i\, e^{-i\omega t}$, where ω is 2π times the frequency of the normal mode in question and $\mathbf{u}_i$ is independent of time. The magnitude of the components of $\mathbf{u}_i$ gives the amplitude of the displacements for the ith atom and the phase of the motion ($\mathbf{u}_i$ may be complex). Finally, we can take advantage of the translational symmetry of the crystal, as we did in the discussion of the electronic structure of solids in Chapter 2. Applying periodic boundary conditions as we did there, we then can show by symmetry (or deduce from direct calculation) that each normal mode can be constructed such that a wave vector $\mathbf{k}$ for that mode relates the motion of all translationally equivalent atoms; that is, if an atom in the undistorted crystal at the position $\mathbf{r}_j$ has an environment identical to an identical atom at $\mathbf{r}_i$, it follows that $\mathbf{u}_j = \mathbf{u}_i\, e^{i\mathbf{k}\,\cdot\,(\mathbf{r}_j - \mathbf{r}_i)}$, with the same $\mathbf{k}$ for all such pairs. This complicated statement means that a normal mode looks like a sound wave of wave number $\mathbf{k}$ and frequency ω_k, so that translationally equivalent atoms at sites $\mathbf{r}_i$ and $\mathbf{r}_j$ have displacements given by

$$\delta\mathbf{r}_i = \delta\mathbf{r}_j e^{i\mathbf{k}\,\cdot\,(\mathbf{r}_i - \mathbf{r}_j)}. \tag{9-1}$$

For the systems we are interested in, there will be at least two kinds of atomic sites, which are not translationally equivalent. In the zincblende structure there are sinc sites and sulphur sites, and a vector $\mathbf{u}(\mathbf{k})$ must be given for each kind of site. Even in the diamond structure, where both zinc and sulphur atoms are replaced by carbon, the two kinds of sites are not equivalent; this is perhaps most readily seen in the two-dimensional analogue, shown in Fig. 8-8, where one type is above a vertical bond and the other below. The construction of the vibrational states is very closely analogous to the construction of electronic states in terms of atomic orbitals. The wave numbers are constructed just as in Eq. (2-3); the components of the displacement vector in Eq. (9-1) are directly analogous to the coefficients in Eq. (2-4). The Brillouin Zones over which the wave numbers run are the same for the same structures and even calculation of frequencies is achieved by the solution of eigenvalue equations of the same form. The use of symmetry to obtain the form of the displacements, Eq. (9-1), gives a very great simplication of the problem. The mode can be written in terms of the six amplitudes describing the components of the displacements at the two types of sites in a zincblende structure. The problem is thus reduced to the solution of six simultaneous equations in six unknowns rather than the $6N_p$ simultaneous equations that would be needed for N_p atom pairs if there were no translational symmetry; the problem becomes comparable to that of obtaining vibrations in a diatomic molecule. We can further simplify the calculation by selecting waves propagating in symmetry directions.

Calculation of the Spectrum

Let us consider a mode in gallium arsenide, which has the zincblende structure; let the wave vector $\mathbf{k}$ lie in a [100] direction. We must write an amplitude vector, $\mathbf{u}_1$, for the gallium atoms so that the displacement of a gallium atom at $\mathbf{r}_i$ is given, as in Eq. (9-1), by

$$\delta\mathbf{r}_i = \mathbf{u}_1 e^{-i(\mathbf{k}\cdot\mathbf{r}_i - \omega_k t)}, \tag{9-2}$$

and we must write an amplitude vector, $\mathbf{u}_2$, for the arsenic atom, to be used in a corresponding expression. In this case, because of high symmetry, there are pure longitudinal modes in which both $\mathbf{u}_1$ and $\mathbf{u}_2$ lie in the same direction as $\mathbf{k}$. This is illustrated in Fig. 9-1.

A simple way to proceed in such a case is to compute the force on a given atom in terms of the relative displacements between it and its neighbors. Such forces are proportional to expressions such as $\mathbf{u}_1 e^{i\mathbf{k}\cdot\mathbf{r}_i} - \mathbf{u}_2 e^{i\mathbf{k}\cdot\mathbf{r}_j}$ for each set of neighbors. These forces are then set equal to the mass of an atom, M_1, times the acceleration $-\omega_k^2 \mathbf{u}_1 e^{i(\mathbf{k}\cdot\mathbf{r}_i - \omega_k t)}$.

Up to this point, the analysis is rigorously correct and general; it is simply a restatement of the Born-Von Karman expansion of the energy in terms of relative displacements—see Eq. (8-17). However, we shall now make a major approximation in taking the force constants from the very simple valence force field that we described in Chapter 8. This will give us a clear and correct qualitative description of the vibration spectra and will even give semiquantitative estimates of the frequencies. Afterward, we shall consider the influence of the many terms that are omitted in this simple model.

Let us consider the displacements specified above and compute the change in energy per atom due to angular and radial distortions, to obtain from these the force on each type of atom. The reader will find it easier to carry out this calculation for himself than to follow the details of a previous calculation. When

206

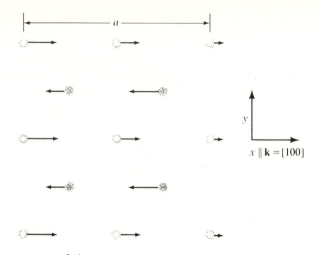

FIGURE 9-1

Displacements for a longitudinal mode propagating in a [100] direction n is in a zincblende crystal.

this is done for both atom types, the phase factors cancel out, leading to two equations,

$$(\kappa - M_1 \omega_k^2)u_1 = \kappa \cos(ka/4)u_2, \quad \Big|$$
$$(\kappa - M_2 \omega_k^2)u_2 = \kappa \cos(ka/4)u_1, \quad \Big| \qquad (9\text{-}3)$$

where

$$\kappa = \frac{64}{9a^2}(C_0 + 12C_1) = \frac{4}{3d^2}(C_0 + 12C_1). \qquad (9\text{-}4)$$

We may easily eliminate u_1 and u_2 and solve for ω_k, to obtain

$$\omega_k^2 = \frac{M_1 + M_0 \pm [(M_1 + M_0)^2 - 4M_1 M_0 \sin^2(ka/4)]^{1/2}}{2M_1 M_0}\kappa. \qquad (9\text{-}5)$$

Notice that we have written a solution for the square of the frequency; the positive and negative solutions correspond to waves propagating in opposite directions. There are, however, double solutions for the magnitude of ω_k, and both are plotted in Fig. 9-2 (labelled LA and LO). There are solutions for arbitrarily large k, but they repeat. In fact, by examining the displacements, in Eq. (9-2), we may see that these large wave numbers, which lie outside the Brillouin Zone, give displacements identical to those of the corresponding mode inside the Brillouin Zone. We initially look at the lower-frequency branch, obtained with the minus sign in Eq. (9-5).

Notice first that ω_k is proportional to k at long wavelengths; this corresponds to

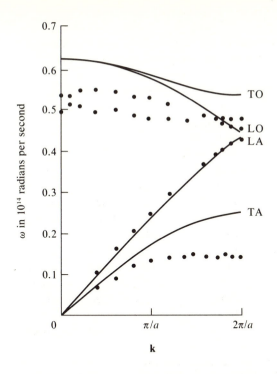

FIGURE 9-2

The vibration spectrum of GaAs, calculated by using the C_0 and C_1 values of Table 8-4 (TO = transverse optical; LO = longitudinal optical; LA = longitudinal acoustical; TA = transverse acoustical). Experimental points are from Dolling and Waugh (1965).

a sound wave of fixed velocity $c_s = \omega_k/k$. This slope agrees with experiment, since the corresponding measured elastic constant, c_{11}, has been used to obtain the C_0 and C_1 values that were given in Table 8-4 and that were used in the calculation of the frequencies. All of the modes in these lower curves are called *acoustical modes*, though at large k the frequency ω_k is no longer linear in k. This corresponds to dispersion of the wave, and to lower values of the group velocity of the sound.

The modes of the upper branch are called *optical modes*; they have frequencies in the infrared and can be excited by light. The optical mode at $k = 0$ is a mode in which all gallium atoms move together and all arsenic atoms move in the opposite direction. Its frequency is given by

$$\omega^2 = \frac{8}{3Md^2}(C_0 + 8C_1),\tag{9-6}$$

where M is twice the reduced mass; $M = 2M_1 M_2/(M_1 + M_2)$. (We have chosen to use M since, for isoelectronic series such as Ge, GaAs, and so on, $M_1 \approx M_2 \approx M$.) We have overestimated this frequency by some 16 percent, which reflects the limited accuracy of the model. The error is larger than, but comparable to, the error in c_{44} computed with the model.

We could have used particular frequencies instead of the elastic constants to determine the force constants in the model; that alternative is of some interest. We have used, for reasons to be discussed, the measured transverse optical frequency at $k = 0$ and the transverse acoustical mode at $k = 2\pi/a$ to obtain alternate values of C_0 and C_1, which are listed in Table 9-1. The differences from the values given

TABLE 9-1

Experimental vibrational frequencies, in units of 10^{14} radians per second, force constants deduced from them, in eV, and ζ, the Kleinman internal displacement parameter.

| Material | Vibrational frequencies | | Force constants | | |
	Transverse acoustical mode, $k = 2\pi/a$	Transverse optical mode, $k = 0$	C_0	C_1	ζ
C	1.52	2.46	21.6	5.71	−0.018
Si	0.282	0.976	49.1	1.07	+0.777
Ge	0.150	0.565	47.2	0.845	+0.812
Sn	0.079	0.377	47.7	0.505	+0.883
GaP	0.201	0.688	44.0	0.984	+0.772
GaAs	0.148	0.509	37.4	0.828	+0.775
InSb	0.070	0.348	41.0	0.398	+0.892
ZnS	0.175	0.515	22.0	0.710	+0.692
CuCl	0.065	0.342	11.6	0.099	+0.904
CuBr	0.068	0.241	8.7	0.179	+0.788
CuI	0.084	0.238	11.3	0.401	+0.668

SOURCES: C, Warren et al. (1967); Si, Dolling (1963); Ge, Brockhouse and Dasannacharya (1963); Sn, Price and Rowe (1969); GaP, Yarnell et al. (1968); GaAs, Dolling and Waugh (1965); InSb, D. L. Price, J. M. Rowe, and R. M. Nicklow; ZnS, Feldkamp et al. (1971); CuCl, Carabatos et al. (1971); CuBr, Prevot et al. (1973); and CuI, Hennion et al. (1972). All data collected by G. Lucovsky.

in Table 8-4 are great, reflecting the limited accuracy of the valence force field in describing the entire range of elastic properties. We shall see in Section 9-B that the main discrepancy comes from the neglect of long-range forces. The discrepancies are accentuated in this table in comparison to vibrational frequencies, which are proportional to the square root of the force constants. The uncertainties are reflected in the internal displacement parameters calculated from the alternative values for C_0 and C_1 and are given in Table 9-1. The differences from the values given in Table 8-4 are a measure of the uncertainty in these values.

Notice that the two longitudinal frequencies at $k = 2\pi/a$ are slightly different. They correspond to high-frequency modes in which the gallium vibrates, with alternate gallium atoms moving in opposite directions but the arsenic atoms not moving at all, and a low-frequency mode in which only the arsenic atoms move. The difference in frequency comes entirely from the small difference in the masses of the two types of atoms. The frequency of these modes has been accurately estimated.

For wave numbers in this direction, there are also transverse modes in which the motion is perpendicular to $\mathbf{k}$. It is clear from symmetry that the two modes are of the same frequency. The calculation of these modes is more complicated. At long wavelengths these modes are describable by the elastic constant c_{44} and are seen therefore to have internal displacements. Thus we must introduce displacement amplitudes in the z-direction as well as in the y-direction. This is not a major

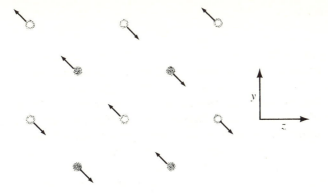

FIGURE 9-3

The displacement of atoms in the transverse acoustical mode at the zone boundary. The x-axis is parallel to **k** in a [100] direction.

problem, but for our purposes it will suffice to look only at the limiting behavior at small k and at $2\pi/a$.

The long-wavelength transverse-wave frequencies are obtainable from elasticity theory and are underestimated by 11 percent owing to the error in the elastic constant. The long-wavelength transverse optical modes will, because of the cubic symmetry, approach the same limiting frequency as the acoustical modes. We note from the experimental curves that this degeneracy is lifted in the real crystal. This is because of long-range Coulomb interactions that we have omitted in this calculation and that will be discussed later in terms of the transverse charge. The modes at $2\pi/a$ are rather simple, at least if the small difference in masses is neglected. We neglect this difference and again take $M_1 \approx M_2 \approx M$. (This is an approximation, but not a bad one for these modes.) The low-frequency mode may be understood by looking along the direction of propagation, say the x-axis. Then planes of atoms are seen to move in unison, as is illustrated in Fig. 9-3.

The frequency of such a mode may be rather simply calculated by constructing the nearest-neighbor vectors from a particular atom, including the displacements **u** for the mode. For a particular atom, this set is given by

$$\left.\begin{aligned}
\mathbf{d}_1 &= \left[\frac{a}{4} - u\sqrt{2}, \quad \frac{a}{4} + u\sqrt{2}, \quad \frac{a}{4}\right], \\[2mm]
\mathbf{d}_2 &= \left[\frac{a}{4}, \quad\quad -\frac{a}{4}, \quad\quad -\frac{a}{4}\right], \\[2mm]
\mathbf{d}_3 &= \left[-\frac{a}{4}, \quad\quad \frac{a}{4}, \quad\quad -\frac{a}{4}\right], \\[2mm]
\mathbf{d}_4 &= \left[-\frac{a}{4} - u\sqrt{2}, -\frac{a}{4} + u\sqrt{2}, \quad \frac{a}{4}\right].
\end{aligned}\right\} \quad (9\text{-}7)$$

For other atoms the signs will be reversed. We can then directly compute the change in length of the four bonds and, by taking dot products, the change in bond angles. We write the total elastic energy per atom, noting that there are equivalent distortions of the bond angles at every atom and that the radial energy in each bond is shared equally between atoms. We can then set the total elastic energy per atom equal to the kinetic energy per atom, $M\omega^2 u^2/(2M)$, and solve for the frequency. The corresponding calculation can also be made for the higher-frequency mode, in which every displacement vector of Fig. 9-3 is rotated (for example, clockwise) by 90°. The two frequencies are given by

$$\omega^2 = \frac{12C_1}{Md^2},\tag{9-8}$$

and

$$\omega^2 = \frac{8(C_0 + C_1/2)}{3Md^2},\tag{9-9}$$

and are shown also in Fig. 9-2 along with a sinusoidal interpolation between them and the small-k modes. The agreement with experiment for the transverse optical mode is comparable to that for the longitudinal optical mode. The transverse acoustical mode—the Brillouin Zone-face mode with $k = 2\pi/a$, in particular—is of very great interest.

We see from Eq. (9-8) that the frequency of this mode depends *only* upon the angular force constant C_1; this is true only at the zone face, not at smaller wave numbers. This is one of the reasons for having chosen the Brillouin Zone-face mode to determine the alternative experimental values of C_0 and C_1 given in Table 9-1. The fact that only C_1 enters the frequency means that all relative displacements of nearest neighbors for this mode are perpendicular to the bonds; it is only the angular rigidity of the lattice that prevents instability in this mode. In just the same sense, it was the constant C_1 alone that prevented the instability of the lattice against the shear distortion shown in Fig. 8-1. We saw theortically that the value of this constant should decrease rapidly with increasing polarity, and the decrease is reflected in Tables 8-4 and 9-1. However, the rigidity against the distortion shown in Fig. 9-3 is far less than that against the distortion shown in Fig. 8-1 (reflected by the smaller value of C_1), so we can expect that if polarity increases until the covalent structure becomes less stable than an ionic structure, the transformation will occur through spontaneous distortion of the kind illustrated in Fig. 9-1. It is called a "soft mode" transformation: the lattice becomes so soft against that deformation that the distortion occurs spontaneously.

9-B   Long-Range Forces

The large difference between the angular force constants C_1 determined in the two ways appears to be the principal defect in application of the valence force field theory to the treatment of covalent solids. The defect is not readily repaired by the consideration of interactions with a few additional neighbors but arises directly from long-range forces. The long-range forces are not the electrostatic contributions that arise in polar crystals and that Martin (1972b) has shown have only small effects; neither is it likely they are from the quadrupole–quadrupole interactions introduced by Lax (1958); rather, they seem to be the long-range forces of electronic origin which were illustrated in Fig. 8-7. It was indicated there that these effects are included in our description of the electronic structure, but not in the valence force field.

We can confirm that this is largely true from calculations of the frequency of the transverse acoustical frequency made by Martin and Chadi (1976), using the special points method. We may deduce values of C_1 from their calculated elastic constants by using Eq. (8-21) and from their calculated transverse acoustical vibrational frequencies by using Eq. (9-8). These are listed in Table 9-2, where we see that values of C_1 deduced from the acoustical mode are much smaller than those deduced from the elastic constant, in rough agreement with experiment but in contrast with the valence force field, which gives them equal values. The experimental values, also given there, show a greater difference than the theoretical ones, but a large portion of the discrepancy has been rectified by accurate treatment of the electronic structure.

There are two further points that should be explained concerning these long-range forces. First, how do they give rise to the difference in the two values of C_1? Second, what parameters of the material do they depend upon?

TABLE 9-2

Angular rigidity force constants C_1, in eV, deduced from the Martin-Chadi (1976) LCAO–special points calculation and from experiment.

	Si	Ge	GaAs	ZnSe
OBTAINED FROM THE ELASTIC CONSTANTS				
LCAO	3.37	2.74	2.32	1.11
Experiment	3.2	2.8	2.3	1.1
OBTAINED FROM THE ZONE-BOUNDARY TA MODE				
LCAO	2.26	1.48	1.40	0.78
Experiment	1.07	0.845	0.828	0.66

SOURCE: Zone-boundary TA mode experimental value for ZnSe from Kunc et al. (1971); all other experimental values from Tables 8-4 and 9-1.

The Weber Model

To see how the long-range forces arise and affect angular rigidity, let us consider a simple one-dimensional example due to Weber (1974) and an analysis of it due to Sokel and Harrison (1976). In the Weber model, each atom is coupled to neighboring bonds with a force constant f. (The atoms are actually thought of as being coupled to neighboring bond *charges*, but the Coulomb charge is not important, and the concept of a bond charge is not useful enough to warrant treatment here.) Likewise, each bond is also coupled to its neighboring bonds by a force constant f'. This is illustrated in Fig. 9-4. It is not difficult to see that it gives a long-range force, analogous to that described in Fig. 8-7. The corresponding Born-von Karman force constant can be computed by displacing one atom, calculating the shifted positions of each of the bonds, and calculating the resulting force on a distant atom. The resulting force constant is easily found to be

$$\frac{\partial^2 E_{tot}}{\partial u_i \partial u_j} = f \, \sin h(\pi\mu/k_0)e^{-\mu|x_i - x_j|}, \qquad (9\text{-}10)$$

where x_i is the equilibrium position of the ith atom, k_0 is the wave number at the Brillouin Zone, $k_0 = \pi/a$, and

$$\cos h(\pi\mu/k_0) = 1 + \frac{f}{f'}. \qquad (9\text{-}11)$$

The force decays exponentially but can be of long range if f'/f becomes large. The same kind of long-range force arises in the **shell model** introduced by Dick and Overhauser (1958) to describe real systems. In both cases, and in the LCAO description, the essential feature is the introduction of electronic degrees of freedom, which can transmit long-range forces.

The vibration spectrum may also be calculated very simply for this model, using essentially the same approach that we described for real systems. It gives

$$M\omega^2 = 2f \, \frac{(f + 2f')\sin^2(\pi k/2k_0)}{f + 2f' \, \sin^2\,(\pi k/2k_0)}. \qquad (9\text{-}12)$$

We see that if f'/f gets large, the dispersion curve becomes quite flat, as was apparent in the experimental transverse mode for GaAs shown in Fig. 9-2. This is illustrated in Fig. 9-5 for two choices of f'/f.

We can in fact fit the speed of sound and the zone-boundary mode to a simple nearest-neighbor force-constant model for the Weber system, just as we did in Table 9-2 for the tetrahedral solids. We obtain force constants

$$\left.\begin{array}{ll} C_1 = \tfrac{1}{2}f + f' & \text{from the elastic limit } (k \rightarrow 0); \\[2mm] C_1 = \tfrac{1}{2}f & \text{from the zone-boundary mode.} \end{array}\right\} \qquad (9\text{-}13)$$

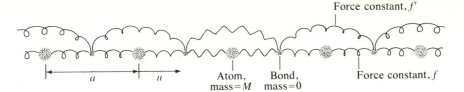

FIGURE 9-4

The Weber model, in which each atom is coupled to its neighboring massless bond by a force constant f, and each bond is coupled to its neighboring bond by a force constant f'. A long-range force between atoms arises through the coupled bonds when an atom is displaced by u.

The essential point is that an extra rigidity against uniform distortions arises from the long-range force, raising the curve at long wavelengths.

Sokel and Harrison used this model and the two limits from Eq. (9-13) to estimate how long the range of the force is that flattens the transverse dispersion curves for the tetrahedral solids. We can do that here immediately by using the C_1 values from Tables 8-4 and 9-1. For the systems considered by Sokel and Harrison, this gives the values k_0/μ, which are proportional to the range of the force, listed in Table 9-3. The meaning of the 2.3 for gallium arsenide, for example, is that the force constant decreases by a factor $e^{-\mu a/4} = e^{-\pi/(2 \times 2.3)} = 0.5$ (for this structure $k_0 = 2\pi/a$) for interactions with each successive plane of atoms, separated by $a/4$ in the [100] direction. It does not require forces of extremely long range to account for the kind of flattening of the dispersion curve that we saw for GaAs in Fig. 9-2.

The main trend in the experimental ranges listed in Table 9-3 is an increase with increasing metallicity. Sokel and Harrison (1976) explained the nature of these forces and the trend by noting that the interaction between atoms can be calculated rigorously in perturbation theory, by Eq. (1-14). Let $|\mathbf{k}_1\rangle$ be an eigenstate

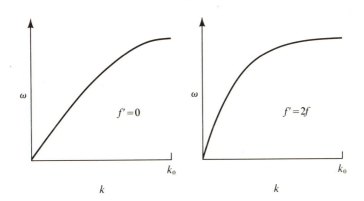

FIGURE 9-5

The dispersion curve for the Weber model, showing the effect of adding the interbond force constant f'.

TABLE 9-3

Ratio of the screening length to the zone-boundary wavelength, based upon the minimum energy gap and effective masses, and an experimental estimate of this ratio, based upon Weber's model of the vibration spectrum.

Material	k_0/k_g	k_0/μ
C	1.7	1.3
ZnS	1.6	1.5
GaP	2.1	2.1
Si	3.1	2.4
GaAs	4.7	2.3
Ge	4.8	2.6
InSb	22.4	2.9

SOURCE: Sokel and Harrison (1976)

from the valence band of the perfect crystal. The system is perturbed by adding a potential $V(\mathbf{R}_i)$ due to the displacement of the ith atom and a potential $V(\mathbf{R}_j)$ due to the displacement of the jth atom. The shift in energy of the eigenstate $|\mathbf{k}_1\rangle$ due to the perturbation is

$$\delta E_1 = E_1 + \langle \mathbf{k}_1 | V(\mathbf{R}_i) + V(\mathbf{R}_j) | \mathbf{k}_1 \rangle + \sum_{k_2} \frac{|\langle \mathbf{k}_2 | V(\mathbf{R}_i) + V(\mathbf{R}_j) | \mathbf{k}_1 \rangle|^2}{E_1 - E_2}.$$

$$(9\text{-}14)$$

The sum can be restricted to conduction-band eigenstates $|\mathbf{k}_2\rangle$, since effects from interaction among occupied states cancel out when we sum Eq. (9-14) over all occupied states. Only the cross-term depends on the displacements of both atoms and therefore contributes to the interaction between them. Thus the interesting term in the change in energy is

$$\delta E_{\text{tot}} = \sum_{k_1, k_2} \frac{\langle \mathbf{k}_1 | V(\mathbf{R}_i) | \mathbf{k}_2 \rangle \langle \mathbf{k}_2 | V(\mathbf{R}_j) | \mathbf{k}_1 \rangle}{E_1 - E_2} + \text{complex conjugate.} \quad (9\text{-}15)$$

The asymptotic form for such an integral at very large separations (large $\mathbf{R}_i - \mathbf{R}_j$) had been obtained by Zeiger and Pratt (1973) in connection with magnetic interactions, and Sokel and Harrison redid it for this case. The essential point is that the coupling between any two sets of states gives a value decaying exponentially with separation, and the range of the force depends upon the energy difference between the states. The longest-range force comes from the states closest to the band gap and their effect can be obtained by expanding the states and energies around the maximum of the valence band and the minimum of the conduction band. The numerator contains a factor $e^{i(\mathbf{k}_1 - \mathbf{k}_2)\cdot(\mathbf{R}_j - \mathbf{R}_i)}$. When the final integral over wave numbers is evaluated by a contour in the complex plane, this is evaluated at a pole occuring at imaginary wave number ik_g, where

$$k_g^2 = 2(m_1 + m_2)E_g/\hbar^2. \tag{9-16}$$

Here m_1 and m_2 are the effective masses and E_g is the band gap. Then the interaction is seen to contain a factor $e^{-k_g|\mathbf{R}_i - \mathbf{R}_j|}$. There are similar terms from other coupled bands, but by Eq. (9-16) we see that they decay more rapidly. There was also a factor of $m_1 + m_2$ in front of the exponential, which reduces the effect of the interaction when the masses are small. Other details of the interaction can be found in Sokel and Harrison (1976).

It is interesting to see how close to the corresponding limiting range k_0/k_g are the ranges k_0/μ that were deduced from the Weber model. Sokel and Harrison evaluated Eq. (9-16); the results are given in Table 9-3. The agreement in magnitude and in trends is gratifying. The very long range interaction indicated for InSb is presumably real but is very weak owing to the tiny effective mass.

In conclusion, the origin and nature of the long-range interactions in semiconductors seem to be fairly well understood, though accurate inclusion of its effects has only been accomplished for simple situations. It is interesting that the same method, Eq. (9-15), can be used to calculate the interaction between atoms in metals, where there is no gap; we shall in fact do just that when we discuss metals. Where there is no gap, the evaluation of the asymptotic form is different, and the evaluation gives an oscillatory form $\cos(2k_F r)/(k_F r)^3$ rather than the exponential decay.

9-C   Phonons and the Specific Heat

Writing the displacements $\delta\mathbf{r}_i$ in terms of the amplitudes of the normal modes corresponds to transforming coordinates to normal coordinates and results in writing the dynamics of the crystal in the form of $6N_p$ independent harmonic oscillators (one for each degree of freedom of the N_p atom pairs). Thus for each wave number $\mathbf{k}$ in the Brillouin Zone for gallium arsenide, there will be six normal modes. A normal coordinate u_k can be associated with each mode. (For simplicity of notation, we let the index $\mathbf{k}$ designate both the wave number and which of the six modes is being considered.) Notice that the solution of the normal mode problem for whichever mode is being considered gives the amplitude vectors $\mathbf{u}_1$ and $\mathbf{u}_2$ for the two atom types in terms of each other or in terms of the normal coordinate u_k, and the energy of the system may be written in terms of the normal coordinate and its time derivative $\dot{u}_k$ in a form $(\frac{1}{2})M\dot{u}_k^2 + (\frac{1}{2})\kappa u_k^2$, to be summed over $6N_p$ values of the index k. Each of these harmonic oscillators can be treated with quantum mechanics just as we treated electrons in Chapter 1. From the energy given above we construct a Hamiltonian in terms of the coordinate u_k and its canonical momentum. Application of quantum mechanics to such macroscopic motions was a bold step when first suggested by Debye (1912), since before then, quantum effects had only been considered on an atomic scale. The application of quantum theory to a harmonic oscillator is an elementary exercise and leads to discrete vibrational energies,

$$E_k = \hbar\omega_k(n_k + \tfrac{1}{2}), \tag{9-17}$$

analogous to the discrete energies of electrons in atoms; n_k is an integer greater than or equal to zero.

Because of this simple form for the energy, the incremental units of energy, $\hbar\omega_k$, are frequently thought of as particles, called *phonons*. If a mode of index **k** is in the n_kth excitation state, as in Eq. (9-17), we say that there are n_k phonons in that mode; the energy of each is $\hbar\omega_k$. The residual energy, $\hbar\omega_k/2$, for $n_k = 0$, is called the *zero-point energy*. Viewed as particles, these phonons are obviously indistinguishable and elementary statistical mechanics shows that they obey Bose-Einstein statistics; that is, in thermal equilibrium at temperature T the average number of phonons $\langle n_k \rangle$ in the mode k is

$$\langle n_k \rangle = \frac{1}{e^{\hbar\omega_k/kT} - 1}, \tag{9-18}$$

where the k in kT is the Boltzmann constant. The derivation of this formula may be found in any elementary book on statistical mechanics. We can immediately write down the total vibrational energy for a crystal in thermal equilibrium,

$$\sum_k E_k = \sum_k \left(\frac{1}{e^{\hbar\omega_k/kT} - 1} + \tfrac{1}{2}\right)\hbar\omega_k, \tag{9-19}$$

and take the derivation with respect to temperature to obtain the specific heat. However, to obtain a value we must perform a sum over the frequencies ω_k of the modes.

This summation can be done rather accurately by fitting a few known vibrational frequencies to a model of the valence force field and then calculating the spectrum by using the Gilat-Raubenheimer (1966) scheme (see Section 2-E). This is much better than the older scheme of sampling over the Brillouin Zone. However, the results of a sampling calculation for gallium arsenide will suffice for our purposes and are given in Fig. 9-6. Notice that the distribution is very strongly peaked at two frequencies, one the doubly degenerate transverse frequency at the zone face and one an optical mode frequency; we can take the transverse optical mode frequency at $k = 0$ as a number characteristic of the optical modes. It is because of this peaking that these two frequencies were tabulated in Table 9-1. Thus it is a very reasonable approximation to replace the entire spectrum of vibration frequencies by $2N_p$ modes at the frequency of the transverse acoustical mode and $4N_p$ modes at the optical frequency;

the evaluation of the sum in Eq. (9-19) using Table 9-1 is trivial. This is called the *Einstein approximation* (Einstein, 1907, 1911).

The approximation did not originate with Einstein, though its use with quantum theory did; he regarded the crystal as a system of $6N_p$ identical oscillators and thus used a single frequency. (Our choice of $2N_p$ and $4N_p$ modes is motivated by Fig. 9-2 and Fig. 9-6.) The particular interest at that time was in understanding, in terms of quantum mechanics, why the specific heat dropped to zero at low temperatures rather than retain its classical value of $6N_p k$ at all temperatures. The quantum expression in Eq. (9-19), and the Einstein approximation to it, explain this feature, though the approximation predicts an even more rapid drop in the specific heat at low temperatures than that observed. This discrepancy was resolved by Debye (1912), who approximated the vibrational frequencies as for sound waves, $\omega_k = c_s k$, with c_s the speed of sound. He took one value for longitudinal sound and another for the transverse branches and restricted k to a sphere

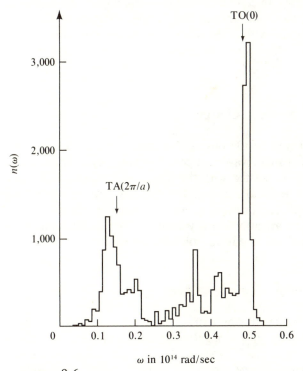

FIGURE 9-6

Distribution of vibrational frequencies in GaAs. The frequencies of the two main peaks are identified with characteristic frequencies of the spectrum. The figure suggests the Einstein approximation of replacing the distribution by two sharp peaks. [After Dolling and Waugh, 1965, p. 19.]

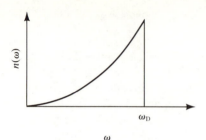

FIGURE 9-7

The distribution of frequencies for one branch of the vibration spectrum, described in the Debye approximation by treating the crystal as an elastic continuum. The Debye frequency is ω_D.

sufficient to give the $6N_p$ degrees of freedom (twice the volume of the Brillouin Zone in the case of gallium arsenide). This more complicated model rectified the defect in the Einstein approximation and led to good quantitative agreement with experiment. The calculation of the Debye spectrum for GaAs is outlined in Problem 9.2. (See Seitz, 1940, for a good account of this subject.)

Indeed, the Debye approximation is more appropriate in any problem where the modes of lowest frequency are important, as they are in thermal properties at low temperatures. In cases were all modes are important, such as in the evaluation of the total zero-point energy, the simpler Einstein model may be preferable. Notice that even within the Debye approximation the frequencies are concentrated near the highest frequency, called the Debye frequency. This is illustrated in Fig. 9-7.

9-D The Transverse Charge

Let us return now to the observed splitting of the optical mode frequencies in polar crystals at small wave numbers, first noted by Lyddane and Herzfeld (1938), and frequently associated with the names of Lyddane, Sachs, and Teller (1941), who gave a theoretical account. The splitting may be understood physically by noting that when atoms of positive and negative charge move with respect to each other, a local polarization density leads to charge accumulation and extra rigidity for longitudinal waves but not for transverse waves.

Definition of the Transverse Charge

We can formalize this by writing the displacements of the two types of atoms along the direction of propagation in the form given in Eq. (9-2). Then, if we imagine that the atoms of type one have a fixed charge $e_T^* e$ and atoms of type two have the opposite charge, we can write the polarization density

$$\mathbf{P}(\mathbf{r}) = (\mathbf{u}_1 - \mathbf{u}_2)e_T^* e\; e^{i(\mathbf{k}\cdot\mathbf{r} - \omega_k t)}/2\Omega_0, \tag{9-20}$$

where $2\Omega_0$ is the volume per atom pair and where we have taken the wave number to be so small that we can neglect the difference in $\mathbf{k}\cdot\mathbf{r}_i$ for the two atoms in the cell and replace both by $\mathbf{k}\cdot\mathbf{r}$. A charge density equal to minus the divergence of the polarization, $-i\mathbf{k} \cdot \mathbf{P}$, arises and is equal to the divergence of the electric field divided by 4π, by Poisson's equation. We must also divide by the dielectric constant to obtain the field felt by each atom since the charge accumulation has given a long-range macroscopic field. (This is the dielectric constant, ε_1, of Eq. (4-22).) We can finally obtain the resulting force on the atom of type one by multiplying this field by the charge $e_T^* e$:

$$\mathbf{F}_1 = \frac{4\pi e_T^{*2} e^2}{(2\Omega_0 \varepsilon_1)} (\mathbf{u}_1 - \mathbf{u}_2) e^{i(\mathbf{k} \cdot \mathbf{r} - \omega_k t)}. \tag{9-21}$$

This force adds directly to the force $-\kappa(\mathbf{u}_1 - \mathbf{u}_2)e^{i(\mathbf{k} \cdot \mathbf{r} - \omega_k t)}$, which enters Eq. (9-3) and from which we obtained the frequency. This can be combined with Eq. (9-6) to obtain the splitting

$$\omega_{LO}^2 - \omega_{TO}^2 = \frac{4\pi e_T^{*2} e^2}{M\Omega_0 \varepsilon_1} ; \tag{9-22}$$

as in Eq. (9-6), M is twice the reduced mass; Ω_0 is the volume per atom.

The atoms cannot really be treated as having simple effective charges, but Eq. (9-22) can be used to define the **transverse macroscopic effective charge** (or simply **transverse charge**), e_T^*; it is defined to give the local polarization induced by relative displacements, as indicated in Eq. (9-20); this polarization is induced in the absence of electric fields, since the fields are accounted for separately by division by ε_1, in obtaining Eq. (9-21). The transverse charge also gives the coupling between transverse lattice vibrations and light; hence the term "transverse." It is a good experimental definition for an effective charge since, if the atoms really had fixed charges, these could be measured directly in this way. Also, the frequencies are easily measured by measuring the reflectance versus wavelength at normal incidence (for transverse frequencies) and at nonnormal incidence (for longitudinal frequencies). (See Kittel, 1967, p. 154.) Values of the experimental transverse charge, in units of the electron charge, have been compiled by Lucovsky, Martin, and Burstein (1971) and are listed in Table 9-4. We immediately notice that these effective charges are quite large in comparison to the effective atomic charges Z^* which we associated with atoms, in Section 5-B. The origin of the discrepancy lies in the "dynamic" contributions, discussed, for example, by Lucovsky et al. (1971). They can be more simply and accurately described in terms of bond orbitals (Harrison, 1974).

Dynamic or Transfer Contributions

Earlier, we obtained the effective atomic charge, $Z^* = 4\alpha_p - \Delta Z$, by adding the charge transfers through the four bonds surrounding each atom. Were the polarities in these four bonds to remain the same as the atoms moved, we would expect the charge to remain fixed at Z^* and for e_T^* to take this value. However, the bond

Effective charges of simple tetrahedral solids.

Compound	Z^*	ζ	e_T^* (Eq. 9-24)	e_T^* (experimental)	e_P^* (Eq. 9-26)	e_P^* (experimental)
SiC	1.02	0.66	1.65	2.57	0.69	—
BN	0.36	0.67	1.17	2.47	0.11	—
AlP	1.03	0.70	2.03	2.28	0.59	—
GaAs	0.93	0.68	1.92	2.16	0.47	−0.47
InSb	1.10	0.78	2.11	2.42	0.81	−0.24
BP	−0.21	0.56	−0.38	—	−0.61	—
BAs	−0.34	0.56	−0.26	—	−0.68	—
AlN	1.33	0.77	2.36	2.75	1.03	—
AlAs	0.92	0.68	1.91	2.3	0.46	—
AlSb	0.82	0.73	1.78	1.93	0.46	−0.22
GaN	1.41	0.79	2.43	3.2	1.13	—
GaP	1.40	0.67	2.43	2.04	0.90	−0.28
GaSb	0.79	0.69	1.74	2.15	0.36	−0.42
InN	1.76	0.81	2.73	—	1.53	—
InP	1.32	0.70	2.35	2.55	0.88	—
InAs	1.22	0.76	2.24	2.53	0.89	−0.13
BeO	0.11	0.80	1.12	1.83	−0.15	0.06
ZnSe	0.99	0.79	1.87	2.03	0.75	0.13
CdTe	1.13	0.87	1.94	2.35	1.01	0.09
BeS	0.53	0.59	−1.32	—	−0.34	—
BeSe	0.47	0.63	1.49	—	−0.13	—
BeTe	0.44	0.56	−1.19	—	−0.37	—
MgTe	1.17	0.82	1.96	—	1.00	—
ZnO	1.15	0.80	1.95	2.09	0.96	1.04
ZnS	1.01	0.82	1.88	2.15	0.82	0.33
ZnTe	0.97	0.79	1.86	2.00	0.73	0.08
CdS	1.21	0.90	1.97	2.77	1.12	0.06
CdSe	1.17	0.90	1.95	2.25	1.08	0.52
CuBr	0.48	0.90	1.04	1.49	0.42	—
AgI	0.53	0.92	1.05	1.40	0.49	—
CuF	0.37	0.92	1.02	—	0.31	—
CuCl	0.47	0.79	1.04	1.12	0.32	0.35
CuI	0.45	0.89	1.04	2.40	0.37	—

SOURCES: Values of ζ from Martin (1972a) or interpolated from Harrison (1974) by using $\zeta = 1 - 0.44(1 - \alpha_p^2)^{3/2}$. Experimental values of e_T^* tabulated by Lucovsky, Martin, and Burstein (1971). Experimental values of e_P^* given by Martin (1972a).

lengths and, therefore, also the covalent energies and polarities change and the charges are redistributed. These redistributions contribute to the transverse charge in a very subtle way that can be most easily seen by imagining a single atom being displaced as illustrated in Fig. 9-8. A direct dipole is produced owing to the charge the atom carries with it. In addition, the covalent energy increases in the shortened bond, making it less polar by transferring charge to the positive atom. Similarly, charge is transferred to the negative atom in the elongated bond. The charge on the displaced atom does not change in the process, but the nearest neighbors become charged in such a way as to add to the dipole moment and therefore to the transverse charge. It is true that there are also changes in the covalent energy due to hybrid misalignment, but they are of the same sign for all four bonds and *do* change the charge localized on the displaced atom, yet do not contribute to the dipole moment to first order in the displacement.

How these transferred charges enter the calculation of optical modes is most easily seen in the two-dimensional analogue to the tetrahedral structures shown in Fig. 9-9. The positive atoms have been moved to the right, giving a polarization in that direction, and the charge transfers are seen to add to the polarization just as when a single atom is displaced. The addition to the polarization occurs by the transfer of electron charge through the crystal, just as in the dielectric response to electric fields, discussed in Chapter 5. For a $k = 0$ mode there is no important change in local charge distribution; the effect of the polarization is only through induced surface charge densities. In longitudinal optical modes of finite wavelength, however, this polarization leads to charge accumulation, as indicated in connection with the derivation of Eq. (9-21), and contributes to the extra force in the case of longitudinal waves and therefore to the transverse charge.

In Fig. 9-9,b is shown the corresponding situation in the piezoelectric effect, which we shall return to. It will be seen that because the atoms are *pulled* into position by their neighboring atoms rather than pushed against them, the charge transfers are reversed and the effective charge is reduced; this can be seen in the figure.

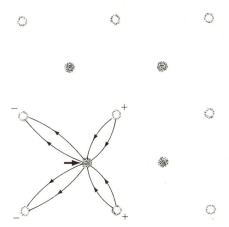

FIGURE 9-8

A positive atom is displaced with respect to its negative neighbors, producing a dipole moment. At the same time the change in the covalent energy in the neighboring bonds causes electron transfer as indicated by the curved arrows and a corresponding charging of the neighbors, which adds to the dipole moment arising from the displacement. It is called a "dynamic" contribution to the effective charge, though it is present in static displacements, and the term "transfer" contribution is preferable.

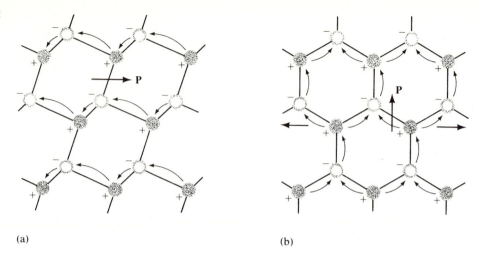

(a)   (b)

FIGURE 9-9

When the positive atoms are displaced with respect to the negative atoms in an optical mode, as illustrated in part(a), they produce a direct polarization owing to the charge Z^*e they carry with them. In addition, the electron transfers illustrated in Fig. 9-8 and indicated here by curved arrows add to the polarization density (**P**) by passing charge through the crystal, giving an effective charge greater than Z^*. In the piezoelectric effect, illustrated in part (b), a lateral elastic strain produces vertical internal displacements but the charge transfers in this case subtract from the dipole, giving a piezoelectric charge less than Z^*.

Calculation of the Transverse Charge

The calculation of e_T^* is made most conveniently in a zincblende structure viewed along a [010] direction, with relative displacement $\mathbf{u} = \mathbf{u}_1 - \mathbf{u}_2$ in a [100] direction, as indicated in Fig. 9-10. The direct polarization per atom pair—appropriate if there are no polarity changes—is clearly $eZ^*\mathbf{u}$. The change in length of the indicated bond AC is $\delta d = u/3^{1/2}$, leading to a change in V_2 of $\delta V_2 = -2V_2u/(d3^{1/2})$, because of the d^{-2} dependence. The polar energy represents the difference in hybrid energy on the two atom types and cannot show a change proportional to the displacement (by symmetry, since the change for reversed displacements must be the same). Thus the polarity change is

$$\delta\alpha_p = \frac{-V_2 V_3\,\delta V_2}{(V_2^2 + V_3^2)^{3/2}} = 2\alpha_p(1 - \alpha_p^2)u/(d3^{1/2}). \tag{9-23}$$

Let the atom C shown in Fig. 9-10 be a positive atom, such as gallium. Then this change in α_p transfers an additional charge $\delta\alpha_p$ to the atom A, as in the derivation of Eq. (5-18). An equal charge is transferred to the neighbor of C above A in the figure and equal charges are transferred from the neighbors on the right onto the gallium. As we indicated, atom C has no net change in its charge so its contribution to the dipole is still $eZ^*\mathbf{u}$, but $2\delta\alpha_p$ electrons are transferred a distance $2d/3^{1/2}$

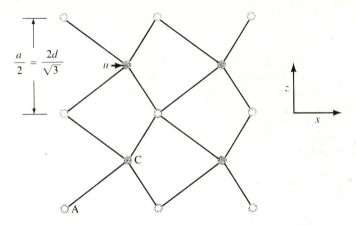

$$\frac{a}{2} = \frac{2d}{\sqrt{3}}$$

FIGURE 9-10

Coordinates for the calculation of the transverse charge.

to the left between neighbors of C, contributing $(8/3)e\alpha_p(1 - \alpha_p^2)\mathbf{u}$ to the dipole. Adding the two contributions and dividing by u and e gives the transverse charge in units of the electron charge,

$$e_T^* = Z^* + (8/3)\alpha_p(1 - \alpha_p^2). \tag{9-24}$$

One might fairly ask if the dipole moments should not be scaled by the same γ that scaled the dipoles in calculations of the dielectric susceptibility; this was in fact done in earlier treatments (see, for example, Harrison, 1974). However, it is not clear how close are the corrections in the two problems; such corrections do not significantly modify the agreement with experiment; hence, it is practical to choose the simpler description without γ. It is neither an error to include the corrections nor an error to omit them.

All the parameters entering Eq. (9-24) were given values earlier. The resulting predictions are listed in Table 9-4. The predictions are not quantitatively accurate, which is not surprising since all of the parameters that entered the calculation were based upon theory or upon experiments in which there was no disruption of tetrahedral symmetry. However, the predictions are semiquantitatively rather good; the large additive corrections, and the general decrease in e_T^* with increasingly polarity, are reproduced, suggesting that the important physical effects have been included.

Eq. (9-24) gives the effective charge within the Bond Orbital Approximation. It would be of interest to calculate corrections in terms of extended bond orbitals, but this has not been done to date.

Notice that in calculating the transverse charge we imagined the motion of the entire metallic sublattice relative to the entire nonmetallic sublattice. However, the derivation remains appropriate if only a single atom is moved, as in Fig. 9-8. The dipole induced locally is obtained by multiplying the displacement by ee_T^*, and the field arising elsewhere in the medium is obtained by dividing the corresponding dipole field by the dielectric constant. Thus it is tempting to think of this as a "true" effective charge. However, we shall now see that the effective charge entering piezoelectricity is very much different, and intuition is probably best served by thinking of Z^* as the true effective charge and remembering that transfer effects are important when there are any distortions of the lattice.

9-E Piezoelectricity

We saw in Section 8-E that the shear strain $\varepsilon_4 = \varepsilon_{yz}$ led to internal displacements of the metallic atoms relative to the nonmetallic atoms in the x-direction; these were written as $\zeta\varepsilon_{yz} a/4$. These internal displacements give a local dipole moment just as they did in the optical vibrational modes. If charges on the atoms were fixed at ee_P^*, the polarization density would be given by

$$P_x = \frac{\zeta a e_P^* e}{4(2\Omega_0)} \varepsilon_{yz}, \tag{9-25}$$

where $2\Omega_0$ is the volume per atom pair. We use this equation to define the piezoelectric charge e_P^* in terms of the measured experimental piezoelectric constant, $e_{14} = \zeta e_P^* e/a^2$, which is the ratio of the local polarization density P_x to the shear strain ε_{yz}. Notice that we have not introduced the dielectric constant in this calculation; it is appropriate not to introduce it if the experiment is performed with external electrodes, for these cancel or shunt out the fields arising from the surface charges produced by the piezoelectric polarization. This is the customary definition; for more general circumstances, the polarization density can be written as $P_i = \Sigma_j e_{ij}\varepsilon_j + \Sigma\chi_{ij}\mathscr{E}_j$, which includes effects of both strain and electric field. In the zincblende structure, $e_{14} = e_{25} = e_{36}$, and all other elements of the piezoelectric tensor vanish.

We can estimate the piezoelectric charge and piezoelectric constant in the Bond Orbital Approximation with no additional assumptions or parameters (Harrison, 1974). Return to the geometry of Fig. 8-5: there we found, as shown in Eq. (8-24), a change in every bond length of magnitude $\delta d = (1 - \zeta)\varepsilon_{yz} d/3$. Proceeding just as in the calculation of e_T^*, we find a change in dipole moment in each bond, of $2e\alpha_p(1 - \alpha_p^2)\delta d$. The components of this change in dipole along the x-direction add for the four bonds surrounding each positive atom but have a direction opposite that of the direct contribution $Z^*e\zeta\varepsilon_{yz} a/4$ arising from the displacement of the atom charge Z^*. Thus the piezoelectric charge in Eq. 9-25 is, in units of the electron charge,

$$e_P^* = Z^* - (8/3)\alpha_p(1 - \alpha_p^2)(1 - \zeta)/\zeta. \tag{9-26}$$

This expression has been evaluated by using the internal displacement parameter of Table 8-4 and, in Table 9-4, the resulting e_P^* are compared with values derived from experimental piezoelectric constants compiled by Martin (1972a). The agreement is not quantitative, though the large cancellation giving values very small in comparison to e_T^* is reproduced. The physical origin of this cancellation was noted in the two-dimensional analogue shown in Fig. 9-9,b. The quantitative errors may arise partly from errors in the internal displacement parameter used and partly from the approximate description of the electronic structure.

9-F   The Electron–Phonon Interaction

Electrostatic Interaction

In our discussion of the transverse charge, we noted that a longitudinal electric field arises which can be obtained from Eq. (9-21) by dividing by $e_T^* e$. These fields are derivable from an electrostatic potential that influences any electron in the crystal. The corresponding potential energy for an electron, obtained by multiplying the potential by the electron charge, $-e$, is given by

$$V(\mathbf{r}) = \frac{4\pi i e^2 e_T^*}{2\Omega_0 \varepsilon_1 k} (u_1 - u_2) e^{i(\mathbf{k} \cdot \mathbf{r} - \omega_k t)}. \tag{9-27}$$

The factor i indicates that $V(\mathbf{r})$ is 90° out of phase with the local displacements. Such an electron potential, arising from phonons in crystals, is called an **electron–phonon interaction**. We saw that electrons may be freed in the crystal when impurities are present and may also be freed by thermal excitation even in the pure crystal. Any such free electrons contribute to the electrical conductivity, but that conductivity will in turn be limited by the scattering of the electrons by lattice vibrations or by defects. We will not go into the theories of such transport properties as electrical conductivity—these are discussed in most solid state physics texts—but will examine the origin of certain aspects of solids such as the electron-phonon interaction, which enter those theories.

For the particular case of longitudinal optical modes, we found in Eq. (9-27) the electrostatic electron–phonon interaction, which turns out to be the dominant interaction with these modes in polar crystals. Interaction with transverse optical modes is much weaker. There is also an electrostatic interaction with acoustic modes—both longitudinal and transverse—which may be calculated in terms of the polarization generated through the piezoelectric effect. (The piezoelectric electron–phonon interaction was first treated by Meijer and Polder, 1953, and subsequently, it was treated more completely by Harrison, 1956). Clearly this interaction potential is proportional to the strain that is due to the vibration, and it also contains a factor of $1/k$ obtained by using the Poisson equation to go from polarizations to potentials. The piezoelectric contribution to the coupling tends to be dominated by other contributions to the electron–phonon interaction in semiconductors at ordinary temperatures but, as we shall see, these other contributions do not contain the divergent factor $1/k$; the piezoelectric coupling becomes dominant when the modes in question are generated ultrasonically (and therefore have much smaller k), rather than thermally.

Deformation Potentials

The origin of other contributions to the electron–phonon interaction can be understood by noticing that a longitudinal mode will change the bond lengths locally; therefore, interatomic matrix elements will be changed and the energies of the electron states will be shifted. Specifically, a local dilatation $\Delta(\mathbf{r})$ (change in volume divided by volume) will, on the average, give a fractional change in bond length one third as large as the dilatation. The fractional change in any interatomic matrix element will be minus two times the fractional change in bond length. In Eq. (3-27) we gave the energies of the bands at the center of the Brillouin Zone in terms of term values and interatomic matrix elements. Let us focus in particular on the maximum energy in the valence band for which the differences in term values and interatomic matrix elements were identified with the covalent and polar energies of Eqs. (4-18) and (4-19). That energy was

$$E = \frac{\varepsilon_p^c + \varepsilon_p^a}{2} - (V_2^2 + V_3^2)^{1/2}, \qquad (9\text{-}28)$$

and the shift in energy owing to the local dilatation becomes

$$\delta E(\mathbf{r}) = -\alpha_c\, \delta V_2 = \tfrac{2}{3}\alpha_c\, V_2\, \Delta(\mathbf{r}). \qquad (9\text{-}29)$$

Similarly, the shift in the triply degenerate conduction-band levels are the negative of this and expressions for the shifts in any other band of interest are derivable from the formulae for the energy.

Such electron–phonon interactions directly proportional to the dilatation are called **deformation potentials**, a concept first introduced by Bardeen and Shockley (see, for example, Shockley, 1950). This is indeed the dominant mechanism for electron–phonon interaction in covalent semiconductors, and the interaction with transverse waves is weaker.

The qualitative validity of Eq. (9-29) can be checked by comparing the measured change in the gap between the energies of the valence and conduction bands at the center of the Brillouin Zone ($\mathbf{k} = 0$) under pressure. (For a discussion, see Paul and Waschauer, 1963). Paul and Waschauer (1963, p. 226) compiled experimental values of $\delta E_0/\Delta$ for Ge, GaAs, GaSb, InP, InAs, and InSb. They range from -3 eV to -9 eV, without conspicuous trends, and where more than one measurement exists for one material, they differ by as much as a factor of two. The predictions based upon equations such as Eq. (9-29) vary from about -2 eV to -4 eV. Thus the physical picture appears to be valid but it is not clear that it is quantitatively useful.

PROBLEM 9-1   *Lattice dynamics*

Consider longitudinal vibrations propagating in the x-direction (in the direction of a Brillouin Zone face) in the hexagonal BN crystal shown in Fig. 9-11. It is clear that for this symmetry there are no internal displacements in the y- or z-directions (the coordinate system is rotated with respect to that in Problem 8-3), so the only amplitudes entering are those for x-direction displacements of the boron and nitrogen atoms. It is clear that angular forces as well as radial forces contribute to the rigidity but you can assume that the nearest-neighbor radial forces dominate, and can take the corresponding C_0 to be that for diamond from Table 9-1. Obtain and plot the spectrum for this direction of propagation; you will obtain both acoustical and optical modes. The zone boundary is at $k = 2\pi/3d$.

The nearest-neighbor force constants are C_0, and you must take care to take appropriate components of relevant relative displacements and of forces when you write $F_i = M_i \ddot{u}_i$ for each atom type.

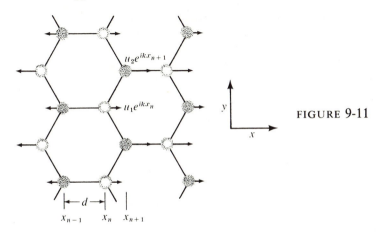

FIGURE 9-11

PROBLEM 9-2   *The Debye spectrum for GaAs*

First consider longitudinal modes. The number of wave numbers allowed by periodic boundary conditions in a volume Ω and having magnitude less than k_D is $(2\pi)^{-3}\Omega 4\pi k_D^3/3$. This follows from Eq. (2-3) and will be derived explicitly following Eq. (15-2). One degree of freedom from each atom is associated with longitudinal modes so the number of wave numbers is equated to twice the number of atom pairs N_p:

$$\frac{\Omega}{(2\pi)^3}\frac{4\pi}{3}k_D^3 = 2N_p.$$

In the tetrahedral structure, $N_p/\Omega = 3(3^{1/2})/(16d^3)$. We may directly obtain the Debye wave number k_D and the Debye frequency $\omega_D = c_s k_D$, where c_s is the speed of sound for longitudinal waves, which may be read from Fig. 9-2. Notice that optical modes are included as part of the Debye spectrum. For transverse modes, the same k_D applies, but c_s is different and there are twice as many modes at each allowed wave number.

For longitudinal modes, we can therefore state that the number of modes with frequency less than ω is $2(\omega/\omega_D)^3$ per atom-pair, or a density of modes of $6\omega^2/\omega_D^3$ modes per atom-pair per frequency-range. Similarly, construct the spectrum for transverse modes and plot the total on the same abscissa as in Fig. 9-6 so that comparison can be made. (That histogram did not have a normalized scale on the ordinate, so you need not worry about the ordinate.) The principal discrepancies are understandable by comparison of the Debye approximation to the spectrum shown in Fig. 9-2.

PROBLEM 9-3   *Calculating the transverse charge*

(a) Calculate the transverse charge for a BN structure by displacing a single boron atom as shown and calculating the resulting electric dipole. Imagine the nitrogen, for example, having one more nuclear charge than carbon, but both π electrons, so Z^* would be zero if α_p were zero. There is a term Z^*, and a transfer term that is due to the change in α_p in the neighboring bonds. The vertical components cancel out, giving a net dipole parallel to **u**. You will do best to determine the sign of each contribution from a sketch.

(b) Obtain the value for e_T^*, using the values of α_p from the solution of Problem 4-2.

(c) Verify that the same result is obtained for a displacement perpendicular to that shown in Fig. 9-12 in the plane of the figure. For displacement out of the plane of the figure, clearly $e_T^* = Z^*$.

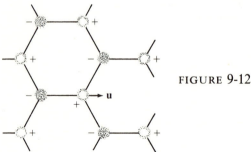

FIGURE 9-12

CHAPTER 10

Surfaces
and Defects

SUMMARY

The principal contribution to the surface energy of a homopolar semiconductor is the energy expended in breaking bonds; thus (111) surfaces are favored energetically, since these have the lowest bond density. There are only half the electrons needed to fill the dangling hybrid states formed at the surface; the states have energy in the gap between the valence and conduction bands. Because these states are only half occupied, the surface undergoes a Jahn-Teller distortion, called surface reconstruction, in which half the surface atoms move into the surface, near the plane of their nearest neighbors, and half move out from the surface to form more nearly perpendicular bonds. This sweeps the dangling hybrid states from the gap but may shift states associated with back bonds into the gap. Reconstruction also eliminates carriers near the surface, pinning the Fermi energy midgap.

A similar reconstruction occurs on the crystallographic (110) surface, a surface that occurs naturally in the polar semiconductors but must be made artificially in the homopolar semiconductors. A fundamentally different reconstruction, involving alternate vacant atomic sites, is expected on the (100) surface. Atoms adsorbed on the (111) face of silicon have different geometries depending upon their valence, with those of valence 3, 4, and 5 forming a bridged configuration over three surface atoms. This configuration appears to have stability sufficient for it to occur spontaneously, once a pure semiconductor has annealed, as a result of diffusion of atoms across the surface, and it produces complex annealed patterns.

Crystalline defects, such as vacancies, also produce dangling hybrids, and the same kind of Jahn-Teller distortion is known to occur; this always tends to sweep states from the gap. The extreme case is the amorphous covalent solid, which, even though disordered, retains an open covalent structure and is semiconducting.

Reconstruction, as well as other effects, shifts the absolute energies of the electrons relative to the vacuum level so that the energy of the valence-band maximum based upon parameters from the Solid State Table is not equal to the experimental photothreshold. However, the total shift seems to result in a reduction of the photothreshold of approximately 3.8 eV for all polar semiconductors, and slightly more for homopolar ones. Thus, with this correction, the absolute energies are meaningful and can be used to estimate heterojunction band steps.

Up to this point we have focused almost entirely on the bulk properties of covalent crystals. Any real crystal has surfaces, but for large systems, say 10^8 atoms long, there are of the order of 10^{24} atoms within the interior and only some 10^{17} on the surface, so that many properties are dominated by the interior. On the other hand, processes such as electron diffraction or catalysis are dominated by surface effects, and it is important that we include some discussion of them. As with other topics, this will necessarily be a cursory view, indicating some central concepts. There is a regular journal, *Surface Science*, exclusively devoted to current developments in the subject. We shall also discuss briefly in this chapter some related concepts concerning crystalline defects and amorphous solids.

10-A   Surface Energy and Crystal Shapes

The fact that crystal surfaces tend naturally to take particular orientations is probably the most familiar attribute of crystals. Diamond tends to form crystals in the shape of an octahedron; so also do the other homopolar tetrahedral solids. This is very easy to understand in terms of the bonding and structure of the crystal. A diamond-structure crystal is viewed along a $[1\bar{1}0]$ direction in Fig. 10-1. If we imagine cutting the crystal at some plane, we can expect the energy required to be proportional to the number of bonds cut. The bond density across the (111) surface plane shown in the figure is $\sqrt{3}/(4d^2)$. (It would be three times this if the plane were moved inside the crystal from the last plane of atoms shown in the figure.) This is smaller than the density of $3/(4d^2)$ on the (100) surface and $3\sqrt{2}/(8d^2)$ on the (110) surface, or the densities on any other planes through the crystal. Thus we may expect the surface of lowest energy, and therefore the natural cleavage plane or growth surface, to be a (111) surface. There are eight such orientations, forming the eight faces of the natural octahedral crystal for the homopolar tetrahedral solids.

Equilibrium Crystal Shape

It is instructive to carry this slightly further. Let us imagine a crystal in the shape of a regular octahedron made up of (111) surfaces. The total surface energy is directly proportional to the surface area and if the octahedron is made irregular (but kept at fixed volume, keeping all (111) surfaces), the distortion will always

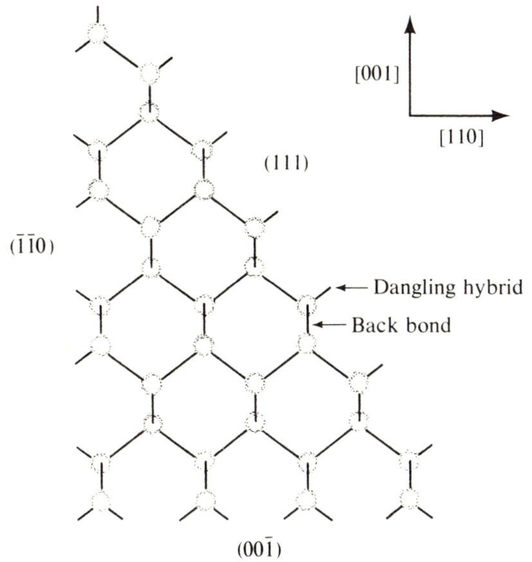

FIGURE 10-1   231

Unreconstructed surfaces in a homopolar semiconductor. The crystal is viewed along a $[1\bar{1}0]$ direction. Atoms connected by diagonal bonds in the figure are nearest neighbors in the same $(1\bar{1}0)$ plane parallel to the plane of the figure. The bonds that appear vertical in the picture make an angle of 54.7° with the plane of the figure. We are viewing each of the indicated surfaces edge-on. We see for the (111) surface, for example, a single dangling hybrid, in the plane of the figure but perpendicular to the (111) surface, originating from each surface atom.

increase the area and therefore the surface energy. The equilibrium geometry is the *regular octahedron*.

Further, imagine that the corners of the octahedron are truncated by (100) planes as shown in Fig. 10-2. If this is done at constant volume, the total surface area will decrease, so if the surface energy were the same on both the (100) and (111) surfaces, the total surface energy would decrease as a result of truncation. Truncation would be energetically favored. However, we have seen that the surface energy is higher on the (100) surface generated by truncation. It is an elementary calculation to write the total surface energy for the two kinds of surfaces, at constant volume, as a function of the extent of truncation s/L (see Fig. 10-2). We can minimize the energy with respect to s/L to obtain

$$\frac{s}{L} = \frac{2}{3}\left(1 - \frac{E_{(100)}}{\sqrt{3}\,E_{(111)}}\right). \qquad (10\text{-}1)$$

Some important features are illustrated by this equation. If $E_{(111)}$ is considerably less than $E_{(100)}$, an unphysical negative value is obtained; no truncation is expected, and the corners should be sharp. If the energies are comparable, quite large cuts are energetically favored. Note also that the shape of the crystal is independent of its size. The zero-order estimates of surface energy discussed earlier give a value for the (100) surface larger than that of the (111) surface by a factor of $\sqrt{3}$ and, therefore, Eq. (10-1) leads to a value of $s/L = 0$. However, any of the corrections to that surface energy, such as will be discussed shortly, could change that and lead to small truncations in the shape that has minimum energy, that is, the equilibrium shape.

The equilibrium shape for any crystal can be obtained by an ingenious construction due to Wulff (1901); for a discussion, see Herring (1953, p. 5). Wulff noted that if ones makes a polar plot of surface energy and constructs planes perpendicular to the polar vector at each point on the surface, the volume con-

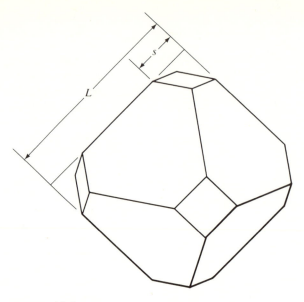

FIGURE 10-2

The natural octahedral shape for a homopolar semi-conductor, with a possible truncation of the corners. Lowest-order theory suggests no truncation for this case, though truncation would occur with slight changes in surface energy.

tained within this set of planes has the equilibrium geometry. It may be characteristic that for simple structures, like those we have discussed here, the variations of surface energy with orientation are very large and the resulting equilibrium geometries tend to be very simple but anisotropic, and that for complex structures the variations are smaller and tend to lead to rather spherical shapes made up of many facets.

Unfortunately there is not the direct relationship between surface energy and crystallography that the foregoing description would suggest. Because it is difficult to obtain equilibrium geometries, particularly in large crystals, the observed geometry tends to be dominated by the growth process of the crystal rather than the energetics itself. An early, but definitive, study of this subject was made by Herring (1953). The author has benefited from discussions of the question with David Turnbull of Harvard University.

Surface Electronic Structure

Let us return to a further description of the electronic structure at the surface of the crystal. Imagine constructing the electron states in a crystal with a surface, just as bulk states were constructed in Chapter 3. Bond orbitals are formed for all the bonds, but those hybrid orbitals directed out of the surface remain unbonded. These orbitals are called *dangling hybrids*. Each retains the energy ε_h, seen in Fig. 3-3 to be greater than the bond energy by V_2^h. (We are interested in total energies, so use of the *hybrid* covalent energy is appropriate.) Cleaving the crystal raises the energy by this amount for both of the electrons in each bond that is cut. One electron is placed in each of the dangling hybrid states created by cleavage of the crystal. The energy change of one electron can be said to contribute to the surface energy of one surface, and the other, to the other surface. Thus we have a contribution to the surface energy on a (111) surface of $V_2^h \sqrt{3}/(4d^2)$, which equals 7600 erg/cm^2 (or dyne/cm) for silicon, ten or more times observed surface energies. There are also other important terms in the calculation of surface energy, just as there were for calculation of the cohesive energy. In addition, we shall see that a reconstruction of the surface can regain a large portion of the lost energy.

The problem is somewhat different for a polar semiconductor. Then two types of dangling hybrids must be dealt with. Those associated with nonmetallic atoms will have lower energy; imagine putting both of the electrons from the broken bond in the dangling hybrid of lower energy. However, notice that cleaving a polar structure on a (111) surface leaves one entire surface of metallic atoms and the opposite surface of nonmetallic atoms. Placing the electrons as suggested leaves an immense surface charge on each of the two surfaces. The resulting charge separation requires too much energy; hence, such a plane is not favorable. In contrast, a (110) surface is half metallic and half nonmetallic and the electrostatic energy gained by this arrangement surpasses the extra energy required to break the bonds; the (110) surface is the cleavage plane in the polar semiconductors. Similar consideration of the (100) face indicates that it also is either purely metallic or purely nonmetallic; however, making a cut at the next plane above or below reverses the charge on the surface, and it would seem that a nonplanar cut with a net (100) orientation is favored over a planar cut.

We should make an important distinction between surface energy for a solid and surface energy for a liquid. To a first approximation the surface energy of the crystal is proportional to the number of bonds that are cut; thus, in this approximation, the surface energy does not change when the crystal is deformed (if that deformation occurs without diffusion of atoms across the surface). The tendency for the crystal to become spherical, as discussed earlier, occurs only through introduction of new crystallographic planes. In contrast, all orientations of a surface plane for a liquid are equivalent and redistribution of the atoms always occurs with deformation of the surface. Then the surface energy will be directly proportional to the surface area and that energy per unit area (or force per unit length) is appropriately called *surface tension*. Using the term to describe a solid is not so appropriate.

10-B   Surface Reconstruction

Let us turn now to a discussion of the detailed structure of surfaces. It has been known for some time that the surfaces of semiconductors are usually of lower symmetry than they would be if the crystal were simply terminated at a plane and all remaining atoms retained their original positions (Schlier and Farnsworth,

1959; Haneman, 1961; and Lander, Gobeli, and Morrison, 1963); this lowering of symmetry is called *reconstruction*. For the ideal unreconstructed crystal it is not difficult to construct primitive surface translations (see Section 3-C for discussion of primitive translations), which are the smallest translations in the plane of the surface by which each atom is replaced by another atom. Let two of these translations be written τ_1 and τ_2. Low-Energy Electron Diffraction (LEED) measurements tell what the actual translational symmetry of a surface is and frequently give a lower symmetry. For example, it may indicate primitive translations of $2\tau_1$ and τ_2; this is called a 2×1 reconstruction. Similarly, an observed translational symmetry of $n\tau_1$ and $m\tau_2$ is called an $n \times m$ reconstruction.

LEED patterns do not tell the nature of the reconstruction, only its symmetry. Thus a series of workers have speculated a variety of patterns of distortion or missing atoms. It has not been possible to choose between these experimentally (except possibly in the case of (110) surfaces of polar semiconductors, to be discussed), so the field has been left in an unsatisfactory state.

One systematic attempt has been made (Harrison, 1976c) to compare the energy of different possibilities theoretically. The discussion here is based upon that study. The qualitative results are in all cases consistent with the experimental information, but not definitively confirmed by it. A general conclusion of the theoretical study was that when distortions occur, these are very much larger than had been previously supposed. More recent studies confirm this, but indicate that the distortions may not be quite as large as indicated by Harrison (1976c).

Jahn-Teller Distortions

We might suspect that the undistorted (111) surface for the homopolar semiconductor described in Section 10-A would be unstable in an important way because we placed a single electron in each of the dangling hybrids. In terms of the energy level diagram we could equally well put two electrons in some hybrids and leave others empty. In such a case, a *degenerate ground state* of the system is said to exist; the Jahn-Teller Theorem (see Jahn and Teller, 1937; also Kubo and Nagamiya, 1969, p. 456) asserts that a system having a degenerate ground state will spontaneously deform to lower its symmetry unless the degeneracy is simply a spin degeneracy. This is easily understood in the context of a (111) surface as follows.

First, imagine that, once all of the material beyond a (111) surface has been removed, the plane of surface atoms might tend to be displaced slightly inward or outward; we could in principle determine this displacement by minimizing the energy of the system with respect to such a displacement. This does not modify the symmetry of the surface; the displacement is not equivalent to a Jahn-Teller effect but is instead called *surface relaxation*. The displacement is expected to be small; perhaps a few percent of the bond distance. Notice in particular that reorientation of the hybrids in the direction of the displaced neighbors has only a small effect since the reorientation is a unitary transformation on the four hybrids at a given atom, which does not change the average hybrid energy.

Having minimized the energy as above, we still have one electron in each dangling hybrid. If we move any atom inward or outward, the energy will increase quadratically with the displacement, since the system has minimum energy. Now let us rearrange the electron so that alternate dangling hybrids have two electrons or are empty. Now, if we move an atom with a doubly occupied hybrid outward, the sum of all energies except that of the added electron will increase quadratically with the displacement, but we will see that the energy of that dangling hybrid decreases *linearly* with displacement. Thus energy can always be gained in the displacement; the only question is how far the displacement will go. Similarly, displacing the atom with the empty hybrid inward will gain an energy linear in the displacement. These *are* Jahn-Teller displacements, and they are inevitable if the model correctly describes the system.

This conclusion appears to be true since a 2×1 reconstruction is seen on freshly cleaved (111) surfaces of silicon (Lander, Gobeli, and Morrison, 1963). Alternate atoms are distinguishable; they form a primitive surface cell of two surface atoms. However, the conclusion is not nearly so persuasive as it seems. First, the Coulomb energy is increased as a result of double occupancy of alternate orbitals, an effect not included in our simple one-electron picture. Second, the dangling hybrid states will inevitably broaden into bands, so the change in energy with distortion will *only* be linear in displacement if the displacements are sufficiently large. Thus the experimental information that preceded these theoretical considerations is essential to the knowledge that the reconstruction will occur. Once the distortion was observed, Haneman (1961, 1968) proposed the alternate in–out distortion described here. It remained only to estimate the magnitude of the displacements, which are not known experimentally.

Magnitude of the Distortion

Consider a (111) surface on a homopolar semiconductor such as silicon, and as illustrated in Fig. 10-1. As the atoms are displaced, first imagine holding the hybrids at their tetrahedral angles (as in Section 8-B) and retaining at the same time a single electron in each dangling hybrid. Then it is possible to estimate the changes in energy by using the valence force field constants discussed in Chapter 8, though in fact the force constants are probably reduced near the surface. Second, rearrange the electrons and allow the hybrids to rotate. For simplicity, freeze the atoms of the second layer at their original positions and move the surface atoms perpendicular to the surface. If we call the outward displacement of the atom u, the change in length of each of the three back bonds (see Fig. 10-1) is seen to be $u/3d$, to lowest order in u. The corresponding change in energy is thus $C_0(u/d)^2/6$, or $9(u/d)^2$ eV for silicon. The angle of rotation of each of these bonds is similarly seen to be $(8/9)^{1/2}u/d$. If only a single surface atom is moved, each of the back bonds is rotated by this angle while the other three bonds at the back atom remain fixed. We see that the change in energy due to the change in the angles between bonds meeting at each back atom is $2C_1(u/d)^2/3$, to be added for each of the three back atoms; this energy would be reduced if we relaxed alternate atoms inward and outward, a correction we can neglect, since $2C_1(u/d)^2/3$ is small in any case. We must also add the angular energy in the three bond angles having apexes at the surface atoms. The evaluation of this energy is also straightforward; it totals $4C_1(u/d)^2$. The total elastic energy of distortion, considerably overestimated, is

$$\delta E_{\text{elast}} = (C_0/6 + 6C_1)(u/d)^2 \equiv \tfrac{1}{2}C(u/d)^2. \tag{10-2}$$

Now let us allow for the addition of an electron to (or the removal of an electron from) the dangling hybrid. We must now add (or subtract) the energy of that electron. It now becomes very important whether or not we rotate the hybrids, since rotation changes the sp mixture and the energy of the hybrid. The calculation requires a little trigonometry. In Fig. 10-3, a dangling hybrid with three back bonds is shown. We construct hybrids $w_\alpha|s\rangle + v_\alpha^1|p_x\rangle + v_\alpha^2|p_y\rangle + v_\alpha^3|p_z\rangle$ along each of the back bonds such that the vector $[v^1, v^2, v^3]$ lies along a bond. Requiring that these be normalized and orthogonal to each other fixes the coefficients. We then require the dangling hybrid $w_{\text{dang}}|s\rangle + v_{\text{dang}}|p_z\rangle$ to be normalized and orthogonal to the others. This gives $w_{\text{dang}} = \sqrt{2}\cot\theta = (3u + d)/2d$, and gives a hybrid energy of

$$\varepsilon_{\text{dang}} = \varepsilon_p - (3u + d)^2 V_1/d^2. \tag{10-3}$$

This remarkably simple result applies for $-d < 3u < d$. It should be added to Eq. (10-2) for a doubly occupied dangling hybrid and subtracted for an empty dangling hybrid. The result is shown for silicon in Fig. 10-4.

Notice that a surface atom with a doubly occupied level is predicted to be displaced almost until the dangling hybrid is completely s-like. The predicted inward displacement of the empty hybrid is not so great; the minimum occurs at

$$u/d = -6V_1/(18V_1 + C), \tag{10-4}$$

or $-0.12d$ for silicon. (Notice that the formula for the full hybrid minimum is obtained by replacing V_1 by $-V_1$ in Eq. (10-4).) Many of the approximations made here tend to underestimate the displacement; therefore, Harrison (1976c) postulated displacements of $-d/3$ and $+d/3$ for the two atoms; these displacements are shown in Fig. 10-5. However, more recent calculations have been made by Chadi (1978), based upon the same essential concepts but with a more careful calculation

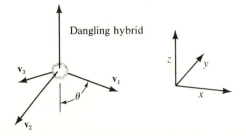

Dangling hybrid

v_3

v_1

θ

v_2

z

y

x

FIGURE 10-3

A (111) surface atom with vectors $\mathbf{v}_\alpha$ in the directions of the three back atoms. The positions of the back atoms remain fixed as the surface atom is displaced a distance u in the z-direction, changing the angle θ which the vectors $\mathbf{v}_\alpha$ make with the normal to the surface.

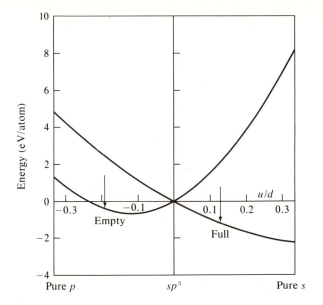

FIGURE 10-4

Calculated energy as a function of the outward displacement u of a (111) surface atom in silicon when the dangling hybrid is doubly occupied or empty. The constitution of the dangling hybrid is indicated at the bottom of the figure. Arrows indicate displacements predicted by Chadi (1978).

of the electronic energy. Chadi constructed a slab of silicon with (111) surfaces; he then calculated and summed the occupied electronic states in detail. This more complete calculation led to minimum energy at the distortions indicated by arrows in Fig. 10-4. It would seem that the displacements are not as large as had been postulated by Harrison but that the physical origin of the reconstruction is as described here and that the displacements are still much larger than had once been supposed. Possibly the error in the calculation came from assuming the back-bond hybrids lie along the back bond, rather than determining their orientation variationally.

The resulting picture of the reconstruction is that of large displacements of alternate atoms as illustrated on the (111) face in Fig. 10-5. The pattern that this leads to over the surface is not easy to predict, but it is observed to be 2 × 1, as illustrated schematically in Fig. 10-6.

It should be noted that the purely theoretical considerations given here could accommodate other patterns of alternate displacements equally well. The most obvious variation can be obtained from Fig. 10-6 by shifting alternate horizontal rows one step to the left. Each shaded circle would still have four empty circles as nearest neighbors and two shaded ones, but the pattern would be 2 × 2, with four atoms per surface-cell. Further elastic and electrostatic calculation would be required to compare the relative energies of these two, or other, variations. On (111) surfaces of annealed silicon a much more complicated 7 × 7 reconstruction is observed. We shall return to a discussion of it in Section 10-D.

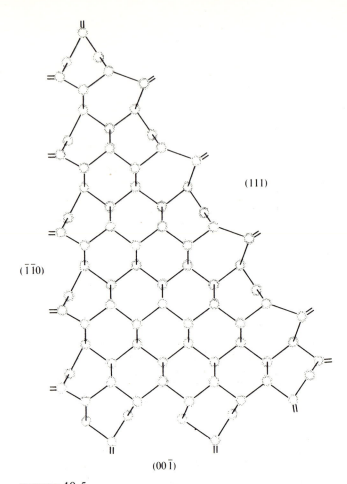

FIGURE 10-5

Reconstructed surfaces in a homopolar semiconductor. The geometry of the crystal before reconstruction was the same as in Fig. 10-1, but here the surface atoms have been allowed to relax. The positions for the corner atoms are not necessarily meaningful. Doubly occupied dangling hybrids are represented by double lines. The reconstruction on a real $(00\bar{1})$ surface is more likely a variation on the reconstruction shown here. One possibility is illustrated in Fig. 10-8.

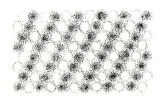

FIGURE 10-6

The pattern of 2×1 reconstruction of the silicon (111) surface, proposed by Haneman (1968). The darkened circles represent surface atoms that have moved into the bulk; the others have moved outward.

(110) Surfaces

Although (111) surfaces have lowest energy, it is possible experimentally to obtain other surfaces on the homopolar semiconductors, notably (110) and (100) surfaces. In considering reconstruction on these surfaces we will assume that the dehybridization energy dominates the distortion, as we found to be so on the (111) surface. Thus we can guess the forms of reconstruction without further calculation.

Let us consider first the (110) surface. The surface in the plane of the paper in Fig. 10-1 is such a surface; the surface atoms are seen to be zig-zag chains of nearest-neighbor atoms lying horizontally in the figure. Such chains are seen edge-on on the ($\bar{1}\bar{1}0$) surface on the left. Each surface atom is bonded to two other surface atoms and to one atom below the surface. The remaining dangling hybrid is not perpendicular to the surface but, as in a (111) surface, there is a single dangling hybrid per atom, and Jahn-Teller-like distortion is expected. The alternate inward–outward displacement of atoms simply rotates the chain for the (110) surface as indicated in Fig. 10-5, where we have assumed sufficient rotation to bring one set into the plane of its nearest neighbors. This may again overestimate the distortion. The dangling hybrid on that atom becomes purely p-like and will be unoccupied. The other atom has 90° angles between bonds, which requires a slight displacement of the chain. The translational symmetry of the surface is not modified by these distortions, but a glide-plane symmetry is removed, so the distortion should be detectable by low-energy electron diffraction.

(100) Surfaces

Now let us examine the (100) surface; a new feature occurs here. This unreconstructed surface has been considered by Appelbaum, Baraff, and Hamann (1975) and we shall simply interpret results that they obtained by detailed calculation. Each (100) surface atom is bonded to only two atoms below the surface and we might at first expect two dangling hybrids, like rabbit ears, coming out of the surface at each atom. However, any linear combination of these two hybrids is still orthogonal to the bonded hybrids, and the appropriate linear combination should be obtained by a variational calculation, just as we obtained the bond orbitals in the bulk semiconductor. This calculation leads to one hybrid—actually a hybrid $|d\rangle = (|s\rangle + |p\rangle)/\sqrt{2}$—perpendicular to the surface directed outward, and to a pure p state perpendicular to the other three and lying in the surface. The state perpendicular to the surface has lower energy than that in the surface, so we may expect the former to be doubly occupied and the latter to be empty, just as in the reconstructed (111) and (110) surfaces. Even without reconstruction there is no degeneracy of the ground state and no Jahn-Teller distortion. It can be expected, however, that the surface atoms will move outward to bring the bond angles near 90°, as indicated at the bottom of Fig. 10-5. This does not change the symmetry of the surface at all, and does not account for the experimental observation of a 2×1 reconstruction for this surface (Schlier and Farnsworth, 1959).

Schlier and Farnsworth have proposed that this reconstruction consists of the movement of adjacent surface atoms together to form bonds between them. But such a reconstruction seems unlikely, since all low-energy dangling hybrids are occupied and all high-energy dangling hybrids are empty and bonding cannot occur without promoting electrons to high-energy hybrids. The energy of such a state has not been calculated but it seems unlikely that it is favorable. Thus, we should consider other possible reconstructions; the alternate vacancy model proposed by Phillips (1973b) is one possibility.

It is remarkable that if, for example, we remove alternate surface atoms, as shown in Fig. 10-7,a, the number of broken bonds is unchanged. Thus, in the

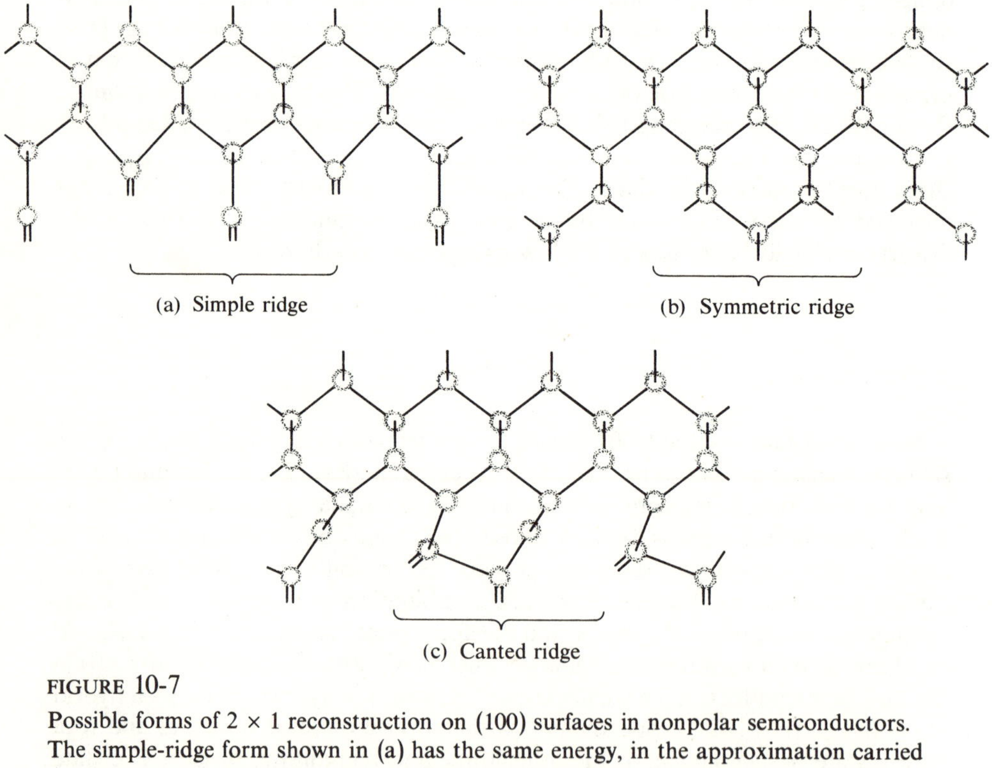

(a) Simple ridge   (b) Symmetric ridge

(c) Canted ridge

FIGURE 10-7

Possible forms of 2×1 reconstruction on (100) surfaces in nonpolar semiconductors. The simple-ridge form shown in (a) has the same energy, in the approximation carried out in the text, as the unreconstructed form given in Fig. 10-1. Removal of the entire set of surface atoms of (a) leads to the symmetric-ridge form shown in (b) which, however, will be unstable against the kind of canting shown in (c). The canted-ridge form has the same calculated energy as the simple ridge of (a) except for possible differences in elastic energy. A third form of approximately the same energy, shown in Fig. 10-8, seems to be favored experimentally. The simple-ridge form is a likely form for polar semiconductors.

absence of relaxation, the surface energy is the same. Further, if we allow the outward relaxation suggested earlier, the change in elastic energy is the same. Thus, we cannot tell which has lower energy. The surface energy per surface atom will equal half the bond energy for the two broken bonds, minus the energy gained in dehybridization and relaxation $(4\varepsilon_h - 2\varepsilon_p - 2\varepsilon_s = \varepsilon_p - \varepsilon_s = 4V_1)$, plus the elastic energy.

An alternative 2×1 reconstruction on (100) surfaces may be obtained by removing all of the surface atoms in Fig. 10-7,a; this is shown in Fig. 10-7,b. Again, the number of broken bonds is not changed but now two atoms with single dangling hybrids are present for each atom with double dangling hybrids, and the reconstruction differs. Each of these single hybrids has [111] orientation (in this case $[\bar{1}\bar{1}1]$ or $[11\bar{1}]$) and they will alternately relax inward and outward as on the (111) surface and as indicated in Fig. 10-7,c. At the same time, the surface atom with the double dangling hybrid will relax downward and an electron will transfer between the atoms with single dangling hybrids. This is also shown in Fig. 10-7,c. Notice that the canted ridge could lean either to the left or to the right, and that a spontaneous electric dipole arises in the plane of the reconstructed surface.

The canted reconstruction is an interesting one. Notice that each ridge acts as a somewhat independent unit (except that they interact electrostatically), and each can cant either way, though the entire length of the ridge perpendicular to the figure must move together. Electrostatic energy favors their canting in the same direction, as in Fig. 10-7,c; notice that if a ridge flips over, the electrons transfer across. Notice also that, since the dangling hybrids on both atoms being brought together are doubly occupied, we do not expect them to bond for the same reason we did not expect the reconstruction proposed by Schlier and Farnsworth to occur.

To compare the surface energies of the canted and simple reconstructions of Fig. 10-7, we must estimate the relaxation energy of the pair of atoms possessing single hybrids. However, that energy is $\varepsilon_p - \varepsilon_s = 4V_1$, just as it was for the surface atom this pair replaced. We find the unreconstructed surface of Fig. 10-5 and the simple and canted 2×1 reconstructions to have the same energy except possibly for the elastic energy difference.

There are an extraordinarily large number of possible reconstructions, all of which have the same energy, so far as has been calculated. The reason is basically the finding, mentioned following Eq. (10-1), that the surface energy for (100) surfaces is $\sqrt{3}$ times that for (111) surfaces; from this it follows that to make (111) facets on a surface which, on the average, is a (100) surface, does not change the energy. We see that the symmetric and canted-ridge configurations of Fig. 10-7 can be thought of as (111) facets formed on the lower surface. We could obviously make a 4×1 reconstruction with ridges twice as high and, to the order we have computed, it would have same energy. It was experiment that required a 2×1 pattern.

It seemed that this reduced the possibilities to those shown in Fig. 10-7. However, Webb notes that a third possibility exists (discussed by Poppendieck et al., 1978) in which the inner row of surface atoms in the simple ridge structure is

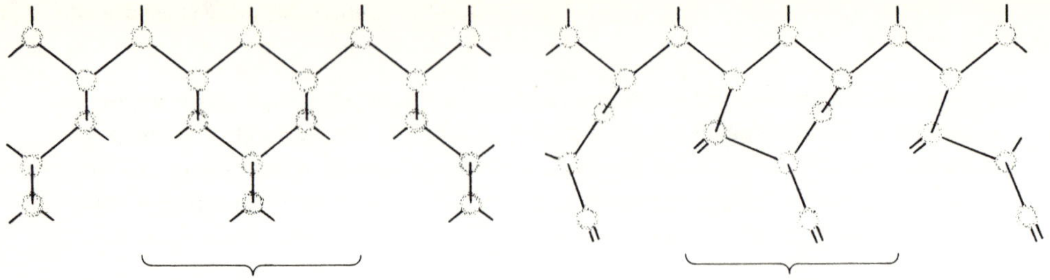

(a) Double ridge   (b) Webb model

FIGURE 10-8

The same number of bonds are broken in making the double-ridge structure of part (a) as were broken in forming the simple ridge of Fig. 10-7. Such a structure will also be canted by reconstruction, as shown in part (b). This, or some variation on it, presently seems to be the most likely form of the silicon (100) surface.

removed. This corresponds to forming trenches twice as deep but of the same width—rows of double vacancies—as shown in Fig. 10-8. This also is a variation on the (111) faceting and has the same approximate energy. Webb's LEED experiments favored this form and, indeed, it is plausible that extending the dielectric medium further down in this way could lower the electrostatic energy of the polar surface and be energetically favored.

The canted-ridge and the Webb reconstructions seem to be favored by an experimental finding communicated to the author by J. E. Rowe, of the Bell Telephone Laboratories, who found a considerable amount of diffuse scattering that might be associated with irregular orientations of the canting of ridges. He also found that the diffuse scattering disappeared when hydrogen was adsorbed. We shall see in Section 10-D that hydrogen, by forming bonds with half-filled dangling hybrids on (111) surfaces, eliminates the corresponding 2×1 reconstruction. Similarly, here the addition of hydrogen should eliminate the canting, returning the system to the symmetric-ridge form as shown in Figs. 10-7,b and 10-8,a.

Polar Semiconductors

We can also expect reconstruction on the surfaces of polar semiconductors. In particular, the distortions on the (110) cleavage plane may be expected to be of the same form as on the (110) surfaces of homopolar semiconductors, as shown in Fig. 10-5, with the nonmetallic atom displaced outward since its hybrid is doubly occupied; the metallic atom is displaced inward with its purely p-like hybrid unoccupied. This is the distortion proposed by MacRae and Gobeli (1966) for essentially the same reasons described here. This appears to have been confirmed by recent analysis of LEED data (Lubinsky, Duke, Lee, and Mark, 1976).

We indicated that a planar (111) surface of a polar semiconductor would have high energy because one face would be purely metallic and the other purely nonmetallic. Nevertheless, there are methods for producing such surfaces experimentally. It is not difficult to see, and not surprising, that nonplanar surfaces are required in order to avoid charge accumulation at such a surface. One can make use of this point to restrict the allowed geometries, and even to select the simplest allowed geometries (Harrison, 1979) but we shall not go into these results here.

10-C   Elimination of Surface States, and Fermi Level Pinning

The spontaneous lowering of the symmetry of a surface by reconstruction is characteristic of covalent solids and is not common in other types of solids (it happens on some transition metal surfaces). It has been interesting to try to understand the origin of this reconstruction. The reconstruction also has important consequences. Before reconstruction, we expect to find dangling hybrid states with energy ε_h, and can expect this energy to be near or in the gap between valence and conduction bands, as indicated by Fig. 3-3,a. We have redrawn part of that energy level diagram in Fig. 10-9, with parameters appropriate to silicon; the band gap shown between the top of the valence band and the conduction band is the experimental one; it has a value somewhat smaller than would be predicted by

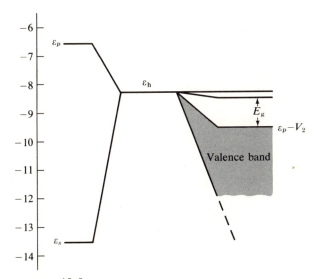

FIGURE 10-9

An energy-level diagram for silicon that gives an energy for dangling hybrids that is slightly above the conduction-band edge. Reconstruction splits these into doubly occupied levels at ε_s, and empty levels at ε_p, both far removed from the gap. The sweeping of levels out of the region of the gap is expected to occur quite generally.

244

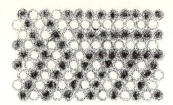

FIGURE 10-10

The 2 × 1 surface reconstruction pattern is presumably divided up into domains of different orientation. Such a pattern allows atoms at the domain boundaries to be converted from one sign to the other without a serious cost in electrostatic energy.

the parameters we are using. This small gap leads to a hybrid state in the conduction band.

The essential effect of reconstruction is to split the dangling hybrid to the energies ε_s and ε_p, both very far from the gap. Again, corrections can be made in the simplified theory, but we would not expect them to modify this result: reconstruction should eliminate all dangling hybrid states from the gap by displacing them deep into the conduction and valence bands.

This suggestion (Harrison, 1976d) is in conflict with traditional theory. It was learned very early, experimentally, that near the surface of semiconductors in contact with metals there are no charge carriers even if the semiconductor is doped to give them. The same has proved true near the surfaces of homopolar semiconductors not in contact with metals; the phenomenon is called *Fermi level pinning*, for reasons we shall see. The absence of charge carriers was explained by Bardeen (1947), who assumed that there were surface electronic states midgap and that any charge carriers at the surface dropped into those midgap states; the explanation involved also a "band bending" effect and the production of a "Schottky barrier," both of which will be discussed shortly. This led to a general conviction that midgap states must be present, even though no one succeeded in proving their existence directly. However, it is now clear that the same reconstruction that eliminates the dangling hybrid states can also eliminate charge carriers at the surface (Harrison, 1976d). Conduction electrons, for example, could pair up, converting an empty high-energy dangling p state to a full low-energy dangling s state by moving the atom to an outward position. If this were to happen at boundaries between surface domains, as illustrated in Fig. 10-10, the displacement would not even cost an appreciable electrostatic energy. In just the same way, the holes near the surface in a p-type semiconductor would be "eaten up" by a modified reconstruction.

The way in which the carriers near the surface are eliminated can be understood in the energy-level diagram of Fig. 10-11. (For details, see McKelvey, 1966, p. 485.) In Fig. 10-11, an n-type semiconductor with a distribution of positively charged donor atoms is considered. Electrons convert dangling hybrid states and produce a negative surface charge. This, combined with the positive charge of the donors that have no negative carriers compensating their charge, produces the potential hill which keeps electrons from the bulk from drifting into the surface region. That potential hill is called a *Schottky barrier* when the semiconductor is in contact with a metal. The region at the surface without carriers corresponds to a Fermi level that is fixed, or pinned, midgap. The thickness of the region without

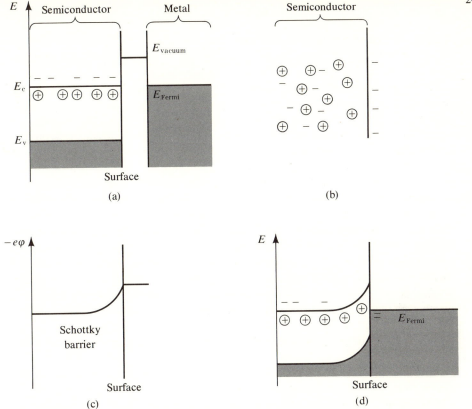

FIGURE 10-11

Imagine that, initially, the valence-band maximum and the conduction-band minimum
remain at the same energies, E_v and E_c, respectively, up to the surface as shown in
part (a). The energy of an electron at rest outside the surface, E_{vacuum}, is higher.
Ionized donors inside the material are indicated by $\oplus$, and the electrons they have
released to the conduction band, by —. If a metal were in electrical contact with the
semiconductor, electrons would be redistributed until the Fermi level, E_F, separating the
occupied from the unoccupied levels in the metal, were near the conduction-band edge of the
semiconductor. Then, if the electrons near the surface were eaten up by surface
reconstruction, as shown in part (b), an electric dipole layer would arise, giving the
potential hill, or Schottky barrier, shown in part (c). The hill or barrier would bend the
bands, as indicated in part (d). One can say that the Fermi level is pinned midgap at the
surface, though it is near the conduction-band edge in the interior.

carriers depends upon the donor density, and can be extremely large on the scale of a lattice distance. The quantitative description of this system is quite straight-forward (McKelvey, 1966); details of it are not essential here.

The mechanism that eliminates carriers is a special case of a general phenomenon—another case is elimination of states in the gap and of carriers in amorphous semiconductors, proposed by Anderson (1975). In general, the mechanism can be described as follows. A conduction electron is an electron in an antibonding state; a system will seek a mechanism to lower its energy by putting that electron in a state of lower energy. Similarly, a hole is an empty bonding state and the system will seek to fill it with an electron. At a surface or in an amorphous semiconductor, the atoms can shift to accomplish this, as we have seen. If foreign atoms are present, as when the semiconductor is in contact with a metal, they will tend to bond in such a way as to eliminate carriers. Only in a semiconductor that is very nearly perfect is the high-energy state corresponding to free carriers preserved. Indeed it is the doped semiconductor possessing free carriers that is the freak—the carrier-lacking region near the surface is the natural state. This same phenomenon is the basis of the familiar fact that colored materials fade in the weather or the sunlight. Reactions and readjustments occur to lower the energy of the occupied states and raise the energy of the empty ones, increasing the gap between the two until there is no light absorption at energies in the visible spectrum.

This mechanism is not so effective in polar semiconductors. The conversion of empty hybrids to doubly occupied hybrids on a GaAs surface would require the double occupation of a gallium hybrid, which is unfavorable because of the polar energy. Indeed, recent experiments (Chye, Babalola, Sukegawa, and Spicer, 1975) indicate that the Fermi level is *not* pinned on surfaces of GaP at the vacuum. Nonetheless, Schottky barriers can arise at GaP–metal interfaces. " Metal-induced surface states " have been proposed as a mechanism (discussed in Section 18-F) but the barriers could well arise simply from incorporation of metal atoms in the semiconductor or vice versa.

The removal of dangling hybrid states from the gap does not preclude, and in fact tends to produce, another kind of surface state in the gap. These are the **back-bond states**. The rehybridization by which the dangling-hybrid-state energy drops to the s level also raises the energy of the hybrid states that form the back bonds to the p level. It is possible that this creates surface states just above the valence-band maximum, but that is not certain. The average increase of the energy of the back-bond hybrids must be sufficiently large in comparison to the metallic energies coupling the back bonds to the interior material that surface states can be created. Photoemission studies (see, for example, Wagner and Spicer, 1972) have detected states just above the valence-band edge, indicating that back-bond surface states are indeed formed. One should think of these states as being pushed up from the initially fully-occupied valence band; these surface states take their electrons with them, and are occupied. Thus they do not influence Fermi level pinning. There is also a possibility that analogous empty states are pulled out from the bottom of the conduction band.

10-D   Adsorption of Atoms and the 7 × 7 Reconstruction Pattern

Bonding with a Dangling Hybrid

The addition of atoms to the surface can make significant changes in the reconstruction pattern and the electron states in the surface because the additional atoms form bonds with the dangling hybrids. This is apparent for the simplest case, atomic hydrogen on a silicon (111) surface, studied by Ibach and Rowe (1974b). A model in which each hydrogen rests above a surface atom and bonds with a dangling hybrid, thus doubly occupying the bond state and removing the tendency for reconstruction, is consistent with their finding that one atomic layer removes the 2 × 1 reconstruction, and with their explanation of the fact.

The energy of the electron state of atomic hydrogen is very low compared to the energy of the dangling hybrid, so the bonding-state energy will be near the hydrogen level and the corresponding antibonding state will be above the dangling hybrid level. The same should be true about the addition of halogen atoms, which should adsorb in just the same way. The adsorption of an alkali metal atom should be similar in all respects, except that since the energy of the alkali metal state is presumably well above the hybrid energy, the energy of the bond formed should be near, and determined by, the energy of the dangling hybrid.

It is also interesting to examine partial coverage; that is, coverage by less than one adsorbed atom per surface atom. Forming a bond with a dangling hybrid forms a neutral surface "molecule" at one of the otherwise polar sites indicated in Fig. 10-6; neutrality can occur at either type of site but electrostatics favors its happening approximately equally on the two. Thus the density of upper and lower hybrid states on the surface should decrease together with increasing coverage. The electrostatic forces will also tend to cause condensation of "surface droplets" of neutral sites, leaving the 2 × 1 reconstruction intact elsewhere.

We see that the effects on the silicon (111) surface of adding atoms from columns 1 and 7 of the Solid State Table are very much the same. This also appears to be true if the added atoms are from columns 2 and 6; likewise if they are from columns 3 and 5: adsorption from columns n and $8 - n$ is similar.

In this regard, oxygen, from column 6, has been studied experimentally by Ibach and Rowe (1974a), though they studied its absorption on annealled surfaces which had undergone a 7 × 7 reconstruction, which we have not so far discussed. Thinking in terms of the 2 × 1 surface, two possibilities for atomic adsorption come immediately to mind. First is a direct bonding of the oxygen to one of the silicon atoms, which has a doubly occupied dangling hybrid; the other is the formation of an *oxygen bridge*, as is present in SiO_2. Ibach and Rowe considered the second possibility, but noted that the distance between nearest silicon neighbors on the surface is 3.84 Å, while the nearest Si—O distance in SiO_2 is only 1.6 Å, so that an oxygen bridge is very unlikely. Indeed, the spectra they saw at coverages of less than a full monomolecular layer bore little resemblence to SiO_2 (though the corresponding SiO_2 spectra did develop at higher coverage). Thus, of the two possibilities, the first is preferred. A third possibility is that the oxygen adsorbs as an O_2 molecule, in which case it presumably becomes O_2^-, quite analogous to adsorption of a chlorine atom. This seems quite likely at present.

Adsorption at Bridging Sites

An added atom from column 5 is expected (Lander and Morrison, 1964) to adsorb above three neighboring silicon surface atoms, forming a bond with each, with the remaining two electrons filling the dangling hybrid. This could be expected for an added phosphorus atom, for example, and similarly an added atom from column 3, such as aluminum, might be expected to take a similar position but leave its dangling hybrid empty. The atomic spacing for bridging in this case is just right, though the angles are quite wrong. There are actually two types of sites over three surface atoms, one directly over a second layer atom and the other not. The latter requires less angular distortion and is preferred. In addition, the fact that the angular force constant, C_1, is so much smaller than the radial force constant, C_0, particularly for polar semiconductors, suggests that semiconductor surfaces can much more readily accommodate to angular distortion.

An added atom from columns 3 or 5 neutralizes the three atoms below it just as added atoms from columns 1 and 7 neutralize the single atoms below them. This would leave the alternate inward–outward reconstruction between such added atoms. We would expect the three back atoms to relax in order to reduce the angular discrepancy. It is possible that they move into the surface, tending to raise the atoms around them. The coverage of such atoms would certainly be limited to one added atom for each three surface atoms, and might well be limited further by the elastic repulsion between the extended distortion patterns.

If the atoms rearrange diffusively we can expect the added atoms to tend to substitute for the surface silicon atoms, thus maintaining the tetrahedral structure and destroying the 2×1 reconstruction. A substituted phosphorus can be thought of as simply another silicon atom with an extra proton in the nucleus. The phosphorus then behaves very much like an added hydrogen atom and stabilizes the surface in just the same way, leaving the surface atoms neutral. An aluminum atom accomplishes the same effect by leaving the dangling hybrid empty. These substitutions, however, require diffusion to be accomplished, and presumably they occur only with annealing.

We turn finally to added atoms from column 4, with the addition of a silicon atom to silicon being the most important case, since it will occur naturally from diffusion across the surface. We can expect the most stable position for an added atom outside the surface to be above three surface atoms, just as for columns 3 and 5. The additional atom bonds with three dangling hybrids, leaving a single dangling hybrid surrounded by an alternate in–out reconstruction. It would be reasonable to expect that such an added atom would have higher energy than those in the undisturbed in–out reconstruction, and that these added atoms would disappear upon annealing. On the other hand, the fact that three dangling hybrids are replaced by one leaves the possibility that the surface energy is lowered by such an addition, in which case a pattern of added atoms will *result* from annealing and a new reconstruction will be made. This is an interesting possibility to explore.

If the energy is indeed lowered by addition of an atom at the surface by means of diffusion across the surface (with the atom presumably coming from some step, or other irregularity on the surface), we might expect to add as many atoms as possible, covering all dangling bonds. On the other hand, there will certainly be a repulsion between such added atoms; the elastic distortion of the substrate will be one mechanism giving a repulsion. Complete coverage of the dangling hybrids would give a 3×2 reconstruction, which has not been observed, at least on silicon, so it appears that a compromise is made between repulsion and the energy gained in adding the atom. It corresponds to that density of added atoms which minimizes the total energy, but since the available sites are fixed by the sublattice, reconstruction must follow some ordered $n \times m$ pattern allowing near-optimum density of added atoms. We cannot predict a 7×7 pattern, though such a pattern is not a surprising result, nor is it surprising that the annealed pattern on germanium—another atom from column 4—is different from that on silicon.

It is premature to speculate on the detailed 7×7 pattern, but an appealing pattern can be obtained by starting with a pattern of vacancies proposed by Lander and Morrison (1963). We do this by adding atoms to each vacant site to obtain the starting planar structure, and then add one atom outside the surface to correspond with each site that Lander and Morrison left vacant. This adds 13 atoms in each primitive surface cell (80 percent coverage), leaving 10 surface atoms uncovered.

10-E   Defects and Amorphous Semiconductors

The simplest defect in a semiconductor is a substitutional impurity, such as was discussed in Section 6-E. There are also structural defects even in pure materials, such as vacant lattice sites, interstitial atoms, stacking faults (which were introduced at the end of Section 3-A) and dislocations (see, for example, Kittel, 1971, p. 669). They are always in small concentration but can be important in modifying conduction properties (doping is an example of this) or elastic properties (dislocations are an example of this).

Let us consider vacant lattice sites, which have been treated in detail by a number of workers (see, for example, Louie, Schlüter, Chelikowsky, and Cohen, 1976; Watkins and Messmer, 1973, p. 133; and an extensive set of references given by the same authors in Messmer and Watkins, 1973). Vacant lattice sites have one electron for each hybrid, and we can expect reconstruction for the same reasons we expected it to occur in surfaces. Jahn-Teller distortion (Section 10-B) is even more certain at vacant sites than in surfaces since there are no band-broadening effects for an isolated vacancy and the Jahn-Teller theorem applies more directly

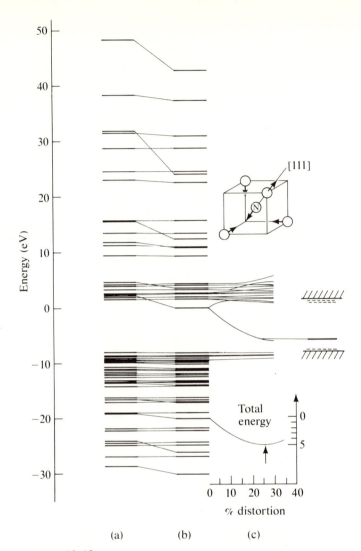

(a)   (b)   (c)

FIGURE 10-12

(a) The energies of the one-electron molecular orbitals for a
35-carbon atom "diamond." (b) The energies with nitrogen as
the central atom, no relaxation. (c) The effect of trigonal Jahn-
Teller distortion is to move the nitrogen and its four nearest
neighbors (see insert). [From Watkins and Messmer, 1973.]

to the resulting degenerate ground state. Watkins and Messmer find such spontaneous distortions lowering the symmetry of the vacancy which, without distortion, had tetrahedral symmetry. They even find such Jahn-Teller distortions for a substitutional nitrogen atom in diamond. Their findings for that case are very interesting and are shown in Fig. 10-12.

Their calculation was based upon the Extended Hückel Approximation described in Section 2-D. However, they took a definite cluster of 35-carbon atoms in the diamond structure and diagonalized the resulting Hamiltonian matrix to obtain molecular orbitals and their energies; this is analogous to making a band calculation, but for a small system rather than an infinite crystal. The resulting set of levels is shown in Fig. 10-12,a. It is like a set of valence and conduction bands, but has also a number of dangling hybrid states due to the surfaces of the cluster. (One can see by counting that there is not the equal number of states above and below the gap that is expected for a system without surfaces.) The dangling hybrid states in their calculation lie within the valence band. Watkins and Messmer next replaced a central carbon atom with a nitrogen atom, making slight shifts in some of the states as shown in Fig. 10-12,b. In particular, a partially occupied set of degenerate levels appear below the conduction-band edge. A Jahn-Teller distortion is to be expected; two forms were considered; that shown in the insert (upper right in the figure) gave the greatest lowering in energy, and the shift in levels resulting from such a distortion is shown in Fig. 10-12,c. It is seen that the degenerate level splits linearly with the distortion, providing the driving mechanism for the distortion.

One aspect of the shift in levels should be commented on. The level which is raised in the distortion soon enters the conduction band and, in detail, the shifts are complex. However, one can see a "ghost" of the shifting level moving through the conduction band, and the shift in total energy varies smoothly, much as we would calculate it to vary if we ignored the presence of the conduction-band states.

Perhaps the most important point is the confirmation of the tendency for levels in the gap to be swept up or down depending upon whether the corresponding states are occupied. When states are partially occupied, they split—again in such a way as to reopen a gap. This universal tendency was discussed in Section 10-C. As indicated there, the concept is important if one is to understand amorphous materials. It might seem that with a loss of crystalline periodicity, the gap separating valence-band and conduction-band states would be lost. That this does not happen was shown in two steps. First, Weaire (1971) showed that in an amorphous material with topological disorder, but with every atom still tetrahedrally surrounded, the gap remains; it does not depend upon long-range order in the system. We saw how such a topologically disordered system could be constructed in Section 3-A. The topological disorder extended only in one direction; it seems certain that when a system is disordered in three dimensions, dangling hybrids and bond-angle distortions must give states in the crystalline gap. The second step was the suggestion that levels are swept out of the gap, equivalent to the formation

of two-electron states suggested by Anderson (1975). The atoms near each dangling hybrid will tend to relax in order to retain the gap. In this regard, amorphous
semiconductors and ordinary glass have properties consistent with the existence
of just the same kind of gap that characterizes their crystalline counterparts.

10-F   Photothresholds and Heterojunctions

In Fig. 10-11 was shown a sharp discontinuity in the energy levels at the interface
with a vacuum. Predicting the size of such discontinuities has been somewhat of a
problem, whether they be at interfaces with the vacuum or between different
solids. The LCAO description of the electronic structure has given us energy levels
in terms of atomic term values which are measured from the vacuum level; the
term value is the negative of the energy required to remove the electron from the
atom and place it at rest an infinite distance away. We might initially hope that
the energies we obtain, such as the maximum in the valence band, would give the
energy required to remove an electron from the corresponding state to infinity; that
particular energy, taking the electron from the highest occupied state, is the *photothreshold*. However, as we noted in introducing the table of atomic term values
(Table 2-2), corrections due to image forces can easily alter the photothreshold,
reducing that value by 1.8 eV or more, depending upon how we make the estimate. In addition, the principal contribution to the overlap interaction of Chapter
7 is a kinetic energy, ultimately, an increase in electronic energy, which will act to
reduce the photothreshold. Finally, reconstructions of the surface, such as we have
been discussing, can introduce dipole layers at the surface, making significant
changes; these will ordinarily act to increase the photothreshold, since we found
that the negatively charged atoms tend to move out from the surface, and positively charged ones tend to move inward. We conclude that the absolute values of
our parameters relative to the vacuum level are not meaningful, and it should
come as no surprise that they do not correspond numerically. The energy of the
valence-band maximum can be rewritten from Eq. (6-2) in terms of the covalent
and polar energies of Eqs. (4-18) and (4-19):

$$E_v = \frac{\varepsilon_p^c + \varepsilon_p^a}{2} - \sqrt{V_2^2 + V_3^2}. \tag{10-5}$$

The value for silicon is -9.5 eV, yet the observed photothreshold is about 5 eV
(Gobeli and Allen, 1962).

Although the absolute values are not meaningful, we may hope that the relative
values, the variation from material to material, are. Table 10-1 lists the magnitudes of the valence-band maximum energy E_v, obtained from Eq. (10-5). These
will be of value in other problems also. Fig. 10-13 shows a plot of the experimental
values of the photothreshold Φ compiled by Ciraci (Harrison and Ciraci, 1974)
against the values of Table 10-1 to which they would correspond in the simplest
picture. To the extent they are systematic, one can say that they differ by an

additive constant, at least for the heteropolar semiconductors. This gives the empirical relation

$$\Phi = |E_v| - 3.8 \text{ eV}. \tag{10-6}$$

In spite of large and very real corrections, the theoretical E_v values give a reasonably good description of the relative values of the observed photothresholds.

It should be noted immediately that of the anticipated corrections, those due to surface dipole layers are different on different crystal faces—owing to different reconstruction, for example. In fact, photothresholds can vary by quantities of the order of an electron volt on different faces. (The energy required to take the electron to infinite distance through different faces cannot differ if both faces are on the same

TABLE 10-1

Valence-band edge E_v, from Eq. (10-5), direct gap E_0, and indirect gap E_g, all in eV, and bond length d.

Material	$-E_v$	E_0	E_g	$d(\text{Å})$	Material	$-E_v$	E_0	E_g	$d(\text{Å})$
C	15.91	—	5.5	1.54	BAs	11.17	—	—	2.07
BN	16.16	—	—	1.57	GaN*	13.66	—	—	1.94
BeO*	16.27	—	—	1.65	BeSe	11.19	—	—	2.20
Si	9.50	4.18	1.13	2.35	ZnO*	15.58	3.40	—	1.98
AlP	10.03	—	—	2.36	CuF	18.41	—	—	1.84
Ge	9.12	0.89	0.76	2.44	InN*	13.00	—	—	2.15
GaAs	9.53	1.52	—	2.45	BeTe	10.00	—	—	2.40
ZnSe	10.58	2.82	—	2.45	AlAs	9.57	2.77	—	2.43
CuBr	11.90	—	—	2.49	GaP	10.00	2.77	2.38	2.36
Sn	8.04	—	—	2.80	ZnS	11.40	3.80	—	2.34
InSb	8.41	0.24	—	2.81	CuCl	13.11	—	—	2.34
CdTe	9.32	1.60	—	2.81	AlSb	8.67	2.5	1.87	2.66
AgI	10.49	—	—	2.80	InP	9.64	1.37	—	2.54
SiC	12.56	7.75	2.3	1.88	MgTe*	9.33	—	—	2.76
BP	11.81	—	—	1.97	CdS	11.12	2.56	—	2.53
AlN*	13.84	—	—	1.89	GaSb	8.69	0.81	—	2.65
BeS	12.05	—	—	2.10	InAs	9.21	0.42	—	2.61
					ZnTe	9.50	2.39	—	2.64
					CdSe*	10.35	1.84	—	2.63
					CuI	10.62	—	—	2.62

SOURCES of data: E_0 and E_g from Phillips (1973a) and Lawaetz (1971), except for AlAs value, estimated by Dingle, Wiegmann, and Henry (1974).

* Wurtzite structure. The three bands are split at Γ. This gives the center of gravity.

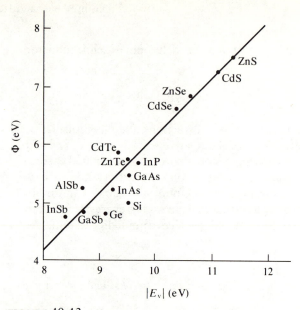

FIGURE 10-13

The experimental photothreshold Φ, plotted against
the energy of the valence-band maximum from
Eq. (10-5). The line corresponds to the empirical
relation, Eq. (10-6). The experimental values are from
Swank (1967), Gobeli and Allen (1962, 1965) and
Fisher (1965, 1966).

specimen, and local electric fields around the specimen must make up the difference. These fields, however, do not influence the experimental result.) All of the
measurements on polar materials were made on (110) surfaces, so this does not
invalidate the use of a single constant in Eq. (10-6). However, the measurements
on silicon and germanium were on (111) surfaces, and we see that they are indeed
displaced from the polar curve. A value of 4.4 eV would give a better fit than does
3.8 eV.

The principal trends among the polar semiconductors can be described as an
increase in photothreshold with increasing polarity and with decreasing metallicity. It may be more precise to say that the principal determinant of the photothreshold is the p-state energy of the nonmetallic atom, with a secondary
influence from the p-state energy of the metallic atom.

Of perhaps even more interest are the discontinuities in bands at a ***heterojunction***, the junction between two semiconductors, which is used in electronic
devices. The predicted discontinuity in the valence-band maximum (Harrison,
1977b) is obtained by simply subtracting the corresponding values from Table 10-1.
The discontinuity in the conduction bands is obtained by making a correction on
each side of the heterojunction for the band gap. In alloys, the values from Table
10-1 can be interpolated, as can the band gaps at a particular point in the Brillouin
Zone. (If there is a change from a direct gap to an indirect gap in the alloy system,

TABLE 10-2   255
Valence-band and conduction-band discontinuities for S–S′ heterojunctions, in eV.

S	S′	$E_v - E_v'$		$E_c - E_c'$	
		Theory	Experimental	Theory	Experiment
Ge	Si	0.38	(0.24 to 0.17)	0.01	−0.12 to −0.19
Ge	GaAs	0.41	(0.36 to 0.76)	−0.35	−0.40 to 0
GaAs	$Ga_{0.8}Al_{0.2}As$	0.01	0.03	−0.24	−0.22
InP	CdS	1.48	1.63	0.29	0.56

SOURCES of experimental values: Si and Ge from Milnes and Feucht (1972), except for $E_v - E_v'$, obtained by the author by using experimental band gaps; GaAs from Dingle, Wiegmann, and Henry (1974); and InP from Shay, Wagner, and Phillips (1976).

as in $Ga_xAl_{1-x}As$, one cannot interpolate the *net* gap.) Not an extensive amount of experimental information has been compiled to date, but that available is given in Table 10-2 and confirms the theory.

A remarkable feature of these predictions was communicated by William Frensley, of the University of California, Santa Barbara, who found that among semiconductor combinations with good matching of the lattice distance is one, InAs–GaSb, for which the valence-band maximum on one side (GaSb) lies above the conduction-band minimum on the other (InAs). This crossing has been experimentally confirmed subsequently by Sakaki et al. (1977), who were motivated by a similar prediction based upon electron affinities.

It is interesting that the earlier method for predicting conduction-band discontinuities by subtracting electron affinities (Anderson, 1960) is equivalent to predicting valence-band discontinuities by subtracting photothresholds. The linearity reflected in Eq. (10-6) indicates that this method should work reasonably well, though it leads to appreciable errors for a junction between homopolar and heteropolar semiconductors. The values in Table 10-1 seem a better basis for estimate in any case. A pseudopotential method was also proposed by Frensley and Kroemer (1976). An analysis of the self-consistency of that method (Harrison, 1977b), indicated that it also should be reasonably reliable as long as the lattice mismatch on the two sides of the heterojunction is not too large, and ordinarily it is not in real heterojunctions.

It seems possible that the same approach will suffice for junctions between semiconductors and insulators, since the latter will be described by the same LCAO parameters. However, the treatment of semiconductor–metal junctions is better handled with pseudopotentials, so will be discussed in Chapter 18.

PROBLEM 10-1   *Surface-state energies*

Construct an energy-level diagram for the GaAs surface, in analogy with that in Fig. 10-9 for silicon, indicating the energies of dangling hybrids before and after full reconstruction.

PROBLEM 10-2   *Heterojunction band discontinuities*

We see from Table 10-1 that GaAs and ZnSe have nearly the same d and might therefore make a good heterojunction. Both the conduction band and the valence band are seen to have discontinuities. We might replace the GaAs by $Al_x Ga_{1-x} As$, giving weighted averages of both E_0 and E_v on that side. At what concentration x can one of the discontinuities be eliminated? Which discontinuity will be eliminated?

CHAPTER 11

Mixed
Tetrahedral
Solids

SUMMARY

Covalent complexes consisting of a central atom surrounded tetrahedrally by four other atoms are expected to be most stable when the central atom is nonmetallic and the outer atoms are oxygen. SiO_2 contains such complexes, each silicon being surrounded by four oxygens, though each oxygen has only two silicon neighbors. To understand the electronic structure of SiO_2 we focus on a bonding unit consisting of an oxygen atom and an sp^3 hybrid from each of its neighboring silicons. The crystal can be constructed as a sum of these units, whereas a crystal cannot be constructed as a sum of SiO_4 complexes, which therefore cannot serve as bonding units.

Energy bands obtained for α-quartz by Chelikowsky and Schlüter (1977) are seen to be related to simpler bands for β-cristobalite that are based upon matrix elements from the Solid State Table. UPS and XPS spectra can be interpreted in terms of the electronic structure of the SiO_2 bonding unit, but the optical spectrum $\varepsilon_2(\omega)$ requires a more complete description. We also study, in terms of the electronic structure of the bonding unit, the angular rigidity of the system at the oxygen and the effective charges associated with coupling of lattice displacements to light.

Various ways of treating the vibrational spectra are indicated, and an approach based upon a simple molecular lattice (having the correct nearest-neighbor configurations but only one silicon per primitive cell) is chosen. A local mode approximation, analogous to the Bond Orbital Approximation for the electronic structure, makes a direct analysis possible and gives a reasonable account of the distribution of modes and of the infrared spectrum. The analysis can be extended directly to the polar counterparts of SiO_2, such as $AlPO_4$, which bears the same relation to SiO_2 that AlP bears to Si.

The zincblende structures we have been studying are the simplest covalent systems. They are made up of simple two-electron bonds, and every bond in the system is identical to every other bond, except near the surface, and we can treat the surfaces separately. Layered compounds, such as graphite and layered BN, which we have given attention to in many of the problems at the ends of chapters, have equally simple two-electron bonds, though they also have one electron per atom in π states. The π electrons occupy states much like those of a two-dimensional metal, and we shall consider them in Problem 15-1. Similarly, filamentary structures such as Se and Te have been treated in problems; conceptually, they require only slight extension of the tetrahedral theory. The first major departure from simple structures is encountered in analyzing mixed tetrahedral solids, such as SiO_2, in which some atoms are tetrahedrally surrounded and some are not. This is an important class of materials, inasmuch as it includes silicates (SiO_2 may be thought of as silicon silicate, $SiSiO_4$), phosphates, sulphates, and so on, in all of which the Si, P, or S is surrounded tetrahedrally by four oxygens. It is desirable to begin this chapter with a discussion of these tetrahedral complexes, which are so familiar in chemistry; however, when we turn to the solid it will be clear that the choice of a complex in which a nonmetallic atom is surrounded by four oxygen atoms does not give an appropriate bonding subunit of the crystal. We wish to see, for example, why it is that SO_4^{2-} is a stable complex while OS_4^{2-} is not, though both complexes have the same number of valence electrons (the superscript refers to a net negative charge, $-2e$). We also wish to see how we may estimate the properties of such complexes.

11-A   Tetrahedral Complexes

Let us consider a central atom A tetrahedrally surrounded by four identical atoms B; a Lewis diagram of this is shown in Fig. 11-1. There are eight electrons around each atom, as in Fig. 5-2, corresponding to a stable arrangement. If the group is neutral, as in carbon tetrachloride, it will be a stable molecule, binding only weakly with other molecular groups. Of more interest will be the case in which this configuration of electrons leads to a net negative charge. Then we can expect it to combine with ions of positive charge to form compounds. In either case, the group is made up of 32 electrons, and each of the four B atoms must have contributed 6 or 7 outer electrons in order for the unit to be neutral or nearly so. A case in which 7 are contributed is CCl_4; a case in which 6 are contributed is SO_4^{2-}.

In generating the electronic structure, we should first construct sp^3 hybrids on the central atom and divide the states on the outer atoms into eight π_p states (p states oriented perpendicular to the bond axes), four σ_p states (oriented along the

FIGURE 11-1
The Lewis diagram for a tetrahedral complex such as CCl_4 or SO_4^{2-}.

bond axes), and four σ_s states. (It could be better to regard these two sets of four states as sp hybrids pointed inward and outward, but that would not significantly modify the model.) These levels are shown in Fig. 11-2. Sixteen pairs of electrons occupy these levels, so all will be filled except the four highest-energy antibonding states. Thus, we should focus on the bonding of the highest two sets of four levels, to define a covalent energy (W_2) and a polar energy (W_3) in terms of them, as indicated in Fig. 11-2. (These have their analogy in the V_2 and V_3 used for the simple tetrahedral solids.) We allow the parameter W_3 to be of either sign. These parameters may be estimated from the Solid State Table, as we shall do shortly for SiO_2.

Next we consider the stability of the complex when each outer atom B contains six outer electrons, as for oxygen and sulphur. We may then ask what change in energy results from separating the complex into four neutral B's and one A (usually charged), with a full electron configuration, as, for example, in $SO_4^{2-} \rightarrow S^{2-} + 4O$. In the separation, the energy of each of the eight electrons involved in the bonding is raised first to the average energy $(\varepsilon_p + \varepsilon_h)/2$, costing the bonding energy $(W_2^2 + W_3^2)^{1/2}$, and then raised to the hybrid energy on atom A, costing an additional energy W_3 per electron. (Since W_3 can have either sign, it could be negative, in which case W_2 and W_3 partly cancel.) For solids, the change in energy resulting from separation would be called cohesive energy. (See Chapter 8; notice that now we have oversimplified the total energy change by ignoring the overlap interaction. This will be adequate for understanding trends; recall from Chapter 8 that even if we incorporate the overlap interaction, only semiquantitative predictions of binding can be made.) Instead of calling the energy change

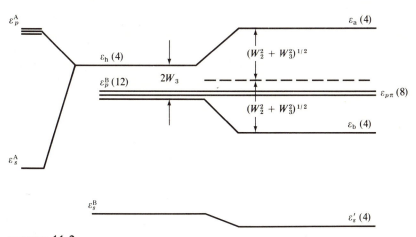

FIGURE 11-2

Energy-level diagram for a tetrahedral complex. At the left are the atomic levels; the energies of the central atom are labelled with a superscript A. The numbers in parentheses indicate the number of orbitals for each level. There are enough electrons in the complex (thirty-two) to fill sixteen orbitals.

cohesive energy for tetrahedral complexes, it is better to call it **heat of atomization.**
We find it to be

$$E_{\text{atom}} = 8[(W_2^2 + W_3^2)^{1/2} + W_3].\tag{11-1}$$

A plot of this energy against W_3 is given in Fig. 11-3, in which W_2 is held constant,
as is appropriate for an isoelectronic series such as SiO_4^{4-}, PO_4^{3-}, SO_4^{2-}, ClO_4^-,
and ArO_4. In assessing stability in this isoelectronic series, the heat of atomization
should be compared with the heat of atomization of some alternative
configuration of atoms, such as $S^{2-} + 2O_2$, which would correspond to $4V_2'$,
based upon a different covalent energy. Such an energy is illustrated by the dashed
line in Fig. 11-3. This is only of qualitative significance since there are other terms
in the total energy that change when a compound is separated into small molecules.

Fig. 11-3 is extremely informative. It tells us that a tetrahedral complex can be
stable only if the outer atoms B are sufficiently electronegative in comparison to
the central A atom. Indeed, because p states, rather than hybrids, are used in the
calculation for the outer atoms, the difference between p state energies, decreased
by a metallic energy, must be sufficiently large. It follows that systems in which the
outer atoms are oxygen, which is the most electronegative element having six outer
electrons, are expected to be the most stable. Furthermore, in the isoelectronic
series in which the outer atoms are oxygen, SiO_4^{4-}, PO_4^{3-}, SO_4^{2-}, ClO_4^-, ArO_4, we
expect a decreasing stability as we move to the right in the series, and indeed the
compound ArO_4 is not stable. On the other hand, replacing the Ar by the heavier
and less electronegative Xe does give a weakly stable molecule. Likewise, the
thiosulphate ion $(S_2O_3^{2-})$, a sulphate ion (SO_4^{2-}) having one outer oxygen
replaced by a sulphur, is also stable, though less so than the sulphate ion. We also
see in Fig. 11-3 that a tetrahedral complex OS_4^{2-} would spontaneously break up
into more stable subgroups since it corresponds to a large negative W_3.

Although the stability against breaking up increases as the central atom is taken
from farther to the left in the periodic table, the net charge on the complex also
increases; hence, such complexes tend to break into individual ions and molecules
(for example, CO_4^{4-} will break into $CO_2 + 2O^{2-}$). Stability of such complexes
comes only with neutralization of the charge, such as by suitably orienting water

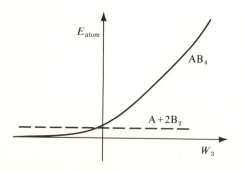

E_{atom}

AB_4

$A + 2B_2$

W_3

FIGURE 11-3

The heat of atomization of the
tetrahedral complex AB_4 as a function of
the polar energy W_3. Also shown is the
heat of atomization of the system made
up of two B_2 molecules and an isolated
A. The tetrahedral complex is stable only
if its heat of atomization is higher than
that of $A + 2B_2$.

molecules in aqueous solution, or by attaching protons (as in H_2SO_4) or positive ions (as in $CaSO_4$). Understanding stability in terms of neutralization of the charge is quite familiar, but understanding how stability is influenced by W_3, as plotted in Fig. 11-3, is not so familiar.

This discussion has been qualitative, and it would be interesting to make it quantitative by introducing W_2 and W_3 from the Solid State Table. However, this has only been done to a very limited extent. The number of possible systems with tetrahedral complexes is immense, and it is not practical to describe all of them completely in this book. Instead, we shall consider only tetrahedra in which the outer atoms are oxygen; in the next section, we shall see how SiO_2 can be described. Then, we shall see how the phosphates, sulphates, arsenates, and so on, can be understood as isoelectronic modifications of SiO_2 and GeO_2, and so on, just as the phosphides, sulphides, arsenides, and so on, were understood as isoelectronic modifications of silicon and germanium. It is interesting that a direct evaluation of W_2 and W_3 for SiO_2, from the Solid State Table, gives 10.95 eV and 2.93 eV, respectively, in rough accord with the values of 10.75 eV and 4.35 eV that S.T. Pantelides has obtained from the spectra. We shall return to that later.

11-B   The Crystal Structure and the Simple Molecular Lattice

Let us consider first SiO_2, which we can generalize to all other mixed tetrahedral solids in much the same way that we could generalize silicon to the other simple tetrahedral solids. In SiO_2, each silicon is tetrahedrally surrounded by four oxygen atoms, forming a regular tetrahedron, as in the complexes discussed in Section 10-A. However, each of these oxygens is bonded to only one other silicon; it has twofold coordination. Thus, each silicon is coupled to four other silicons, through its neighboring oxygens, as in the simple tetrahedral solids, so that topologically it is also a tetrahedral network, except that an oxygen is associated with each silicon–silicon " bond." In all forms of SiO_2, the Si—O—Si system is bent at the oxygen, typically with an angle less than the 180° straight configuration by 36°. (There was long thought to be a form, β-cristobalite, in which the silicons were arranged as in diamond and the oxygens were colinear with the silicon–silicon separation, but that form apparently does not occur. Nevertheless, because of its simplicity, it is frequently used as a theoretical model, and we will find it useful here.) There are many different crystalline structures for SiO_2, with α-quartz being the most familiar; it contains three silicon atoms (and six oxygen atoms) per primitive cell. SiO_2 also forms a glass, called vitreous silica, which is a tetrahedral network analogous to amorphous silicon. Presumably, the ease with which an amorphous form in SiO_2 is produced arises from the bend in the Si—O—Si unit, which makes it possible for a random network to form while retaining angles close to the tetrahedral values at the silicon; the angles are retained by various rotations of the Si—O—Si units. The formation of glasses is still further facilitated by the addition of sodium (for example, Na_2O or Na_2CO_3), as Phoenician merchants learned when they discovered that blocks of natron used to support their cooking pots on Mediterranean beaches fused with the sand in the heat of the fire, and formed glass. According to the Roman author Gaius Plinius Secundus, whose *Natural History* was written 2000 years ago, this was how man first learned to make glass. Sodium ions can heal any remaining dangling oxygen atoms or remaining silicon hybrids, reducing the working temperature of the glass. Ordinary commercial glass contains sodium or calcium for this purpose (this is called soda-lime glass).

The properties we shall discuss are relatively unaffected by differences between the crystal structures, even if the structure is an amorphous glass. The essential feature is the twofold-fourfold coordination; bond angles and details of the connectivity are not so important. Thus, it would make sense to pick the simplest structure for calculations; however, even the simplest structure is complicated, and so, instead of selecting a real crystal structure, it will be simpler to generate a conceptual model that is topologically possible and mathematically tractible, though it is not a geometric possibility. We construct a *simple molecular lattice* that has one formula unit (SiO_2) per primitive cell, but where each atom has the same coordination as in the real crystal. This is illustrated in Fig. 11-4.

We must exercise some caution when using ficticious lattices. A simple molecular lattice for SiO_2 inevitably has the topology of a two-dimensional system, as is apparent in Fig. 11-4, and calculation of some properties can give results that are qualitatively incorrect. To construct a three-dimensional molecular lattice for this coordination, we require at least Si_2O_4, which would then have the topology of

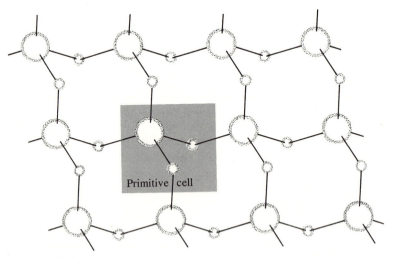

FIGURE 11-4

The simple molecular lattice for SiO_2. The large circles represent silicon atoms; the small, oxygen atoms. There is one SiO_2 per primitive cell. The coordination of each atom is the same as in real SiO_2, but the connectivity and symmetry have been simplified.

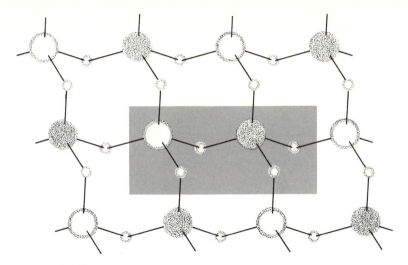

FIGURE 11-5

The simple molecular lattice for $AlPO_4$. To make this, beginning with the simple molecular lattice of SiO_2, from Fig. 11-4, we remove a proton from alternate silicon nuclei (open circles), making them aluminum, and place the proton in the other silicon nuclei (shaded circles), making them phosphorus. The primitive cell has become twice as large but the structure is the same as that of SiO_2 and isoelectronic with it.

the diamond structure. However, for our needs, the simple molecular lattice will suffice.

It will help here to mention the structure of polar counterparts to SiO_2. These are constructed exactly as aluminum phosphide is constructed from silicon in the simple tetrahedral solid. The process is illustrated for the simple molecular lattice in Fig. 11-5, which shows how the structure for SiO_2 can be transformed to the structure for aluminum phosphate. Transferring an additional proton leads to magnesium sulphate. Indeed, the counterpart of each AB tetrahedral semiconductor, ABO_4, is possible in principle. The structures may be obtained from Wyckhoff (1963), and if the structure has twofold-fourfold coordination, it can be analyzed by the methods outlined here.

11-C The Bonding Unit

Let us now construct a representation of the electronic structure of SiO_2. There have been many early studies of SiO_2, principally aimed at interpreting various experimental spectra; these have been reviewed recently by Ruffa (1968, 1970). More recently, studies based upon calculations for large clusters of atoms have been made by Reilly (1970), Bennett and Roth (1971a,b), Gilbert et al. (1973), and Yip and Fowler (1974). Most recently, a full, self-consistent pseudopotential calculation on quartz was made by Chelikowsky and Schlüter (1977). Here, we shall

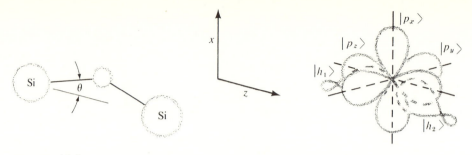

FIGURE 11-6

A bonding unit in SiO_2, showing the coordinate system and the orbitals that are included.

follow an analysis based upon simple bonding units, made by Pantelides and Harrison (1976), and make use of the results of the full calculation by Chelikowsky and Schlüter.

Given the simple molecular lattice or any realistic structure for SiO_2, we might write linear combinations of the four orbitals on each atom. However, to adhere to the simplicity achieved in our discussion of simple tetrahedral solids, we should perform a unitary transformation to obtain bonding orbitals, in terms of which the occupied states can be approximately described. At this stage, an intuitive understanding of the bonds must be incorporated. Some transformations are not allowed (for example, if they lead to nonorthogonal orbitals or to the wrong number of orbitals), but even among the allowed ones are good and bad choices. Following Pantelides and Harrison (1976), let us choose a **bonding unit** consisting of two silicon hybrids and an intervening oxygen. (See Fig. 11-6.) We can carry out a variational calculation based upon the orbitals in this bonding unit and select orbitals from the resulting set to represent the valence bands; thus, the bonding unit in mixed tetrahedral solids and the bond in simple tetrahedral solids are counterparts. Before proceeding further with the silicon dioxide bonding unit, some general comments about the procedure are needed.

It is imperative that the bonding unit be stoichiometric. This means that a direct addition of the bonding units must give a representation of the electronic structure of the entire system. In the bonding unit we have taken, addition of one quarter of each of the two silicons (two hybrids) to one oxygen makes $Si_{1/2}O$, which is equivalent to the formula SiO_2. We shall later wish to make a transformation from the orbitals of our bonding unit to band states, and this cannot change the number of states. This may seem obvious, but it is not uncommon to see studies of electronic structure where, for example, an SiO_4 cluster has been regarded as a bonding unit; the electron states for such a cluster have then been constructed and loosely put into correspondence with the states of the solid. Such a treatment is unambiguously wrong; any attempt at a careful formulation on this basis will fail. It is possible, in principle, to make a formulation with one silicon and half of the orbitals on the surrounding four oxygens, but this approach will not be as successful, even though it will not be wrong in the sense that using a full SiO_4 cluster for a bonding unit is. A similar point will be made concerning vibrational modes, in Section 11-D.

States in the Bonding Unit

For the SiO_2 bonding unit shown in Fig. 11-6, the orbitals that are to be included are shown to the right. The oxygen s state is so low in energy in comparison to the others that we simply consider it doubly occupied and regard it as a core state, so it is not shown. It is possible to correct for this approximation in the end and such an estimate, described at the end of Section 11-E, suggests that significant angular forces arise from the s state. We will also see an appreciable s-band width in the accurate bands in Fig. 11-9,a. In a more complete calculation the oxygen s states should be included. There remain the two hybrids and the three oxygen p states. For the purpose of obtaining the simplest description of the bonding, we discard all interatomic matrix elements except those between each silicon hybrid and the p states on the oxygen toward which the hybrid is directed. (For a detailed description of the energy bands it is necessary to introduce additional matrix elements, just as we added matrix elements for the simple tetrahedral solids in Chapter 6; Pantelides made a detailed description of SiO_2 in this manner—see Pantelides and Harrison, 1976.) Indeed, an oxygen p state can be decomposed into two terms, one oriented along the $Si-O$ axis and one perpendicular to it. The matrix element between the hybrid and the perpendicular p state vanishes by symmetry, and taking the matrix element between the hybrid and a p state oriented parallel to it as $-W_2 = -\frac{1}{2}(V_{sp\sigma} + \sqrt{3}V_{pp\sigma})$, we obtain $\langle h_1|H|p_x\rangle = -W_2 \sin\theta$ and $\langle h_1|H|p_z\rangle = -W_2 \cos\theta$ and $\langle h_1|H|p_y\rangle = 0$. (The signs depend upon suitable choice of the signs of the orbitals.) We also take the energies of all of the oxygen p states to be the same, and to be lower by $2W_3$ than the value $\langle h_1|H|h_1\rangle$ for the hybrids. From the Solid State Table we obtain $W_2 = 10.95$ eV, and $W_3 = 2.93$ eV for SiO_2. This gives values for all of the matrix elements of the five-by-five Hamiltonian matrix obtained by seeking the minimum-energy orbitals for the bonding unit. This corresponds to the matrix based upon the coefficients in Eq. (3-11) for the simple tetrahedral case. It is a submatrix of the total Hamiltonian matrix shown in Fig. 3-4 for the simple tetrahedral case.

Diagonalization of a five-by-five matrix can require numerical computation, but in this case it is simplified by the symmetry of the system. Because of the symmetry of the Hamiltonian when undergoing reflection in the y- and z-planes (planes given by $y = 0$ and $z = 0$, respectively), the eigenstates of the bonding unit can be taken as either even or odd under reflection in these planes. The only orbital that is odd under reflection in the y-plane is $|p_y\rangle$, so it is an eigenstate of the Hamiltonian matrix; it is "nonbonding" and has energy $-W_3$. The fact that this is an eigenstate of energy $-W_3$ is also obvious from Fig. 11-7. The remaining states are even under y-reflection but can be even or odd under z-reflection. There are bonding and antibonding combinations of each. The bonding combinations are shown in Fig. 11-8. Their energies may be obtained very simply. For example, the state B_z is written

$$|B_z\rangle = u_0|p_z\rangle + u_h(|h_1\rangle - |h_2\rangle)/\sqrt{2}. \qquad (11-2)$$

$$
\begin{array}{ccccc}
h_1 & h_2 & p_x & p_y & p_z
\end{array}
$$

$$
H = \begin{pmatrix}
W_3 & 0 & -W_2 \sin \theta & 0 & -W_2 \cos \theta \\
0 & W_3 & -W_2 \sin \theta & 0 & +W_2 \cos \theta \\
-W_2 \sin \theta & -W_2 \sin \theta & -W_3 & 0 & 0 \\
0 & 0 & 0 & -W_3 & 0 \\
-W_2 \cos \theta & +W_2 \cos \theta & 0 & 0 & -W_3
\end{pmatrix}
\begin{matrix}
h_1 \\ h_2 \\ p_x \\ p_y \\ p_z
\end{matrix}
$$

FIGURE 11-7
The Hamiltonian matrix for the SiO_2 bonding unit.

This reduces the problem to two equations in two unknowns, as in Eq. (3-11). The eigenvalue for B_z, and the eigenvalue from the corresponding treatment of B_x, are

$$
\begin{aligned}
\varepsilon_{B_z} &= -(2W_2^2 \cos^2 \theta + W_3^2)^{1/2}; \\
\varepsilon_{B_x} &= -(2W_2^2 \sin^2 \theta + W_3^2)^{1/2}.
\end{aligned}
\tag{11-3}
$$

By analogy with the treatment of the bond orbitals in Chapter 3, we can also write down the corresponding bond orbitals:

$$
\begin{aligned}
|B_z\rangle &= \left(\frac{1 + \beta_{pz}}{2}\right)^{1/2} |p_z\rangle + \left(\frac{1 - \beta_{pz}}{4}\right)^{1/2} (|h_1\rangle - |h_2\rangle); \\
|B_x\rangle &= \left(\frac{1 + \beta_{px}}{2}\right)^{1/2} |p_x\rangle + \left(\frac{1 - \beta_{px}}{4}\right)^{1/2} (|h_1\rangle + |h_2\rangle),
\end{aligned}
\tag{11-4}
$$

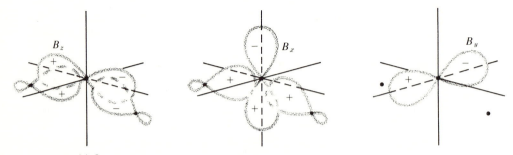

FIGURE 11-8
Bond orbitals for the SiO_2 bonding unit. All are doubly occupied. The antibonding combinations A_z and A_x, obtained by reversing the sign of the p states in B_z and B_x, are empty.

with partial polarities given by

$$\begin{aligned}\beta_{pz} &= W_3/(2W_2^2 \cos^2 \theta + W_3^2)^{1/2}; \\ \beta_{px} &= W_3/(2W_2^2 \sin^2 \theta + W_3^2)^{1/2}.\end{aligned} \Bigg\} \qquad (11\text{-}5)$$

We now immediately construct an effective charge Z^* for the oxygen by adding twice the squared amplitudes for the occupied orbitals, $6 + \beta_{pz} + \beta_{px}$, and subtracting the core charge of six to obtain

$$Z^* = \beta_{pz} + \beta_{px}, \qquad (11\text{-}6)$$

which comes out to be 0.72 for SiO_2.

As in the simple tetrahedral solids, the bond orbitals are not eigenstates because matrix elements exist between them and bond orbitals on adjacent sites. We can in fact use the values for the matrix elements between two hybrids on the same silicon atom, $-V_1$—see Eq. (3-5)—to obtain the matrix elements between adjacent bond orbitals in SiO_2:

$$\begin{aligned}\langle B_z | H | B_z' \rangle &= \pm \frac{1 - \beta_{pz}}{4} V_1; \\[2mm] \langle B_z | H | B_x' \rangle &= \pm \frac{(1 - \beta_{pz})^{1/2}(1 - \beta_{px})^{1/2}}{4} V_1; \\[2mm] \langle B_x | H | B_x' \rangle &= -\frac{1 - \beta_{px}}{4} V_1.\end{aligned} \Bigg\} \qquad (11\text{-}7)$$

These matrix elements are needed to obtain the energy bands. For silicon, $V_1 = 1.41$ eV.

Extension of the calculation to the polar counterparts of SiO_2, such as aluminum phosphate, is also quite direct and can be made without the introduction of any new parameters. The energies of the two hybrids in the bonding unit now differ, $\langle h_2 | H | h_2 \rangle - \langle h_1 | H | h_1 \rangle = 2V_3$, but the hybrid energies are just those given in Table 2-2. We must generalize the bond orbital by allowing different coefficients for $|h_1\rangle$ and $|h_2\rangle$ in Eq. (11-2), and must solve a cubic rather than a quadratic equation.

11-D   Bands and Electronic Spectra

Bands for SiO_2

It is a somewhat formidable undertaking to make an energy-band calculation for a crystal such as α-quartz with nine atoms per primitive cell; a full calculation has only recently been completed by Chelikowsky and Schlüter (1977), who used a self-consistent pseudopotential method. The results of this calculation are shown

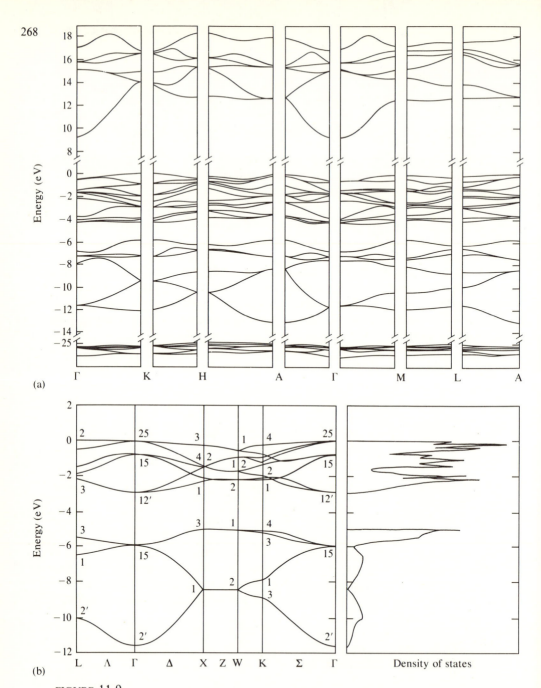

(a)

(b)

FIGURE 11-9

In (a) are shown the energy bands of α-quartz obtained by Chelikowsky and Schlüter (1977) from a self-consistent pseudopotential calculation. In part (b) are shown the bands of the simpler β-cristobalite structure (Pantelides and Harrison, 1976) from an LCAO calculation and a Bond Orbital Approximation; that calculation was restricted to the valence bands. The corresponding density of states is shown to the right. Because the symmetry of the structures is different, different symmetry points are indicated below each figure. [Part (a) after Chelikowsky and Schlüter, 1977; part (b) after Pantiledes and Harrison, 1976.]

in Fig. 11-9,a. Chelikowsky and Schlüter took the zero of energy at the valence-band maximum (which is not at Γ in their calculation, nor is this surprising for p-like bands in a complex structure). The set of bands near -25 eV are associated with the oxygen s states; the bands extending from -6 eV to -12 eV may be associated with B_z and those from zero to -4 eV with the B_x and B_y states. The conduction bands arising from the antibonding state, A_x, begin at approximately $+9$ eV. This interpretation of the bands is confirmed by plots of electron density from the various bands (Chelikowsky and Schlüter, 1977). There are just enough bands at negative energy to accommodate the 36 electrons per cell from oxygen atoms and 12 electrons from silicon atoms.

The oxygen s bands are quite narrow, as expected, suggesting that one could treat them as core states, though they are broad enough to give some worry about whether they *should* be so treated. It is striking that the B_x and B_y bands are also rather narrow and well mixed, as they would be if there were no bending of the bonds at the oxygen (that is, were $\theta = 0$). Then the only matrix elements coupling these oxygen p states and their neighbors would be π-bonding matrix elements through the silicons, or matrix elements between second neighbors. Thus the density of states is quite similar to that which would be obtained by using the β-cristobalite structure discussed earlier, which was used by Pantelides and Harrison (1976) in their discussion of spectra. (It is topologically the same as the silicon structure, but an oxygen is inserted in the center of each bond.) The valence bands obtained by Pantelides and Harrison for the β-cristobalite structure (by using the Bond Orbital Approximation) are shown in Fig. 11-9,b; they are in fair quantitative accord with the more complete calculation. They have one advantage in that the Brillouin Zone is the same as for the simple tetrahedral solids, so the computer program for the density of states (see Section 2-E) could be directly applied to these bands, giving the density of states shown to the right of the bands. Chelikowsky and Schlüter used, instead, a limited grid, sampled, and smoothed the results.

If $\theta = 0$, as in β-cristobalite, the orbitals in the bonding unit become particularly simple and are illustrated in Fig. 11-10. The B_x orbital can be thought of as

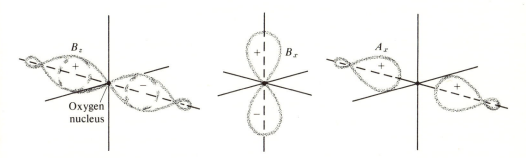

FIGURE 11-10

Orbitals of the bonding unit when it is taken to be straight ($\theta = 0$), as in β-cristobalite. This approximation simplifies the understanding of the SiO_2 spectra. The dark dot at the center of each combination of orbitals is the oxygen nucleus.

a pure oxygen p state, as can the B_y orbital oriented perpendicular to it. It may in fact be useful to think in terms of the states $(B_x \pm iB_y)/2^{1/2}$, which have cylindrical symmetry around the bonding unit and can be thought of as electrons orbiting in opposite directions. Correspondingly, the antibonding state, A_x, can be thought of as an even sum of silicon hybrid states with no contribution from the oxygen. Viewed from the oxygen, the symmetry is that of the $3s$ oxygen state, the s state of the next highest shell. Chelikowsky and Schlüter in fact chose to describe the conduction-band states in that way.

The B_z orbital can be thought of as a bonding combination of the $p\sigma$ state from the oxygen and silicon hybrids. In SiO_2, the wave function of this orbital enters the calculation of the bands as did the silicon bond orbital in tetrahedral silicon; the wave function is odd around the bond center, which changes the symmetry of the wave function but does not affect the resulting bands. Thus B_z bands in β-cristobalite have the same topology as the silicon bands, as can be seen in Fig. 11-9,b, and in fact Pantelides and Harrison (1976) used the silicon V_1 matrix elements to compute the coupling between adjacent bond orbitals in calculating the band structure for β-cristobalite.

Electronic Spectra

Chelikowsky and Schlüter (1977) calculated a crude, smooth density of states from the α-quartz bands of Fig. 11-9,a, and their results are shown in Fig. 11-11. The peaks can be identified by comparison with corresponding peaks in β-cristobalite, from Fig. 11-9,b, and with the bond orbitals. The three peaks corresponding to highest energy, A, B, and C, are from the B_x and B_y bands. Peaks D and E are from the upper and lower parts of the B_z bands, and F is from the oxygen s bands. The area under the F peak should be the same as the total of the D and E peaks, and the area under peaks A, B, and C should equal that under peaks D, E, and F, though they appear not to be equal in the figure. Three plots of spectra obtained by experiment, which, at least in simple terms, measure the density of states, are given for comparison and appear to be in essential agreement. (See Chelikowsky and Schlüter, 1977, and Pantelides and Harrison, 1976, for discussions of additional spectra.)

Of particular interest are the optical spectra. Chelikowsky and Schlüter calculated the joint density of states for direct transitions (which would be proportional to ε_2 were the dipole matrix elements all equal)—see Section 4-A—with the result shown at the bottom of Fig. 11-12. It bears little resemblance to the experimental ε_2 curve (uppermost in the figure), for a number of reasons. The prominent peak at 10.4 eV appears to be an exciton peak (See Section 6-E), as had been suggested earlier by Platzoder (1968) on the basis of observed temperature dependence. Pantelides and Harrison took this peak to result from interband transitions, since it lay at an energy above the photoconductivity threshold of 9 eV (DiStephano and Eastman, 1971b); that would rule out the possibility that the peak represents a simple exciton, but not that it represents an excitonlike

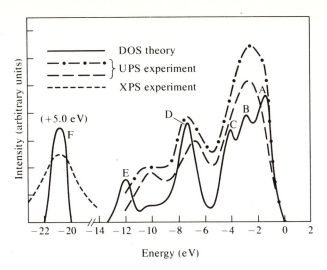

FIGURE 11-11

The density of states (labelled "DOS theory") for the
α-quartz energy bands calculated by Chelikowsky and
Schlüter (1977), compared with ultraviolet
photospectroscopy ("UPS") data obtained experimentally
by Ibach and Rowe (1974c) (broken line) and DiStefano
and Eastman (1971a) (dashed-dotted line), and X-ray
photospectroscopy ("XPS") data obtained experimentally
by DiStefano and Eastman (1971a). A shift of 5 eV was
made in the oxygen *s* peak, in comparison to the bands
of Fig. 11-9,a. [After Chelikowsky and Schlüter, 1977.]

resonance. Such a resonance should not appear in a plot of the *one-electron*
density of states and does not appear in the lowermost plot of the figure.

The calculated joint density of states in fact rises smoothly through the energy
of the next two peaks in the experimental ε_2 plot. Only when Chelikowsky and
Schlüter incorporated calculated dipole matrix elements (the middle curve,
labelled $\varepsilon_2(\omega)$ theory in Fig. 11-12), did those two peaks show up. Variations of
the matrix elements with energy presumably arise from interference between the
dipole matrix elements for *different* bonding units in the primitive cell, in which
case the structure in ε_2 should not be interpreted in terms of a *single* bonding unit.

These considerations argue against the interpretation of the optical spectra in
terms of the bonding unit used by Pantelides and Harrison (1976) to obtain
parameters W_2 and W_3. Nonetheless, we see by comparison between the bands of
Pantiledes and Harrison and those of Chelikowsky and Schlüter (Fig. 11-9) that
the matrix elements of Pantelides and Harrison were approximately correct. This
was really not accidental since they considered many properties simultaneously
and only used optical properties to fix exact final values. In fact, the use of larger

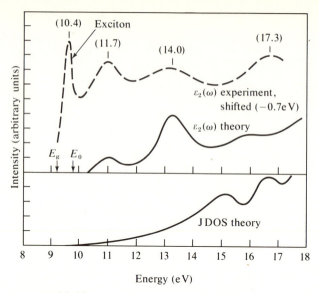

FIGURE 11-12

The lower panel shows the calculated joint density of states for α-quartz. It is quite different from the experimental $\varepsilon_2(\omega)$ shown in the upper panel (Klein and Chun, 1972). Inclusion of dipole matrix elements gives the theoretical $\varepsilon_2(\omega)$ plot shown in the upper panel; it duplicates some of the features of the spectrum found by Klein and Chun, but not the exciton peak. Theoretical indirect and direct gaps are indicated by arrows. [After Chelikowsky and Schlüter, 1977.]

W_3 gave generally better results than the value from the Solid State Table: Pantelides and Harrison used values of 10.75 and 4.35 eV for W_2 and W_3, respectively. A recalculation of these without overlap S gives 8.95 and 4.35, respectively. The present treatment will make use of values 10.95 and 2.93, respectively, obtained by using the expressions given earlier and the Solid State Table. It would be preferable to obtain matrix elements by fitting an LCAO calculation of the bands to those given in Fig. 11-9,a, but that has not yet been done. The values obtained from the Solid State Table for SiO_2 and GeO_2, by use of

$$\left.\begin{array}{l} W_2 = \tfrac{1}{2}(V_{sp\sigma} + \sqrt{3}\,V_{pp\sigma}) = 3.73\,\dfrac{\hbar^2}{md^2}, \\[4mm] W_3 = \tfrac{1}{2}\left(\dfrac{3\varepsilon_p^c + \varepsilon_s^c}{4} - \varepsilon_p^a\right), \end{array}\right\} \tag{11-8}$$

are given in Table 11-1. These will be used in the calculation of other properties of SiO_2.

Before leaving the spectra, we should note that though the general band positions and widths are not so sensitive to details of crystal structure, nor even to the

TABLE 11-1

Values of parameters for the electronic structures
of SiO_2 and GeO_2, obtained from the Solid
State Table.

Parameter	Value for SiO_2	Value for GeO_2
d	1.61 Å	1.74 Å
θ	18°	25°
W_2	10.95 eV	9.38 eV
W_3	2.93 eV	2.88 eV
β_{px}	0.52	0.46
β_{pz}	0.20	0.23
Z^* (Oxygen)	0.72	0.69

bending of the bonds at the oxygen, one feature of the electronic structure *is*
influenced by details of the geometry of the structure: that is the position of peaks
and valleys within individual bands. These can be understood qualitatively in
terms of quantization conditions around closed rings of atoms, since indeed the
phase of every band state must change by an integral multiple of 2π around every
ring in the crystal. A structure consisting of rings each made up of four silicons, as
in Fig. 11-4, will give a much different set of peaks than a structure made up of
predominantly five-member or six-member rings (see Fig. 11-13). This point was
emphasized first, apparently, by Weaire and Thorpe (1973, p. 295). See also
Friedel (1978). Randomness in amorphous structures tends to wash out such
peaks and valleys by providing rings of all sizes.

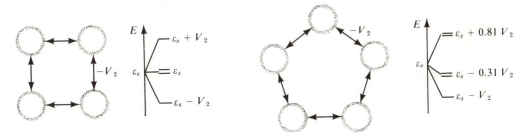

FIGURE 11-13
The density of states is affected by the evenness and oddness of the number of atoms
in constituent rings, as can be illustrated by calculating the states for small isolated
rings made up of s states, each state being coupled to its neighbor by $-V_2$; the energies
of the states for rings such as these were calculated in Problem 2-1. The solution for a
ring of four atoms is shown at the left in the figure; at the right, for a ring of five.
Doubly drawn energy levels are degenerate. Notice that none of the energies for the
odd-membered ring are at ε_s; changing the sign of V_2 inverts the entire energy-level
diagram. Such differences are reflected in the final density of states when the rings are
combined to form a lattice.

Dielectric Susceptibility

We can directly calculate the dielectric susceptibility by following the procedure that led to Eq. (4-28) for the dielectric susceptibility of the tetrahedral solids; that is, we calculate the polarizability of the bonding unit. We do *not* introduce a scale factor γ; therefore, we obtain a direct prediction of the corresponding contribution to the susceptibility. We may expect it to be rather good since scaling is necessary principally in the tetrahedral solids for high-metallicity systems with narrow gaps between bands.

The calculation for the special case of $\theta = 45°$ is exactly equivalent to what it is in tetrahedral solids, since B_x and B_z become degenerate; we may use even and odd combinations of those wave functions, which corresponds to treating the bond between a silicon hybrid and a σ-oriented p state of oxygen as a two-electron bond. The result is the same as Eq. (4-28), with the factor γ^2 omitted, V_2 and V_3 replaced by W_2 and W_3, α_p replaced by $\beta_x = \beta_z$, and with the electron density N replaced by four times the density N_O of oxygen atoms for the four electrons per bonding unit.

The calculation for θ not equal to 45° is slightly more cumbersome. For fields in the z direction, we must include coupling between orbitals B_z and A_x, and between orbitals B_x and A_z. For fields in the x direction, we must couple B_z to A_z and B_x to A_x. Fields in the y direction cause no polarization in this approximation. The resulting dielectric susceptibility averaged over field direction is

$$\chi = \frac{N_O e^2 d^2}{3W_3} \left\{ \frac{2\beta_z \beta_x (1 - \beta_z \beta_x)\cos^2 \theta}{\beta_z + \beta_x} + [\beta_z(1 - \beta_z^2) + \beta_x(1 - \beta_x^2)]\frac{\sin^2 \theta}{2} \right\}.$$

$$(11-9)$$

This formula for χ is not very sensitive to changing θ. Substituting values from Table 11-1, and using the oxygen-atom density of $N_O = 0.053$ per Å³, gives $\chi = 0.058$. The experimental dielectric constant of 2.40, for quartz, corresponds to a considerably higher dielectric susceptibility of 0.111. This immediately suggests that other contributions to the susceptibility must be important. An upward scaling of σ matrix elements, as by the factor γ used in the simple tetrahedral solids, is probably not appropriate. Nor does it now seem that the intra-atomic coupling of oxygen states discussed by Pantiledes and Harrison (1976) is responsible: it is likely that the additional contributions to susceptibility come principally from the coupling of π states on the oxygen with π-oriented p states on the neighboring silicons. The matrix elements necessary to calculate this contribution are available, but the calculation is not elementary, since the orbitals cannot be separated into bonding units the way they can be in simple covalent solids, and the corresponding values for W_3 (2.38 eV) and W_2 (2.90 eV) are comparable to each other; consequently, the treatment we use in discussing ionic solids in Chapter 14 is inaccurate. The π-state contributions have not been pursued further.

Another general result of the analysis is interesting. If we compare different allotropic forms of SiO_2, we expect the matrix elements and bond lengths to remain essentially the same and the principal variation in susceptibility to be from the proportionality to the electron density. The refractive index is equal to the square root of the dielectric constant, which is given by $1 + 4\pi\chi$, so Pantelides and Harrison (1976) suggested that the refractive index should vary with density ρ among the allotropic forms as $n = (1 + C\rho)^{1/2}$. This is much more plausible than the dependence usually assumed, $n = A + B\rho$, and it depends only upon a single parameter; it can be fitted for quartz and used to predict n directly for the other forms, whereas two parameters must be fitted for the usual form. The two forms work equally well.

11-E   Mechanical Properties

We now turn to the dependence of SiO_2 band energy upon the positions of the atoms, and to questions related to effective charges. An appropriate first step would be the calculation of the overlap interaction between oxygen and silicon atoms, as described in Chapter 7, but this has not yet been done; instead, we must rely on the observed spacing and measured angular rigidity. We wish also to know if the overlap interaction between the silicons is important. Their separation is over 3 Å, greater than that in tetrahedral silicon by 30 percent; consideration of Fig. 7-1 would suggest that the interactions between neutral silicons at 3 Å are quite unimportant. We need to return to the topic of Coulomb interactions.

Within the SiO_2 bonding unit, only the orbitals B_x and B_z change when the bonding unit is distorted, and the sum over the energies of the four electrons occupying them gives

$$E_{bond} = -2(2W_2^2 \cos^2 \theta + W_3^2)^{1/2} - 2(2W_2^2 \sin^2 \theta + W_3^2)^{1/2}, \quad (11\text{-}10)$$

which has a minimum value at $\theta = 45°$, corresponding to a 90° angle at the oxygen. At this minimum, the two bond energies are equal, and we could take even and odd combinations of wave functions, corresponding to each silicon hybrid bonded to a single p state on the oxygen. Formation of individual two-atom bonds is thus favored energetically; and this is the origin of the bending of the bonding unit. We could in fact use the same mechanism to explain the tendency for the hydrogens to go to 90° angles in the central hydrides (H_2O, NH_3, H_2S, and so on), though as we indicated in Section 1-C, it is better to treat hydrogen differently than other atoms are treated.

Coulomb Forces

Why is the angle at the oxygen 144° rather than 90°? The answer that immediately comes to mind is that the Coulomb repulsion between the two silicon atoms (each of charge $2Z^*$) may open up the angle (this mechanism can be used to account for the larger-than-90° angle in water). However, the question is very sticky, because we have already implicitly included *some* interaction between these ions as a part of the electron energy. Furthermore, the potential due to one charged silicon will affect the electron energy on the other, and therefore, we can expect a change in angle to change W_3, also. The dilemma is illustrated by the fact that Z^*, as given in Eqs. (11-6) and (11-5), depends upon angle, and we will deduce a different force if we use $(2Z^*e)^2/(2d \cos \theta)^2$ than if we differentiate $(2Z^*e)^2/2d \cos \theta$ with respect to θ. The correct procedure is to incorporate the dependence of W_3 upon angle and Z^* into the calculation of the electronic states, minimize the total energy with respect to Z^* at each angle, and then evaluate that total energy as a function of angle. That is a highly intricate task, but it has been carried through by Pantelides and Harrison, (1976), leading to minimum energy at θ equal to 40°, in very poor agreement with the angle that is observed, 18°. The Coulomb repulsion between the two silicons failed to open up the angle, contrary to expectations; Pantelides and Harrison assumed that the overlap interaction between the silicons was responsible.

Two things have developed since that calculation was made. First, the overlap interaction was evaluated and appears much too small to account for the angle (Chapter 7), and second, the covalent and polar energies have been related to pseudopotentials (Chapter 18), suggesting that W_3 is much less sensitive to distortions than one would guess from interatomic Coulomb interactions. We shall therefore not repeat the full analysis made by Pantelides and Harrison (1976) but extract only the parts that now seem most relevant.

First, we shall address the question raised earlier, since it remains true that Z^* depends upon angle: how should interatomic Coulomb forces be calculated? It can be seen from the full analysis that the correct result is obtained from the force equation, $(2Z^*e)^2/(2d \cos \theta)^2$, not from differentiation of the energy with respect to angle. The reason depends upon the fact that the energy is minimum with respect to variations in Z^* at constant angle.

Using this equation directly is a questionable procedure, since a silicon atom is subjected not only to a force from its neighboring silicon atoms but also from the oxygen atoms to which it is connected. Since these atoms are oppositely charged, we can expect some cancellation of the force; we can also expect that a correct result will be obtained only by taking a sum over the entire lattice or by evaluating the total electrostatic energy (for fixed Z^*) of the lattice as a function of angle. Such electrostatic sums converge slowly. Hence, finding values for them is difficult. In Chapters 13 and 19 we shall discuss methods for performing them and give values for particular structures, but the calculation we need for SiO_2 has not yet been done. An attempt at finding an approximate value was made (Pantelides and Harrison, 1976) by summing over small clusters in each of which the charge vanishes; this gave a reduction of the force by a factor of four. Indeed, it was principally this small value that led to the large discrepancy in equilibrium angle. We shall see here that even using the large value gives poor results if we use the new parameters (and the correspondingly smaller Z^*) of Table 11-1. It is nevertheless appropriate to indicate how such a calculation is carried through.

Let us take the torque on each Si—O axis to be $(2Z^*e)^2 \, d \sin \theta/(2d \cos \theta)^2$.

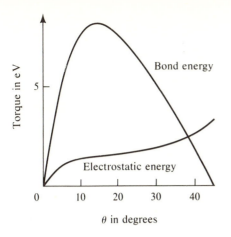

FIGURE 11-14

Two contributions to the torque on each leg of an SiO_2 bonding unit. The crossing point is a prediction of the angle in the equilibrium bonding unit. The observed angle is 18°.

This is then equated to $(1/2) \, \partial E_{bond}/\partial\theta$ (a change in θ rotates both Si—O axes by $\delta\theta$), with E_{bond} as defined in Eq. (11-10), to obtain the equilibrium angle. Notice that we cannot simply add the electrostatic energy to Eq. (11-10) and then minimize the energy, because Z^* depends upon angle. The magnitudes of the torques are readily calculated and are plotted in Fig. 11-14. (The calculation also gives Z^*, which varies from 1.19 at $\theta = 0$ to 0.52 at 45°.) The crossing is seen to occur at 38°, in poor agreement with the observed 18°. A similar result is obtained for GeO_2. Perhaps, as in the simple tetrahedral solids, the principal error lies in the Bond Orbital Approximation itself and a more complete analysis is needed.

An additional appropriate correction is the inclusion of the oxygen s state. Its coupling with the B_x state can be found with matrix elements from the Solid State Table. It is seen to lower the s state energy by about 4 eV and by itself would open up θ by some 2°.

We can also obtain an estimate of the force constant from such curves. This can be done in terms of the angle φ (in radians) between the two Si—O axes, $\varphi = \pi - 2\theta$. Then, the elastic energy is written as

$$E_{elast} = \tfrac{1}{2}C'_1(\delta\varphi)^2 \tag{11-11}$$

for each bonding unit. The value of C'_1, proportional to the difference in slope of the two curves, is 25 eV at the crossing. This value is too large, which is not surprising in view of the large discrepancy in the point of crossing.

It is not clear whether the discrepancies here come from errors in the parameters used or from the approximations used. The physical origins of the bending and the rigidity are presumably those given here. In Section 11-F we shall add an angular force at the silicon, and we shall discuss the vibrational spectra.

11-F Vibrational Spectra

Because of the polar nature of the mixed tetrahedral solids, there is stong coupling between electromagnetic radiation and lattice vibrations. This has allowed direct experimental study of the vibrational spectra through examination of infrared absorption (direct conversion of photons into phonons) or **Raman spectra** (in which one photon is absorbed and a second, of lower energy, is emitted, a phonon also being emitted at the same time). We shall discuss both the spectra and the coupling here, as we did when treating the simple tetrahedral solids. A full analysis is somewhat intricate and we shall merely outline the main features of a simplified analysis.

The lattice vibrations for the simple tetrahedral lattice were studied in Section 9-A. The state of the distortion of the lattice was specified by giving the displacement δr_i of each atom. We then made a transformation to normal coordinates u_k, each corresponding to a normal mode frequency $\omega(\mathbf{k})$, and these were plotted as a function of $\mathbf{k}$ in Fig. 9-2. There were three curves for each atom in the primitive cell. We see immediately that there will be difficulties in complex structures; in quartz there are 27 sets of modes, and even in the simple molecular lattice there are 9. This complexity suggests that one should proceed by computer. One such approach was taken by Bell, Bird, and Dean (1968). They took a large cluster

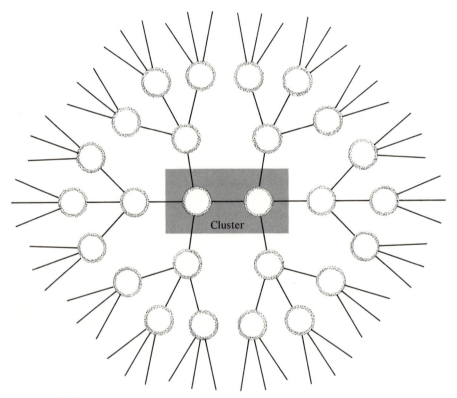

FIGURE 11-15

In the cluster–Bethe-lattice method, a central cluster is treated exactly and the surrounding medium is replaced by a successively branching network containing no closed paths.

of atoms, a simple valence force field, and calculated the mode frequencies directly. Another such approach is the use of a method combining cluster techniques and the Bethe lattice method, called the *cluster–Bethe-lattice method*. It was applied to the calculation of vibration spectra by Joannopoulos and Pollard (1976), Laughlin and Joannopoulos (1977), and Sen and Thorpe (1977). It was applied to the problem of calculating the electronic density of states by Yndurain and Joannopoulos (1976). In this method one focuses on some cluster, which may be as small as one wishes, and then replaces the rest of the system by a continually branching lattice (a Bethe lattice), as illustrated in Fig. 11-15. By this approximation, any paths through the lattice which close on themselves are neglected; the procedure enables one to deal with an infinite system while reducing the complexity of the calculation essentially to that of the starting cluster. This seems a very appropriate way to proceed, and it can be made as accurate as one wants, or as simple as one wants, by suitable choice of cluster size. A very extensive cluster analysis was undertaken by Joannopoulos; it also included study of surface vibrations.

We choose here instead an analytic formulation based upon the simple molecular lattice, which will immediately be generalized. This will bring the most important concepts to light and provide an interpretation of the spectrum. It can also provide the starting point for a quantitative application of the cluster–Bethe-lattice method; use of the Bethe lattice should improve accuracy and reduce the computation required in comparison to the direct cluster technique.

Vibrations of the Simple Molecular Lattice

We initially consider the simple molecular lattice and only the modes with $\mathbf{k} = 0$, those modes at Γ in the Brillouin Zone. These are, in fact, just the modes that are coupled to infrared light. (Light is of sufficiently long wavelength to be regarded as having $\mathbf{k} = 0$.)

The nature of the corresponding nine modes is immediately clear. There are three acoustical modes, such as we studied in the simple tetrahedral solids. Their frequencies go to zero at $\mathbf{k} = 0$. There are six optical modes, but in these, all silicon atoms move together exactly in phase at $\mathbf{k} = 0$, since there is just one silicon atom per primitive cell in the simple molecular lattice. The six modes correspond to motion of the two oxygens in the cell relative to the silicons. Such local descriptions of the modes were made by Saksena (1940) and more recently by Kleinman and Spitzer (1962). In fact, Kleinman and Spitzer thought in terms of oxygen displacements in the x, y, and z directions shown in Fig. 11-16 for the bonding unit. These would indeed be the nomal modes were the structure sufficiently symmetric and were only a single oxygen atom allowed to move. Our formulation will be closely related to this insight.

We could make a unitary transformation on the atomic displacements to obtain coordinates that correspond to the kinds of local modes envisaged by Kleinman and Spitzer, three for each kind of atom in the crystal. As in the treatment of

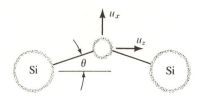

electronic states, these unitary transformations do not lead to solutions of the problem (that is, they do not lead to normal modes) but lead to an equivalent set of dynamical equations, which, if solved, would lead to exactly the same normal modes obtained by solution of the initial equations. As in the treatment of electronic states, the point is to put the problem in a form where approximate solutions can be obtained. The transformation to local modes must lead to modes associated with a molecular unit (for our purposes, to SiO_2) or some multiple of it. (In this case, as for the electronic case, the multiple is $\frac{1}{2}$.) Thus, the vibrations of a tetrahedral molecule (or a SiO_4 group) have little relevance to the problem of finding normal modes for SiO_2.

In the simple molecular lattice, at $\mathbf{k} = 0$, the problems of finding the acoustical modes and the optical modes are independent. Even at $\mathbf{k} \neq 0$ they would be independent problems if the ratio of the silicon to the oxygen mass were sufficiently large. In the end we shall see that the lowest optical mode frequencies overlap the acoustical mode frequencies, so to separate the problems is not entirely valid. However, for the present, let us separate the problems.

Having made the separation, we see that generalization to real structures, and even to amorphous (glassy) structures, is immediate. The three acoustical modes per silicon atom are treated together in the Debye approximation, taking the Debye wave number large enough to account for all of the degrees of freedom of the silicon atoms. Then we turn to the vibrations of the oxygen atoms in an essentially immobile silicon lattice. We make an approximate treatment of the coupling between these vibrations; the treatment does not depend upon a specific structure, and therefore we can think of the structure as being that of glass. The primitive cell taken is so large it becomes the specimen itself, and all modes are $\mathbf{k} = 0$ for the tiny Brillouin zone corresponding to the huge primitive cell. Thus we are considering the entire spectrum.

The displacements of the ith oxygen atom are given by using the coordinate system of Fig. 11-16, $\mathbf{u}^i = [u_x^i, u_y^i, u_z^i]$, and the kinetic energy is written as $\frac{1}{2}M\dot{u}^{i2}$ summed over all oxygen atoms. The mass M should be a reduced mass but for a large mass ratio it is approximately equal to the oxygen mass and we shall use that value. The potential energy is a quadratic form in the displacements, which could be evaluated for any model or from experiment. There are diagonal terms such as $(\kappa_{xx}^{ii}/2)u_x^{i2}$, which could be evaluated by displacing a single oxygen and holding all other atoms fixed. There are also off-diagonal terms coupling the displacements of different atoms.

In terms of these parameters, the acceleration of each atom can be written in terms of the forces on it (or, the Lagrangian for the system can be constructed, and Lagrange's equation can be written for each component of each $\mathbf{u}^i$). These provide a set of simultaneous equations of the form

$$-M\ddot{\mathbf{u}}^i = \sum_j \overleftrightarrow{\kappa}^{ij}\mathbf{u}_j, \tag{11-12}$$

where the right side represents a sum over all displacements coupled to the displacement appearing on the left. For a normal mode, $\ddot{\mathbf{u}}$ can be replaced by $-\omega^2\mathbf{u}$, and this is formally the same as the set of equations for the electronic structure, Eq. (3-18), with the eigenvalue E_k replaced by $M\omega^2$, and with matrix elements of the Hamiltonian replaced by coupling coefficients $\overleftrightarrow{\kappa}^{ij}$. The solution would be obtained by constructing a matrix in analogy with Fig. 3-4, though it would be termed the *dynamical matrix*, and by diagonalizing it. However, just as for the calculation of the electronic structure, we proceed in an approximate manner.

By constructing the dynamical matrix in this way, we have represented the optical modes as individual local oxygen vibrations, three for each oxygen, coupled to each other through the off-diagonal elements of the dynamical matrix. The diagonal elements correspond to *local oxygen mode* frequencies ω, given by

$$\left.\begin{aligned} M\omega_x^2 &= \kappa_{xx}^{ii}, \\ M\omega_y^2 &= \kappa_{yy}^{ii}, \\ M\omega_z^2 &= \kappa_{zz}^{ii}. \end{aligned}\right\} \tag{11-13}$$

These are the same for every oxygen if the environment of each oxygen is the same. Notice that there is one of each type of local mode for every oxygen present, and that fact does not depend upon the assumption of a simple molecular lattice.

Of the three local oxygen mode frequencies, ω_z is expected to be largest, since it involves large bond-stretching deformations (See Fig. 11-16). The frequency ω_x is expected to be next largest, since it reflects bond stretching to a lesser extent, and ω_y should be the smallest, since it reflects only angular distortion, and we have always found the angular spring-constants to be softest. Furthermore, we may expect the modes to be only weakly coupled, since the coupling between different modes arises from angular forces only. Actually, there is an additional *dynamic coupling* between optical modes, because the silicon mass is not infinitely greater than the oxygen mass. Thus, the modes do involve a small motion of the silicon atoms, which couples adjacent modes. That coupling is of order $M^2\omega^2/M_{\text{Si}}$. It should be added to the corresponding κ^{ij} coupling. The weakness of the total coupling suggests an approximation analogous to the Bond Orbital Approximation for electronic states: *we neglect the coupling between the three different kinds of local oxygen modes in comparison to their frequency differences.* This **local mode approximation** leads immediately to three sets of optical frequencies, each set centered at the local mode frequency given in Eq. (11-13).

It is possible to make a valence force field analysis on this basis, by introducing radial forces between nearest neighbors and angular forces at the silicon and at the oxygen. For this, the model described earlier by Kleinman and Spitzer (1962) is appropriate. The three force constants are fitted by matching the highest three observed infrared-active modes with the frequencies ω_x, ω_y, and ω_z. Then, these constants are used to calculate elastic constants and to produce a Debye approximation to the acoustical modes. The coupling between optical modes of the same type can be used to calculate the breadth of individual peaks. This is done by the method of moments, used for electronic states, for example, by Lannoo and Decarpigny (1974). The point can be seen by returning to Eq. (11-12) to see that the normal modes are eigenvectors of the matrix $(1/M)\overset{\leftrightarrow}{\kappa}$, with eigenvalues ω_k^2. In the local mode approximation, there is an N_O-by-N_O matrix giving the values for all the modes made up of displacements of, for example, the x-type for N_O values of **k**. Similarly, the matrix $(\overset{\leftrightarrow}{\kappa}/M - \omega_x^2 \overset{\leftrightarrow}{I})^2$ has eigenvalues equal to $(\omega^2 - \omega_x^2)^2$, and the trace of that matrix, divided by the number of modes, gives the second moment of the distribution. Using unitarity, one can write it in terms of the force constants, $\langle(\omega^2 - \omega_x^2)^2\rangle = M^{-2}\Sigma_j \kappa_{ij}\kappa_{ji}$, where the sum is over all $j \neq i$ for atoms that couple with the ith atom. (The terms with $j = i$ were cancelled by the $-\omega_x^2 = -\kappa_{ii}/M$ terms.) The distribution for each type of mode can be approximated by a Gaussian of the form $N(\omega) = Ae^{[-\alpha(\omega^2 - \omega_x^2)^2]}$. The results depend very much on how we treat the dynamic coupling and on how we average over angles in treating the force-constant coupling. Therefore, the results presented in Fig. 11-17 should be regarded as schematic. They are compared there with results, by Bell, Bird, and Dean (1968), of a cluster calculation and a simpler valence-force field.

The general conclusions probably are independent of the details. The two higher-frequency peaks, derived from ω_x and ω_z, can be reasonably well described in a local mode approximation, but the lowest mode, derived from ω_y, is so mixed with the acoustical modes that it should be treated together with the acoustical modes in a more complete calculation.

11-G   Coupling of Vibrations to the Infrared

An electric dipole arises from any displacement of an atom in SiO_2, and this gives rise to a coupling of the modes with electromagnetic radiation. In Chapter 9 this coupling was characterized by a transverse charge that, when multiplied by the atomic displacement, gave the induced dipole. (Recall that if this is done in such a way that electric fields are generated in the system, an additional polarization, equal to the susceptibility times the local electric field, should be added.) We wish to use the same definition here, but notice that because of the lower symmetry the transverse charge will depend upon the direction of the displacement. In particular, for the coupling with the optical modes, we will need transverse charges e_x^*, e_y^*, and e_z^* for the oxygen, for displacement in the three directions indicated in Fig. 11-16. These have been calculated (Pantelides and Harrison, 1976), who included the electrostatic corrections discussed in Section 11-E. We will not include those corrections here, in order to be consistent with the analysis carried out in Section 11-E, and will examine the differences between results obtained in this way and those of the more complex analysis.

The transverse charge for displacement in the y-direction is simplest. One sees immediately from Fig. 11-16 that displacement in this direction causes no changes in bond length to first order, nor does it change the bond angle to first order. Therefore, no transfer of charge is caused by the displacement, unlike what was described for the tetrahedral solid discussed in Section 9-D, and so e_y^* is simply the Z^* given in Eq. (11-6).

Displacement in the x-direction changes both the angle θ and the bond lengths d and, therefore, causes a transfer of charge between the oxygens and silicons. A change in bond length causes a change in W_2 so that $\delta W_2 / W_2 = -2\,\delta d/d$, like what was described for the simple tetrahedral solids. We neglect any change in W_3 proportional to displacement. Doing so was rigorously valid by symmetry in the simple tetrahedral solids, but that is not the case here. Pantelides and Harrison made estimates of the change in W_3, as indicated at the beginning of Section 11-E, with the result that the effect of changes in W_3 was overestimated, so we shall neglect such changes altogether, here, but will describe the results of Pantelides and Harrison later. The first step is to differentiate Z^* with respect to d and θ, using the assumptions just made about W_2 and W_3, and by using Eqs. (11-5) and (11-6). We immediately obtain

$$\frac{\partial Z^*}{\partial d} = \frac{2}{d}[\beta_{px}(1 - \beta_{px}^2) + \beta_{pz}(1 - \beta_{pz}^2)] \tag{11-14}$$

and

$$\frac{\partial Z^*}{\partial \theta} = \beta_{px}(1 - \beta_{px}^2)(-\cot\theta) + \beta_{pz}(1 - \beta_{pz}^2)\tan\theta. \tag{11-15}$$

The displacement u_x indicated in Fig. 11-16 causes a change $\delta d = u_x \sin\theta$ and a change $\delta\theta = u_x \cos\theta/d$. The change in dipole due to this transfer is $\delta Z^*\, d \sin\theta$, corresponding to a total transverse charge of

$$e_x^* = Z^* + (3\sin^2\theta - 1)\beta_{px}(1 - \beta_{px}^2) + 3\sin^2\theta\beta_{pz}(1 - \beta_{pz}^2). \tag{11-16}$$

This differs considerably from the Pantelides-Harrison prediction of transverse charge for x-displacement, which was dominated by changes in W_3.

Displacements in the z-direction give somewhat different effects. The two changes in bond length are of opposite sign and so are the changes in angle between the Si—O and Si—Si axes. Thus there are two changes in Z^*, each of half the magnitude given in Eq. (11-14) and (11-15) and of opposite sign. Thus there is no net change in the charge on the oxygen atom, but half the charge given in Eqs. (11-14) and (11-15) is transferred between the two silicons. Multiplying this by $2d \cos \theta$ gives the change in dipole due to the transfer and a transverse charge of

$$e_z^* = Z^* + 3 \cos^2 \theta \beta_{px}(1 - \beta_{px}^2) + (3 \cos^2 \theta - 1)\beta_{pz}(1 - \beta_{pz}^2). \quad (11\text{-}17)$$

One other transverse-charge parameter can be evaluated. That parameter, δp_x, gives the change in dipole in the x-direction due to the change in distance R between the two silicons in the bonding unit (the distance between the oxygen and the Si—Si axis is held fixed). Its evaluation is closely related to the calculation of e_z^* leading to Eq. (11-17) and can be written as

$$\frac{R}{p_x} \frac{\delta p_x}{\delta R} = 1 - \frac{e_z^*}{Z^*}. \quad (11\text{-}18)$$

These expressions differ from those given by Pantelides and Harrison (1976), partly because W_3 is held constant here and partly because a different method of calculation has been used which corresponds to slightly different approximations (Pantelides and Harrison used perturbation theory and coupling of bonding and antibonding states). In addition, different values for W_2 and W_3 are used here. The values as obtained by the calculations outlined here (and the Pantelides-Harrison values, given in parentheses) for e_x^*, e_y^*, e_z^* are 0.50 (0.45), 0.72 (1.02), and 2.08 (2.74). The values for $(R/p_x) \, \delta p_x / \delta R$ are -1.89 and (-1.37), respectively. The new values for e_i^* are listed in Table 11-2. Values for the four parameters for GeO_2 have also been estimated from these formulae; these values are 0.64, 0.69, 1.90, and -1.76.

It is interesting that these transverse charges depend so strongly upon direction; this dependence comes entirely from the transfer charge. Kleinman and Spitzer

TABLE 11-2
Experimental frequencies and predicted coupling constants for the local oxygen modes in SiO_2.

Mode	ω_x	ω_y	ω_z
$\omega_i/2\pi c$ (cm^{-1})	778	495	1080
$M \, d^2\omega_i^2$ (eV)	93.0	37.6	179.2
e_i^*	0.50	0.72	2.08
$4\pi\rho_i$	0.02	0.11	0.19

SOURCE: Frequencies are from Kleinman and Spitzer (1962).

(1962) introduced a model to account for the differences in absorption intensities. The study by Pantelides and Harrison indicated that the parameters fitted by Kleinman and Spitzer gave *magnitudes* similar to the estimates obtained theoretically but a different sign for e_z^*. Only the magnitudes of e_z^* enter the calculation of absorption intensity, so the theory and experiment are in agreement, and the sign obtained theoretically is probably correct.

It is convenient to write the coupling of the modes in terms of the dimensionless coupling strength, which is the contribution of each type of mode to the static electric susceptibility. Let us imagine an electric field $\mathscr{E}$ making an angle φ with the oxygen displacement in the ith local oxygen mode. We now displace that single oxygen by $\delta x_i = u_i$. The energy of the resulting dipole in the field is $-(-ee_i^*u_i)\mathscr{E} \cos \varphi$. The elastic energy, obtained by using Eq. (11-13), is $M\omega_i^2 u_i^2/2$, so the total is minimized, with $u_i = (-ee_i^*)\mathscr{E} \cos \varphi/M\omega_i^2$, and a component of dipole along the field equal to $e^2 e_i^{*2}\mathscr{E} \cos^2 \varphi/M\omega_i^2$. We average over orientation of the field and sum over the N_O oxygen atoms per unit volume to obtain the contribution to the susceptibility, defined to be the coupling strength for the ith set of optical modes ($i = 1, 2, 3$):

$$\rho_i = \frac{e^2 e_i^{*2} N_O}{3M\omega_i^2}. \tag{11-19}$$

The contribution of these modes to the **dielectric constant** is the sum over $4\pi\rho_i$. This calculation neglects the coupling between modes, as did the calculation of the frequencies ω_i, and each mode is active in the infrared. When the coupling of the modes is included, the frequencies split and the strength of coupling with the infrared is redistributed among the modes. Values for SiO_2 are given in Table 11-2. Summing the three $4\pi\rho_i$ gives 0.32. This result is inevitably of the right order of magnitude. Vitreous silica has a static dielectric constant of 3.75 and an optical dielectric constant of about 2.3, which suggests a sum for SiO_2 of 1.5; a sum for quartz of about 2.0 is suggested. The discrepancy between these and our value of 0.32 is significant and means that we will in general underestimate the infrared intensities, proportional to ρ_i, by a factor of about five. This appears to be an accumulation of errors from the various approximations made.

Table 11-2, combined with Fig. 11-17, now provides the information necessary to give a preliminary description of the absorption spectra. The frequencies of the three peaks have been taken from experiment. The total intensity for each peak and the breadth of each peak are predicted by the model; each has been fitted by a Gaussian distribution curve. The acoustical modes, in the simplest picture, do not contribute to the infrared absorption intensity, so we obtain a simple sum of three Gaussian distributions, as illustrated in Fig. 11-18. For comparison, an experimental curve is given in the lower part of the illustration. As noted earlier, the absolute values of the absorption are underestimated but the vertical scale has been adjusted to make the two curves comparable. The accord between the two is sufficiently good to suggest that the local oxygen mode description of the spectrum is meaningful. The major differences between the density of modes and the

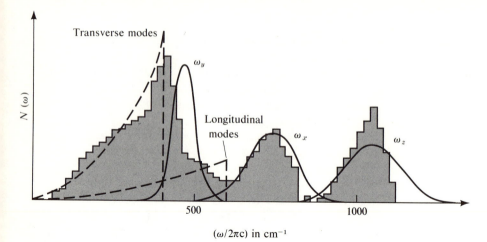

FIGURE 11-17

The distribution of vibrational modes in SiO_2. The smooth curves correspond to use of a local mode approximation for optical modes and a Debye approximation for acoustical modes (three per silicon atom). The histogram given for comparison is from a cluster calculation (Bell et al., 1968). The frequency is customarily specified in wave numbers by giving the reciprocal of the wavelength of the corresponding light, $\omega/2\pi c$, with c the speed of light in cm/sec. The scale on the ordinate is fixed by the conservation of degrees of freedom. In all cases force constants were fitted to experiment.

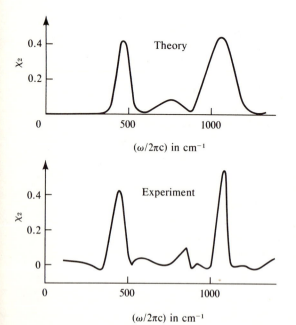

FIGURE 11-18

Imaginary susceptibility in the infrared, χ_2, plotted against the frequency. In the theoretical curve no effect is included for acoustical modes and the relative intensities of the three oxygen local mode peaks are scaled by the appropriate value for $4\pi\rho_i$ from Table 11-2. (The absolute scale was chosen to fit the experimental curve.) The experimental curve is from Miler (1968).

spectrum are explained. The very strong mixture of the low-frequency mode with the acoustical branches has not had an immense effect on the shape of the spectrum. The suppression of the intermediate-frequency peak is caused by the small effective charge predicted by the theory described here. However, notice at the same time that the agreement with experiment is not impressive. The three frequencies have been fitted and the absolute vertical scale has been adjusted by a factor of two. A full calculation, possibly using the cluster–Bethe-lattice method, will be necessary for a more complete description. The calculation will require use of effective charges, such as we have given. Such a study might include Raman scattering as well as infrared absorption. (For a discussion of the Raman spectra of silica, see Hass, 1970.)

PROBLEM 11-1   *The bonding unit*

The treatment of the electronic structure of the SiO_2 bonding unit becomes much simpler if the bond angle at the oxygen is 90° rather than the 144° that in fact is found in SiO_2. Then σ-oriented p states can be constructed so that there are independent left and right bonds in addition to the π-oriented p states.

(a) What is the relation between the polarity β of each bond and the β_{pz} and β_{px} given in Eq. (11-5)? These bond orbitals can be the basis for an elementary band calculation, carried out in part (b) of this problem.

(b) If only nearest-neighbor matrix elements are included, no matrix elements couple the three orbitals shown in Fig. 11-19. However, matrix elements $-V_1^h$ couple silicon hybrids on the same atom. If all other matrix elements (except V_1, W_2, and W_3) are discarded, the bands become independent of $\mathbf{k}$. Obtain the corresponding bands that result from B_l and B_r (and the corresponding A_l and A_r) and B_y, all based upon the Solid State Table. Compare the bands with the band edges shown in Fig. 11-9,a, measuring energies from that of B_y.

Note: The bond orbitals sharing an atom are coupled by $W_1 = -(1 - \beta)V_1/2$. Use the Bond Orbital Approximation by neglecting couplings between A's and B's, so that each set reduces to a four-by-four Hamiltonian matrix having eigenvalues relative to equal diagonal matrix elements of $-W_1$ (threefold degenerate) and $3W_1$ (nondegenerate).

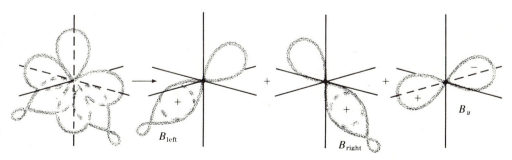

FIGURE 11-19

PROBLEM 11-2   *The transverse charge*

For oxygen, e_T^* depends upon the direction of displacement (Table 11-2), but if the environment of the silicon is assumed to have tetrahedral symmetry, its e_T^* will not. Use values of e_T^* from Table 11-2 to estimate the transverse charge for the silicon in SiO_2.

PART III

CLOSED-SHELL SYSTEMS

TWO-ELECTRON BONDS between neighboring atoms characterize covalent solids. Since atoms with a single valence s state and three valence p states can form only four such bonds, each atom in a covalent structure is surrounded by four other atoms at most and the structures are inevitably open. That is, if you imagine that the diamond structure is constructed from rigid balls at the atomic sites, as discussed in Section 3-A, you will see that there is room in the interstices to accommodate an equal number of additional balls. The electronic structure and the properties of covalent solids are dominated by these two-electron bonds.

If the angular rigidity of the structure is sufficiently reduced, either because of an increased bond polarity or because of a reduction of covalent energy itself (as the bond lengths are increased), the structure will collapse to a more closely packed structure in the sense of the rigid balls just mentioned. The formation of two-electron bonds between neighbors is no longer possible and a completely fresh view of the electronic structure and the bonding should be adopted. It is undesirable to let the success of the theory of covalently bonded solids force the same concepts onto unrelated electronic structures. In fact, when we consider the open and close-packed electronic structures in terms of similar pseudopotentials, we shall see that totally different approximations are appropriate for each.

Notice that the difference between the two electronic structures is a *consequence* of their different crystal structures. It is true that a two-electron bond keeps silicon in a tetrahedral structure, but if silicon is forced into a close-packed structure by pressure or heat, it becomes as metallic as aluminum and should be thought of as such. Once a structure has become close-packed, the solid will be metallic or insulating depending upon the atoms making up the structure. Recall the boundary between the two classes of materials shown in the schematic phase diagram of Fig. 2-6. Although materials lie close to the boundary, usually the distinction between metals and insulators is very clear. We shall discuss the insulating materials first, and then the metallic materials. In the intermetallic compounds—those that are intermediate between the metals and the insulators—we may expect the concepts applicable for both metals and insulators to be relevant.

Ionic solids are the largest class of insulators (though some covalent solids, such as SiO_2 or diamond, are also insulating). They consist of ions, atoms that have gained or lost electrons. Of particular interest in any study of solids is the ion having an inert-gas configuration, such as the configuration of He, Ne, Ar, Kr, or Xe. Such configurations, called closed-shell configurations, are extremely stable, so atoms tend naturally to arrange themselves to give such configurations. Their stability corresponds to a large gap in energy between occupied and empty states and, therefore, to insulating properties and a transparent or whitish appearance. The general tendency for materials to bleach when exposed to oxygen is a consequence of the tendency for ionic solids to form with a large gap between energy bands, as well as the tendency for states to be removed from the gap, as discussed in Chapter 10.

A special case of closed-shell configurations are the ions of zero charge, the inert gases themselves. The inert-gas solids may not be of great intrinsic interest, but they provide the best starting point for the understanding of ionic solids, so we discuss them first. We shall then see how the ionic solids can be understood in terms of transfer of protons between nuclei, much as we understood the polar covalent solids in terms of transfer of protons between the nuclei of the homopolar solids.

CHAPTER 12

Inert-Gas
Solids

SUMMARY

Neon, and the elements directly below it in the periodic table or the Solid State Table, form the simplest closed-shell systems. The electronic structure of the inert-gas solid, which is face-centered cubic, is essentially that of the isolated atoms, and the interactions between atoms are well described by an overlap interaction that includes a correlation energy contribution (frequently described as a Van der Waals interaction). The total interaction, which can be conveniently fitted by a two-parameter Lennard-Jones potential, describes the behavior of both the gas and the solid. Electronic excitations to higher atomic states become excitons in the solid, and the atomic ionization energy becomes the band gap. Surprisingly, as noted by Pantelides, the gap varies with equilibrium nearest-neighbor distance, d, as d^{-2}.

Neon, and the elements below it in column 8 of the Solid State Table have very similar properties. We call them *normal inert gases* and our discussion will center on them. For a good discussion of inert-gas crystals, see Kittel (1976, p. 76). Helium, above Ne in column 8, is also an inert gas, but with an outer configuration of s^2, rather than s^2p^6. Helium is an extremely interesting material in its own right, partly because of the strong quantum-mechanical effects it shows (superfluidity, if the nucleus contains four nucleons, and something analogous to superconductivity, if it contains three); we shall discuss helium only to the extent that it has relevance to other solid types.

We may think of the inert gases as neutral atoms with their electronic structure only weakly perturbed by their neighbors. The atoms attract each other weakly through the so-called *Van der Waals interaction*, which is a manifestation of the correlation energy mentioned in Chapter 7 and discussed in Appendices A and C. One way of understanding the origin of this interaction between well-separated atoms is to imagine that quantum-mechanical fluctuations produce a dipole moment in an otherwise spherically symmetric atom. This produces a field at the second atom that decreases with the distance d to that atom, as d^{-3}. That atom is polarized by the field, giving a dipole moment also proportional to d^{-3}, so the interaction energy between atoms, the dot product of the field and the dipole, varies with separation as d^{-6}. A way of calculating this interaction (see Kittel, 1971, 4th Edition, Appendix) is to treat each atom as a harmonic dipole oscillator (of frequency equal to the first excitation energy divided by $\hbar$). The dipole–dipole coupling splits the oscillator frequencies, and the decrease in the net zero-point energy of the oscillators is the Van der Waals interaction.

An interaction decreasing as d^{-6} may seem very short ranged, but it is the only interaction between neutral atoms that remains at separations sufficient that the electron distributions do not overlap. When the atoms are brought close enough to overlap, the Van der Waals interaction is more conveniently thought of as coming from the change in correlation energy of the electrons as the internuclear separation changes. It is included in this way in the overlap interaction discussed in Chapter 7 and Appendix C.

Their weak interatomic interaction is responsible for the condensation of the normal inert gases into solids. Atoms of normal inert gas are brought together until the repulsive terms in the overlap interaction prevent further contraction. The attraction favors a close-packed structure, and all of the normal inert gases form face-centered cubic lattices. These two contributions to the total interaction will remain almost the same in the ionic crystals, but with added Coulomb interactions, so it is desirable to understand all of these contributions with some care.

12-A   Interatomic Interactions

We wish to discuss interatomic interactions at two levels; first, from the detailed though approximate quantum-mechanical calculation by Gordon and Kim; second, in terms of the parameterized model of Lennard-Jones, which will be useful for approximate calculations in the ionic solids as well as the inert-gas solids.

The first overlap interaction calculated by Gordon and Kim (1972) was actually calculated for inert gases. The result was very similar to that subsequently obtained by Sokel (Harrison and Sokel, 1976) for silicon and shown in Fig. 7-1. The overlap interaction in the figure (not the total interaction, containing bonding energy) appears to rise monotonically with decreasing d, but there is actually a very shallow minimum at large d, as there is in Gordon and Kim's calculation for the inert gases. Part of the results of the Gordon and Kim calculation for neon— the region near the minimum—is shown in Fig. 12-1. Notice that the scale on the

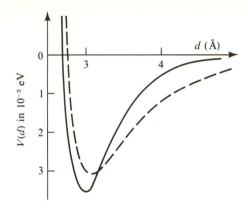

FIGURE 12-1

The interatomic interaction for neon. The solid line is the overlap interaction calculated by Gordon and Kim (1972). The dashed line is the Lennard-Jones interaction with parameters determined by Bernardes (1958).

ordinate is in millielectron volts, so that this portion of the curve would lie within the line widths on the scale used in Fig. 7-1. Notice also that the minimum occurs at very large spacing on the scale given in Fig. 7-1.

The calculated overlap interaction was compared with molecular beam experiments that tested small-d values where the interaction energy was from one to ten electron volts; the agreement was remarkably good. Gordon and Kim also tested the overlap interaction near the minimum by comparing the position and depth of the minimum they predicted with values determined experimentally. That comparison is given in Table 12-1. Except for helium, the predictions are in good accord with experiment and the description may be regarded as adequate for our purposes. At larger separation, the calculated overlap interaction rises to zero too rapidly, since the approximation used to calculate the correlation energy does not give the d^{-6} term that must be present at sufficiently large d; it gives only an exponentially decaying attraction. This, however, will not be important in our discussions. What is important is that the origins of both the repulsive and the

TABLE 12-1

Predictions of Gordon and Kim (1972) for the position d_{min} and depth ε of the minimum, in the interaction between inert-gas atoms, compared with experiment.

	d_{min} in Å		ε in millielectron volts	
Inert gas	Theory	Experiment	Theory	Experiment
He	2.49	2.96	3.9	1.0
Ne	2.99	3.03	3.5	3.9
Ar	3.63	3.70	10.9	12.2
Kr	3.89	3.95	15.5	17.1

SOURCES for experimental values: He and Ne, from Siska et al. (1971); Ar, from Parson and Lee (1971); and Kr, from M. Cavallini, M. G. Dondi, G. Scoles, and U. Valbusa, quoted in Gordon and Kim (1972).

attractive parts of the interaction are quantitatively understood; they arise from kinetic energy and correlation energy, respectively, and rather reliable, tabulated interactions can be found if they are needed. The same interaction gives a good description of the solid as well as the gaseous phase.

Of somewhat more use here is an approximate description of the interactions in terms of the well-known Lennard-Jones interaction (Lennard-Jones, 1924, 1925). This is a two-parameter model that Lennard-Jones fitted to properties of the gaseous phase. This form contained an attractive term proportional to d^{-6}, in accord with the Van der Waals form. The repulsive term rises more rapidly, and a term proportional to d^{-12} is commonly taken, though most results are not highly sensitive to the exact exponent. The Lennard-Jones overlap interaction is then written

$$V_o(d) = 4\varepsilon[(\sigma/d)^{12} - (\sigma/d)^6]. \tag{12-1}$$

The choice of parameters is a convenient one. The parameter σ is the value of d at which V_o equals zero. By differentiation with respect to d we see immediately that the minimum occurs at $d = 2^{1/6}\sigma = 1.12\sigma$, and the value at the minimum is $-\varepsilon$. A more recent fit of the two parameters to experimentally determined deviations of the gaseous phase from the ideal-gas laws has been made by Bernardes (1958). Bernardes's values are given in Table 12-2. The form of the interaction is not the same as that calculated by Gordon and Kim, as can be seen in Fig. 12-1, where the Lennard-Jones interaction (calculated with parameters from Table 12-2) is plotted along with the Gordon and Kim interaction. However, a comparison can be made by taking the value from the crossing and ε from the minimum in the Gordon and Kim overlap interaction. The corresponding values are given in Table 12-2. The comparison can be regarded as another experimental test of the Gordon and Kim overlap interaction as it applies to properties of the gaseous

TABLE 12-2
Experimental Lennard-Jones parameters compared with estimates based upon the overlap interaction (Gordon and Kim, 1972).

Inert-gas atom	Experiment		Gordon/Kim	
	σ in Å	ε in 10^{-3} eV	σ in Å	ε in 10^{-3} eV
He	—	—	2.20	3.9
Ne	2.74	3.1	2.71	3.5
Ar	3.40	10.4	3.28	10.9
Kr	3.65	14.0	3.48	15.5
Xe	3.98	20.0	—	—

SOURCE of experimental values: Bernardes (1958).

TABLE 12-3   295

Nearest-neighbor distance and cohesion predicted from the Lennard-Jones potential, compared with experiment.

Inert-gas solid	d (Å)		Cohesive energy (eV per atom)	
	Theory, 1.12σ	Experiment	Theory, 6ε	Experiment
Ne	3.08	3.13	0.019	0.02
Ar	3.82	3.76	0.062	0.080
Kr	4.10	4.01	0.084	0.116
Xe	4.47	4.35	0.120	0.17

SOURCE of experimental values: Kittel (1967).

phase. We now wish to turn to the inert-gas solid, and for that purpose we shall use the experimental values of σ and ε.

Under the influence of the interatomic interaction, the normal inert gases form a face-centered cubic crystal at low temperatures. If we neglect the effect of all but the nearest-neighbor interactions, we would predict a nearest-neighbor distance equal to the position of the minimum in V_0 (at 1.12σ) and a heat of atomization equal to ε times the number of nearest neighbors (12) divided by two. These predictions are compared with experiment in Table 12-3. This comparison tests the validity of interactions determined from the gaseous phase, for solids. Again the agreement is remarkable. It is quite simple to include more distant interactions (see Problem 12-1); this modifies the results by several percent and improves the agreement with experiment.

Knowing the interactions, one can proceed to calculate any of the other mechanical properties, and one can expect reasonably good results. Kittel (1976) indicates in particular that results obtained for the bulk moduli are quite good, and that they are improved by the inclusion of quantum effects associated with the zero-point oscillations. A study of the full vibrational spectra of Ar and Kr has been made by Grindlay and Howard (1965).

12-B Electronic Properties

To the extent that the electronic structure is describable in terms of independent atoms, the properties of inert-gas solids are easily understandable and not so interesting. There are, however, one or two points that should be made. The optical absorption spectra of isolated atoms consists of sharp lines that correspond to transitions of the atom to excited states, and to a continuous spectrum of absorption beginning at the ionization energy and continuing to higher energy. The experimental absorption spectra of inert-gas solids (Baldini, 1962) also show fairly sharp lines corresponding to transitions from the valence p states to excited s

states and to a continuum of states at higher energies, and are thus consistent with description in terms of independent atoms.

The spectra may also be described in the language of solid state theory. The atomic excited states are the same as the excitons that were described, for semi-conductors, at the close of Chapter 6. They are electrons in the conduction band that are bound to the valence-band hole; thus they form an excitation that cannot carry current. The difference between atomic excited states and excitons is merely that of different extremes: the weakly bound exciton found in the semiconductor is frequently called a *Mott-Wannier exciton*; the tightly bound exciton found in the inert-gas solid is called a *Frenkel exciton*. The important point is that the excitonic absorption that is so prominent in the spectra for inert-gas solids does not produce free carriers and therefore it does not give a measure of the band gap but of a smaller energy. Values for the exciton energy are given in Table 12-4.

The band gap itself measures the energy required to produce a mobile carrier of current, the counterpart of ionization of the atom. For inert-gas solids, the spectrum in this energy-range displays a peak (Baldini, 1962) that appears to be understandable in terms of resonant states, which we shall not discuss here. (For one discussion, see Rössler and Schütz, 1973.) The band gap is inevitably inter-mediate between the energy required to form an exciton and the energy required to ionize the free atom. To understand this, we can imagine removing an atom from the crystal, ionizing it, and then returning the ion and the electron separately to the crystal; it is clear that more energy will be gained in returning the ion to the crystal than was required to remove the neutral atom, and we can expect a positive electron affinity for the solid, so both effects reduce the resulting gap energy in the solid.

One measure of the gap is the resonant energy listed in Table 12-4, which is seen to be slightly less than the free-atom ionization energy. Pantelides (1975c) noted

TABLE 12-4

Excitation energies for the inert-gas solids. The energy of the lowest exciton peak is E_x. A resonant energy that may be associated with the band gap is E_r, and the ionization energy of the isolated atom is E_i. All energies are in electron volts.

Inert-gas solid	$d(Å)$	E_x	E_r	E_i	$29.2\ \hbar^2/md^2$
Ne	3.13			21.56	22.7
Ar	3.76	12.1	14.3	15.76	15.8
Kr	4.01	10.2	12	14.00	13.9
Xe	4.35	8.4	9.4	12.13	11.8

SOURCES: d, from Kittel (1971); E_x, from Baldini (1962); E_r, from Rössler and Schütz (1973); and E_i, from Moore (1949, 1952).

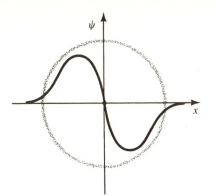

FIGURE 12-2

The valence *p* state of an inert-gas atom, viewed as a wave having a wavelength equal to the atomic diameter.

the remarkable fact that the gaps in inert-gas solids vary with nearest-neighbor distance as d^{-2}, just as in covalent solids. This carries over to ionization energies of isolated atoms, which are a crude measure of the band gaps, as illustrated by the last column of Table 12-4, which lists $\eta_g h^2/(md)^2$, with η_g taken as 29.2 to fit the argon value. This differs from the value given by Pantelides largely because he used a different definition for the parameter d.

We saw in Section 2-D that a d^{-2} dependence of an energy gap is a natural result for the solid, but it may seem surprising that the ionization energy of an isolated atom should be related to an interatomic distance in the solid. The result can be made plausible by thinking of the atom as a sphere and the valence *p* state as a wave with wavelength equal to the diameter d, as illustrated in Fig. 12-2. Thus the kinetic energy of the state becomes $(h^2/2m)(2\pi/d)^2$. By the virial theorem, the potential energy is twice the negative of this, and leads to a binding energy—or ionization energy—of $\eta_g h^2/(md^2)$, with η_g equal to $2\pi^2 = 19.7$, in crude agreement with the experimental value of 29.2.

Pantelides (1975c) also discussed the valence bands for the inert-gas solids, indicating that they consist of a narrow *p* band and an *s* band, which may be taken as completely sharp. He gave a universal width for the *p* band, of $\eta_v h^2/(md)^2$, with $\eta_v = 4.2$. (Again, his numerical value was different because of a different definition of d.) Presumably the conduction bands, corresponding to electrons added to the crystal, would be quite like free-electron bands.

Properties such as the dielectric susceptibility may be thought of as a superposition of the effects of individual atoms, but one should recognize that corrections can be made for this model. The most obvious correction in the case of the dielectric susceptibility is an accounting for the effect of the field arising from neighboring dipoles, the local field effects that we mentioned at the beginning of Chapter 4 but did not treat in detail for covalent solids. Local field effects are not difficult to include. (See, for example, Mott and Gurney, 1953, pp. 10ff.) In addition, it is clear from an expression such as Eq. (4-5), which is applicable to all solids, that having a band gap smaller than the atomic ionization energy would give an increase in the polarizability of the atoms even if the valence wave functions were exactly as in the free atom. We shall not explore this for the inert-gas solids, but make the point since it becomes a very important effect in the ionic solids.

The diamagnetic susceptibility of the inert-gas solids is dominated by the Langevin term (see Section 5-E), is readily calculated for the atom, and can be multiplied by the density of atoms to obtain values in good agreement with experiment. (See, for example, Kubo and Nagamiya, 1969, p. 438.)

Just as inert-gas crystals may be thought of as being composed of independent atoms, many crystals are made up of essentially independent molecules, loosely held together by Van der Waals forces or, for example, electric dipole–dipole interactions. The principal bonding is within the molecular unit. Such structures are characterized by wide spacing between molecules that can exist also in the gaseous phase. The simplest case is solid molecular hydrogen, which has a close conceptual relationship to solid helium (which exists at high pressure), though it does not show the interesting properties that condensed helium does. There are as many molecular crystal types as there are molecular types, but we shall not explore them further.

PROBLEM 12-1   *Second neighbor interactions*

Estimate the total contribution to the cohesion from the second-neighbor interactions in a face-centered cubic inert-gas solid. If you set the spacing d at the minimum of the Lennard-Jones interaction, that is, $d = 2^{1/6}\sigma$, the result can be written as a fraction of the contribution from nearest neighbors.

PROBLEM 12-2   *The Lennard-Jones binding*

If a pair of inert-gas atoms held together by the Lennard-Jones interaction rotates, the centrifugal force $MR\omega^2$ will increase their separation. The maximum frequency of rotation will be at the separation R such that the attractive force $\partial V_o/\partial R$ divided by MR is maximum. Determine that frequency and the corresponding separation for neon, using experimental σ and ε from Table 12-2.

CHAPTER 13

Ionic
Compounds

SUMMARY

Ionic compounds may be thought of as consisting of inert-gas atoms with modified nuclear charges such that the crystal itself is neutral. They form close-packed structures in which alternate positive and negative ions give low electrostatic energy. In addition to the Coulomb interaction, there is an overlap interaction well described by approximations made by Gordon and Kim. The overlap interaction is very nearly the same in an isoelectronic series such as Ar_2, KCl, CaS, and can thus also be obtained approximately even from measurements made on the appropriate inert gas. The combination of electrostatic energy and overlap interaction gives a good account of the spacing, cohesion, the bulk modulus, and even the angular rigidity of the crystal. The edge of the overlap interaction is sufficiently hard that ionic radii can be defined, and they are useful for estimating lattice distances, but at the same time, it is sufficiently soft that the ionic radii vary from structure to structure and provide an unreliable guide to structural stability.

13-A   The Crystal Structure

Varieties of Ionic Compounds

Ionic compounds include all insulating compounds that form close-packed structures, that is, solids in which the constituent atoms have more than four nearest neighbors. Transition-metal compounds are a special case, which we discuss in Chapter 19. In the present chapter we shall focus upon ions that have inert-gas configurations and think of them as inert-gas atoms with a modified number of nuclear charges. In Fig. 13-1 the portion of the Solid State Table that gives such ions is redrawn from Fig. 2-7. The block of atoms Be, B, Mg, and Al was included, though when these atoms occur in AB compounds (for examples, BeO, BeS, and BN), they usually form tetrahedral covalent solids (MgO, MgS, and MgSe are exceptions—they form in the rocksalt or NaCl structure); in other circumstances (for example, BeF_2) they can form ionic crystals.

The elements in Fig. 13-1 form ions of charge varying from -3 to $+3$. There are also compounds for which it is useful to think of ions of charge ± 4 forming, as in TiC, but the elements for these compounds are not included in Fig. 13-1. There are also many compounds containing ions that differ from a closed-shell configuration only by some number of d or f electrons. These are the transition-metal compounds that will be discussed in Chapter 19. A particularly simple special case are the compounds AgF, AgCl, and AgBr, all of which form in the rocksalt structure. These contain a closed $4d$ shell in addition to a kryptonlike core. These have some of the simplicity of the ionic compounds discussed in this chapter, but the d shells are important to dielectric and bonding properties, as illustrated by the ion distortion which was discussed in Chapter 8.

We should also mention another class of ionic solids, the **ten-electron solids**, such as PbS, PbSe, and PbTe (in which lead contributes four and S, Se, or Te contributes six electrons per molecular unit). These form in the same structure as NaCl and can be thought of, by analogy, as consisting of closed-shell S^{2-}, Se^{2-}, or Te^{2-} ions and Pb^{2+} ions that have fully-occupied s states and completely empty p states; the Pb^{2+} is closed-shell in almost the sense that He is; hence, the same approaches that are developed here for other ionic crystals are appropriate to the ten-electron solids. Another such set of compounds are TlCl, TlBr, and TlI, which form in the same structure as CsCl. Still another ionic compound without an inert-

		5	6	7	8	9	10	11			
				1 H	2 He	3 Li	4 Be	5 B			
	7 N	8 O	9 F	10 Ne	11 Na	12 Mg	13 Al				
	15 P	16 S	17 Cl	18 Ar	19 K	20 Ca	21 Sc				
	33 As	34 Se	35 Br	36 Kr	37 Rb	38 Sr	39 Y				
	51 Sb	52 Te	53 I	54 Xe	55 Cs	56 Ba	57 La				
	83 Bi	84 Po	85 At	86 Rn	87 Fr	88 Ra	89 Ac				
$Z =$	-3	-2	-1	0	$+1$	$+2$	$+3$				

FIGURE 13-1

The principal elements that form closed-shell ions with the same electron configurations as the inert gases of column 8. The charge Z associated with the ion is indicated along the bottom of the figure.

gas configuration is the hyperoxide of sodium, NaO_2, which forms essentially the same structure as NaCl but with the covalently bonded O_2^- molecular ion functioning exactly like the Cl^- ion in NaCl.

A more significant generalization is required for *mixed ionic–covalent solids*, such as the feldspar minerals. β-Eucryptite, $LiAlSi_3O_8$, can be understood by thinking first of mixed-tetrahedral SiO_2 (or Si_4O_8), which is a covalent solid, and then modifying that structure to obtain β-eucryptite. One out of every four silicon atoms will be replaced by an aluminum atom. For each aluminum atom, a lithium atom will be needed to provide the missing electron (aluminum having one less electron than silicon), so that the same covalent bonding is present in the $AlSi_3O_8^-$ network. That network, however, will now be charged, forming a giant " ion " with compensating Li^+ ions in the interstices. By thinking of mixed ionic–covalent solids in this way, it is relatively clear how the mathematical approaches developed in Chapter 11 and in this chapter can be used to treat the electronic structure and properties of such systems. They will not be treated here separately, however.

We shall concentrate instead on the simple ionic solids. These we shall think of as being made up of inert-gas atoms interacting through a predominantly repulsive overlap interaction but having net charges that cause them to form close-packed structures of alternate positive and negative charges. Two differences between inert-gas solids and ionic solids are important: in ionic solids the Coulomb attraction is much larger than the Van der Waals interaction, so interatomic spacing will be considerably smaller; and the binding together of the ions is orders of magnitude stronger in the ionic solids than in the inert-gas solids. The electrostatic energy of the interacting ions is so important in the properties of ionic solids that we shall want to consider it at the outset, but it will be helpful to specify three of the most important structures for the ionic solids first.

The Crystal Structures for Simple Ionic Solids

The simplest ionic solids consist of equal numbers of positive and negative ions with charges of equal magnitude. The ions form either the rocksalt or the cesium chloride structure because these two structures have such low electrostatic energy. The cesium chloride structure was shown in Fig. 2-1. It consists of a repeating, simple cubic arrangement of cesium ions with a chlorine ion at every cube center. Thus the positive and negative ions form individual simple cubic lattices interpenetrating each other. The more common structure is the rocksalt structure shown in Fig. 13-2; this is seen to be a simple cubic arrangement of alternate positive and negative ions, with each species by itself forming a face-centered cubic lattice. The zincblende structure of the covalent solid is also made up of two interpenetrating face-centered cubic lattices, but with lattices separated by [111] $a/4$ rather than [100] $a/2$. Since the primitive lattice for the rocksalt lattice is face-centered cubic (see Section 3-C), as was the primitive lattice of the zincblende structure, the

Brillouin Zone is the same for both zincblende and rocksalt, and is as shown in Fig. 3-6.

The structures for ionic crystals that are *not* made up of equal numbers of oppositely charged ions are much more varied, but in essence all may be thought of as consisting of closely packed ions arranged to give a low electrostatic energy. We shall only discuss one of these, the *fluorite structure*, which illustrates the main features of these crystals. A complete listing of these structures, including helpful ways for understanding them, is given by Wyckoff (1963). CaF_2, or fluorite, forms a structure in which the calcium ions are in a face-centered cubic structure. Notice that there are two kinds of interstitial sites in the face-centered cubic structure where the flourine ions can be placed. First, there are sites at the center of cube edges (and at the cube center) that are each surrounded by six calcium ions arranged as the corners of a regular octahedron. It is easy to see that there is one such *octahedral* site for each calcium; filling them forms the rocksalt structure that has just been discussed. Second, there are also sites displaced from each calcium site by [111] $a/4$, which are surrounded by four calcium ions arranged as the corners of a regular tetrahedron. There are two *tetrahedral* sites for each calcium. The zincblende structure can be obtained by filling half of these. In CaF_2, all of the tetrahedral sites are filled, as illustrated in Fig. 13-3.

It is important to recognize that though the fluorine ions are tetrahedrally surrounded, each calcium ion has eight fluorine neighbors; this makes an ionic—not a covalent—structure. Any attempt to construct a reasonable bonding unit leads to a unit that is the fluorine ion.

Compounds with more complex formulae must form more complex structures, and these occur in extraordinary variety (see Wyckoff, 1963). All of the ionic

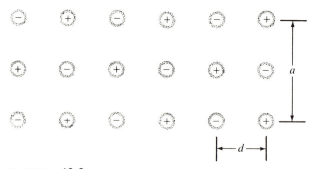

FIGURE 13-2

The rocksalt crystal structure viewed in a [100] direction. Additional planes of ions, displaced successively from the plane of the figure by [100] d, have positive and negative ions interchanged so that each positive ion has six nearest-neighbor negative ions, and vice versa. The unit cube edge is $a = 2d$; the primitive lattice translations are of length $a/\sqrt{2}$, in [110] directions, and the interion spacing is d.

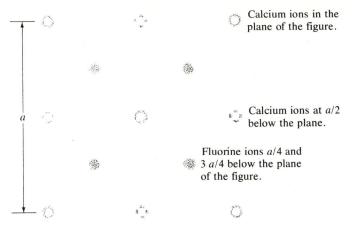

Calcium ions in the plane of the figure.

Calcium ions at $a/2$ below the plane.

Fluorine ions $a/4$ and $3\,a/4$ below the plane of the figure.

FIGURE 13-3

A unit cube of the fluorite structure. The unshaded circles represent calcium ions and form a face-centered cubic lattice. There are eight fluorine ions occupying all of the tetrahedral sites in the unit cube.

structures, however, are characterized by a closeness of packing and an alternation of charge, which minimizes the electrostatic energy. In the same way, those systems containing water of crystallization utilize the electric dipole of the water molecule to lower the electrostatic energies.

13-B Electrostatic Energy and the Madelung Potential

The *electrostatic energy* of an ionic crystal is defined relative to the energy of ions that are separated from each other at infinite distance. It is calculated by treating the ions as spherically symmetric, which gives exactly the same energy as treating the ions as if they were point charges at the nuclear positions in the solid. We shall see that the ions in the crystal "soften" somewhat (see Section 14-C) such that they have effective charges slightly less than the integral values given in Fig. 13-1. However, when we realize that the electrostatic energy is a part of the energy associated with bringing ions together from infinite spacing, and that any charge redistribution occurs at the very last step of this assembly, it is clear that the electrostatic energy should be defined in terms of the integral ionic charges Z of Fig. 13-1.

The calculation of the electrostatic energy is completely straightforward, in principle. For example, the electrostatic energy for the sodium chloride crystal is obtained in the following way.

1. Number each ion with an index i.
2. Sum the electrostatic interaction energy $(\pm)e^2/r_{ij}$ for all other ions (indexed j), using a plus or minus depending upon whether the ith and jth ions are of the same or opposite sign.
3. Sum over all i.

4. Divide by two since each pair has been counted twice (for example $i = 6$, $j = 7$ and $i = 7$, $j = 6$).

5. Divide by the number of ion pairs N_p to obtain the electrostatic energy per ion pair, called the *Madelung energy*,

$$E_{\text{electro}} = \frac{1}{2N_p} \sum_{i,j}{}' \frac{(\pm)e^2}{r_{ij}}. \qquad (13\text{-}1)$$

It should be realized that taking the sum over all ions in a crystal is a nontrivial task, because the sum converges very slowly. This is a classic problem addressed early in this century by Madelung (Madelung, 1909); for a recent discussion, see Brown (1967, p. 93 ff).

It is convenient to write the result of such a calculation in general form. Let us imagine equal numbers of ions of charge Ze and $-Ze$; the energy for these will be proportional to Z^2. Since all interaction energies scale inversely with distance, and therefore with nearest-neighbor distance, d, it is convenient to write the total electrostatic energy of the crystal, divided by the number of ion pairs, in the form

$$E_{\text{electro}} = -\alpha Z^2 e^2/d, \qquad (13\text{-}2)$$

where α is called the *Madelung constant*. The Madelung constant is dimensionless, of order one, and depends only upon the configuration of ions, not the magnitude of the charge nor the absolute spacing, both of which have been extracted in Eq. (13-2). Methods for precise calculation of the Madelung constant have been reviewed by Brown (1967). Shortcuts can be found, also, if one only wants to calculate changes in Madelung constants resulting from ionic displacement. Such a calculation is illustrated for the perovskite structure at the end of Chapter 19.

It is convenient also to generalize the Madelung constant to structures with more than one *magnitude* of charge (as in CaF_2). This can be done for the case of a solid with formula $A_{n_1}B_{n_2}$ in terms of a reduced Madelung constant (discussed, for example, by Johnson and Templeton, 1961). Charge neutrality requires that the magnitudes of the two charges, Z_1 and Z_2, satisfy $n_1 Z_1 = n_2 Z_2$. Then the reduced Madelung constant, α, is defined such that the electrostatic energy per molecular unit of $n_1 + n_2$ atoms is given by

$$E_{\text{electro}} = -\alpha(n_1 + n_2)Z_1 Z_2 e^2/2d. \qquad (13\text{-}3)$$

Here d is the nearest-neighbor distance in the structure. Notice that this reduces to Eq. (13-2) for the case $n_1 = n_2 = 1$ and that the value of α does not change if we define a molecular unit to be an integral number of primitive cells of a simpler structure; for any definition that does not meet these requirements, the values of the corresponding constant will vary significantly from one structure to another. For this definition they do not, as can be seen in Table 13-1, where values obtained by Johnson and Templeton (1961) for a number of structures are tabulated.

Since the calculation is purely arithmetic, the values can be (and have been) obtained to very high precision. Use of more significant figures than are given in Table 13-1 is probably misleading, however, because there are always other uncertain terms in the energy, which vary from structure to structure. Johnson and Templeton also give electrostatic energies for spinel and some other tertiary compounds, but they cannot be written in terms of the reduced Madelung constant just defined.

Madelung constants can be directly related to the potential at one ionic site due to all other ions. In fact, the potential at a site of type one (taken for convenience to be a positive ion) can be written

$$\varphi_1 = -\alpha Z_2 e/d, \tag{13-4}$$

and the potential at the second site can be written as $\varphi_2 = \alpha Z_1 e/d$. (Adding a constant to both equations does not change the electrostatic energy because of charge neutrality.) We can verify that the potential in Eq. (13-4) is correct by multiplying it by the charge of the ion $(Z_1 e)$ and completing the calculation of electrostatic energy to obtain Eq. (13-3).

The meaning of this electrostatic *potential* is much less clear than the meaning of the electrostatic *energy*; that energy corresponds to the work done in bringing the charges from infinity. At the last stage, when the ions come into contact, there

TABLE 13-1
Madelung constants for ionic crystals.

Compound	α	Compound	α
$AlCl_3$	1.40	βSiO_2	1.47
αAl_2O_2	1.68	SiF_4	1.25
$BeCl_2$	1.36	$SrBr_2$	1.59
BeO	1.64	$TiCl_2$	1.45
$CaCl_2$	1.60	TiO_2 (anatase)	1.60
CaF_2	1.68	TiO_2 (brookite)	1.60
$CdCl_2$	1.50	TiO_2 (rutile)	1.60
CdI_2 (Bozorth)	1.46	UD_3	1.64
CdI_2 (Hassel)	1.46	V_2O_5	1.49
$CsCl$	1.76	YCl_3	1.41
Cu_2O	1.48	YF_3	1.59
$LaCl_3$	1.54	Y_2O_3	1.67
La_2O_3	1.63	ZnO	1.65
MgF_2	1.60	ZnS (cubic)	1.64
$NaCl$	1.75	ZnS (hexagonal)	1.64

SOURCE: Johnson and Templeton (1961).

is some rearrangement of charge that may affect the local potentials but that has little effect on the total work done. Indeed, by the time Na and Cl ions are put together in a sodium chloride structure, the charge density corresponding to the superposition of free ion densities is only very slightly different from what would be obtained by a superposition of free *atom* densities. A weak octahedral distortion of the sodium 3s atomic charge cloud makes it resemble a chlorine ionic charge cloud.

The ambiguity can be eliminated if we imagine trying to calculate the change in the energy of one of the core states as the atom becomes part of an ionic crystal; this is called a **chemical shift** in the core energy. For a sodium core, the energy due to interaction with neighboring ions can be immediately calculated by using the Madelung constant, but we must subtract the interaction with the sodium valence electron that was removed from the atom when the crystal was formed. That term will very nearly cancel the Madelung potential, from Eq. 13-4; the Madelung potential has a value typically of eight electron volts, and observed chemical shifts have values typically from one to three electron volts. If one wishes to model some property, such as chemical shifts, using the Madelung potential, it is probably more appropriate to use the effective charges Z^* that will be introduced later than the charge Z that is used for the electrostatic energy. However, because of the cancellation, the errors and uncertainties may be large.

We have discussed Madelung energies and potentials only in connection with ionic crystals, though clearly, given a covalent crystal structure and effective charges, we can define Madelung potentials for covalent structures also. Indeed, a number of covalent structures are included in Table 13-1. We did not need these when we discussed the cohesive energy of the covalent solids, since there we chose a different path for assembling the crystal. We brought together neutral atoms, with no electrostatic energy gain (except for the contribution to the overlap interaction), and then wrote the change in energy as that associated with bond formation occuring at the observed spacing. *The two paths lead to the same energy change*, and we simply choose the one which, for each system, is most conveniently estimated.

The mention of Madelung potentials in covalent solids again points to an uncertainty in calculation of properties when the crystal is distorted. One contribution to the polar energy V_3 will be the difference between the Madelung potentials for the two atomic species in a polar covalent solid. When a crystal distortion changes the Madelung potential, we might expect a corresponding change in V_3. We discussed this point in Section 4-C and noted that pressure experiments did not seem to support it. We also noted in Chapter 11 that an intricate analysis of such shifts in the mixed tetrahedral solids (Pantiledes and Harrison, 1976) no longer seems warranted. Perhaps this is not so surprising in hindsight. The cancellation mentioned earlier in connection with chemical shifts may make itself felt in the distortion, and the pseudopotential representation (Chapter 18) of the origin of the polar energy suggests some insensitivity to distortion. Thus, any use of the Madelung potential other than in the calculation of cohesion should be considered questionable.

13-C   Ion–ion Interactions

At this point we shall restrict our discussion to the simple ionic solids of the AB type (equal numbers of oppositely charged ions), taking KCl as the prototypic system. It has the rocksalt structure. The interionic interactions are obtained by adding a Coulomb interaction $\pm Z^2 e^2/d$ to the overlap interaction. The Coulomb interaction causes the system to form an ionic structure and reduces the interion spacing to a value limited by the repulsive terms in the overlap interaction.

Interactions in Molecules

As in the inert gases, we wish to discuss the interionic interactions at two levels: first, at the microscopic level, which corresponds to the calculations of Gordon and Kim, and second, in terms of the phenomenological Lennard-Jones interaction. We look first at the interion interaction itself, and compare this with the experimental properties of the diatomic molecule. Gordon and Kim (1972) recalculated the overlap interaction for potassium and chlorine ions, among other systems, including the Coulomb interaction between them. From the position and depth of the minimum in the interaction, they predicted the properties of the molecule given in Table 13-2. The agreement with the experimental values is good; indeed, it is typical for the systems they considered. We conclude that the interionic interactions are well understood in terms of this conceptually simple, though computationally complicated, description.

We next consider the interactions at the second level, in terms of Lennard-Jones potentials. We considered the electronic structure of a system like KCl by beginning with the inert gas Ar and transferring protons between alternate nuclei to produce K^+ and Cl^-. Certainly, the extra proton will reduce the size of the K^+ ion in comparison to Ar, but the Cl^- ion will be correspondingly expanded, and we may hope that the K^+-Cl^- overlap interaction will not be changed too much.

TABLE 13-2
Predicted and experimental parameters for the diatomic molecule, KCl. The distance d_0 is the separation giving a minimum in the interion interaction, V_0.

Parameter	d_0 in Å	$-V_0(d_0)$ in eV
Gordon and Kim (1972)	2.56	4.97
Lennard-Jones	2.76	4.86
Experiment	2.66	5.0

SOURCES: The Lennard-Jones prediction is based upon the experimental parameters for Ar given in Table 12-1. Experimental values are from Smitt (1971).

Thus, we add a Coulomb interaction $-e^2/d$ to the Lennard-Jones overlap interaction for argon and again predict the spacing and binding of the KCl molecule. That result is also given in Table 13-2. Considering that the experimental input was from properties of the gaseous phase of argon and that the prediction is for KCl, the agreement is indeed remarkable. It is this kind of agreement—found when using empirical input from widely different sources—that tests the conceptual basis of the description.

The convenience of the Lennard-Jones form and the similarity between it and the form tabulated by Gordon and Kim, evidenced by Fig. 12-1 for the inert gases and by Table 13-2 for KCl, suggests that we use the Lennard-Jones form in our calculations of the mechanical properties of the ionic crystals. That still leaves us with several choices for determining the parameters σ and ε. In some ways the most appealing choice is to use inert-gas values from Table 12-2, preferably Bernardes's experimental values rather than the values calculated by Gordon and Kim, for the ions. For skew compounds containing elements from different rows we would use intermediate values. The most plausible choice seems to be the average of the two rows for σ and the geometric mean for ε. This scheme can be tested for the inert gases by comparing the interactions calculated by Gordon and Kim for the cross-interactions between different types of inert-gas atoms with the average based upon their calculations and listed in Table 12-2. This comparison is made in Table 13-3. The estimates are quite good, and are better than other choices, such as also using an average for ε.

The appeal of this approach is in the small number of parameters which need be put into the calculation (those of Table 12-2, which are also given in the Solid State Table) as well as in the remoteness-of-origin of these parameters relative to the mechanical properties of the ionic crystals being studied. We use that approach though it is quite crude. Notice, in particular, that it predicts the same properties for complementary skew compounds such as NaCl and KF, which we shall see is far from true. Much higher accuracy could be obtained by fitting σ and ε to the observed spacing and bulk modulus for each compound. That choice is better

TABLE 13-3
Lennard-Jones parameters calculated by Gordon and Kim (1972), compared with estimates based upon Table 12-2.

Atoms	σ in Å		ε in 10^{-3} eV	
	Gordon/Kim	Average	Gordon/Kim	Geometric mean
HeAr	2.78	2.74	5.8	6.5
NeAr	3.09	3.00	4.9	6.2
NeKr	3.27	3.10	5.1	7.4
ArKr	3.40	3.38	12.7	13.0

than using a separation energy, which is dominated by the Madelung potential. The resulting Lennard-Jones interaction could then be used to predict the elastic shear constants and other mechanical properties of the compounds.

Let us then turn to the ionic solids themselves. Kim and Gordon (1974) recalculated the total energy for the alkali halides that form rocksalt structures and thereby computed from first principles the lattice spacing, the *separation energy* (the energy per ion pair required to separate the solid into isolated ions—this comes from the theory more naturally than does the cohesive energy, which is relative to isolated neutral atoms), and the bulk modulus. For KCl, the agreement of the values for the three properties with experimental values is typical of the calculations. The calculated (and in parentheses, the experimental) values for KCl are 3.05 (3.15) Å, 175 (166) kcal/mole, and 2.3 (1.9) $\times 10^{11}$ dyne/cm^2. Again we may say that the interactions are quite well understood in terms of the microscopic theory. We shall return to the interpretation of these properties in terms of simple models in Section 13-D.

For *accurate* modeling of the properties we would take a form for the overlap interaction given by Eq. (12-1), add the Madelung energy, and fit the observed lattice spacing and bulk modulus, to obtain the parameters σ and ε for each compound. For best accuracy the experimental parameters should have values taken at a temperature of absolute zero (even if one wishes to interpret experiments at finite temperature since, for example, the lattice distance changes with temperature though the interaction presumably does not). For each compound, the values obtained for the parameters σ and ε are significantly different from values obtained from the inert gases. We should not be fooled by the fact that good agreement is found for separation energy in any compound, because that is heavily dominated by electrostatic energy. (The Madelung contribution to the separation energy is 8.0 eV per pair for KCl; the total separation energy found by experiment is 7.4 eV.) It is also presumably as reasonable to use an exponential representation of the overlap interaction. (See, for example, Kittel, 1967, p. 93, or Tosi, 1964. Both list the relevant parameters for the alkali halides.) Here, the same form used for the inert gases has been chosen for simplicity and internal consistency, and the inert-gas parameters from Table 12-2 will be used for the rest of our study. Other forms of the overlap interaction could be used, and for the Lennard-Jones form, more accurate parameters can be determined from the properties of the compound. We shall list the experimental properties from which the parameters can be determined, in Section 13-D.

13-D   Cohesion and Mechanical Properties

We have seen that the nearest-neighbor distance in ionic solids is determined by the joint action of the Coulomb interaction and the predominantly repulsive overlap interaction that we approximate with the Lennard-Jones form. First, let us explore the properties of KCl, an ionic solid with ion charge $Z = 1$. We have seen that the Coulomb energy per ion pair in the face-centered cubic structure is

$-1.75\ e^2/d$ for ions having $Z = 1$. In that structure, each ion has six nearest neighbors, so the energy per ion pair from the overlap interaction is six times $V_o(d)$. That overlap interaction is given in Eq. (12-1); with K and Cl both neighbors to argon, we use $\sigma = 3.4$ Å and $\varepsilon = 10.4 \times 10^{-3}$ eV; these are experimental values given for argon in Table 12-2 and the Solid State Table. We have a separation energy (a positive number) per ion pair,

$$E_{sep} = 1.75e^2/d - 24\varepsilon[(\sigma/d)^{12} - (\sigma/d)^6]$$
$$= 7.41\eta - 0.250(\eta^{12} - \eta^6)\ \text{eV}, \tag{13-5}$$

with $\eta = \sigma/d$. By setting the derivative with respect to η equal to zero, we find the maximum at $\eta = 1.12$, thus predicting the equilibrium d to be 3.05 Å (compared to the experimental value of 3.12 Å at the absolute zero of temperature, $T = 0$), and a separation energy of 7.82 eV per pair (compared to 7.36 eV at $T = 0$). It is good agreement, considering that no data from ionic solids were used, and it confirms that a description based upon the overlap interaction, with Coulomb interaction added, is essentially correct.

The spacing for KCl is much smaller than the spacing for solid argon ($d = 3.76$ Å in Ar) and the binding is two orders of magnitude larger, both because of the Coulomb interaction. The prediction of the separation energy is not sensitive to the overlap interaction; the electrostatic contribution alone was 8.27 eV and the overlap contribution -0.45 eV. The spacing in the KCl solid is somewhat larger than in the KCl molecule since, though the electrostatic contribution is only increased from e^2/d to $1.75e^2/d$, the overlap contribution is increased by a factor of six. It is important to bear in mind that *cohesive* energy is defined as the energy required to separate the crystal into isolated neutral atoms and that it differs from the *separation* energy by the energy gained in transferring the electrons from the negative ions to the positive ions and thereby neutralizing them. For alkali halides, this gain is the ionization energy of the alkali atom minus the electron affinity of the halogen.

We can also estimate the bulk modulus of KCl by differentiating the separation energy twice with respect to d. Doing so gives $d^2\partial^2 E_{sep}/\partial d^2 = -1.09$ eV at the minimum-energy separation. This corresponds to a value for the bulk modulus of $B = (1/18d)\ \partial^2 E_{sep}/\partial d^2$ or 3.4×10^{11} dyne/cm^2 (see Table 13-4 for appropriate formulae) compared to the experimental value, at room temperature, of 1.7×10^{11} eV. This is pushing the model rather far and the result is only semiquantitative. Better results for the bulk modulus are obtained by fitting σ and ε to d and the separation energy.

Although the simple model does give a semiquantitative account of bond length, cohesion, and compressibility for KCl, it is less useful to list the results of doing the arithmetic for the other alkali halides than merely to list the experimental parameters, as in Table 13-4. Predictions from the simple model will be approximately as accurate as the KCl results are, and more accurate models can be fitted to the experimental parameters if one wishes. Extension of the model to ions

TABLE 13-4

Properties of alkali halides in the rocksalt structure, at room temperature except where specified.

Compound	d in Å	Bulk modulus B in 10^{11} dyne/cm²	Separation energy E_{sep} in eV per ion pair	
			$T \approx 293$ K	$T = 0$ K
LiF	2.014	6.71	10.52	10.71
LiCl	2.570	2.98	8.63	8.76
LiBr	2.751	2.38	8.24	—
LiI	3.000	1.71	7.71	—
NaF	2.317	4.65	9.31	9.46
NaCl	2.820	2.40	7.93	8.04
NaBr	2.989	1.99	7.53	7.57
NaI	3.237	1.51	7.08	7.04
KF	2.674	3.05	8.24	8.44
KCl	3.147	1.74	7.20	7.36
KBr	3.298	1.48	6.88	6.91
KI	3.533	1.17	6.51	6.56
RbF	2.815	2.62	7.87	—
RbCl	3.291	1.56	6.91	—
RbBr	3.445	1.30	6.62	—
RbI	3.671	1.06	6.29	—

SOURCES of data: E_{sep} is the energy per ion pair required to separate the crystal into isolated ions; $B = \Omega_p \, \partial^2 E_{sep}/\partial\Omega_p^2 = (1/18d) \, \partial^2 E_{sep}/\partial d^2$, and $\Omega_p = 2d^3$ is the volume per pair; values are from Tosi (1964), except values from $T = 0$. Values for $T = 0$, from Kittel (1967).

of higher charge and to more complex structures is immediate, though we can expect larger errors to result when Z is greater than one.

One feature that should be noted in passing is that the bulk modulus varies approximately as d^{-3} in contrast to covalent solids, in which it varies as d^{-5}. (We shall see that in metals it also varies as d^{-5}.) As a consequence, the bulk modulus for the alkali halides can be given the universal value of 6.7 eV per ion pair. It is not clear what the significance of this empirical rule is.

We can also examine general distortions of ionic solids. The rocksalt structure cannot be made stable by radial nearest-neighbor interactions alone; the stability is provided by the Madelung energy. Stability may also be enhanced by second-neighbor overlap interactions, but these contributions should be negligible. In any case, both the overlap interaction and the Coulomb interaction are central-force interactions, and it can be proved that any crystal of cubic symmetry held in equilibrium by the action of central forces alone will have elastic constants satisfying $c_{12} = c_{44}$. This is one of the **Cauchy relations**. (See, for example, Born and Huang, 1954, p. 136ff; also Wallace, 1972, p. 104.) Such a relation is very

valuable in studying bonding properties, since experimentally determined deviations from this relation give very direct evidence that noncentral forces are present
in the crystal. Experimental elastic constants for a number of cubic ionic crystals
are given in Table 13-5. We see immediately that the elastic constants satisfy the
relation $c_{12} = c_{44}$ quite well in the alkali halides, both in the rocksalt structure
and in the cesium chloride structure. However, for the other compounds there are
significant deviations. These deviations are traditionally associated with
covalency, and in fact, when, in Chapter 19, we add matrix elements between
nearest neighbors to our calculation of rigidity, these matrix elements being analogous to the covalent energy V_2, we will find angular forces arising in fourth order.
We call these a "chemical grip" and will evaluate them for the alkali halides in
terms of the electronic structure. Using the approximate forms for the band gaps
and matrix elements from the Solid State Table, we obtain a correction to c_{44} of
$14.8 \times 10^{11}/d^5$ (in erg/cm^3 if $d(\text{Å})$ is substituted) for all alkali halides, and no
correction to c_{12}. We see that this is of the right general magnitude to explain the
differences between c_{12} and c_{44}, but it is clear that other factors affect the difference also. For one thing, the particular chemical grip considered in Section 19-F
can only contribute positively to $c_{44} - c_{12}$, but some differences are observed
to be negative. This is also consistent with the differences in the experimental and
electrostatic values for c_{12}, which we shall describe next. We shall return to a
detailed treatment of the chemical grip in Chapter 19 but ignore it for the present,
since it is a small term.

For most distortions, both overlap interactions and Madelung terms contribute
to the elastic constants. However, for the distortion ε_4 associated with c_{44} in the
rocksalt structure, only the Madelung term enters. (There are no changes in
nearest-neighbor distance to first order in the strain.) Thus the experimental
values can be compared with the Madelung term, and by virtue of the Cauchy
relation, c_{12} should take the same value. That contribution has been calculated by
Kellerman (1940) and is $0.348Z^2e^2/d^4$. Values of this expression are given in Table
13-5 for comparison with experimental values for c_{12} and c_{44}. For the alkali
halides the scale of the agreement is roughly that for which the Cauchy relations
themselves are satisfied. Thus, combined with the theory given for the bulk modulus, $B = (c_{11} + 2c_{12})/3$, a semiquantitative theory of the elastic properties of the
alkali halides, based only on inert-gas data, is rather complete. The same may be
expected for the alkali halides in the CsCl structure. However, the agreement is
quite poor for the other ionic solids, which apparently cannot be described without a more careful treatment of chemical grips of the type discussed in Chapter 19.

Given this description of the forces, there is no difficulty in carrying out the
calculation of the full vibration spectrum for an ionic solid in the manner
described in Chapter 9. At least for the compounds we have considered here the
structures are sufficiently simple that the complications which arose in calculating
the spectra for the mixed tetrahedral solids are not present. The elastic constants
describe the low-frequency lattice vibrations and there is no reason to expect
the description at high frequencies to be either worse or better. (Some discussion
of the vibration spectra of ionic crystals is given, for example, by Wallis, 1965; a
number of properties related to anharmonicity are described by Cowley, 1971.)

TABLE 13-5
313
Elastic properties of cubic ionic crystals.

Structure and compound	$d(\text{Å})$	10^{11} erg/cm^3			
		c_{11}	c_{12}	c_{44}	$0.348Z^2e^2/d^4$
ROCKSALT STRUCTURE $(Z = 1)$					
LiF	2.01	11.2	4.5	6.32	4.91
LiCl	2.57	4.94	2.28	2.46	1.84
LiBr	2.75	3.94	1.87	1.93	1.40
LiI	3.00	—	—	—	0.99
NaF	2.32	9.7	2.44	2.81	2.77
NaCl	2.82	4.85	1.25	1.27	1.27
NaBr	2.99	3.97	1.06	0.99	1.00
NaI	3.24	3.03	0.89	0.734	0.73
KF	2.67	6.56	1.46	1.25	1.58
KCl	3.15	4.05	0.66	0.629	0.81
KBr	3.30	3.46	0.56	0.515	0.68
KI	3.53	2.75	0.45	0.369	0.52
RbF	2.82	5.52	1.40	0.925	1.27
RbCl	3.29	3.56	0.60	0.46	0.68
RbBr	3.45	3.14	0.48	0.383	0.57
RbI	3.67	2.56	0.36	0.280	0.44
CESIUM CHLORIDE STRUCTURE $(Z = 1)$					
CsCl	3.57	3.64	0.92	0.80	—
CsBr	3.71	3.07	0.84	0.75	—
CsI	3.95	2.45	0.67	0.63	—
HIGHER-VALENCE ROCKSALT STRUCTURE					
MgO	2.11	29.2	9.1	15.4	16.2 $(Z = 2)$
TiC	2.16	50.0	11.3	17.5	59.0 $(Z = 4)$
TEN-ELECTRON CESIUM CHLORIDE STRUCTURE					
TlCl	1.53	0.76	—	3.32	4.01 $(Z = 1)$
TlBr	1.48	0.756	—	3.45	3.78 $(Z = 1)$
TEN-ELECTRON ROCKSALT STRUCTURE					
PbS	2.97	12.7	2.98	2.48	4.12 $(Z = 2)$
PbTe	3.23	10.7	0.77	1.30	2.95 $(Z = 2)$
FLOURITE STRUCTURE					
CaF$_2$	4.7	3.39	—	2.36	16.4
SrF$_2$	4.30	3.13	—	2.51	12.35
BaF$_2$	4.06	2.53	—	2.68	9.01
PbF$_2$	4.72	2.45	—	2.57	8.88

SOURCE: Data are from Landolt-Börnstein (1966).

13-E   Structure Determination and Ionic Radii

Since the nearest-neighbor distance is determined principally by the overlap interaction, and the separation energy is dominated by the electrostatic energy, we might at first think that we could predict the structure simply by comparing Madelung constants from Table 13-1. For AB compounds, the CsCl structure has the highest value for α and would be the predicted structure. The first difficulty, however, is that the nearest-neighbor distance depends upon the structure: recall that the spacing is larger for the rocksalt structure than it is for the diatomic molecule. In particular, though the Madelung constants are nearly the same for the rocksalt and the CsCl structure, the former has six nearest neighbors and the latter has eight, so CsCl can be expected to have a larger spacing. It is very simple to rework the calculation following Eq. (13-5) for the modified Madelung constant and the number of nearest neighbors in the CsCl structure, to obtain the predicted equilibrium spacing d_0 and the Madelung energy. The results are given in Table 13-6 for the rocksalt, CsCl, and zincblende structures, obtained with parameters for potassium chloride.

The results are just as anticipated, with the cesium chloride structure having the largest spacing. It is in fact sufficiently large that the electrostatic energy is higher than for the rocksalt structure; the differences between nearest-neighbor distances are of more consequence than the differences between Madelung constants. We may compare the total separation energies, using these spacings and Eq. (13-5), to see that indeed, as expected, the prediction switches from the cesium chloride structure to the rocksalt structure. That might be of some comfort, since rocksalt is the correct structure for KCl, the parameters of which have been used in Table 13-6. However, the analysis leads to the same prediction for all alkali halides, including the cesium halides, which occur in the cesium chloride structure rather than the rocksalt structure. Thus the analysis fails to make a distinction between the two structures, a point made long ago by Hund (1925). We should regard the energy difference between the rocksalt and cesium chloride structures as being too small to be reliably given, at least by the simple theory. At the same time, the energy difference between an ionic structure and a covalent structure (the zincblende structure carried along also in Table 13-6) *is* sufficiently large that the

TABLE 13-6
Estimated nearest-neighbor distance, Madelung energy, and E_{sep} (Eq. 13-5), for KCl in three different structures.

Structure	d_0 at minimum (in Å)	$-\alpha e^2/d_0$ (in eV)	E_{sep} (in eV)
Rocksalt	3.04	-8.27	7.79
Cesium chloride	3.11	-8.13	7.74
Zincblende	2.95	-8.00	7.48

conclusion, from Table 13-6, that the covalent structure is not stable is meaningful.

There is a second difficulty with the simple theory just given and with the simple calculation. We have assumed that only the overlap interaction between nearest neighbors—that is, between a negative and a positive ion—is important in establishing crystal spacing. That is certainly appropriate if the two ion types are of similar size, since the overlap interaction drops rapidly with distance. Indeed, if we estimate the ion size from the size of the inert-gas atoms, a simple geometrical calculation confirms that in ionic structures the nearest-neighbor interactions always establish the crystal spacing. By transferring a proton to form the ions, however, the metallic ion is made smaller and the nonmetallic ion is made larger, so it might be that the spacing is established predominantly by the nonmetallic ion, with the metallic ions "rattling around" in the interstices. In particular, one might expect this to be so for the lithium compunds (though maybe not for LiF). The idea that the halide-ion contacts might set the spacing has been used to rationalize the different structures of the alkali halides. (For example, it has been argued that the cesium ion has a size comparable to halide ions, so that the spacing is established by the sum of the cesium and halide radii in both structures; the larger Madelung constant for the CsCl structure forces that structure. Alkali ions other than cesium are much smaller than halide ions, so the structure is determined by the contacting halides; the rocksalt structure gives the smaller alkali–halide ionic spacing and that structure is favored.) However, the idea appears to offer the wrong explanation. It cannot be made to work if one assumes reasonable ion sizes, and the complete description of solids made up of ions with variable ionic radii, given earlier, indicates that the difficulty lies in predicting the cesium structures, not the others.

All of the foregoing analysis argues against assuming that ions have rigid radii, particularly when analyzing subtle differences in structure. However, Table 13-6 indicates that at the equilibrium spacing, separation of ions in KCl varies only by 5 percent between the zincblende and CsCl structures and only by about 2 percent between the CsCl and rocksalt structures. Thus, on that scale of accuracy, it may be possible to define radii for the ions which can simply be added to give nearest-neighbor distances in various structures. The idea is very old; tables of such radii have been made by Pauling and by Zachariasen. A table formed by Zachariasen is shown in Table 13-7. These radii have relevance only to ionic crystals. Attempts have been made to develop tables for other crystals; for example, "covalent radii" can be developed for the tetrahedral solids discussed earlier and "metallic radii" can be defined for metals. The ionic radii are of particular value since they can be useful in studying the many complex structures in which many different constituent elements form crystals. For this purpose, corrections depending upon the number of nearest neighbors can also be obtained from tables (see, for example, Kittel, 1967.)

Notice that each row in Table 13-7 corresponds to a single electronic configuration, the same as that of the inert gas that appears within the shaded column. There is a steady decrease in the ionic radii in each row as we move to the

TABLE 13-7
Ionic radii in Å for closed-shell configurations, due to Zachariasen.

			He	Li 0.68	Be 0.30	B 0.16					
O 1.46	F 1.33	Ne	Na 0.98	Mg 0.65	Al 0.45	Si 0.38					
S 1.90	Cl 1.81	Ar	K 1.33	Ca 0.94	Sc 0.68	Ti 0.60					
Se 2.02	Br 1.96	Kr	Rb 1.48	Sr 1.10	Y 0.88	Zr 0.77	Nb 0.67				
Te 2.22	I 2.19	Xe	Cs 1.67	Ba 1.29	La 1.04	Ce 0.92					
Po 2.30	At 2.27	Rn	Fr 1.75	Ra 1.37	Ac 1.11	Th 0.99	Pa 0.90	U 0.83			

SOURCE: Kittel (1967).

right, corresponding to increasing nuclear charge; that point was discussed earlier. The average radius for two ions of equal and opposite charge in each row does not vary so greatly, however; it shows a mild decrease in radius with increasing valence difference. This may correspond to a reduction in the spacing from the increased Coulomb attraction against a rather constant overlap interaction.

Changes in ionic radii have a particularly dramatic effect in *superionic conductors*. (See Huggins, 1975, for a review; see also the proceedings of a conference on fast-ion transport in solids, Van Gool, 1973.) Copper iodide at elevated temperatures is a superionic conductor. At low temperatures copper iodide has a zinc-blende structure, with the iodine atoms lying in a face-centered cubic arrangement and the small copper atoms lying in the tetrahedral intersticies. This is a highly polar compound and the energy difference would not be great if the coppers were in octahedral interstices, thus forming the rocksalt structure. At elevated temperatures, a transformation occurs in which the copper sublattice "melts" and the copper becomes highly mobile, diffusing as ions between the rigid framework of iodine ions. More commonly, the crystal structure of the nonmetallic ions changes as the lattice of metallic ions melts; in both AgI and CuBr (at 146° C and 485° C, respectively), the nonmetallic ions transform to a body-centered cubic arrangement, but again the noble-metal ion lattice melts. There is a wide variety of such superionic conductors, all characterized by small ions diffusing through a rigid framework. In many, the small ion is a noble metal, but other small ions such as lithium or fluorine diffuse in other superionic conductors.

PROBLEM 13-1 *Estimating lattice distances*

The fluorite structure of CaF_2 was given in Section 13-A. Use the appropriate inert-gas overlap interaction from Table 13-3 (use the values based upon Table 12-2) and the electrostatic energy from Table 13-1 to estimate the nearest-neighbor distance of CaF_2 and compare with experiment. You can ignore the d^{-6} term if you like; it makes only a one-percent correction. Compare also with the sum of ionic radii.

PROBLEM 13-2   *Estimate of bulk modulus*

By using the same interactions mentioned in Problem 13-1, estimate the bulk modulus from $\partial^2 E_{sep}/\partial d^2$, evaluated at the observed bond length. (Notice that you need to recalculate the ion density for the fluorite structure.) Notice the relative size of the contributions to the bulk modulus made by the d^{-12}, d^{-6}, and electrostatic terms. Compare the total with the observed bulk modulus, $B = (c_{11} + 2c_{12})/3$.

CHAPTER 14

Dielectric Properties of Ionic Crystals

SUMMARY

The band gap in the electronic structure of ionic crystals is associated with the energy required to remove an electron from a negative ion and place it on a positive ion, rather than an ionization, as in the inert gases. Nevertheless, Pantelides found that the band gap varied approximately as d^{-2} for each valence difference, as in the inert gases. The interatomic matrix elements, as reflected in the width of the negative-ion p band, showed the same dependence. The optical absorption is dominated by the band gap and by the corresponding interionic transitions. Correspondingly, the dielectric susceptibility is dominated by interionic oscillator strengths, and cannot be described in terms of the traditional picture of independent polarizable ions, but Pantelides's theory of the susceptibility, based upon the d^{-2}-dependent band gap, works well and has fewer parameters. Although the bands in the ionic compounds are rather simple, the electronic properties themselves are complicated by excitonic, polaron, and self-trapping effects, which are characteristic of these solids.

The coupling between positive and negative ions that is essential to the interionic transitions also reduces the total charge associated with the ions, an effect we call softening of the ion. Simple theory gives a softening of 50 percent for all of the alkali halides in the rocksalt structure, a value that should not depend upon pressure; the softening is greater when the valence difference is greater than one (for example, CaS) and requires extension of the simple theory. The same coupling leads to charge redistribution when the lattice is distorted, and a predicted transverse charge of 1.16 for the rocksalt alkali halides, in reasonable accord with experiment. Though the effective ionic charge and transverse charge are useful for description of the appropriate properties, it is the integral valence charge that enters calculations of the cohesive energy and the current carried in ionic diffusion.

14-A Electronic Structure and Spectra

We think of the ionic solids as made up of closed-shell ions, so that to a first approximation, the electronic structure is like that of the inert-gas solids. Two important differences arise, however, from the transfer of protons to make ions: first, the electronic states on different ions are not the same and, second, the spacing is sufficiently reduced that there are important effects from the matrix elements between states on adjacent ions. It will be best to begin with the simplest description, as we have in other systems, and introduce complications as we go; one reason for this is that many properties are understandable without the full complexity of the true electronic structure.

Let us again take KCl as a prototypic system and center our discussion on the alkali halides. It appears that most of the concepts necessary for understanding the entire range of ionic solids arise already in these simple systems.

We begin with atomic argon, shown on the extreme left in the energy-level diagram in Fig. 14-1. The $3s$ and $3p$ levels are fully occupied and the excited state of lowest energy is the $4s$ level. Above that is the ionization energy of argon,

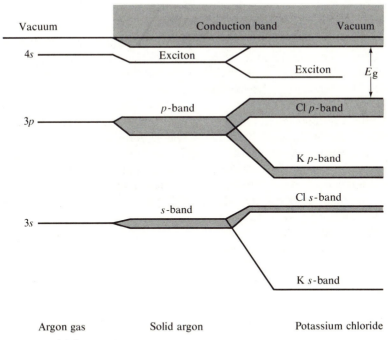

FIGURE 14-1

Formation of the energy-band structure of KCl. We start with argon atoms, and then put them in a simple-cubic crystal structure. Protons are then transferred between neighboring nuclei to form potassium and chlorine ions.

labelled "vacuum" level; excitation to this level is necessary for current to flow. We next place the argon atoms in a simple cubic lattice with nearest-neighbor distance equal to the nearest-neighbor distance in KCl, considerably smaller than the spacing in the face-centered cubic structure of real argon. Each of the levels is thereby broadened into a band. The energy required to free an electron from the argon atom (but not from the solid) is slightly reduced, corresponding to a conduction-band edge slightly below the vacuum level, as we indicated in the Chapter 13. This reduces the energy from the top of the $3p$ band to the bottom of the conduction band—this is now the band gap for the solid. A localized state remains, derived from the atomic $4s$ state, which is the excitonic state also discussed in the Chapter 13.

Finally, we transfer protons between adjacent nuclei, converting the atom which gives up the proton to a chlorine ion, and converting the atom which receives it to a potassium ion. This splits each of the bands into a chlorine band and a potassium band. The exciton level is also split; the exciton deriving from the chlorine is lifted into the conduction band, as shown.

The Band Gap

The most important parameter of the electronic structure is the band gap, which is indicated by E_g in Fig. 14-1. Experimental values for this parameter were compiled by Poole, Jenkin, Liesegang, and Leckey (1975), and are given in Table 14-1. These authors have made a critical study of the data and its interpretation; see also Poole, Liesegang, Leckey, and Jenkin (1975).

Since we associate the valence band with the anion p states and the conduction band with the cation s states, the most natural first estimate of the band gap is the difference between the two atomic term values, obtained from the Solid State Table. This gives the lower sets of values in parentheses in Table 14-1. Indeed, they are about the correct magnitude and describe the strong trend with atomic number of the halogen ion. They do not correctly describe the much weaker trend with the atomic number of the alkali ion. The principal trend in the band gaps, then, is the decrease with increasing size of the halogen ion, associated with its decreased electronegativity.

One should also check for a variation of the band gap with d^{-2}; and at this stage it should come as no surprise that when Pantelides (1975c) did check, he found such a variation. We see from Table 14-1 that the agreement of such a universal form,

$$E_g = \eta_g \hbar^2/(md^2), \tag{14-1}$$

for the alkali halides is perhaps even better than it was for the term-value difference; Eq. (14-1) correctly gives both trends mentioned above. It will also prove convenient for the treatment of other properties. The reason two such different views can simultaneously be correct is probably the same as it was for the inert gases, and was illustrated in Fig. 12-2.

TABLE 14-1   321

Experimental band gaps in the alkali halides, in eV.
In parentheses are shown, first, values from Pantelides's
formula, $9.1 \, \hbar^2/md^2$, and second, values of $\varepsilon_s^c - \varepsilon_p^a$
from the Solid State Table.

	Li	Na	K	Rb	Cs
F	13.6	11.6	10.7	10.3	9.9
	(17.1)	(12.9)	(9.7)	(8.7)	—
	(11.5)	(11.9)	(12.8)	(13.1)	(13.4)
Cl	9.4	8.5	8.4	8.2	8.3
	(10.5)	(8.7)	(7.0)	(6.4)	(5.4)
	(6.8)	(7.2)	(8.1)	(8.4)	(8.8)
Br	7.6	7.5	7.4	7.4	7.3
	(9.2)	(7.8)	(6.4)	(5.8)	(5.0)
	(5.7)	(6.1)	(7.0)	(7.3)	(7.6)
I	—	—	6.0	6.1	6.2
	(7.7)	(6.6)	(5.6)	(5.1)	(4.4)
	(4.5)	(4.8)	(5.8)	(6.0)	(6.4)

SOURCE: experimental values are from Poole, Jenkin, Liesegang, and
Leckey (1975).

NOTE: CsCl, CsBr, and CsI are in the CsCl structure.

For the alkali halides, Pantelides found a value $\eta_g = 9.1$, giving the predictions
listed in parentheses in Table 14-1. The one parameter does reasonably well
for all alkali halides, though clearly a much better fit can be made by letting
η_g differ for each alkali metal. In particular, values of η_g equal to 7.7, 9.1,
10.5, 11.4, and 13.3 do very well for Li, Na, K, Rb, and Cs, respectively. It is
interesting that, though the cesium compounds form a different structure, they do
not appear particularly out of line with the others. Pantelides also gave empirical
parameters for the divalent compounds (chalcogenides) and trivalent compounds
(pnictides) in the rocksalt structure. These are listed in Table 14-2. The agreement
with experiment for these compounds is comparable to that for the alkali halides.
(For a comprehensive tabulation of gaps, see Strehlow and Cook, 1973.) The
apparent insensitivity of the gaps to structure in the alkali halides would offer
hope that Table 14-2 might be of value for other structures also. Gaps for fluorite-
structure crystals (Table 14-3) do not show such simple systematics, though the
alkali-chalcogenide gaps can be predicted to within about 20 percent of the
experimental values, with $\eta_g = 2.8$. The structure of the alkali-chalcogenides is
called *antifluorite*, since it is the fluorite structure with all ion charges changed
in sign.

A second important parameter of the electronic structure is the *photothreshold*,
the energy required to take an electron from the top of the valence band, out of the

TABLE 14-2

Band and band-gap parameters given by Pantelides (1975c); η_g can be used to obtain the band gap and η_v to obtain the width of the upper valence band (nonmetallic p band).

Z	η_g	η_v
1	9.1	3.1
2	5.3	4.1
3	1.6	5.1

crystal; that is, to move an electron to the vacuum level. This parameter was also studied by Poole, Jenkin, Liesegang, and Leckey (1975), and, as suggested by Fig. 14-1, it is generally less than an electron volt in excess of E_g. Thus, as in the tetrahedral solids, it is less than the LCAO estimate of ε_p^a, but by a rather constant amount. They also studied the *electron affinity*, the threshold energy minus the gap.

TABLE 14-3

Band gaps (E_g, in eV) and nearest-neighbor distance (d, in Å) for compounds having a fluorite or antifluorite structure. The E_g values in parentheses are $2.8\,h^2/md^2$.

FLUORITE STRUCTURE

	CaF_2	BaF_2	CdF_2
E_g	10.0	9.07	6.0
d	2.36	2.7	2.33

ANTIFLUORITE STRUCTURE

	Na_2S	Na_2Se	Na_2Te
E_g	2.4 (2.7)	2.0 (2.5)	2.3 (2.1)
d	2.83	2.95	3.17

	K_2S	K_2Se	K_2Te
E_g	2.1 (2.1)	1.8 (1.9)	1.9 (1.7)
d	3.20	3.33	3.53

SOURCES: E_g from Strehlow and Cook (1973); d from Wyckoff (1963).

The Energy Bands

Poole, Liesegang, Leckey, and Jenkin (1975) have reviewed published band calculations for the alkali halides and tabulated the corresponding parameters obtained by various methods. Pantelides (1975c) has used an empirical LCAO method that is similar to that described for cesium chloride in Chapter 2 (see Fig. 2-2), to obtain a universal one-parameter form for the upper valence bands in the rocksalt structure. This study did not assume only one important interatomic matrix element, as we did in Chapter 2, but assumed that all interatomic matrix elements scale as d^{-2} with universal parameters. Thus it follows that all systems would have bands of exactly the same form but of varying scale. That form is shown in Fig. 14-2. Rocksalt and zincblende have the same Brillouin Zone and symmetry lines, which were shown in Fig. 3.6. The total band width was given by

$$W_v = \eta_v \hbar^2/md^2. \tag{14-2}$$

Pantiledes's values of η_v are given in Table 14-2. He went on to study the relation between these bands and the zincblende bands (Pantiledes, 1975c), as discussed in Chapter 6. In addition, he has carried out the corresponding analysis for the valence bands in the fluorite structure, to obtain universal bands for that system (unpublished). A more recent band calculation for fluorite itself has been made by Albert, Jouain, and Gout (1977).

We may rather easily compare these predictions with those we would obtain with the atomic matrix elements from the Solid State Table by carrying out an elementary LCAO calculation for the rocksalt structure, in analogy with that for CsCl in Section 2-A. A limited discussion will be given here, and later, a more general discussion will be given in connection with ion softening. We again consider wave numbers in a [100] direction and anion p orbitals oriented in the same direction; this will lead to the band labelled Δ_1 in Fig. 14-2. However, there will be no width to the band unless we also include cation s states, since the Solid State Table includes only nearest-neighbor matrix elements and anions do not have anions as nearest neighbors. An anion p state only has nonzero matrix elements between itself and the two neighbors in the [100] and [$\bar{1}$00] directions (this can be seen by symmetry); therefore the problem reduces to that of alternate s and p states along a line, each coupled to its neighbor by a matrix element $V_{sp\sigma}$. The solution is elementary, giving energies (relative to midgap) of $\pm[(E_g/2)^2 + (2V_{sp\sigma}\cos kd)^2]^{1/2}$. The lower band is the p band. Its width (which from Fig. 14-2 is about equal to the total band width) may be obtained by subtracting values at $k = 0$ and $k = \pi/(2d)$; to lowest order in $V_{sp\sigma}/E_g$, the width is $4V_{sp\sigma}^2/E_g$. It is convenient to write the width of the p band in this form, which corresponds both to the perturbation-theoretic formula, Eq. (1-14), and to the way we will treat ion softening later in the chapter. Substituting values for $\eta_{sp\sigma}$ and η_g then gives the corresponding $\eta_v = 1.48$. This is smaller by a factor of about two than the value given by Pantelides. In its very simplest form, this description of the bands is only semiquantitative, perhaps because of neglect of the cation p states. We shall not explore the question further.

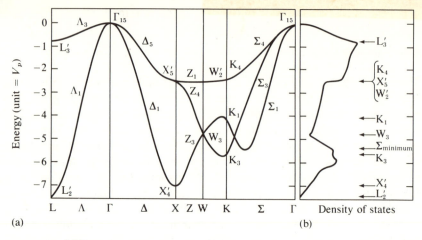

FIGURE 14-2
The universal form of the upper valence bands, and the density of states, found by Pantelides (1975c) for crystals of the rocksalt structure. The energy unit for the ordinate, V_p, is a second-neighbor matrix element which entered his fittings; the total width W_v is 7.5 V_p. [After Pantiledes, 1975c.]

The Spectra

The energy-level diagram, Fig. 14-1, and the parameters given in Tables 14-1 and 14-2, are the necessary ingredients for understanding the various spectra of the ionic crystals of the rocksalt structure. The interpretation of such spectra has not been straightforward, however, because of the difficulty in distinguishing transitions to exciton states from transitions to the conduction bands, and because of uncertainties in identifying peaks because of the variations in the intensity of absorption in different experiments. We illustrated this difficulty in the mixed tetrahedral solids with Fig. 11-12, showing a number of different spectra. Very recently, studies of angle-resolved photoemission on NaCl have been made by Himpse and Steinmann (1978), who appear to have surmounted these difficulties and given rather complete and unambiguous experimental determination of the bands for that system.

The optical absorption spectra were obtained earlier and interpreted for a very large number of ionic crystals. We have made use of such data in obtaining the numbers given in Tables 14-1 through 14-3. Reference was made in particular to the recent ultraviolet photoelectron-emission studies of the alkali halides by Poole et al. (1975), who reviewed other optical studies as well. We have noted the trends that are present in the electronic structure and that are reflected in these studies.

The interpretation of the X-ray spectra has also caused problems. The spectra have been determined from experiments in which an X-ray is absorbed, lifting an electron from a core state to some excited state. That excited state may be in the conduction band or it may be a lower energy state with the electron bound to the core hole. Different workers have interpreted the same peaks in different ways, leaving the field explained unsatisfactorily. Recently, Pantelides (1975b) included a careful determination of the energy of the conduction-band minimum in his interpretation of the X-ray spectra for the alkali halides, and could therefore systematically distinguish between core-hole excitons and conduction-band excitations. He found that the behavior was significantly different depending upon whether the electron was excited from the alkali core or the halogen core. In particular, the excitation spectra from the alkali-ion cores were dominated by excitonic effects. He attributed this to a lack of screening of the hole; it may be a more direct consequence of the fact that the conduction-band states themselves are predominently centered on the alkali ions, giving a more strongly bound excitonic state. Excitonic effects were weaker for excitations from the halogen-ion cores, which require greater separation of the excited electron and the core hole, but there were exciton effects in all cases. Even when the excitations were to levels above the conduction-band edge, the effect of the potential from the core hole was sufficient that the observed intensity, as a function of energy, did not strongly reflect the density of conduction-band states. A similar difficulty arose in our discussion of the optical spectrum of SiO_2.

Effects of Lattice Distortion

A little more should be said about the nature of the electronic excitations themselves. We have noted that binding of electrons to holes has a major effect on the absorption spectra of the ionic solids. There are other deviations from simple band behavior that are characteristic of ionic solids, though these effects are unimportant in covalent solids. The first is the effect of the lattice distortion due to an electron in a conduction-band state. That band is quite free-electron-like, though the probability density for the state is weighted more on the metallic ions than on the nonmetallic ones. An electron in that state draws neighboring positive ions towards it and repels neighboring negative ions, through direct Coulomb interaction. The combined state, consisting of an electron and its associated distortion, is called a *polaron*. (For some discussion of polarons, see Kittel, 1963, p. 137ff; for a more extensive discussion, see Schultz, 1956.) The polaron is mobile in the solid, but its effective mass is enhanced by the inertia of the displaced ions. The enhancement of the mass can be written

$$m_{\text{polaron}} = m[1 + (1/6)\alpha], \tag{14-3}$$

where α is a dimensionless coupling constant defined by Fröhlich, Pelzer, and Zienau (1950). (It is assumed here that the effective mass of the band itself is simply m.) It turns out that α is just twice the number of optical lattice-vibrational quanta (phonons) that would on the average arise if the electron itself were annihilated. (The dropping of the electron into a core state, for example, leaves a distorted lattice behind, and therefore, some vibrational energy.) The calculation leading to Eq. (14-3) assumed weak coupling; that is, it assumed that α is small compared to one. Actual estimates of α listed by Kittel (1963) range from 5.25 for

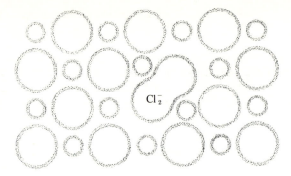

FIGURE 14-3

Two Cl^- ions (represented by large circles) move together, with a hole, to form a Cl_2^- ion. This "self-trapped hole", also called a V_K center, is immobile.

LiF to 6.4 for RbCl, indicating the necessity for the use of strong-coupling theory for a quantitative understanding of the alkali halides.

The effects of lattice distortion on a hole in an alkali halide are even stronger. The hole in fact becomes immobilized and is called a *self-trapped hole*. The nature of the distortion that does this is rather easily visualized: the hole is a missing electron on, for example, a chlorine ion. That ion and a neighboring chlorine ion then move together to form a Cl_2^- molecular ion, as shown in Fig. 14-3. The displacement of the ion immobilizes the hole. The hole could of course equally well lie on another ion pair, but to move it requires the rearrangement of the ions. Thus the rate at which a self-trapped hole can tunnel to neighboring sites is on the scale at which ions, rather than electrons, tunnel through potential barriers and is ordinarily completely negligible.

The lowering in energy corresponds to the rising of the hole energy-level into the gap. Thus it is possible for electronic excitations to occur with energy less than E_g, and correspondingly, for absorption peaks in the optical spectrum to occur at frequencies in the visible range. These color the otherwise transparent crystal and are called *color centers*. (For a discussion of color centers, see Brown 1967, p. 329ff.) The particular one illustrated in Fig. 14-3, called the V_K center, was identified first by Castner and Känzig, (1957). A more familiar center is the F center, in which an electron is trapped by the excess positive charge owing to a halogen-ion vacancy. The electronic state in an F center resides principally on the six neighboring alkali-metal ions, not within the vacancy itself.

14-B Dielectric Susceptibility

It would be quite natural in terms of our picture of the electronic structure to represent the dielectric response of an ionic crystal as the sum of the dielectric responses of the individual ions, and this is the way in which these properties have traditionally been understood. (See, for example, Kittel, 1967, p. 384.) Recently, however, Pantelides (1975a) has pointed out that this representation is not consistent with the view that the principal peaks in the optical absorption spectra correspond to transitions from valence-band states concentrated on the nonmetallic ion to conduction-band states concentrated on the metallic ion. Recall that in Section 4-A we saw that the same oscillator strengths that determine the optical absorption intensity enter the dielectric susceptibility. The assumption of independent polarizable ions corresponds to the assumption that the principal optical absorption comes from the excitation of electrons to excited states of the same ion. This was defensible in the inert-gas solids (though we noted that the gap was slightly reduced in those solids), but in the ionic crystal the nonmetallic ion electronic levels are greatly raised and the important excited levels (for exciton levels as well as for lower conduction-band levels) are dominated by the states on metallic ions; see Fig. 14-1. Pantelides noted in fact that a critical study of the analysis of experiments in terms of the independent-ion model did not support the model. The model *appeared* to work for the alkali halides, but this was by fitting 16 experimental numbers with 8 adjustable parameters and the systematic variation made this fitting possible. Little success was had with other compounds.

Calculation of the Susceptibility

Universal matrix elements were not available at the time of Pantelides's study, but now we may use them for a direct prediction of the susceptibility in the alkali halides. This is closely analogous to the calculation of susceptibility for the tetrahedral solids, done in terms of bond dipoles in Section 5-A by means of Eq. (5-7), and the corresponding calculation for the mixed tetrahedral solids. Since now we do not have independent two-electron bonds, however, the susceptibility must be formulated differently; the final result will be equivalent to the extreme polar limit of the tetrahedral solids.

In our starting picture of the electronic structure, each occupied state was an atomic state on the nonmetallic ion. There are, however, matrix elements between each of these states and the empty states on the neighboring ions, and these modify the occupied states in a way which can be calculated by perturbation theory, from Eq. (1-16):

$$\psi(\mathbf{r}) = \psi_i(\mathbf{r}) + \sum_j \frac{H_{ji}\psi_j(\mathbf{r})}{H_{ii} - H_{jj}} + \cdots \tag{14-4}$$

Thus a fraction of the charge, equal to the square of the coefficient of the state $\psi_j(\mathbf{r})$, is transferred to the state $\psi_j(\mathbf{r})$ on the neighboring ion. We shall sum this up carefully in the next section to obtain the total "softening" of the ion; for now let us look only at the terms that will be of interest in obtaining the susceptibility. In particular, if we are to apply a field in the x-direction, along a cube edge of the NaCl structure, we will be interested in any transfer of charge from a p state on a given halogen ion to the s state on the alkali ion to its right (the positive x-direction) or to its left (the negative x-direction). That transfer will only come from the x-oriented p state by way of the interatomic matric element $V_{sp\sigma}$ and, from Eq. (14-4), will be given by $V_{sp\sigma}^2/(\varepsilon_s^c - \varepsilon_p^a)^2$. We have written the energy difference explicitly, since an applied field $\mathscr{E}$ in the x-direction will modify it by $e\mathscr{E}d$, giving $\varepsilon_p^c - \varepsilon_p^a = E_g \pm e\mathscr{E}d$. Thus the field transfers a charge $\pm 2(V_{sp\sigma}^2/E_g^3)e\mathscr{E}d$

(to first order in $\mathscr{E}$) and induces a dipole, due to the coupling between this pair of states, of $2e^2 d^2 \mathscr{E} V^2_{sp\sigma}/E^3_g$. There is such a contribution from the coupling to the right and to the left for each spin for each of the four halide ions per unit cube of edge $2d$. This corresponds to an optical susceptibility of

$$\chi = 4e^2 V^2_{sp\sigma}/(E^3_g d). \tag{14-5}$$

Notice the proportionality of χ to $V^2_{sp\sigma}$; this indicates that Eq. (14-5) is entirely an interatomic contribution, as suggested by Pantelides. It could be evaluated by using the term-value difference from the Solid State Table for E_g, or the approximate formula in d^{-2} from Eq. (14-1) for E_g, or the observed E_g. We look first at the result of using the values from the Solid State Table; these values are listed in Table 14-4 along with the experimental values. The agreement is quite erratic and

TABLE 14-4
Theoretical and experimental optical susceptibility χ for ionic crystals in the rocksalt structure. Theoretical values are from Eq. (14-5) with $E_g = \varepsilon^c_s - \varepsilon^a_p$. The two compounds in each row are isoelectronic.

Compound	Optical susceptibility, χ		Compound	Optical susceptibility, χ	
	Theory	Experiment		Theory	Experiment
LiF	0.23	0.07	BeO	Wurtzite	—
LiCl	0.32	0.14	BeS	Zincblende	—
LiBr	0.39	0.18	BeSe	Zincblende	—
LiI	0.51	0.22	BeTe	Zincblende	—
NaF	0.11	0.06	MgO	0.71	0.16
NaCl	0.17	0.10	MgS	2.40	0.33
NaBr	0.21	0.13	MgSe	3.96	0.39
NaI	0.29	0.17	MgTe	Wurtzite	—
KF	0.04	0.06	CaO	0.21	0.18
KCl	0.07	0.10	CaS	0.53	0.28
KBr	0.08	0.11	CaSe	0.72	0.33
KI	0.11	0.14	CaTe	1.10	0.42
RbF	0.03	0.07	SrO	0.13	0.20
RbCl	0.05	0.10	SrS	0.31	0.27
RbBr	0.05	0.11	SrSe	0.42	0.31
RbI	0.08	0.14	SrTe	0.69	0.38
AgF	0.11	—	CdO	0.60	0.41
AgCl	0.34	0.26	CdS	Wurtzite	—
AgBr	0.52	0.32	CdSe	Wurtzite	—

SOURCE of experimental data: Van Vechten (1969a).

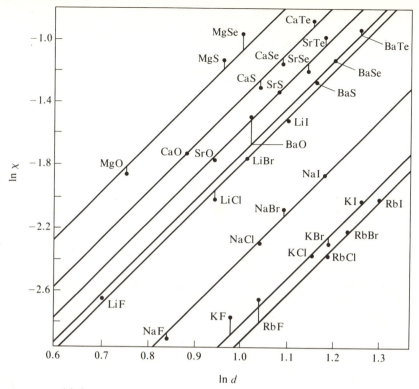

FIGURE 14-4

Plot of $\ln \chi$ versus $\ln d$ for rocksalt structure compounds. The diagonal lines correspond to $\chi = (d/d_\chi)^3$, with d_χ as given in Table 14-5. [After Pantelides, 1975a.]

becomes bad when $V_{sp\sigma}$ is comparable to the gap $\varepsilon_s^c - \varepsilon_p^a$, as in the lithium and magnesium compounds. Otherwise the general magnitudes are correct and the principal trends are reproduced. Considering the fact that none of the input comes from dielectric properties of the compounds, this strongly supports Pantelides's finding that the principal effects are interatomic, rather than intra-atomic as in the traditional model.

If, instead of using the term-value difference of Eq. (14-5) for E_g, we use Eq. (14-1), giving E_g proportional to d^{-2}, we obtain $\chi = d/(29.5 \text{ Å})$ for the alkali halides (with a different coefficient for other compounds), giving a value near 0.1, with only weak trends from one alkali halide to another. Pantelides (1975a) checked this by plotting $\ln \chi$ against $\ln d$ for compounds in the rocksalt structure, as shown in Fig. 14-4. These clearly indicate an experimental proportionality to d^3 rather than to d, with a coefficient depending upon the metallic ion involved; that is,

$$\chi = (d/d_\chi)^3, \tag{14-6}$$

with d_χ depending upon the metallic ion. The values extracted by Pantiledes are given in Table 14-5 and provide an empirical formula for the susceptibility. It is

TABLE 14-5
Parameters d_χ (in Å) entering the
empirical formula for the
susceptibility of ionic solids in
the rocksalt structure; $\chi = (d/d_\chi)^3$.

Li	4.9	Mg	3.8
Na	6.0	Ca	4.3
K	6.9	Sr	4.6
Rb	7.2	Ba	4.8

SOURCE: After Pantiledes (1975a).

interesting that a proportionality of the susceptibility to d^3 for ionic solids corre-- sponds to a universal matrix element of p_x for each metallic ion (see Eq. (4-5)), so Eq. (14-6) is analogous to the empirical finding in Section 4-B of a universal matrix element of p_x for all tetrahedral solids. It is not clear whether much meaning can be attached to the empirical formula. It is preferable to interpret the susceptibilities in terms of the theoretical formula, Eq. (14-5), and the values from the Solid State Table. Some of the discrepancy between theory and experiment is removed if we use experimental band gaps from Table 14-1 rather than term-value differences. However, discrepancies remain.

Susceptibility for the Fluorites

A corresponding analysis can also be applied to the fluorite structure. Focusing on a Ca ion in CaF_2, we note that the s state is coupled by $V_{sp\sigma}$ to a p state on each of its eight fluorine neighbors. Taking again a [100] field, the charge is transferred along the field a distance of $d/3^{1/2}$ rather than d, which was the distance the charge was transferred for the two neighbors in the rocksalt structure. Furthermore, the density of calcium ions is four per cube of edge $4d/3^{1/2}$, rather than four per cube of edge $2d$. These modify the leading factor in Eq. (14-5), giving

$$\chi = 2\sqrt{3}\, e^2 V_{sp\sigma}^2/(E_g^3 d) \tag{14-7}$$

for the fluorite or antifluorite structure. This may be readily evaluated, just as the expression in Eq. (14-5) was for the rocksalt structure. The results are compared with experiments in Table 14-6. We see that they are reasonably good for the fluorites; similarly, it was noted in connection with Table 14-3 that the band gaps were well given by the term-value differences for the fluorites. In Table 14-6, the one antifluorite is very poorly given, as are the band gaps based upon term values; however, using the more accurate band gap from Table 14-3 makes the agreement even worse. This is presumably because in that one case $V_{sp\sigma}$ is

TABLE 14-6

Theoretical and experimental optical susceptibilities for fluorite and antifluorite structures. Theoretical values are from Eq. (14-7), with $E_g = \varepsilon_s^c - \varepsilon_p^a$.

Compound	$d(\text{Å})$	$(\varepsilon_s^c - \varepsilon_p^a)$ in eV	χ from Eq. (14-7)	χ Experimental
		FLUORITE STRUCTURE		
CaF_2	2.36	11.58	0.09	0.08
SrF_2	2.51	11.99	0.06	0.09
$SrCl_2$	3.02	7.31	0.10	0.14
BaF_2	2.68	12.54	0.04	0.09
$BaCl_2$	3.18	7.86	0.06	0.16
		ANTIFLUORITE STRUCTURE		
Li_2O	2.00	8.65	0.47	0.14

SOURCE of experimental values: Pantelides (1975a).

nearly one-half E_g while in the others it is smaller by a factor of about 6; the ionic limit is quite inaccurate for this case.

Again, the susceptibility that is discussed here corresponds to that measured at frequencies that are high compared to the vibrational frequencies but low compared to $E_g/\hbar$; this is the optical susceptibility. The refractive index n of the crystal is given by the square root of the corresponding dielectric constant

$$n = \sqrt{\varepsilon_1} = \sqrt{1 + 4\pi\chi}. \tag{14-8}$$

At lower frequencies, the displacement of the ions contributes to the susceptibility; we turn to the contributions of these displacements next.

14-C Effective Charges and Ion Softening

As in the covalent solids, the meaning of the static charge on an ion in an undistorted ionic crystal depends upon the model used. A recent discussion of the charge distribution in ionic crystals and of the difficulty in defining effective charges is given by Jennison and Kunz (1976). For ionic crystals, however, one model is natural to the LCAO context, just as was true for covalent solids. In particular, our starting picture of inert-gas atoms having transferred protons leads to a charge of $+e$ for a potassium ion in KCl, of $+2e$ for a calcium ion in CaS, and so on. If the charges were that large for ionic solids and remained that large as the ions were moved, we would call the ions "hard" and the corresponding integral charges would enter the various properties of the solid. However, there must be matrix elements between the occupied orbitals on the chlorine and the empty orbitals on the potassium, so there is a reduction of the charge on the chlorine. We call that reduction *softening of the ion*, and it will change as the ions are moved.

Ion Softening

We may estimate the ion softening in perturbation theory as in Section 14-B, using the expansion of the wave function to first order in the interatomic matrix elements, Eq. (14-4), which we rewrite

$$\psi(\mathbf{r}) = \psi_i(\mathbf{r}) + \sum_j \frac{H_{ji}\psi_j(\mathbf{r})}{H_{ii} - H_{jj}} + \cdots \tag{14-9}$$

The terms omitted are of order $(H_{ij}/|H_{ii} - H_{jj}|)^2$; they are neglected in lowest-order perturbation theory. Thus, if a p state on a chlorine ion in KCl is coupled to an s state on its neighboring potassium ion by a matrix element H_{ij} (which we will relate to $V_{sp\sigma}$) and the difference in energy between the two states $H_{jj} - H_{ii} = E_g$ is large compared to the coupling, the probability that an electron in the perturbed state lies on that potassium ion is $(H_{ij}/E_g)^2$.

Now we may calculate the effective ionic charge on the potassium in the rock-salt structure. That charge, in units of the electronic charge, is $+1$ minus the contributions $(H_{ij}/E_g)^2$ from the electrons in p states on the six neighboring chlorine ions. Focusing on the potassium ion, we orient the p states on the neighboring chlorines parallel or perpendicular to the interion distance as in Fig. 14-5; only those with parallel (σ) orientation have nonzero matrix elements with the potassium s state. Those atomic matrix elements are obtained from the Solid State Table; thus, $V_{sp\sigma} = 1.84\hbar^2/md^2$. There are contributions to the charge on the potassium from the corresponding spin-up and spin-down electrons from each of the six neighboring chlorine ions, leading to an effective ionic charge on the potassium of

$$Z^* = Z - 12\left(\frac{V_{sp\sigma}}{E_g}\right)^2 = 1 - 12\left(\frac{1.84}{\eta_g}\right)^2 = 0.51. \tag{14-10}$$

In the last step the value of η_g was taken from Table 14-2 to obtain a simple universal result.

This charge may be used in a first estimate of the electrostatic fields and field gradients in a crystal; in Chapter 19 we shall discuss the importance of such potentials in splitting the atomic d states in some transition-metal compounds, or, as it is called, *crystal-field splitting*. The charge should *not* be used in the evaluation of the Madelung contribution to the cohesive energy. We saw in Section 13-B that the appropriate charge there is Z rather than Z^*. We might at first be surprised that the softening is as large as it is. Recall, however, that in the covalent

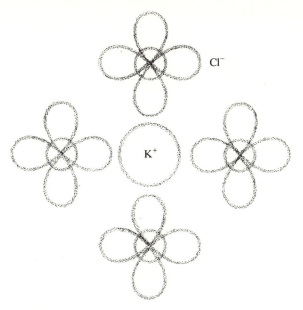

FIGURE 14-5

The p states on the six neighbors of a potassium ion in KCl contribute charge to the potassium ion, softening its charge from 1 to about 0.51. The same 50 percent softening is expected in the other alkali halides.

solids, when we transferred one proton to make Ge into GaAs, the corresponding electron transfer was *twice* as large, corresponding to correction terms in Eq. (14-10) larger than Z.

The most remarkable aspect of Eq. (14-10) is that the dependence upon bond length—and therefore, the dependence upon which alkali halide is being considered—has cancelled out. Another consequence is that the ion softening does not depend upon pressure. This explains why theories of the crystal-field splitting and its pressure dependence that are based upon hard ions have been successful; the 0.51 factor can be absorbed in an undetermined scale parameter, depending upon the shape of the orbitals being split, since the factor 0.51 does not change with distortion.

The calculation of Z^* for the cesium chloride structure is also quite direct and is discussed in Problem 14-2. However, the perturbation theory that we have used becomes inaccurate when we go to the rocksalt structure with $Z = 2$ or 3, or to the fluorite structure. The assumption of weak coupling restricts us to small percentages of softening. For $Z = 2$ and $\eta_g = 5.3$ from Table 14-2, Eq. (14-10) leads to a softening of 72 percent, and even larger values are obtained for $Z = 3$ and for the fluorites.

There is no difficulty in defining an effective charge in terms of the LCAO electronic structure even when the softening is large. Ultimately the band states can be written as linear combinations of atomic orbitals, with coefficients $u_{k\alpha}$, as in Eq. (3-17). If the band states have been calculated, we may select a constituent atom from, let us say, the nth column of the Solid State Table. Then, if we select just those values of α which correspond to atomic orbitals on the constituent atom, we may sum over all valence-band states $\mathbf{k}$ and over the orbitals α to obtain

$$Z^* = n - 2\Sigma_{\alpha,\,\mathbf{k}}\, u_{\mathbf{k}\alpha}^* u_{\mathbf{k}\alpha}, \tag{14-11}$$

where the factor of two is for spin. Eq. (14-10) was an approximate evaluation of this expression.

The Transverse Charge

We turn now to a matter of more direct physical relevance, the local electric dipole moment induced when a particular ion is displaced by some vector $\mathbf{u}$. The transverse charge e_T^* is defined to be the magnitude of that dipole moment divided by the displacement (and by the magnitude of the electronic charge). We saw in Eq. (9-22) that e_T^* is directly related to an observable splitting between the longitudinal and transverse optical-mode frequencies, so that this is a quantity that can be compared with experiment.

We will make the calculation, using perturbation theory, in the spirit of the calculation for covalent solids illustrated in Fig. 9-8. Consider a potassium chloride crystal with a single potassium ion displaced in the [100] direction, as illustrated in Fig. 14-6. The arrow represents a displacement of magnitude u. If the matrix elements all remained the same under the displacement, we would expect a dipole moment of Z^*eu toward the right, with Z^* given by Eq. (14-10). It is clear from symmetry that the matrix elements between the potassium s state and the two p states above and below the potassium, as well as those states disposed perpendicular to the plane of the figure, change only to second order in u (they change by the same amount for positive or negative u), so they contribute to the dipole only to third order in the displacement and do not contribute to e_T^*. One may also see that the changes in matrix elements between the potassium s state and the other p states on those chlorine ions also contribute only to third order. Similarly, the energy of the potassium s state can change only to second order in u (and if all potassium ions were displaced, the chlorine energies would shift only to second order in u), so we may neglect any change in E_g. However, the changes in matrix elements between the potassium s state and the p states to the right and the left *do* contribute. Noticing that $V_{sp\sigma}$ is proportional to d^{-2}, we find a change $\delta V_{sp\sigma} = -2V_{sp\sigma}u/d$. This transfers $4(V_{sp\sigma}/E_g)^2 u/d$ electrons for each spin to the potassium ion from the chlorine ion on the right. Similarly, the same number are transferred from the potassium ion to the chlorine on the left, leaving the charge on the potassium unchanged. However, $8(V_{sp\sigma}/E_g)^2 u/d$ electrons have been transferred $2d$ to the left, adding $16(V_{sp\sigma}/E_g)^2 eu$ to the dipole moment. Combined with the direct contribution of $[1 - 12(V_{sp\sigma}/E_g)^2]eu$ from Eq. (14-10), this leads to a transverse charge of

$$e_T^* = 1 + 4(V_{sp\sigma}/E_g)^2. \qquad (14\text{-}12)$$

This value, in contrast to Z^* itself, can be directly compared with experiment. For the alkali halides having a rocksalt structure, Eq. (14-12) applies, and we predict $e_T^* = 1.16$. This can be compared with the experimental values collected in Table 14-7. The agreement is rough, but the table gives no evidence of important trends which have been missed. A recalculation for the cesium chloride structure (Problem 14-2) gives a slightly higher prediction, of 1.22, but the experimental values for corresponding compounds are even higher, at 1.3.

To the extent that perturbation theory is appropriate, Eq. (14-12) can also be used for the alkaline-earth oxides, leading to a prediction of $e_T^* = 1.49$. Indeed, the observed values are also higher, with 1.77, 1.96, and 2.11 for MgO, CaO, and SrO, respectively (Lucovsky, Martin, and Burstein, 1971). Perhaps, as was suggested for the band structure, the numerical discrepancy arises from effects of the

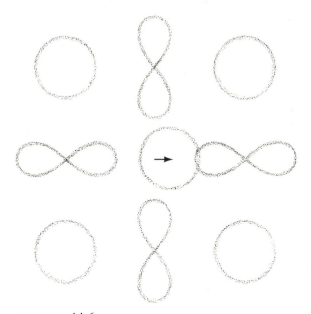

FIGURE 14-6

A displaced potassium ion in KCl. Only the p states oriented parallel to the ion separation are shown on the nearest-neighbor chlorine ions. The displacement not only moves the charge on the potassium ion but also transfers electrons from the chlorine on the right to that on the left.

TABLE 14-7

Experimental values of e_T^* for the alkali halides.
The simple ion-softening theory predicts $e_T^* = 1.16$
for all but the cesium compounds and 1.22 for
CsCl, CsBr, and CsI. (See Problem 14-2.)

	Li	Na	K	Rb	Cs
F	0.85	1.03	1.17	1.28	—
Cl	1.23	1.11	1.13	1.16	1.31
Br	1.28	1.13	1.13	1.15	1.30
I	—	1.25	1.17	1.17	1.31

SOURCE: Lucovsky, Martin, and Burstein (1971).

cation p states. The same approach can, of course, be applied directly to the fluorite structure. In this case, a value $e_T^* = 1 + (8/3)(V_{sp\sigma}/E_g)^2$ is obtained, which cannot be reconciled with the values less than one obtained by Samara (1976), who obtained e_T^* values of 0.86, 0.88, and 0.92 for CaF_2, SrF_2, and BaF_2, respectively. Further study is necessary in these other structures.

The e_T^* values discussed here can be combined directly with a model for the interion interactions, to obtain the lattice contribution to the static susceptibility, just as this was obtained for covalent solids in Section 9-D.

14-D Surfaces and Molten Ionic Compounds

The surfaces of ionic crystals are much simpler than those of the covalent solids. There does not appear to be evidence for the symmetry-breaking reconstructions that were discussed in Section 10-B. The surfaces are much like the bulk crystal, with all ions beyond a particular plane removed. The main part of the surface energy is the increase in electrostatic energy per ion due to termination of the crystal. The electrostatic surface energy was first calculated by Madelung (1918, 1919), who obtained $0.0422\, Z^2 e^2/d^3$ for the (100) rocksalt surface. Note that it is again the full ion charge, not the effective charge Z^*e, which enters the calculation, for the same reasons that Z entered the calculation of cohesive energy in Section 13-B. This is presumably the smallest electrostatic surface energy, which explains why the (100) surfaces are the natural cleavage and growth planes. Seitz (1940, p. 97) discusses corrections due to the overlap interaction and indicates that they are surprisingly large.

The liquid state of ionic compounds seems also to be rather simply analyzed. The change from crystal to liquid is much less dramatic in ionic solids than in covalent solids, since the ionic crystal is already close-packed. The liquid is presumably a somewhat random, but locally neutral, conglomeration of ions. Most of the studies of ionic compounds made in this part of the book have not depended greatly on details of structure and so remain appropriate in the liquid. It should be noted, however, that for studies of transport and diffusion, one should

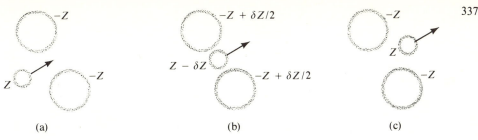

FIGURE 14-7

A diffusing ion transports Z electron charges, even though at various stages of its motion we may think of it as having an effective charge $Z^* = Z - \delta Z$.

take the magnitude of the ion charge to be Z and not Z^*. This is most easily seen by imagining a single ion moving past a collection of other ions, or past just two, as illustrated in Fig. 14-7. Although when an ion encounters other ions it may take on an effective charge, these differences disappear after the encounter and the net transport of charge is as if the ion had retained its charge of Z.

The melting process itself has been the subject of a number of studies. A very recent one, by Pietronero (1978), reviews these briefly and adds an interesting suggestion. Pietronero assumes that at zero temperature the spacing in the crystal is set by nearest-neighbor repulsions but that with increasing temperature vibrations of the larger ions (usually nonmetallic) may bring second neighbors into contact; this is taken as the condition for melting. By setting the vibrational energy equal to kT one obtains a relation between the typical excursion of the ion at the melting temperature, $(kT_m/C_0)^{1/2}$, and the separation between second-neighbor hard spheres. This gives a linear relation between the square root of the melting temperature and the ratio of ionic radii, in good accord with experiment. However, to make a quantitative fit it is necessary to take different ionic radii for the second-neighbor interaction than those used for the first-neighbor interaction.

PROBLEM 14-1   *Estimates of band gaps*

Use atomic term values from the Solid State Table to estimate E_g for the fluorite and antifluorite systems listed in Table 14-3.

PROBLEM 14-2   *Ion softening and e_T^**

Recalculate the effective ionic charge Z^* and transverse charge e_T^* for alkali halides in the CsCl structure, using the η_g from Table 14-2.

PART IV

OPEN-SHELL
SYSTEMS

IN INTRODUCING CLOSED-SHELL SYSTEMS it was noted that when the stability of the two-electron bond is not sufficient to hold the covalent solid in its open structure, a close-packed structure forms. This requires a new view of the electronic structure and its relation to the bonding properties. We have discussed the situation of particular stability in which the atoms and electrons rearrange to form the closed-shell configurations that occur in the inert-gas atoms. The tendency to do this is illustrated by the plot of atomic term values given in Fig. 1-8. Electrons shift from the high-energy levels corresponding to the left end of each period to the low-energy levels corresponding to the right, emptying some shells and filling others.

Such rearrangements are not always possible. If, for example, every atom in the system is an aluminum atom, there can be no such closing of shells. We wish to turn now to such systems in which covalent bonding is not maintained and in which closed-shell systems are not formed. The first case we shall discuss is that of simple metals, such as aluminum. We shall then turn to the more complex systems containing elements in the wings of the Solid State Table. Those to the left are the transition metals; those to the right are the rare earths (lanthanides) and actinides. The additional atomiclike d and f states that are partially filled in these systems give them a special character. We shall see that there are systems that are covalent, ionic, and metallic in exactly the sense we have discussed in earlier chapters, but with effects of d and f states added in. We shall see that there are also systems in which the d states are of central importance in the bonding properties. Strictly speaking, the covalent solids themselves are open-shell systems, as are molecular crystals of N_2, O_2, and so on. Their bonding properties were treated separately, however, in Part II.

The simple metals were the first system for which there was a relatively complete quantitative theory of the bonding properties in the sense presented in this text. There has therefore been time for a number of refinements and a variety of simplified versions to be developed. It is appropriate here to select a single approach, but some discussion of the relative merits of the others is desirable. We shall therefore begin with a summary of the history of this field and then outline current developments.

CHAPTER 15

Simple Metals

SUMMARY

After outlining the history of the theory of simple metals, we formulate free-electron states for the valence electrons, giving the model that has sufficed for a wide range of electronic properties. We then evaluate the total energy for this model, incorporating an extra kinetic energy for the free electrons because of the presence of core electrons. This term, called the repulsive term in the pseudopotential, compensates the strong Coulomb interaction with the metallic ions to produce a weak net interaction, which is approximated here by Ashcroft's simple empty-core pseudopotential, specified by a single empty-core radius for each element. The combined free-electron and electrostatic energy is minimized with respect to electron density to predict the equilibrium density, bulk modulus, and binding energy.

Metals such as sodium, magnesium, and aluminum are in some ways the simplest form of condensed matter. They appear in close-packed structures, and the valence electrons, only weakly perturbed by the presence of the ions, form a gas in the solid. The perturbation can readily be calculated and gives a simple and direct theory of the bonding properties. The theory of the bonding properties was developed in the early 1960's and has not changed fundamentally since then. (For a full account, see Harrison, 1966a. A more recent review is given by Heine and Weaire, 1970.) A relatively complete account of these systems can be given that is much more compact than the account given for the more complex covalent solids. Though some crucial developments are recent, the history of the theory is quite old.

It might seem surprising, after discussing a broad range of solids in terms of LCAO electronic structure, that we should suddenly switch to an electron gas as a basis for discussing metals. The basic reason was given in Section 2-C, where it was said that in metallic systems the dominant feature is the metallic energy in comparison to the covalent energy, and that just the reverse is true in covalent systems. The metallic energy is the matrix element that broadens the levels into bands; this broadening can also be thought of as the kinetic energy of the electrons in the bands. Treating this dominant attribute in metallic systems as free-electron kinetic energy has proved particularly simple and accurate.

This does not mean that the LCAO approach of the type we have used is incorrect or not useful. Recent applications of LCAO theory, based only upon electron orbitals that are occupied in the free atom, have been made to the study of simple metals (Smith and Gay, 1975), noble-metal surfaces (Gay, Smith, and Arlinghaus, 1977), and transition metals (Rath and Callaway, 1973). In fact, the LCAO approach seems a particularly effective way to obtain self-consistent calculations. The difficulty from the point of view taken in this book is that, as with many other band-calculational techniques, LCAO theory has not provided a means for the elementary calculations of properties emphasized here, but pseudo-potentials have.

The origin of the difficulty can be quickly seen. If we consider aluminum, with a nearest-neighbor distance of 2.86 Å, we estimate a $V_{ss\sigma}$ value of 1.31 eV. This is not large, but with each atom having twelve neighbors, this broadens the s band, in the simplest description, by $12 \times 1.31 = 15.69$ eV. This is three times as great as $\varepsilon_p - \varepsilon_s$. The p band is broadened even more and all atomic states become thoroughly mixed. Out of this mixture emerge the free-electron bands. It has only been possible to make the limited-basis LCAO calculation work by including the small (off-diagonal) matrix elements and overlaps with a very large number of distant neighbors. (Actual atomic wave functions and potentials were used in the calculations referred to in the preceding paragraph, rather than matrix elements fitted to other band calculations.) Even then, to obtain the total energy, one must sum over only the occupied portion of the band. Unitarity cannot be used and one must make a machine calculation of the density of states in order to perform the sum. We shall nonetheless return to an LCAO description of the bands when we consider transition metals in Chapter 20. The LCAO approach can provide useful insight into the nature of metals, just as we shall see in Chapter 18 that the pseudopotential approach can provide useful insight into the nature of covalent solids.

15-A   History of the Theory

As early as 1900, the German physicist Paul Drude recognized that many of the properties of metals could be understood if one assumed that the electrons within the metal were free. With the development of quantum mechanics in the 1920's, it became clear why such an assumption might be justified, and what seemed a relatively complete description of metals quickly developed, reflected in *The Theory of the Properties of Metals and Alloys* by N. F. Mott and H. Jones in 1936. In the 1950's, L. Onsager and A. B. Pippard recognized that the Fermi surface— the surface separating occupied and unoccupied states in wave number space— could be experimentally determined, and that it constituted an aspect of the electronic structure independent of the detailed models used to describe metals. The Fermi surface became a focus for experimental work on metals and for realistic energy-band calculations done with the newly invented electronic computers, and culminated in 1960 in the Fermi Surface Conference (see Harrison and Webb, 1960). Fermi surfaces were known in considerable detail for a number of simple metals; perhaps more important, knowledge of Fermi surfaces made it clear that the nearly-free-electron picture, which had seemed quite naive till then, gave a remarkably accurate description of the electronic structure of the simple metals. The residual interaction was associated with pseudopotentials, such as had been used for semiconductors by Phillips and Kleinman (1959), and was treated in perturbation theory (Harrison, 1963) to give the rather full account of the theory of simple metals that will be given here.

The pseudopotential had its roots in the orthogonalized plane-wave method for band calculations given by Herring (1940). The relation of this approach to a general form of pseudopotential was given by Austin, Heine, and Sham (1962), and a method for optimizing it was given by Cohen and Heine (1961). The use of that optimized form for a study of the properties of metals (Harrison, 1966a) has continued and the calculations have been refined to a remarkable accuracy (see, for example, Hafner and Nowotny, 1972, and Hafner and Eschrig, 1975). During this period there have been a series of parallel developments. They have been reviewed by McLaren (1974), and we shall indicate the high spots.

Early in the development of pseudopotential theory, an alternative formulation, the ***model potential method***, was given by Abarenkov and Heine (1965). In this method, the deep potential at the atom core was replaced by a flat potential that gave the same electron scattering. Animalu and Heine (1965) and Animalu (1966) evaluated the potentials for all of the simple metals, giving parameters that have not really been superseded since. (These parameters appear also in Harrison, 1966a.) A subsequent optimization of the model potential by Shaw (1968) did significantly improve the accuracy of the method for the calculation of bonding properties, however, and additional refinements have been made by Appapillai and Williams (1973).

Some of the motivation for refinement of the methods has been a search for an understanding of what determines the crystal structure of individual metals. The first serious attempt at predicting structures was directed at Na, Mg, and Al (Harrison, 1964), and it succeeded completely for these three metals. Subsequent studies (reviewed by Harrison, 1966a, p. 189ff) suggested that these metals were a fortuitous choice, however. More recent, careful studies seem to have confirmed that second-order pseudopotential theory is not adequate to determine the structures correctly. One contrary example is the work of Hafner and Nowotny (1972), which correctly gave nine out of nine structures for pure metals. An extension of second-order pseudopotential theory to alloys was made by Hafner (1976). It may be too early to know the significance of this most recent work. The rationalization of structures by Heine and Weaire (1970), which was based on qualitative second-order pseudopotential theory, has seemed unconvincing. One hesitates to regard the crude theory as meaningful when the more careful treatments prove unreliable.

The apparent inadequacy of second-order theory, at least for determination of structure, led to an exploration of higher-order terms. The contribution of third-order terms to the energy was first evaluated by Lloyd and Sholl (1968). The application of third-order theory to magnesium by Brovman, Kagan, and Holas (1971) indicated that indeed the contributions were significant. A study of the form of the multibody interactions that these higher-order terms lead to (Harrison, 1973b) also suggested that they may dominate the determination of structure. Bertoni et al. (1973) found that third-order terms are essential to an understanding of the details of the vibration spectrum of the carbon-row metals (beryllium in particular), but are much less important for magnesium and the heavier elements. Further refinements of the higher-order theory have been made by McLaren and Sholl (1974) and by Paasch, Felber, and Eschrig (1972).

The general conclusion from these studies has been that the simple second-order theory that will be given here provides a meaningful account of most of the bonding properties of the simple metals. The determination of specific crystal structures seems to be a particularly subtle matter. It appears that in solids in general, it may be possible to use theory to predict whether a system should form a covalent, ionic, or metallic structure, but it may not be possible to use theory to choose from among possible structures in each category. In some other isolated situations—the vibration spectrum of beryllium being an example—the simple theory also proves inadequate.

As refinements of pseudopotential theory have emerged, some simpler approximate models have also developed. Apparently the first was the point-ion model (Harrison, 1963c, described in Harrison, 1966a, pp. 55ff and 298ff). This was followed up by the empty-core model of Ashcroft and Langreth (1967), which has been used also by Shyu and Gaspari (1967, 1968). It is in fact the empty-core model that we shall use here. Smoother forms, designed to give better convergence, have been proposed by Krasko and Gurskii (1969) and by Brovman, Kagan, and Holas (1970). The idea in all of these models was to fit the simple forms to experiment or to accurate calculations, and then to treat other properties. A comprehensive discussion of this approach has been given by Cohen and Heine (1970). Another motivation for the introduction of approximate models has been to allow one to proceed beyond second-order theory. For example, the extra complexity of third-order theory prohibited the use of rigorous pseudopotential operators. Even now, new models are proposed, but they are little more than new algebraic fits of the same function and are of limited interest.

Our general approach will be to proceed with the simplest theory and add refinements, as was done for other systems. In particular, we first discuss the free-electron theory, which by itself accounts for many of the properties of simple metals, and then introduce the pseudopotential, which allows a treatment of the rest of the bonding properties. Most of the things that will be discussed are covered more completely in *Pseudopotentials in the Theory of Metals* (Harrison, 1966a) or in Heine and Weaire (1970).

15-B   The Free-Electron Theory of Metals

The central assumption of free-electron theory is that each atom gives up its valence electrons to the metal, and the states of these electrons are unaffected by the metallic ions formed from the atoms thus stripped of their electrons. The number Z of electrons each atom gives to the metal is unambiguous; it is the number in excess of the last inert-gas shell or in excess of the last completed d shell, whichever is less. These electrons form a uniform electron gas in the metal. We may thus proceed to a discussion of such a gas and obtain the consequences for the properties of the metal. In Section 16-F we shall introduce the modification of the electron states caused by the metallic ions, describing the influence of those ions by a pseudopotential.

The Geometry of Wave Number Space

We may think of a free-electron gas as having a vanishing potential (or equivalently, a constant potential, since we can measure energies from that potential level). The Hamiltonian becomes simply $-\hbar^2\nabla^2/2m$, and the solutions of the time-independent Schroedinger equation, Eq. (1-5), can be written as plane waves, $e^{i\mathbf{k}\cdot\mathbf{r}}$. We must apply suitable boundary conditions, and this is most conveniently done by imagining the crystal to be a rectangular parallelepiped, as shown in Fig. 15-1. Then we apply periodic boundary conditions on the surface, as we did following Eq. (2-2). The normalized plane-wave states may be written as

$$|\mathbf{k}\rangle = \Omega^{-1/2}e^{i\mathbf{k}\cdot\mathbf{r}}, \tag{15-1}$$

and the components of the wave numbers are restricted by the periodic boundary conditions to satisfy

$$\left.\begin{array}{l} k_x = 2\pi n_x/L_x, \\[4pt] k_y = 2\pi n_y/L_y, \\[4pt] k_z = 2\pi n_z/L_z, \end{array}\right\} \tag{15-2}$$

in analogy with Eq. (2-3); n_x, n_y and n_z are integers.

The allowed wave numbers form a closely spaced grid in wave number space, as illustrated in Fig. 15-2, with one allowed wave number for each cell of size

$$\delta k_x\,\delta k_y\,\delta k_z = \frac{2\pi}{L_x}\frac{2\pi}{L_y}\frac{2\pi}{L_z} = (2\pi)^3/\Omega.$$

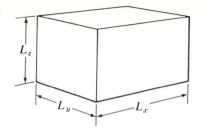

FIGURE 15-1

A metal in the shape of a rectangular parallelopiped. Periodic boundary conditions are applied on the surface. The total volume is written $\Omega = L_x L_y L_z$.

This corresponds to a density of states, in wave number space, of $\Omega/(2\pi)^3$. The energy of each state is given by $\hbar^2 k^2/2m$, so the state of the gas that is of lowest energy will be obtained by putting one electron of each spin in each state within a sphere, the **Fermi sphere** in wave number space, with a radius k_F chosen to be just large enough that all electrons are accommodated. If there are N electrons per unit volume, or a total of $N\Omega$ electrons, the product of the sphere volume, times two for spin, times the density of states in wave number space, must equal $N\Omega$:

$$\left(\frac{4\pi k_F^3}{3}\right) \cdot 2 \cdot \frac{\Omega}{(2\pi)^3} = N\Omega. \tag{15-3}$$

The volume of the system cancels out, so the Fermi wave number does not depend upon the size of the system, only upon the density of electrons. We may solve immediately for k_F:

$$k_F = (3\pi^2 N)^{1/3}. \tag{15-4}$$

The electron density is given by the number Z of valence electrons per atom divided by the volume per atom, Ω_0. Values of k_F for the simple metals are given in Table 15-1.

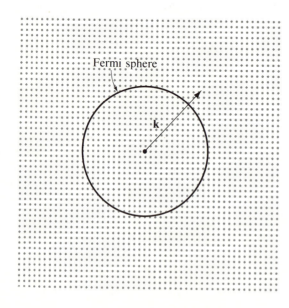

Fermi sphere

k

FIGURE 15-2

The states allowed by periodic boundary conditions form a fine grid in wave number space. The lowest energy state of the system is obtained if pairs of electrons are placed in the states of smallest **k**, thus occupying all states within a sphere called the Fermi sphere.

TABLE 15-1

The simple metals, their principal crystal structures, and Fermi wave numbers k_F, in Å^{-1}. Arrangement is by column and row of the Solid State Table. The values of k_F shown here also appear in the Solid State Table.

1 ($Z = 1$)		2 ($Z = 2$)		3 ($Z = 3$)		4 ($Z = 4$)		9 ($Z = 1$)		10 ($Z = 2$)	
								Li			
								hcp	1.126		
		Be				C		Na		Ca	
		hcp	1.942			dia	2.759	hcp	0.909	fcc	1.113
		Mg		Al		Si		K		Sr	
		hcp	1.374	fcc	1.752	dia	1.809	bcc	0.733	fcc	1.019
Cu		Zn		Ga		Ge		Rb		Ba	
fcc	1.361	hcp	1.590	—	1.661	dia	1.743	bcc	0.686	bcc	0.983
Ag		Cd		In		Sn		Cs			
fcc	1.204	hcp	1.412	—	1.506	white	1.633	bcc	0.633		
						dia	1.520				
Au		Hg		Tl		Pb					
fcc	1.206	—	1.363	hcp	1.461	fcc	1.580				

NOTE: Only the three most common metallic structures are listed: hcp = hexagonal close-packed, bcc = body-centered cubic, and fcc = face-centered cubic; values for the diamond structure will be discussed in Chapter 18.

It is important to recognize that there is no flexibility in the choice of Z. There are no corrections to the integral values, and even though, for example, lead may form compounds as a doubly charged ion (the ten-electron compounds mentioned earlier), the true bands in the metal resemble the free-electron bands for $Z = 4$, not for $Z = 2$. On the other hand, the values of k_F given in Table 15-1 do not really warrant the precision indicated there; they depend upon pressure and temperature, and where real accuracy is required, they should be redetermined for the circumstances at hand, from Eq. (15-4).

Properties of a Free-Electron Gas

The treatment of the electrons as a free-electron gas ignores the effect of the ions and is in some sense like replacing the ions with a uniform positive background that cancels the charge on the electrons. We will not make that approximation here in calculating the total energy, but will compute the potential energy of the uniform electron gas in the presence of real ionic potentials. However, the approximation in which the background is kept uniform has proved sufficiently important to be given a name; such an approximation is called a *jellium model*. The jellium model can be extended to the vibrational properties by allowing the positive background, or jelly, to deform as a continuum, and in the jellium model it is assumed that the jelly has no rigidity of its own, only that arising from the

Coulomb energy. Thus the only parameters required to specify the jellium model are the density of electrons and the mass density of the jelly. The jellium model is an approximation to the free-electron treatment used here; at a number of points it will be of interest to note the results that are obtained with the jellium model.

The theory of the electronic properties of the simple metals that has been built from simple free-electron theory is extraordinary. It extends to thermal properties such as the specific heat, magnetic properties such as the magnetic susceptibility, and transport properties such as thermal, electrical, thermoelectric, and galvano-magnetic effects. This theory is discussed in standard solid state physics texts (see, for example, Harrison, 1970) and will not be discussed here. As a universal theory for all metals, it is not sensitive to the electronic structure; it depends only upon the composition of the metals through simple parameters such as those of Table 15-1, or upon electron scattering times, which are ordinarily regarded as empirical parameters. We focus in this chapter upon bonding properties and upon those features of the free-electron theory which are relevant to them.

We may note, in passing, that the properties of a free-electron gas are not as different from the properties of a covalent solid as one might at first think. A *classical gas* (as opposed to the electron gas, with its restriction of two electrons per state) has a specific heat of $(3/2) k$ per electron (k is the Boltzmann constant). A covalent solid has no available excited states with energy less than the gap E_g, so this specific heat is suppressed at temperatures $T \ll E_g/k$. Similarly, in a metal, only electrons within kT of the Fermi energy can find excited states available to them, and the specific heat is suppressed by a factor of the order of kT times the number of states per unit energy at the Fermi energy. The suppression turns out to be by a factor kT/E_F, of order 0.004 at room temperature. We calculated the magnetic susceptibility of covalent solids in Section 5-E, finding both paramagnetic and diamagnetic contributions. Similarly, in the free-electron gas, there are paramagnetic contributions (from the spin moment) and diamagnetic contributions and, as in the covalent solids, they are of order Ne^2d^2/mc^2 with d the nearest-neighbor distance. It is principally in the electrical conductivity itself (and other transport properties) that the behavior is fundamentally different.

The properties of electrons having the Fermi wave number are obtainable immediately from Table 15-1. The kinetic energy can be obtained immediately from Table 15-1: $\hbar^2 k_F^2/2m = 7.62 \times 1.94^2/2 = 14.34$eV for Be. The Fermi velocity, $\hbar k_F/m$, is about 1 percent of the speed of light and a thousand times the speed of sound. Of more interest for an understanding of bonding properties is the *average* kinetic energy of the electrons,

$$\left\langle \frac{\hbar^2 k^2}{2m} \right\rangle = \frac{\int_0^{k_F} (\hbar^2 k^2/2m) 4\pi k^2 dk}{\int_0^{k_F} 4\pi k^2 dk} = \frac{3}{5} \frac{\hbar^2 k_F^2}{2m}. \qquad (15\text{-}5)$$

This kinetic energy, varying as the 2/3 power of the electron density—see

Eq. (15-4)—is the term that keeps the metal from collapsing, just as it was the electronic kinetic energy in the overlap interaction that prevented the collapse of the covalent and ionic solids. The forces pulling the metal together are the electrostatic attraction between the free electrons and the compensating, positively charged ions, just as the Coulomb interaction between ions pulled the ionic solid together and the potential energy associated with V_2 pulled the covalent solid together.

Let us then obtain the total energy of the simple metal in terms of the free-electron model. As in the other solid types, the total energy consists of kinetic energy of the valence electrons, potential energy (the potential energy of interaction between electrons and ions, electrons and other electrons, and among ions), and exchange and correlation corrections to the energy of the electron gas. As in other solids, the core electrons are taken to be constituents of ions which do not change in the formation of the solid. However, we shall see that the influence of the core electrons on the valence electrons is important and somewhat subtle.

15-C   Electrostatic Energy

Let us look first at the potential energy terms, calculated in the context of a free-electron model in which the electrons form a gas of Z electrons per ion of uniform charge density. The ions are spherically symmetric (to the extent that they are the same as in the isolated atom) and to a very good approximation the charge distributions on neighboring ions can be taken as nonoverlapping; each carries a net charge of Z. If they were indeed *point ions*, all of the potential energy terms would correspond to the electrostatic energy of point-positive charges imbedded in a uniform, compensating, negative background. We shall discuss that energy first, and then discuss the corrections for the finite size of the ions.

The calculation in the approximation that considers ions to be points, called the electrostatic energy of the metal, is straightforward but tricky, as was the calculation of the electrostatic energy in the ionic crystals. The method was given by Fuchs (1935, 1936). There are frequently shortcuts for obtaining the energy for a particular arrangement in terms of the known value for the energy of another arrangement (see Harrison, 1966a, p. 165ff; see also Section 19-G for an example). The electrostatic energy per ion is customarily written in the form

$$\cdot \; E_{\text{electro}} = -\frac{1}{2}\frac{Z^2 e^2}{r_0}\,\alpha, \tag{15-6}$$

where r_0 is the radius of a sphere with volume equal to the atomic volume Ω_0 and α is the dimensionless constant characteristic of the crystal structure. Like the Madelung constants of the ionic structures, α depends only on the geometry of the arrangements; charge magnitudes and distances have been factored out. A value in electron volts is obtained by letting $e^2 = 14.40$ eV-Å, as given in the Solid

TABLE 15-2

Geometric coefficients α for the
electrostatic energy of several structures.

fcc	1.79
hcp (ideal close packing)	1.79
bcc	1.79
Diamond	1.67
Simple cubic	1.76

SOURCE: Harrison (1966a).

State Table. The atomic sphere radius may be obtained in angstroms with the formula

$$r_0 = (9\pi Z/4)^{1/3}/k_F, \tag{15-7}$$

derived from Eq. (15-4). Thus, for Be, this gives $r_0 = 1.25$ Å and $E_{\text{electro}} = -23.1\alpha$ eV.

The principal crystal structures that the simple metals take (see Table 15-1) are the face-centered cubic (fcc) and hexagonal close-packed (hcp) structures described in Section 3-A, and the body-centered cubic (bcc) structure, which is the same as the CsCl structure except that the different species of ions in the CsCl structure are replaced by the same metallic ion in the simple metal structure. Values of α for these three metallic structures, for the diamond structure, and for the simple cubic structure are given in Table 15-2. As with the Madelung constants, values of α are actually known very precisely (see, for example, Harrison, 1966a) and the body-centered cubic value is higher than the hexagonal close-packed or face-centered cubic values by one part in 10^4. However, this is such a small difference that it is quite irrelevant to the determination of structures among those three. Yet, that the values are higher for the three metallic structures than for the nonmetallic structures *is* the reason why the three more closely packed structures are the ones that occur. The other structures observed (e.g., Hg and In) are similarly close-packed; so, in fact, is the structure of the liquid metal.

It is interesting that if we estimate the electrostatic energy by taking a single ion and its share of the background as a uniform sphere of radius r_0, we may readily calculate the electrostatic energy to obtain $\alpha = 9/5$. That this is so close to the value of 1.79 for close-packed structures suggests that there is only weak interaction between the neutral cells of metallic structures. Also interesting is that if the electrostatic contribution is carried out cell by cell (Hall, 1979), there are additional contributions to α, of the order of -20 percent, which depend on the cell shape (and therefore on the surface of a finite crystal). It is likely that in a conducting solid these effects are eliminated by charge redistribution.

15-D The Empty-Core Pseudopotential

The fact that the ions are not points—that instead the core electrons lie outside the nucleus—means an additional potential well is localized at the core. (Notice that outside the core the potential is exactly $-Ze^2/r$, as in the point-ion approximation.) Let us call that difference in potential between the point-ion and the true core potential $v_c(\mathbf{r})$ at each core; each core shifts the energy of each valence electron by $\int \psi^*(\mathbf{r})v_c(r)\psi(\mathbf{r})\,d^3r$, giving a total shift in the energy of the crystal, per ion, of

$$E_{\text{core}} = \int_{\text{one core}} Nv_c(r)\,d^3r, \tag{15-8}$$

where N is again the uniform density of the valence electrons. The other terms in the core energy are the same as in the free atom and need not be considered.

We turn next to the kinetic energy of the electrons. In Eq. (15-5) was given the kinetic energy for the uniform valence-electron gas, and we might at first think that this could be directly added to the kinetic energy of the electrons in the cores. This would not be consistent with the way we calculated kinetic energy for the overlap interaction in Chapter 7, however, and would not be correct. In Chapter 7 we computed the kinetic energy locally in terms of the two-thirds power of the *total* electron density at each point. Let us write the valence density N and the core density $N_c(\mathbf{r})$; notice that $[N + N_c(\mathbf{r})]^{2/3} > N^{2/3} + N_c(\mathbf{r})^{2/3}$. Thus, even in the free-electron approximation, we should include extra kinetic energy for the valence electrons in the region of the core. This is the subtle way in which the core electrons influence the valence electrons and give rise to what is called the **repulsive term** in the pseudopotential representing the interaction between the cores and the valence electrons. We shall formulate the effect in sufficient detail here to obtain a convenient form for the pseudopotential; a careful quantum-mechanical derivation is given in Appendix D.

Combining Eqs. (15-4) and (15-5), we may write the average kinetic energy of the electrons at a point in terms of the electron density at that point; the density is $N + N_c(\mathbf{r})$ when we include both valence electrons and core electrons:

$$\left\langle \frac{\hbar^2 k^2}{2m} \right\rangle = \frac{3}{5}\frac{\hbar^2}{2m}(3\pi^2)^{2/3}[N + N_c(\mathbf{r})]^{2/3}. \tag{15-9}$$

To obtain the total electronic kinetic energy we simply multiply by the electron density and integrate over the crystal:

$$E_{\text{ke}} = \frac{3}{5}\frac{\hbar^2}{2m}(3\pi^2)^{2/3}\int [N + N_c(\mathbf{r})]^{5/3}\,d^3r. \tag{15-10}$$

In the region between cores, $N_c(\mathbf{r})$ is zero and the contribution is equivalent to Eq. (15-5). Within the cores, the core density may be very much larger than the valence electron density and we may expand Eq. (15-10) in N/N_c, keeping only the first two terms:

$$E_{\text{ke}} = \frac{3}{5}\frac{\hbar^2}{2m}(3\pi^2)^{2/3}\int N_c^{5/3}(\mathbf{r})\,d^3r + \frac{\hbar^2}{2m}(3\pi^2)^{2/3}\int N N_c^{2/3}(\mathbf{r})\,d^3r. \tag{15-11}$$

352

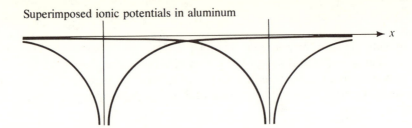

Empty-core model of the pseudopotential

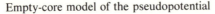

FIGURE 15-3
Superimposed ionic potentials in aluminum, and the empty core
approximation, which includes the effects of the repulsive kinetic
energy term.

The first term is simply the kinetic energy in the isolated ion, which is the same as
for the free atom. The second term is the interesting one, because it gives a shift in
the total energy exactly as if a potential given by

$$V_R = \frac{\hbar^2}{2m} (3\pi^2)^{2/3} N_c^{2/3}(\mathbf{r}) \tag{15-12}$$

had been introduced. Eqs. (15-11) and (15-8) can be compared to see this. The
term V_R is the repulsive term in the pseudopotential for the approximation of
kinetic energy calculated locally. The sum of the core potential and V_R is called a
pseudopotential since it contains terms that are not a potential, though it enters the
theory mathematically as a potential. We could substract the contribution that we
would have obtained in this region if there were no core electrons—it would be of
the form of Eq. (15-12) with $N_c^{2/3}N$ replaced by $N^{5/3}$ and would be of higher
order—and add it to the free-electron energy of Eq. (15-5) to obtain the total
kinetic energy.

This repulsive term tends to cancel the true potential of the core itself, a feature
noted very early in the development of the pseudopotential theory that we have
described. Ashcroft (1966) took advantage of this feature in proposing the *empty-
core model* of the pseudopotential. In this model, the repulsive term of Eq. (15-12)
is combined with the Coulomb potential of a point ion and the core potential v_c of
Eq. (15-8) to give a potential due to each ion, which is approximated by

$$\begin{aligned} w^0(r) &= 0 && \text{for } r < r_c; \\ w^0(r) &= -Ze^2/r && \text{for } r > r_c. \end{aligned} \tag{15-13}$$

Values for r_c are given in the Solid State Table. Their origin will be discussed in
Chapter 16. Here we have uŝed w rather than v; this is customary for pseudopotentials, and the superscript zero indicates that we have not added the screening
corrections.

This might seem a very arbitrary form, but the core radius can be chosen such that the integral of the total energy over the core gives the exact result, and then it is no approximation at all in the present calculation. We shall also see a mathematical justification for such a form in Appendix D. This has turned out to be a very convenient model and we shall use it here. When we go beyond the free-electron approximation, we shall include screening, that is, changes in the potential owing to redistribution of the valence charge density, but for the present we simply superimpose the empty-core potentials from Eq. (15-13), as illustrated in Fig. 15-3.

15-E   Free-Electron Energy

The combined effect of the potential energy and the extra kinetic energy has been to add a potential $+Ze^2/r$ for $r < r_c$ to each of the point-ion potentials we used for the evaluation of the electrostatic energy. Thus the empty-core contribution to the total energy per ion is simply

$$E_{ec} = \int_0^{r_c} N \frac{Ze^2}{r} d^3r = 2\pi Ze^2 r_c^2 N. \tag{15-14}$$

Finally, corrections to the energy of the electron gas are needed because of the correlated motion of the electrons; these corrections were mentioned at the beginning of Chapter 7 and are discussed more fully in Appendix C. The principal contribution is the exchange energy given by

$$E_x = -3Ze^2 k_F/(4\pi) \tag{15-15}$$

per ion. For beryllium, for example, it is given by $-3 \times 2 \times 14.4 \times 1.94/4\pi = -13.3$ eV.

The free-electron kinetic energy, Eq. (15-5), of $(3Z/10)\hbar^2 k_F^2/m$ per ion may be combined with the empty-core energy of Eq. (15-14) and the exchange energy of Eq. (15-15) to obtain what is called the *free-electron energy*; note that it depends upon the density but is otherwise independent of the arrangement of the ions. We may add this to the electrostatic energy of Eq. (15-6) to obtain the total energy per ion of the metal in the free-electron approximation. We have

$$E_{tot} = \frac{3Z\hbar^2 k_F^2}{10m} + \frac{2Z}{3\pi} e^2 r_c^2 k_F^3 - \frac{3Ze^2 k_F}{4\pi} - \frac{Z^2 e^2 \alpha k_F}{(18\pi Z)^{1/3}} . \tag{15-16}$$

Eq. (15-16) is a theoretical expression for the total energy per ion in terms of k_F and r_c. We specify r_c for each element from other sources, and then we can minimize E_{tot} with respect to k_F, to predict the equilibrium k_F and, therefore, also the equilibrium density. Thus, in the discussion of equilibrium density, compressibility, and total energy in this section, the only parameters needed for each system are r_c and the valence. (A similar study was made much earlier by Ashcroft and Langreth, 1967, who used the empty-core pseudopotential but included additional terms and additional adjustments.) We will see that the origin of the r_c values is quite independent of the volume dependence of the energy of the metal and does not make use of any data from the metal; the comparison of the predictions with experiment provides a true test of the empty-core model and the free-electron theory.

15-F   Density, Bulk Modulus, and Cohesion

To predict the equilibrium density, we simply write $dE_{tot}/dk_F = 0$, using Eq. (15-16), and solve the corresponding quadratic equation in k_F to obtain the values given in Table 15-3. These correspond to direct predictions of the equilibrium volume for the metal from microscopic atomic theory. The comparison with the experimental values in parentheses shows that the magnitudes are essentially correct and that all important trends are included. Lithium and beryllium are not as well given as the heavier elements, a discrepancy that has become familiar because it was found to be so also in the other solid types discussed. The discrepancies grow also for the heavy elements of high Z. The noble metals could have been included in this table, but there are large uncertainties in the choice of parameters and it seems clear that for the study of bonding properties, the noble

TABLE 15-3

Test of the minimum of E_{tot} (Eq. 15-16) as a prediction of equilibrium lattice spacing. The first values are predicted k_F in $Å^{-1}$, based on r_c values from the Solid State Table. Values in parentheses are experimental values of k_F from the Solid State Table.

2	3	4	9	10
			Li 0.89 (1.13)	
Be 1.53 (1.94)	–		Na 0.85 (0.91)	
Mg 1.29 (1.37)	Al 1.68 (1.75)		K 0.72 (0.73)	Ca 1.10 (1.11)
Zn 1.51 (1.59)	Ga 1.72 (1.66)		Rb 0.64 (0.69)	Sr 0.91 (1.02)
Cd 1.40 (1.41)	In 1.64 (1.51)	Sn 1.91 (1.63)	Cs 0.59 (0.63)	Ba 0.68 (0.98)
Hg 1.40 (1.38)	Tl 1.72 (1.46)	Pb 1.97 (1.58)		

metals should be included with the transition metals rather than the simple metals. In Problem 15-1 we recalculate the equilibrium density for diamond structures and find comparable agreement, including the lower density for diamond structures.

We may also obtain an immediate estimate of the bulk modulus,

$$B = \Omega^2 \, \partial^2 E_{tot}/\partial\Omega^2 = (1/9)k_F^2 \, \partial^2 E_{tot}/\partial k_F^2, \tag{15-17}$$

from Eq. (15-16), evaluated at the minimum or at the observed spacing. Notice that, strictly speaking, the last two expressions in Eq. (15-17) are only equivalent at the minimum, but we shall use the final expression in tabulating results at the observed spacing. From Eq. (15-17), one directly obtains the value in electron volts per ion. This must be multiplied by the ion density, and the units changed, to obtain

$$B = (0.0275 + 0.01102k_F r_c^2)k_F^5 \times 10^{12} \text{ erg/cm}^3, \tag{15-18}$$

where again, k_F in Å^{-1} and r_c in Å are to be substituted.

It is interesting to notice that for small core radii we obtain a universal constant times k_F^5, varying as the inverse fifth power of the interatomic distance, just as we noted empirically for the tetrahedral solids in Table 7-1. This is the first point at which we have derived such a dependence for the bulk modulus; the dependence given in Table 7-1 was simply an empirical fit. The corresponding relation between the covalent and metallic solids will continue to develop as we proceed with other properties. The relation to the closed-shell systems is not so close; recall that the bulk modulus for the ionic solids varied as d^{-3}. In fact, $k_F r_c^2$ is not small in Eq. (15-18) but is relatively constant, so that the d^{-5} dependence is maintained. Values obtained by substituting k_F and r_c from the Solid State Table are listed in Table 15-4, along with experimental values. Again the agreement and trends are impressive in view of the extreme simplicity of the free-electron approximation and empty-core model.

The magnitude of E_{tot} at the minimum is the **binding energy** of the metal, the energy required to separate the metal into isolated ions and isolated electrons. Eq. (15-16) is in fact not a bad estimate of that energy. Direct substitution of values in Eq. (15-16) gives, for example, 5.3, 21.6, and 52.2 eV per ion for sodium, magnesium, and aluminum at the observed spacing. This is in fair agreement with the observed values of 6.3, 24.4, and 56.3 eV, respectively. (Experimental values were given by Ashcroft and Langreth, 1967. They included other contributions in the prediction of E_{tot} and obtained better agreement with experiment.)

The binding energy, however, is not of as much interest as the **cohesive energy** which we have discussed for other systems; the cohesive energy is the energy required to separate the metal into neutral atoms. It is the binding energy minus the total ionization energy, the energy required to remove all three valence electrons in aluminum, for example. The cohesive energies for sodium, magnesium, and aluminum are 1.1, 1.5, and 3.3 eV, respectively. They are very small and, in

TABLE 15-4
The first value for each element is the bulk modulus (in 10^{12} erg/cm^3), obtained from $B = (1/9)k_F^2 \, \partial^2 E_{tot}/\partial k_F^2$, Eq. (15-18), evaluated at the observed k_F. Experimental values are in parentheses.

2	3	4	9	10
			Li 0.24 (0.11)	
Be 2.7			Na 0.074 (0.064)	
Mg 0.54 (0.35)			K 0.031 (0.031)	Ca 0.22 (0.17)
Zn 0.89 (0.56)	Ga 1.16		Rb 0.026 (0.025)	Sr 0.19
Cd 0.52 (0.45)	In 0.66 (0.37)	Sn 1.04 (0.50)	Cs 0.020 (0.016)	Ba 0.28
Hg 0.47 (0.25)	Tl 0.57 (0.44)	Pb 0.82 (0.43)		

SOURCE of experimental values: *International Critical Tables* (1928).

fact, are of the order of the errors in our calculated binding energy, so that we could not estimate them by subtracting the experimental ionization energies. It would be necessary to calculate the energies of the atoms with approximations similar to those used in calculating the binding energy. Then there would be hope that in calculating the difference the errors would cancel. Replacing the valence electrons in the atom by a uniform spherical electron gas (of ionic radius r_0, determined variationally) is not a good enough approximation. In thinking through that calculation, notice that one obtains just the same value as in the metal (except for the slight change in electrostatic constant α), so that it leads to no cohesive energy at all. The calculation requires a much more careful treatment of the atom itself than is appropriate here.

One way of understanding the origin of the cohesive energy in metals is in terms of the broadening of atomic levels into bands as the atoms are brought together, as was shown in Fig. 2-3. The coupling of the atomic levels tends to broaden the bands such that the average energy does not change, but since the bands are only partially occupied, only the lower energy states are occupied and energy is gained; this is analogous to the occupation of only bonding states when there is bonding–antibonding splitting. Since cohesion depends upon the change in energy in going from atoms to the solids, a method of calculation is needed which gives both the energy of the atom and that of the solid. One such method is the cellular method, or Wigner-Seitz method, discussed in Section 20-D. The use of this method and a comprehensive discussion of the theory of cohesion in metals has been given by Brooks (1963). It is also convenient to include here a table of experimental values for the simple metals—Table 15-5.

In terms of the free-electron approximation, which neglects the modification of

TABLE 15-5

The cohesive energy (energy of atomization) of the simple metals, in eV. Values of some semiconductors and semimetals are also included.

2	3	4	5	9	10
				Li 1.63	
Be 3.32	B 5.77	C 7.37		Na 1.11	
Mg 1.51	Al 3.39	Si 4.63		K 0.934	Ca 1.84
Zn 1.35	Ga 2.81	Ge 3.85	As 2.96	Rb 0.852	Sr 1.72
Cd 1.16	In 2.52	Sn 3.14	Sb 2.75	Cs 0.804	Ba 1.90
Hg 0.67	Tl 1.88	Pb 2.03	Bi 2.18		

SOURCE: Kittel (1976), p. 74.

the electron states by the metallic ions, we have obtained not only a description of a wide range of electronic properties of the metal, but also a semiquantitative theory of the equilibrium density, bulk modulus, and binding energy. Although we have left the states as free-electron plane waves, we have included a shift in their energies owing to the ion-core potentials and an extra kinetic energy owing to the presence of the core electrons. These last two effects were approximated by an empty-core model pseudopotential, but since the parameter describing that model came ultimately from atomic data rather than metallic data, we may still regard the description as microscopic theory. We could even go on to treat some other properties meaningfully in terms of the free-electron theory, such as the elastic shear constants; in the alkali metals, at least, these constants are dominated by the electrostatic energy. However, these other effects are generally so intimately related to the changes in the electron states that we do best to consider those changes first.

PROBLEM 15-1   *Two-dimensional free-electron bands*

We may approximate the π bands in graphite by a two-dimensional free-electron band. The density of states in two-dimensional wave number space is $2A/(2\pi)^2$, where A is the area of the system, obtained just as the $2\Omega/(2\pi)^3$ for three dimensions. There is one π electron per atom, so you may find the Fermi energy E_F just as in three dimensions. This is to be compared with half the π-band width (there are two atoms per primitive cell), which can be read directly from the figure in Problem 3-3.

PROBLEM 15-2   *Free-electron theory of covalent solids*

Calculate the Fermi wave number for C, Si, Ge, and Sn by minimizing the energy in Eq. (15-16) for both a face-centered cubic structure and a diamond structure. The values of k_F given for C, Si, and Ge in the Solid State Table are for an electron density corresponding to the diamond structure. That for Sn is for the metallic structure (not face-centered cubic but a structure of similar packing density). Compare the predicted ratio of k_F values for the two structures in tin by noticing that the observed specific gravity is 5.75 in the diamond structure and 7.31 in the metallic white-tin structure.

PROBLEM 15-3 d^{-5}-*Dependence of the bulk modulus*

Use Eq. (15-18) to estimate the bulk modulus of C, Si, and Ge in the diamond structure and compare with the experimental values from Table 7-1. Check the d^{-5}-dependence of theory and of experiment by considering B/k_F^5.

Electronic Structure
of Metals

SUMMARY

The lattice pseudopotential, no matter how weak, can cause diffraction of the electrons, modifying their orbits in a magnetic field. The corresponding changes can be represented as a reassembled Fermi sphere, called the Fermi surface for nearly free electrons. As the pseudopotential is increased, there is additional distortion of the Fermi surface and of the energy bands near the Bragg planes for the perfect crystal. The pseudopotential also changes the charge distribution, causing screening of the pseudopotential, which is calculated in perturbation theory. In first order the calculation can be done atom by atom, leading to screened-pseudopotential form-factors, which can be used also for crystals with defects. In such cases the electrical resistivity that the defects cause can also be calculated in perturbation theory.

The free-electron approximation described in Chapter 15 is so successful that it is natural to expect that any effects of the pseudopotential can be treated as small perturbations, and this turns out to be true for the simple metals. This is *only* possible, however, if it is the pseudopotential, not the true potential, which is treated as the perturbation. If we were to start with a free-electron gas and slowly introduce the true potential, states of negative energy would occur, becoming finally the tightly bound core states; these are drastic modifications of the electron gas. If, however, we start with the valence-electron gas and introduce the pseudopotential, the core states are already there, and full, and the effects of the pseudopotential are small, as would be suggested by the small magnitude of the empty-core pseudopotential shown in Fig. 15-3.

16-A   Pseudopotential Perturbation Theory

Let us systematically discuss the effect of the pseudopotential in terms of the empty-core form, Eq. (15-13). We first write the total pseudopotential in the metal as a superposition of the individual pseudopotentials $w^0(\mathbf{r} - \mathbf{r}_i)$ centered at the ion positions $\mathbf{r}_i$,

$$W^0(\mathbf{r}) = \sum_i w^0(\mathbf{r} - \mathbf{r}_i). \tag{16-1}$$

Since we wish to treat it as a perturbation upon the free plane-wave states, it will always enter the calculation as a matrix element between two such plane-wave states, as in Eq. (1-14). It is most convenient to write such a matrix element between a state of wave number $\mathbf{k}$ and a state with wave number $\mathbf{k} + \mathbf{q}$ as

$$\langle \mathbf{k} + \mathbf{q} | W^0 | \mathbf{k} \rangle = \int \frac{e^{-i(\mathbf{k}+\mathbf{q}) \cdot \mathbf{r}}}{\Omega^{1/2}} W^0(\mathbf{r}) \frac{e^{i\mathbf{k} \cdot \mathbf{r}}}{\Omega^{1/2}} d^3r = \frac{1}{\Omega} \int e^{-i\mathbf{q} \cdot \mathbf{r}} W^0(\mathbf{r}) \, d^3r. \tag{16-2}$$

This matrix element is the Fourier component of the pseudopotential with wave number equal to the difference $\mathbf{q}$. In the more complete pseudopotential theory of Appendix D, the pseudopotential becomes an operator, so that $W_0(\mathbf{r})$ and $e^{i\mathbf{k} \cdot \mathbf{r}}$ cannot be interchanged in the final step of Eq. (16-2), and the matrix element depends upon $\mathbf{k}$ also. This is not a major complication, but we shall utilize the simpler form given in the last step of Eq. (16-2), called the *local approximation to the pseudopotential.*

The Form Factor

Next, we use the expanded form of the pseudopotential, Eq. (16-1), to rewrite the matrix element as

$$\langle \mathbf{k} + \mathbf{q} | W^0 | \mathbf{k} \rangle = \frac{1}{\Omega} \sum_i \int e^{-i\mathbf{q} \cdot \mathbf{r}} w^0(\mathbf{r} - \mathbf{r}_i) \, d^3r$$

$$= \frac{1}{\Omega} \sum_i e^{-i\mathbf{q} \cdot \mathbf{r}_i} \int e^{-i\mathbf{q} \cdot (\mathbf{r} - \mathbf{r}_i)} w^0(\mathbf{r} - \mathbf{r}_i) \, d^3r. \tag{16-3}$$

In writing the first form, we interchanged the sum and the integral; in the second, we multiplied by $e^{-i\mathbf{q} \cdot \mathbf{r}_i} e^{+i\mathbf{q} \cdot \mathbf{r}_i} = 1$. This allows a very important factorization of the matrix element, which is familiar in diffraction theory. Within the integral, $\mathbf{r} - \mathbf{r}_i$ has become a dummy variable. Thus $\mathbf{r} - \mathbf{r}_i$ may be replaced by $\mathbf{r}$, and the integral is then seen to be independent of the ion position. The remaining sum depends only upon the ion positions and is completely independent of the form of the pseudopotential. Thus, we may write the matrix element in the factored form

$$\langle \mathbf{k} + \mathbf{q} | W^0 | \mathbf{k} \rangle = S(\mathbf{q}) w_q^0, \tag{16-4}$$

where the *structure factor*

$$S(\mathbf{q}) = \frac{1}{N_a} \sum_i e^{-i\mathbf{q} \cdot \mathbf{r}_i} \qquad (16\text{-}5)$$

depends only on the positions of the N_a ions making up the solid, and the *pseudo-potential form factor*

$$w_q^0 = \frac{1}{\Omega_0} \int e^{-i\mathbf{q} \cdot \mathbf{r}'} w^0(r') \, d^3r' \qquad (16\text{-}6)$$

depends upon the ion type. It depends upon the ion positions only through the atomic volume, $\Omega_0 = \Omega/N_a$. It may be evaluated immediately for the empty-core pseudopotential of Eq. (15-13), if we are cautious. There is a difficulty of convergence at large distances, which is eliminated if we replace $w^0(r')$ by $w^0(r')e^{-\kappa r'}$ in evaluating the integral, and then let κ become zero afterward. When we treat screening, we shall see that a factor of this form, with nonzero κ, is appropriate, but here it is simply a trick. Taking polar coordinates with a z-axis along $\mathbf{q}$ and integrating first over angle, we obtain

$$\begin{aligned}
w_q &= \frac{1}{\Omega_0} \int_{r_c}^{\infty} \int_0^{\pi} e^{-iqr'\cos\theta} \left(\frac{-Ze^2}{r'} \right) e^{-\kappa r'} 2\pi \sin\theta \, d\theta r'^2 dr' \\
&= \frac{-4\pi Ze^2 \cos qr_c}{\Omega_0(q^2 + \kappa^2)}.
\end{aligned} \qquad (16\text{-}7)$$

We see that indeed a convergent answer would be obtained by letting κ go to zero. However, let us anticipate the effect of screening, which eliminates the Coulomb field at large distances with a nonzero κ, and plot Eq. (16-7) with parameters appropriate to aluminium (see Fig. 16-1); because we choose to do this, we eliminate the superscript zero from w_q^0 in Eq. (16-7); lack of the zero indicates that w_q is a screened pseudopotential.

The plot of Eq. (16-7) shown in Fig. 16-1 is the conventional form for such plots; both ordinate and abscissa are dimensionless. We have plotted values from the model potential (discussed in Appendix D) as points for comparison; they are the result of a full calculation giving tables for all the simple metals (listed also in Harrison, 1966a, p. 309). We shall see that the screening calculation requires that w_q approach $-(2/3)E_F$ at small q, with E_F the Fermi energy, so both curves approach that limit. We have chosen r_c such that the two curves cross the horizontal axis first at the same point. Corresponding values for r_c are listed for all of the simple metals and for some other elements in Table 16-1. We shall see that most properties depend principally, or only, upon the values of the form factor for $1.4 < q < 2k_F$, and the two curves are extremely close in that domain. (An exception will be in the extension to covalent solids that is made in Chapter 18, where values at $q = 1.1 \, k_F$ are important and the errors in the model become significant.) Both curves are somewhat artificial at larger q, and as long as the form factor

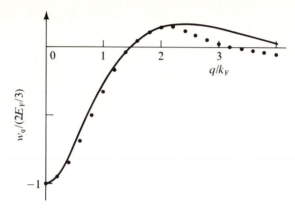

FIGURE 16-1

A plot of the form factor for the empty-core pseudopotential, from Eq. (16-7), for aluminum. For comparison, the points give the model potential (Animalu and Heine, 1965). The choice of r_c is such that the two curves cross the axis at the same point.

TABLE 16-1

Core radii for the empty-core model, in Å. These are listed also in the Solid State Table.

2	3	4	5	6	7	9	10
						Li 0.92	
Be 0.58	B 0.44	C 0.37		0 0.42		Na 0.96	
Mg 0.74	Al 0.61	Si 0.56	P 0.51	S 0.47	Cl 0.5	K 1.20	Ca 0.90
Zn 0.59	Ga 0.59	Ge 0.54	As 0.51	Se 0.50		Rb 1.38	Sr 1.14
Cd 0.65	In 0.63	Sn 0.59	Sb 0.56	Te 0.54		Cs 1.55	Ba 1.60
Hg 0.66	Tl 0.60	Pb 0.57	Bi 0.57				

SOURCES of values: Most were chosen such that $w_q = 0$ at the same q as in the model potential due to Animalu and Heine (1965) and Animalu (1966). Values of q at the first crossing of the horizontal axis were taken from Cohen and Heine (1970, p. 235). The exceptions are O, S, and Cl, which were obtained from fitted pseudopotentials (Cohen and Heine, 1970), and Rb and Cs, from Ashcroft and Langreth (1967).

approaches zero rather smoothly at large q, the detailed form is not very important. Given such a curve or the value of r_c for a given material such that the curve can be generated by using Eq. (16-7), we have all the information about the interaction between the electrons and the ions that is needed. By using perturbation theory and the structure factors corresponding to any situation of interest, we may directly calculate the properties.

Obtaining r_c from Atomic Parameters

An alternative method of obtaining r_c was suggested by Ashcroft (1966), in which r_c was adjusted in the unscreened empty-core pseudopotential given in Eq. (15-13), so that the bound valence s state had an energy equal to that required to remove an s electron from the atom experimentally. Although this gave a value in good accord with the values of Table 16-1, the values in Table 16-1 are preferable since they are more directly related to metals than to single atoms. The procedure for obtaining values from the ionization energy by using atomic parameters is given in Problem 16-2.

Ashcroft (1966) noted that these values of core radii for metallic ions are roughly equal to the ionic radii (see Section 13-E) that determine the spacings in ionic crystals. The comparison may be made from the Solid State Table, where values of r_i are listed, as well as values of r_c. This uniqueness of "ion size" is gratifying even though very crude.

It should be noticed that curves like those shown in Fig. 16-1 are defined only for $q \neq 0$. The $q = 0$ value is best obtained by combining all of the terms in the energy, as we did in the discussion in Chapter 15. Notice, however, that in the discussion of cohesion in Chapter 15 we used the same r_c that we use here for the discussion of w_q for $q \neq 0$. This remarkable feature appears to be special to the empty-core pseudopotential.

It is appropriate also to comment further on the origin of the pseudopotential. The manner in which it was introduced here, and more particularly, the manner in which it is introduced in Appendix D, might suggest that it is intrinsically a quantum effect. In fact, it may be thought of as a classical effect in which energetic electrons will move *faster* where the potential is deep and will therefore have reduced probability density where the potential is deep. By restricting consideration to valence electrons, attention can be focussed on those energetic electrons that are effectively repelled by an attractive potential. On the other hand, the *total* electron density (including core electrons) is higher where the potential is deep.

One should avoid being led by the discussion of cancelling terms into believing that it is the smallness of the values of w_q that makes the perturbation theory valid; that is not the case. We may estimate the size of the form factors in the absence of the repulsive term by setting $r_c = 0$ in Eq. (16-7). That does increase the magnitude of the form factors, but not in a qualitative way. The essential point is the alternation of sign as a function of q so that no bound states arise; that is what makes the perturbation theory a convergent procedure. This may be seen more directly in the real-space sketch of the pseudopotential of Fig. 15-3. The deep part of the potential that forms the core states has been eliminated.

16-B   Pseudopotentials in the Perfect Lattice

Let us now turn to the structure factors of Eq. (16-5), to determine them first for the perfect crystal. What we do here is formulate the diffraction theory for crystal lattices, since the interaction of the electron waves with the crystal is a diffraction phenomenon. A perfect crystal is characterized by a set of lattice translations $\mathbf{T}$ that, if applied to the crystal, take every ion (except those near the surface) to a position previously occupied by an equivalent ion. The three shortest such trans-lations that are not coplanar are called *primitive translations*, τ_1, τ_2, and τ_3, as indicated in Section 3-C. For the face-centered cubic structure, described also in Section 3-A, such a set is $[011]a/2$, $[101]a/2$, $[110]a/2$. The nearest-neighbor dis-tance is $d = a\sqrt{2}/2$. Replacing one of these by, for example, $[01\bar{1}]a/2$, would give an equivalent set. For a body-centered cubic lattice, such a set is $[\bar{1}11]a/2$, $[1\bar{1}1]a/2$, and $[11\bar{1}]a/2$, and the nearest-neighbor distance is $a\sqrt{3}/2$. For each of these struc-tures, the position of every ion in the crystal, starting from an ion at the origin, can be given by some lattice translation

$$\mathbf{T} = n_1\tau_1 + n_2\tau_2 + n_3\tau_3, \tag{16-8}$$

where n_i are integers. In the hexagonal close-packed structure, as in the diamond or zincblende structure, such a set of vectors gives only half the ions; a second ion is positioned with respect to each of these by a vector τ, which is not a lattice translation: the crystal is said to have two ions per primitive cell. The statement of the ion positions given by Eq. (16-8) is very convenient for the evaluation of structure factors.

Suppose we let the crystal be a parallelepiped, as in Fig. 15-1, but not neces-sarily rectangular, with edges $N_1\tau_1$, $N_2\tau_2$, and $N_3\tau_3$. This is illustrated in Fig. 16-2. If there is one ion per primitive cell, there are $N_a = N_1 N_2 N_3$ ions in the entire crystal. This geometry for the crystal is the most convenient one, and clearly the properties we discuss do not depend upon the exact shape of the crystal, so we may as well use this.

We now apply periodic boundary conditions upon the crystal surfaces, as we did in Eq. (15-1), but now the condition on the wave number for each plane-wave state becomes

$$\left.\begin{aligned} \mathbf{k} \cdot N_1\tau_1 &= 2\pi m_1, \\ \mathbf{k} \cdot N_2\tau_2 &= 2\pi m_2, \\ \mathbf{k} \cdot N_3\tau_3 &= 2\pi m_3, \end{aligned}\right\} \tag{16-9}$$

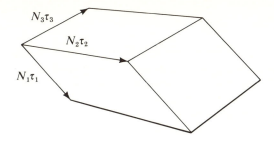

$N_3\tau_3$

$N_2\tau_2$

$N_1\tau_1$

FIGURE 16-2

A convenient choice of crystal geometry based upon the primitive translations τ_1, τ_2, and τ_3. There are $N_1 N_2 N_3$ primitive cells in the crystal. Periodic boundary conditions are taken on the crystal surfaces.

rather than Eq. (15-2), to which Eq. (16-9) reduces if the τ_i are mutually perpendicular; the m_i are integers. It may be verified by substitution that Eq. (16-9) will be satisfied if **k** is given by

$$\mathbf{k} = (m_1/N_1)\mathbf{k}_1 + (m_2/N_2)\mathbf{k}_2 + (m_3/N_3)\mathbf{k}_3, \tag{16-10}$$

with

$$\left.\begin{array}{l} \mathbf{k}_1 = (2\pi\tau_2 \times \tau_3)/\tau_1 \cdot (\tau_2 \times \tau_3); \\[4pt] \mathbf{k}_2 = (2\pi\tau_3 \times \tau_1)/\tau_2 \cdot (\tau_3 \times \tau_1); \\[4pt] \mathbf{k}_3 = (2\pi\tau_1 \times \tau_2)/\tau_3 \cdot (\tau_1 \times \tau_2). \end{array}\right\} \tag{16-11}$$

This then gives the grid of allowed **k**, in analogy with Eq. (15-2), and as illustrated in Fig. 15-2.

The difference between the wave numbers for any two of these plane waves can also be written in this form, so the structure factors of Eq. (16-5) can be written explicitly as

$$S(\mathbf{q}) = \frac{1}{N_a} \sum_i \exp\{-i[(m_1/N_1)\mathbf{k}_1 \cdot \mathbf{r}_i + (m_2/N_2)\mathbf{k}_2 \cdot \mathbf{r}_i + (m_3/N_3)\mathbf{k}_3 \cdot \mathbf{r}_i]\}. \tag{16-12}$$

If there is a single ion per primitive cell, we may write every $\mathbf{r}_i$ as

$$\mathbf{r}_i = n_1\tau_1 + n_2\tau_2 + n_3\tau_3, \tag{16-13}$$

with $0 \le n_1 < N_1$, $0 \le n_2 < N_2$, and $0 \le n_3 < N_3$. The sum can be factored into three sums,

$$S(\mathbf{q}) = \frac{1}{N_1} \sum_{n_1=0}^{N_1-1} e^{-i2\pi n_1 m_1/N_1} \frac{1}{N_2} \sum_{n_2=0}^{N_2-1} e^{-i2\pi n_2 m_2/N_2} \frac{1}{N_3} \sum_{n_3=0}^{N_3-1} e^{-i2\pi n_3 m_3/N_3}. \tag{16-14}$$

Each of these sums is a geometric series of the form

$$(1/N_i) \sum_{n_i=0}^{N_i-1} (e^{-i2\pi m_i/N_i})^{n_i} = (1 - e^{-i2\pi m_i})/(1 - e^{-i2\pi m_i/N_i}). \tag{16-15}$$

This is identically zero unless m_i is an integral multiple of N_i; then by direct evaluation of the sum it is seen to be unity. A vanishing structure factor is obtained unless this is satisfied for all three sums; that is, for $\mathbf{q}$ given by

$$\mathbf{q} = m_1' \mathbf{k}_1 + m_2' \mathbf{k}_2 + m_3' \mathbf{k}_3, \qquad (16\text{-}16)$$

where again the m_i' are integers. Such special wave numbers are called *lattice wave numbers*. The smallest ones, given in Eq. (16-11), are called primitive lattice wave numbers; they were introduced in Section 3-C and are illustrated for the simple cubic lattice in Fig. 16-3.

If there are two ions per primitive cell, there is an additional term, differing from the first by $e^{-i\mathbf{q}\cdot\mathbf{\tau}}$, and the structure factor becomes

$$S(\mathbf{q}) = (1 + e^{-i\mathbf{q}\cdot\mathbf{\tau}})/2, \qquad (16\text{-}17)$$

rather than unity for each lattice wave number $\mathbf{q}$.

This completes the specification of the pseudopotential as a perturbation in a perfect crystal. We have obtained all of the matrix elements between the plane-wave states, which are the electronic states of zero order in the pseudopotential. We have found that they vanish unless the difference in wave number between the two coupled states is a lattice wave number, and in that case they are given by the pseudopotential form factor for that wave number difference by Eq. (16-7), assuming that there is only one ion per primitive cell, as in the face-centered and body-centered cubic structures. We discuss only cases with more than one ion per primitive cell when we apply pseudopotential theory to semiconductors in Chapter 18. Then the matrix element will be given by a structure factor, Eq. (16-17),

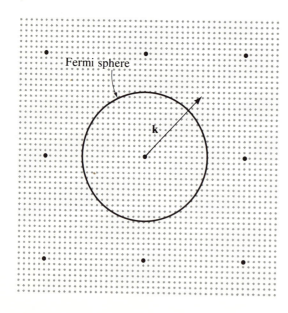

FIGURE 16-3

The grid of wave numbers allowed by periodic boundary conditions, as in Fig. 15-2. Also shown are several lattice wave numbers (heavier dots). The relative spacings and geometry correspond to a simple cubic lattice of 6859 ions (see Problem 16-1). A sample wave number $\mathbf{k}$ is shown.

times the pseudopotential form factor. In considering distorted crystals, or crystals with defects, it will be necessary to reevaluate the structure factor, starting from the defining equation, Eq. (16-5), but that is best postponed until the occasion arises.

How to proceed with these matrix elements will depend upon which property one wishes to estimate. Let us begin by discussing the effect of the pseudopotential as a cause of diffraction by the electrons; this leads to the *nearly-free-electron approximation*. The relation of this description to the description of the electronic structure used for other systems will be seen. We shall then compute the screening of the pseudopotential, which is necessary to obtain correct magnitudes for the form factors, and then use quantum-mechanical perturbation theory to calculate electron scattering by defects and the changes in energy that accompany distortion of the lattice.

16-C   Electron Diffraction by Pseudopotentials

The lattice wave numbers defined in Eq. (16-16) are the *reciprocal lattice vectors* familiar in diffraction theory. Only if the change in wave number resulting from the diffraction is equal to a lattice wave number can a wave, whether it be an X-ray or an electron, be diffracted; otherwise the wavelets scattered by the different ions interfere with each other and reduce the diffracted intensity to zero. Only for diffraction by $\mathbf{q}$ equal to a lattice wave number do the scattered waves add in phase. Thus a wave having wave number $\mathbf{k}$ can only be diffracted to final states of wave number $\mathbf{k}'$ that can be written as $\mathbf{k}' = \mathbf{k} + \mathbf{q}$, where $\mathbf{q}$ is a lattice wave number. Furthermore, the diffracted wave will have the same frequency as the incident wave if it is an X-ray, or the same energy if it is an electron, from which it follows that $|\mathbf{k}'| = |\mathbf{k}|$. Combining the two conditions, $|\mathbf{k} + \mathbf{q}|^2 = k^2$ gives the *Bragg condition* for diffraction,

$$\mathbf{k} \cdot \mathbf{q} = -q^2/2. \tag{16-18}$$

This may be stated as the condition that an electron can only be diffracted if its wave number lies on a plane bisecting some lattice wave number. (We have used the fact that if $\mathbf{q}$ is a lattice wave number, then so is $-\mathbf{q}$.) Such planes, called *Bragg planes*, are illustrated in Fig. 16-4 for the simple cubic lattice.

In ordinary diffraction phenomena, waves of various frequencies, for example, X-rays, are brought in from outside a crystal, and most are transmitted, because they do not happen to satisfy the Bragg condition. Those that do satisfy Eq. (16-18) may be diffracted and emerge in the direction of $\mathbf{k} + \mathbf{q}$, giving the familiar Laue spots. In the metal, we are interested in electron waves that are propagating *within* the crystal; most do not satisfy the Bragg condition and so are not diffracted; thus the free-electron theory of Chapter 15 is quite appropriate for almost all electron states. However, by applying fields, we can deflect the electrons (the acceleration of electrons by applied fields was discussed in Section 2-B), thereby bringing many electrons, one after the other, to the Bragg planes, to make the effects of the diffraction felt. The study of such effects in metals is called *Fermiology*; though it is not central to our studies, the subject has connections with other parts of our discussion. Therefore, the subject will be introduced.

Only quite small *electric* fields can be applied in a metal, because of the high conductivity; in the time an electron moves before colliding with a defect or with the surface it can only be accelerated very slightly. Unless it happens to lie very close to a Bragg plane it will not be affected by the diffraction. Since most electrons are thus "unaware" of the lattice, conductivity can to a large extent be treated in the free-electron approximation, as we indicated earlier.

On the other hand, *magnetic* fields that are very strong *can* be applied. In the presence of a uniform magnetic field, free electrons are deflected into circular or helical orbits, with a characteristic frequency of $\omega_c = eH/mc$, called the *cyclotron frequency*. In this formula, ω_c is given in radians per second if e is written as 4.8×10^{-10} electrostatic units; H is given in gauss; m is 9.1×10^{-28} grams, and c is 3×10^{10} centimeters per second or 1.76×10^{10} radians per second at one kilogauss. In pure metals at low temperatures and in high magnetic fields, the time an electron will travel before being scattered by a defect can be sufficiently long that it would complete many orbits in the field. The electron wave number is rotating at the same rate, and we see from Fig. 16-4 that the **k** shown there would cross Bragg planes many times. Because of the periodic potential, the electron is diffracted whenever it encounters a Bragg plane, so in the crystal it will not complete the circular orbit. The behavior of the electron is not difficult to understand. Consider an electron with the wave number **k** in the plane of the Fig. 16-4. Let a magnetic field be applied perpendicular to the figure such that the velocity of the electron, and the wave number, rotate counterclockwise. This is illustrated in Fig. 16-5,a. When **k** reaches the upper Bragg plane at *A*, the wave number changes by the negative of the lattice wave number of which that Bragg plane is the bisector;

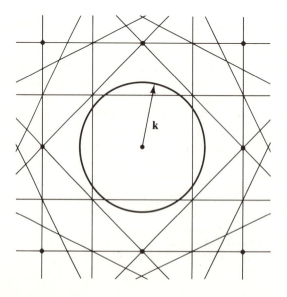

FIGURE 16-4

The lattice wave numbers and Bragg planes for the system shown in Fig. 16-3. We imagine, though, a much larger crystal, so the mesh of wave numbers allowed by periodic boundary conditions becomes very fine and is not shown. This does not change the lattice wave numbers. The wave number **k** shown does not satisfy the Bragg condition.

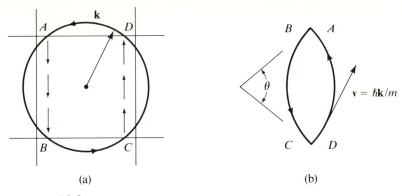

(a)   (b)

FIGURE 16-5

A uniform magnetic field perpendicular to the plane of the figure causes the
wave number of the state to rotate as shown in Part (a). When it reaches A
it is diffracted to B and continues to C; it is diffracted to D and again
continues to complete its orbit. The motion of the electron in the real crystal
is shown in part (b); the relative scales depend upon the size of the field.

it changes abruptly to the wave number at B. It then continues rotating at the
same rate to C, diffracts to D, and then returns to its initial wave number.

The time to complete this cycle is just the total arc through which it moved (2θ)
divided by 2π times the time required for a free electron to complete its circular
orbit in that field; it has a cyclotron frequency higher than that of a free electron.
We may also see how this electron has moved within the crystal itself, by observing that the velocity at any time is just $\hbar k/m$ at that time. The orbit in real space is
shown in Fig. 16-5,b, and is seen to be made up of similar arcs rotated 90° from
those shown in part (a). The real orbit must of course be continuous. The size of
the orbit in real space will be smaller if the field is larger but will always extend
over very many lattice distances. We might similarly construct orbits for other
electrons moving (for example) initially directly to the right in Fig. 16-5,a, or for
electrons moving initially with a velocity 45° from the vertical (or horizontal).
Similarly, an orbit for an electron moving with some component of velocity
along the field will move in an orbit taking it out of the plane perpendicular to
the field. Since there are techniques for experimentally studying the orbit shapes
in the crystal, this diffraction description can be accurately verified, and has been
(see Harrison and Webb, 1960, or Cracknell, 1969). The behavior of the electrons
and their orbit shapes can be very complex, though diffraction is basically quite
simple; furthermore, our discussion can be generalized immediately from the
simple cubic to the face-centered and body-centered cubic structures.

16-D   Nearly-Free-Electron Bands and Fermi Surfaces

We should now relate what we have found in this chapter to the energy-band
description introduced initially in Section 2-A; the crystal used there also had the
translational symmetry of the simple cube. There we also defined wave numbers
for the states but restricted their domain to a Brillouin Zone; similarly, in Section

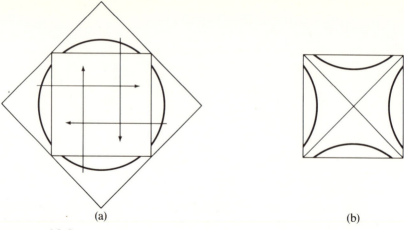

(a)   (b)

FIGURE 16-6

The displacement of regions of wave number space so that each state is represented by a wave number inside the Brillouin Zone equivalent to its wave number outside. The "reduced-zone scheme" in part (b) gives the second band Fermi surface in the ordinary band description in the Brillouin Zone.

3-C we took wave numbers only within the appropriate Brillouin Zone, calling any wave number differing from a wave number within the Brillouin Zone by a lattice wave number an *equivalent wave number*. The nearly-free-electron description has led us to associate wave numbers outside the Brillouin Zone with most of the states—this is called the *extended-zone representation* and is the most convenient for most studies in metals. To relate the resulting free-electron bands to those restricted to the Brillouin Zone, notice first that the region of wave number space containing $k = 0$ and bounded by the nearest Bragg planes in every direction is the Brillouin Zone, a simple cube in the simple cubic lattice, as shown in Fig. 16-4. Then the energy band, $E = \hbar^2 k^2 / 2m$, for k in a region outside the Brillouin Zone, must be replotted as a function of the equivalent wave number $k - q$ that lies within the Brillouin Zone. This was precisely the procedure used to obtain the nearly-free-electron bands shown in Fig. 2-2,b. In this way, the states in the triangular regions *outside* the Brillouin Zone and the segments of the Fermi sphere that are contained in them are translated *into* the Brillouin Zone, as shown in Fig. 16-6. The translation results in what is known as the *reduced-zone scheme*. It is not difficult to see that the states translated from these regions become the second band in the Brillouin Zone and that the second-band "Fermi surface" consists of six segments of a sphere at the six faces of the zone.

Since states on opposite faces of the zone are entirely equivalent, one may repeat the bands and the Fermi surface by constructing identical zones around every lattice wave number, as shown in Fig. 16-7. In this representation, called the *periodic-zone scheme*, the second-band Fermi surface is seen to consist (again, for the simple cubic structure) of three closed lens-shaped segments. The advantage of this representation is that all discontinuous jumps in wavenumber have been eliminated; the wave numbers connected by diffractions have been plotted together. Thus the electron orbit described in Fig. 16-5 corresponds to the cross-section of some piece of Fermi surface and has exactly the same shape. In fact, every orbit that can occur in a magnetic field corresponds to some Fermi surface cross-section in the periodic-zone scheme. This has made the periodic-zone scheme very useful for the study of electron orbits in metals.

Construction of Free-Electron Fermi Surfaces

It is in fact rather easy to construct Fermi surfaces in the periodic-zone scheme. Notice that the Fermi surface is made up entirely of segments of spheres centered at the lattice wave numbers. Thus one may proceed directly by constructing Fermi spheres around every lattice wave number. It can then be seen that the second-band Fermi surface, for example, is the surface that separates regions of wave number space that lie within two spheres from regions that lie within only one. Having made the construction, one may pick a Brillouin Zone wherever one wishes to give the Fermi surface for the corresponding band, since by making the periodic construction one regains the entire space of surfaces. For the simple cubic example in the figures, one could then display the second-band Fermi surface as in Fig. 16-7 and the first-band Fermi surface separating regions within one sphere from those within none. (The latter is a diamond-shaped region with concave sides.)

Exactly the same procedure is used for other structures. For aluminum, one

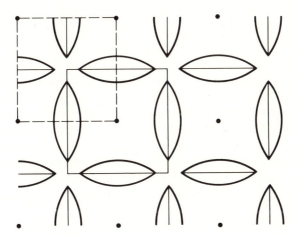

FIGURE 16-7

The second-band Fermi surface from Fig. 16-6 displayed in the periodic zone scheme. A Brillouin Zone can be constructed at any point in order to specify the Fermi surface; two possibilities are shown as dashed-line and solid-line squares.

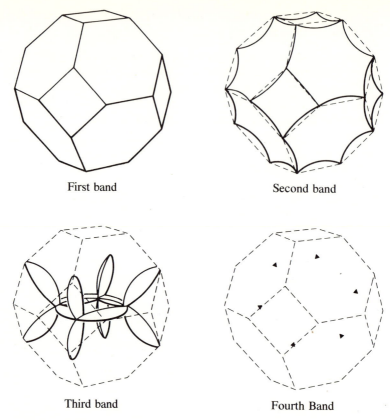

First band

Second band

Third band

Fourth Band

FIGURE 16-8

The Fermi surface of aluminum. The position of the Brillouin Zones has been shifted in the third and fourth bands, in analogy with the dashed-line square shown in Fig. 16-7.

constructs the wave number lattice for the face-centered cubic structure. Since there are three electrons per atom (and per primitive cell) there are enough electrons to fill one and one-half bands or a Fermi sphere of 1.5 times the volume of the Brillouin Zone. One constructs the spheres and counts the number of spheres which contain any given **k** to find the Fermi surface. The result is shown in Fig. 16-8. The first band is completely full and has no Fermi surface. However, there is Fermi surface in the next three. Notice that the construction is purely geometrical. Parameters of the material set the scale, but the shapes are precisely given and depend only upon valence and crystal structure.

Corrections to Nearly-Free-Electron Surfaces

Although the nearly-free-electron construction is absolutely precise, it gives only an approximate description of the electron orbits in the real crystal. The approximation lies in the assumption that pseudopotentials are of sufficient

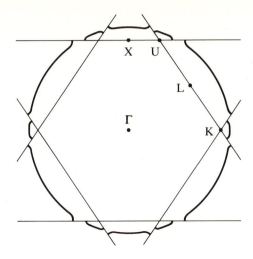

FIGURE 16-9

A (110) section of the Fermi surface of aluminum in the extended zone scheme, showing the distortion at the Bragg planes. [After Harrison, 1966a.]

weakness that an electron does not "see" them unless its wave number lies exactly on the Bragg plane. Then the magnitude of the pseudopotential does not influence the orbit shapes; only the geometry of the Bragg planes does. We shall see in Chapter 17 that electrons *near* the Bragg planes are also perturbed. The consequence is that the electron begins diffracting before it reaches the Bragg condition, Eq. (16-18), and the orbit shapes are rounded at the corners; the sharp points shown in Fig. 16-5,b do not occur in the real crystal. Exactly the same rounding of corners occurs in the Fermi surface, so the experimental study of Fermi surfaces has also given information about the strength of the diffractions and therefore the magnitudes of the pseudopotentials. Although we shall not discuss this in detail, it should be noted that these experimental studies have provided the most complete test of the pseudopotential theory being presented in this book.

The rounding off of the Fermi surface in the periodic-zone scheme also corresponds to a distortion of the free-electron sphere in the extended-zone scheme. That distortion, calculated for aluminum, is shown in Fig. 16-9. It has the effect of reducing the total area of free surface and, as we shall see, in the case of covalent solids the same effect eliminates the entire surface.

16-E Scattering by Defects

Before discussing screening, let us examine the effect of lattice imperfections on the electronic diffraction. The simplest defect is a vacancy, a missing ion at the site r_0. Any such change in atomic arrangement requires us to reevaluate the structure factors, which requires application of Eq. (16-5). We rewrite the sum with a prime to indicate the omission of the term $i = 0$, and write $N_a - 1$ to indicate the decrease in the number of ions by one:

$$S(\mathbf{q}) = \frac{1}{N_a - 1} \sum{}' e^{-i\mathbf{q} \cdot \mathbf{r}_i}. \qquad (16\text{-}19)$$

Consider first the evaluation for $\mathbf{q}$ equal to a lattice wave number from the perfect lattice. Every phase factor in the sum is still unity, but there are now only $N_a - 1$ terms, so the structure factor is still unity. There is nevertheless a slight reduction in the matrix elements because the atomic volume, now $\Omega/(N_a - 1)$, has been slightly increased. We neglect this change when doing the scattering calculation, though in calculating the total energy to obtain the energy to form a vacancy (Harrison, 1966a, p. 204ff) we cannot. For calculation of the total energy it is preferable to form a vacancy by rearranging the atoms at constant total volume in forming the vacancy. The removal of an atom, as done here, is simpler and more convenient in scattering where the same precision is not required.

Consider next the evaluation for $\mathbf{q}$ differing from a lattice wave number. Then, a sum over *all* r_i gives zero and the sum over all but $\mathbf{r}_0$ gives

$$S(\mathbf{q}) = -e^{-i\mathbf{q} \cdot \mathbf{r}_0}/(N_a - 1) \cong -1/N_a. \qquad (16\text{-}20)$$

(In the last step we have taken our origin at the vacancy site; the phase factor would not affect the final result in any case, since the scattering is proportional to the magnitude of the matrix element squared.)

Thus the introduction of the vacancy has given a small matrix element, $-w_q/N_a$, coupling *every* pair of states. This is no longer called diffraction but *scattering*. We did not discuss scattering theory in Chapter 1 and here it will suffice simply to outline the calculation and give the result. When one electron state is coupled to a continuum of other states, as in this case, we may calculate the probability per unit time that an electron in that one state will make a transition to each of the others. We may then add up those probabilities to obtain the total scattering rate; actually this becomes an integration over all states of the same energy or, for an electron with the Fermi energy, an integration over the area of the Fermi surface, A_{FS}. Transport theory tells us that of more interest is the scattering rate weighted by $1 - \cos \theta$, where θ is the change in angle in the scattering. The result of this calculation is to give a weighted scattering rate (see, for example, Harrison, 1970, p. 194ff.),

$$\frac{1}{\tau} = \frac{\pi \Omega}{\hbar} \frac{dN}{dE} \frac{1}{A_{FS}} \int |\langle \mathbf{k} + \mathbf{q} | W | \mathbf{k} \rangle|^2 (1 - \cos \theta) \, dA_{FS}$$

$$= \frac{\Omega k_F m}{2\pi \hbar^3} \int_0^\pi |\langle \mathbf{k} + \mathbf{q} | W | \mathbf{k} \rangle|^2 (1 - \cos \theta) \sin \theta \, d\theta; \qquad (16\text{-}21)$$

dN/dE is the density of electron states, per unit energy and per unit volume, at the Fermi energy, equal to $3N/(2E_F)$. The integral over the Fermi surface has become an integral over scattering angle θ, through use of the cylindrical symmetry of the factors that enter the integral, as indicated in Fig. 16-10, and the matrix element similarly depends only upon angle. From elementary trigonometry, $q^2 = 2k_F^2(1 - \cos \theta)$, so w_q, from Eq. (16-7), is also a simple function of angle. We have returned to the extended-zone representation where the Fermi surface is a sphere. Here, as in almost any calculation one carries out, the extended-zone representation is the most convenient. It would be an intricate geometrical task to carry out this integral over the Fermi surface in the reduced-zone scheme, but if correctly done, the integration would give the same answer.

We may substitute the structure factor for the vacancy from Eq. (16-20) and write Ω/N_a as the atomic volume. The result may also be more meaningful if we postulate some number of vacancies N_v and add the scattering rates for each, to obtain

$$\frac{1}{\tau} = \frac{\Omega_0 k_F m}{2\pi\hbar^3}\left(\frac{N_v}{N_a}\right)\int_0^\pi w_q^2(1 - \cos\theta)\sin\theta\,d\theta. \tag{16-22}$$

We then see that the result is proportional to the fraction of sites, N_v/N_a, which are vacant, and is independent of the size of the system, as it should be.

The scattering due to vacancies is not in itself of much importance, but exactly the same approach is applicable to a very wide variety of problems. For example, the scattering due to a substitutional impurity gives the same result, except that w_q in Eq. (16-22) is replaced by the difference $w_q - w_q'$ between the form factors for the host and the impurity. Similarly, one can allow atoms to be displaced from their equilibrium positions—near an impurity or elsewhere—and recalculate the structure factor to obtain the scattering rate. (A number of such examples are discussed by Harrison, 1966a, Chapter 4.)

To allow for these more general circumstances, we rewrite Eq. (16-22) in terms of the structure factor. At the same time, we may convert from $1/\tau$ to resistivity. Transport theory leads to an expression for the resistivity in terms of the scattering rate, $\rho = m/(Ne^2\tau)$, with N again the electron density, Z/Ω_0. We may combine this with Eq. (16-22) to give the resistivity as

$$\rho = CN_a\int_0^2 |S(\mathbf{q})|^2 w_q^2 x^3\,dx; \tag{16-23}$$

C, equal to $3\pi m\Omega_0/(8\hbar e^2 E_F)$, is given for the simple metals in Table 16-2. In writing this form we have changed variables from θ to $x = q/k_F$.

The lattice distortions of particular interest are the phonons. We shall return to them and to the electron–phonon interaction in Chapter 17.

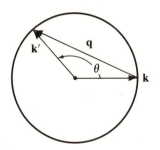

FIGURE 16-10

Owing to a vacancy or other defect, an electron in a state of wave number $\mathbf{k}$ on the Fermi sphere may be scattered to some state of wave number $\mathbf{k}'$ elsewhere on the sphere.

TABLE 16-2

Values of C entering the formula of Eq. (16-23) for the resistivity due to defects in the simple metals. Values are in microohm-centimeters per squared electron volt.

1	2	3	4	9	10
				Li 27.21	
	Be 3.58			Na 79.8	
	Mg 20.12	Al 8.95		K 233	Ca 57.7
Cu 10.59	Zn 9.71	Ga 11.70		Rb 325	Sr 90.0
Ag 19.51	Cd 17.62	In 19.15	Sn 16.98	Cs 485	Ba 107.6
Au 19.32	Hg 19.60	Tl 22.24	Pb 20.0		

16-F   Screening

We developed the pseudopotential, using the free-electron approximation, in such a way that the valence electrons were included as a uniform negative charge density. The electron states are in fact distorted by the pseudopotential, and the corresponding charge density is modified; this results in corrections to the pseudopotential. This is called screening of the pseudopotential. A complication arises inasmuch as we must know the pseudopotential in order to obtain the charge density, and we must know the charge density in order to obtain the pseudopotential; we therefore require a self-consistent solution. Such solutions are frequently obtained by iteration, but here we may use the smallness of the pseudopotential to keep only terms of first order in the pseudopotential, and we find we can solve directly for the screening. The best way to proceed is to calculate the distortion of the electron states by using quantum-mechanical perturbation theory, but it will be adequate for our purposes to use an approximation to this approach that leads to a simpler form. This will also provide a clearer physical picture of the screening. After obtaining that form, we shall give the result of the full quantum-mechanical calculation.

Notice first that in obtaining the matrix elements of the pseudopotential, Eq. (16-2), we have obtained a Fourier expansion of the pseudopotential. Writing $W_q^0 = \langle \mathbf{k} + \mathbf{q} \,|\, W^0(\mathbf{r}) \,|\, \mathbf{k} \rangle$, we may take the inverse transform of Eq. (16-2) to write

$$W^0(\mathbf{r}) = \sum_q W_q^0 \, e^{i\mathbf{q} \cdot \mathbf{r}}. \tag{16-24}$$

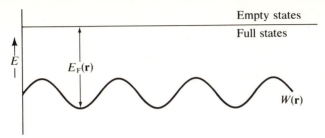

FIGURE 16-11

The redistribution of electrons in a metal, due to a sinusoidal
perturbation $W(\mathbf{r})$. Where the potential is low, the local
Fermi energy and therefore the local electron density is high.

To first order in the pseudopotential, each Fourier component will cause a charge
redistribution with the same spatial dependence, so that each Fourier component
is an independent problem; we can compute the screening of each component
separately.

The Fermi-Thomas Approximation

Let us focus on a particular Fourier component of wave number $\mathbf{q}$. Before
screening, that component was $W_q^0 \, e^{i\mathbf{q}\cdot\mathbf{r}}$. Let us write the screened value $W_q e^{i\mathbf{q}\cdot\mathbf{r}}$.
Our goal is to determine W_q. If the wavelength were sufficiently long that we could
treat each region as having a constant potential, the electrons would redistribute
themselves until the highest energy of occupation would be the same in every
region, much as water redistributes itself so that the surface everywhere is at the
same level. This is illustrated in Fig. 16-11. Notice that it is the total potential
$W(\mathbf{r})$ that is relevant. Then the local Fermi kinetic energy must also vary sinu-
soidally to compensate for the variation of the potential; that is,

$$E_F(\mathbf{r}) + W(\mathbf{r}) = \text{Constant.} \tag{16-25}$$

This leads to a local Fermi wave number, $k_F(\mathbf{r})$, given by $\hbar^2 k_F(\mathbf{r})^2/2m = E_F(\mathbf{r})$, and
from Eq. (15-4), a local electron density $N(\mathbf{r}) = k_F(\mathbf{r})^3/(3\pi^2)$. This is called the
Fermi-Thomas approximation. The essential assumption that is required is that the
potential does not vary greatly over the distance corresponding to the electron
wavelength.

The second important approximation is to treat the variation in $W(\mathbf{r})$ as being
small compared to E_F, so that any terms higher than first order in $W(\mathbf{r})$ can be
neglected. Then, working back to obtain the first-order term in $N(\mathbf{r})$, we obtain

$$\delta N = \frac{3k_F^2}{3\pi^2}\,\delta k_F = \frac{k_F}{\pi^2}\frac{m}{\hbar^2}\,\delta E_F$$

$$= -\frac{k_F m}{\pi^2 \hbar^2}\,\delta W \tag{16-26}$$

Here N, k_F, E_F, and W are all functions of position. This gives us the electron density in terms of the potential. We may now use Poisson's equation to write the screening potential, $\delta W(\mathbf{r}) - \delta W^0(\mathbf{r})$, in terms of the electron density fluctuation,

$$-\nabla^2[\delta W(\mathbf{r}) - \delta W^0(\mathbf{r})] = 4\pi e^2 \, \delta N(\mathbf{r}), \tag{16-27}$$

and then solve for $W(\mathbf{r})$ in terms of $W^0(\mathbf{r})$. Notice that all terms vary as $e^{i\mathbf{q}\cdot\mathbf{r}}$, so that $-\nabla^2$ can be replaced by q^2. Then, substituting Eq. (16-26) in Eq. (16-27) and solving for δW gives

$$\delta W = \delta W^0/(1 + \kappa^2/q^2), \tag{16-28}$$

where the **Fermi-Thomas screening parameter** κ is given by

$$\kappa^2 = 4e^2 k_F m/(\pi\hbar^2). \tag{16-29}$$

Using the values for e^2 and $\hbar^2/m$ from the Solid State Table, we obtain $(\kappa/k_F)^2 = 4 \times 14.40/(\pi \times 7.62 k_F)$. For beryllium, for example, with $k_F = 1.94$ Å^{-1}, this gives $\kappa/k_F = 1.114$.

The result is very simple. Each Fourier component is reduced by a **dielectric function** (*function* rather than *constant* since it depends upon q) that is given by

$$\varepsilon(q) = 1 + \kappa^2/q^2, \tag{16-30}$$

in direct analogy with the dielectric screening in semiconductors described in Section 5-C. Notice that this is just the form that was used in replacing q^2 by $q^2 + \kappa^2$ in the denominator of Eq. (16-7), as we indicated it would be. We may also confirm immediately that as q approaches zero, w_q approaches $-2E_F/3$, as shown in Fig. 16-2.

A More Accurate Calculation of Screening

A more complete theory of screening is based upon quantum-mechanical perturbation theory, which does not assume that the potential varies slowly with distance but does assume that it is small. This calculation, which follows quite closely the perturbation-theoretic calculation of the energy which we carry out in Chapter 17, gives

$$\varepsilon(q) = 1 + \frac{me^2}{2\pi k_F \hbar^2 \eta^2}\left(\frac{1-\eta^2}{2\eta}\ln\left|\frac{1+\eta}{1-\eta}\right| + 1\right), \tag{16-31}$$

where $\eta = q/2k_F$. (For a derivation, see, for example, Harrison, 1966a, p. 46ff.) This form approaches κ^2/q^2 at small q, as does Eq. (16-30), and approaches 1 at large q, as does Eq. (16-30), so the simpler form will suffice for most purposes. However, there are two relatively important differences. First, at large q, Eq. (16-31) approaches $1 + 16me^2 k_F^3/(3\pi\hbar^2 q^4)$ and approaches 1 as q^{-4} rather than q^{-2}. We shall see that this makes a difference in convergence when we calculate the total energy. Second, Eq. (16-31) contains a term varying as $(q - 2k_F) \ln (q - 2k_F)$ near $q = 2k_F$; this is a logarithmic singularity with an infinite negative slope at that point. It arises because, for $q < 2k_F$, the potential can scatter an electron to a final state of the same energy, enhancing the screening. This singularity can become prominent under some circumstances, particularly when we transform to interactions or charge densities in position space, as will be seen.

PROBLEM 16-1   *Free-electron Fermi surface*

A central cross-section of the Fermi sphere for aluminum in the extended-zone scheme was shown in Fig. 16-9. By a study such as that illustrated in Fig. 16-5, you can identify two orbit types, which correspond to cross-sections of the surfaces for the second and third bands, shown in Fig. 16-8. The topology of the Fermi surface is the same for lead but, with four electrons per atom in lead, the sphere is larger in comparison to the zone than that for aluminum. Estimate the area of cross-sections of the surfaces for the second and third bands for lead, in units of $(2\pi/a)^2$, with a the cube edge.

It is most convenient to begin by constructing a (110) plane in wave number space, with the wave number lattice as shown in Fig. 16-12. The electron density is $16/a^3$, because there are four atoms per cube and four electrons per atom, so the Fermi wave number can be obtained in units of $2\pi/a$. Either a graphical solution or an approximate geometrical solution is adequate. These areas are directly measured in the de Haas van Alphen effect. The experimental values (Gold, 1958) are 1.00 and 0.11 times $(2\pi/a)^2$ for the second band and third band, respectively.

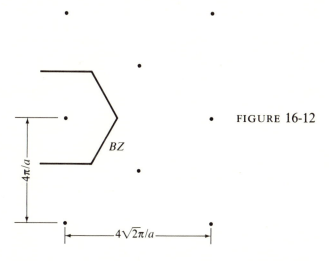

FIGURE 16-12

PROBLEM 16-2   *Obtaining pseudopotential core radii from atomic levels*

The lowest-lying s state for the empty-core pseudopotential is illustrated in Fig. 16-13,a. Its energy E can be computed by numerical integration and obtained as a function of Z and r_c. We shall make an approximate solution.

Following Schiff (1968, p. 93ff), an $l = 0$ wave function is written as

$$\psi(r) = - \left[\frac{1}{8\pi} \left(\frac{2Z}{na_0} \right)^3 \frac{(n-1)!}{n(n!)^3} \right]^{1/2} e^{-\rho/2} L_n^1(\rho), \tag{16-32}$$

where a_0 is the Bohr radius (0.529 Å) and $\rho = 2Zr/(na_0)$. For $r_c = 0$, these are the hydrogen-atom wave functions, L_n^1 is an associated Laguerre polynomial, and the energy is $E_n = -Z^2 e^2/(2a_0 n^2)$ for every integer n.

Solutions of the empty-core problem can be obtained for this particular set of energies by fitting the above solutions, beyond the last node, to the solution $A\rho^{-1} \sinh(\rho/2)$ which holds for $r < r_c$. The L_n^1 are polynomials containing terms up to order $n-1$, but the last two $(\rho^{n-2}$ and $\rho^{n-1})$ dominate the fit at r_c. If we in fact keep only these two terms, we can simply let n become nonintegral (corresponding to intermediate energies) and obtain a

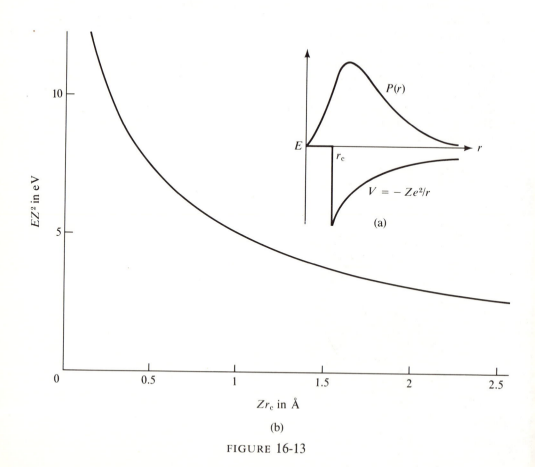

(b)

FIGURE 16-13

condition for the fit at $\rho_c = 2Zr_c/(a_0 n)$. Essentially, this interpolates for intermediate energies and turns out to be exact at $n = 2$. The condition is

$$\rho_c = n(n - 1) + n\rho_c^{-1}[\rho_c - (n - 1)^2](1 - e^{-\rho_c}), \qquad (16\text{-}33)$$

which can be solved by iteration and plotted as E/Z^2 against Zr_c, as in Fig. 16-13,b. Approximate solutions for $Zr_c > 2.5$ Å are obtained as in part (d) of this problem. In Fig. 16-13,b, E is the ionization potential, that is, the energy required to remove the last s electron to produce an ion of charge Z. It is customarily written as NaI for sodium ($Z = 1$), MgII for magnesium ($Z = 2$), AlIII for aluminum ($Z = 3$), and so on, and is tabulated in standard handbooks, for example, in the Chemical Rubber Company Handbook (Weast, 1975, p. E-68).

(a) For monovalent metals, values of E can be identified as the ε_s given in the Solid State Table. Use these and the curve in Fig. 16-13,b to estimate r_c for all alkali metals (metals in column 9) and compare with the r_c given in the Solid State Table. (The values of r_c in the Solid State Table were obtained from fit to the Animalu-Heine form factors.) The table gives no r_c value for francium; presumably a better prediction could be made by plotting the result from the form factor against our estimate here and extrapolating the resulting curve for francium.

(b) Values of E for NaI, MgII, and AlIII (Weast, 1975, p. E-68) are 5.14 eV, 15.03 eV, and 28.44 eV, respectively. Estimate r_c for each of these from the curve in Fig. 16-13,b and compare with the Solid State Table.

(c) It is interesting that values for r_c obtained by using AlI (and again, the energy required to remove an s electron rather than a p electron) and AlII are 0.72 Å and 0.53 Å, respectively; the predicted value of r_c is not so sensitive to ionization state as one might have expected. One may therefore estimate values for the rare earths from the ionization potentials (Weast, 1975, p. E-68). These are typified by sumarium. SmI is 5.6 eV; SmII is 11.2 eV. Ignore the uncertainty as to which state is emptied to treat these as $Z = 1$ and $Z = 2$ transitions to estimate r_c. Values from the first ionization potential (for example, SmI) were used to obtain the r_c given for the f-shell metals in the Solid State Table.

(d) For Zr_c/a_0 greater than 3 the correction $e^{-\rho_c}$ in the formula for the condition for the fit at Eq. (16-33) can be dropped and the resulting equation solved for ρ_c. Use this formula to obtain r_c for Al with the AlIII ionization potential given in part (b).
Note: Part (d) is complicated and may be of limited importance.

PROBLEM 16-3   *Calculation of structure factors and resistivity*

Consider a perfect crystal with the single atom, equilibrium position $\mathbf{r}_0$, displaced by $\mathbf{u}$. Evaluate $S(\mathbf{q})$ for $\mathbf{q}$ not a lattice wave number.

(a) Expand it to lowest order in $\mathbf{u}$. Evaluate the corresponding $S^*(\mathbf{q})S(\mathbf{q})$ and average it over the direction of $\mathbf{q}$. This average is relevant to scattering rates, with u^2 replaced by an averaged value, $\langle u^2 \rangle$, for each atom.

(b) If we think of the vibration spectrum of potassium in the Einstein approximation, discussed in Chapter 9, we will associate $kT/2$ at high temperatures with the kinetic energy in each degree of freedom of this atom, $(M/2)\omega_E^2\langle u^2 \rangle$, with ω_E the representative frequency. An appropriate value for potassium may be read from Fig. 17-4, $\omega_E = 2\pi\nu \sim 12 \times 10^{12}$ sec^{-1}, and there are $3N_a = 3N$ such scatterers in a unit volume.

Thus, Eq. (16-23) may be used to estimate the high temperature resistivity of potassium. An analogous calculation, using the Debye approximation, is carried out in Problem 17-2.

For simplicity, approximate w_q by $(-1.37 + 0.44x^2)$ eV; fit to give $-2E_F/3$ at $x = 0$ and to vanish at $qr_c = \pi/2$.

The linear temperature dependence is appropriate, but the actual value at room temperature agrees only in order of magnitude with the observed 7 microohm-centimeters.

CHAPTER 17

Mechanical Properties
of Metals

SUMMARY

The perturbation of electron states by the pseudopotential modifies the total energy by a term called the band-structure energy, which is obtained to second order in the pseudo-potential; it is essential to an understanding of the bonding properties, and it can be directly evaluated for any specified positions of the metallic ions. It may also be thought of as contributing to an effective interaction between ions; the interaction shows Friedel oscillations at large separations.

The vibration spectra of simple metals are directly calculable in terms of the band-structure energy and the electrostatic energy, and are well described, particularly for low-valence metals. Pseudopotential theory also can be used to calculate the electron–phonon coupling constant, which gives values for high-temperature resistivity, renormalization of the electronic specific heat, and superconducting critical temperatures in metals. The theory of the resistivity of liquid metals is very simple, but though the theories of surface energies, work functions, and chemisorption also follow immediately, these have required major computation.

As in the other solid types, the entire range of structural, elastic, and vibrational properties are determined by the electronic structure. Likewise, as in other systems, the density, bulk modulus, and cohesion are considered together as a separate problem and, for the metals, were treated in Chapter 15. We have given a reasonably simple description of the electronic structure of simple metals in Chapter 16, and can now use it to treat the more detailed aspects of the bonding properties.

17-A   The Band-Structure Energy

In examining how changes in the electron states caused by the pseudopotential change the total energy of the electron gas, it is best not to use the Fermi-Thomas approximation, used in Section 16-F, but to compute the energy of the electrons directly by applying perturbation theory. For a particular electron of wave number $\mathbf{k}$, Eq. (1-14) directly gives

$$
\begin{aligned}
E_{\mathbf{k}} &= \frac{\hbar^2 k^2}{2m} + \langle \mathbf{k} | W | \mathbf{k} \rangle + \sum_{\mathbf{q}} \frac{\langle \mathbf{k} | W | \mathbf{k} + \mathbf{q} \rangle \langle \mathbf{k} + \mathbf{q} | W | \mathbf{k} \rangle}{(\hbar^2/2m)(k^2 - |\mathbf{k} + \mathbf{q}|^2)} \\
&= \frac{\hbar^2 k^2}{2m} + w_0 + \sum_{\mathbf{q}} \frac{S^*(\mathbf{q})S(\mathbf{q})w_q^2}{(\hbar^2/2m)(k^2 - |\mathbf{k} + \mathbf{q}|^2)}.
\end{aligned} \tag{17-1}
$$

In the final form, the matrix elements are factored into structure factors and form factors. Eq. (17-1) should be summed over all occupied states. The first two terms have already been included in the free-electron energy, so the final term is the correction to that energy, which was obtained in Chapter 15.

The shift in energy of levels relative to each other will mean that some redistribution of the occupation of levels will be required to obtain the ground state. In particular, levels inside Bragg planes are lowered in energy and levels just outside are raised, so that the Fermi surface becomes distorted at the planes, as was illustrated for aluminum in Fig. 16-9. It can be shown (Harrison, 1966a, p. 88ff) that the energy is obtained correctly to second order in the pseudopotential by ignoring the distortion of the Fermi surface and summing Eq. (17-1) over the undistorted Fermi sphere. This is not obvious, particularly since the energies given in Eq. (17-1) are seen to diverge when $\mathbf{k}$ is too close to a Bragg plane. (These are just the planes for which the energy denominator vanishes.) That the distortion can be ignored is proved by showing that the correct second-order energy is obtained by taking principal parts in the integrations across a single Bragg plane.

Thus, to obtain the effect of the pseudopotential on the energy, we sum the final term in Eq. (17-1) over $k < k_F$, multiply by two for spin, and divide by N_a to obtain the energy shift per ion, called the *band-structure energy*:

$$
\begin{aligned}
E_{bs} &= \frac{2}{N_a} \sum_{\mathbf{q}, \mathbf{k}} \frac{S^*(\mathbf{q})S(\mathbf{q})w_q^2}{(\hbar^2/2m)(k^2 - |\mathbf{k} + \mathbf{q}|^2)} \\
&= \sum_{\mathbf{q}} S^*(\mathbf{q})S(\mathbf{q}) \frac{2}{N_a} \sum_{\mathbf{k}} \frac{w_q^2}{(\hbar^2/2m)(k^2 - |\mathbf{k} + \mathbf{q}|^2)}.
\end{aligned} \tag{17-2}
$$

The second step is an extremely important one, since everything in the sum over $\mathbf{k}$ is independent of the arrangement of ions. When we evaluate it as a function of q for a particular metal, that same function will be applicable to all properties of the metal, with the ion positions entering only through the structure factors as they did in the electronic properties.

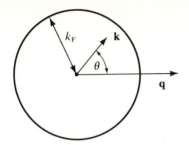

FIGURE 17-1

The coordinate system used for the calculation of the energy wave number characteristic; **q** remains fixed as we integrate over **k**.

Before evaluating the sum we must make one other correction. In simply adding electron energies we make the same double-counting error for the Coulomb interactions that was discussed at the beginning of Chapter 7. We may correct the error here by replacing one of the pseudopotential form factors in Eq. (17-2) with the unscreened form factor, that is, by replacing w_q^2 with $w_q w_q^0$ (see Harrison, 1966a, 53ff). If one thinks about the double counting, it will be seen that this result is plausible, though it is not an obvious result.

We may now proceed with the evaluation of the sum over **k** in Eq. (17-2) by converting the sum to an integral, using the coordinate system of Fig. 17-1. The $w_q w_q^0$ factor which replaced w_q^2 does not depend upon **k**, so it may be taken out of the sum:

$$\frac{2}{N_a} \sum_{k<k_F} \frac{1}{k^2 - |\mathbf{k} + \mathbf{q}|^2} = \frac{2\Omega}{(2\pi)^3 N_a} \int_0^{k_F} dk 2\pi k^2 \int_0^\pi \frac{\sin\theta\, d\theta}{-2kq\cos\theta - q^2}$$

$$= \frac{\Omega_0}{(2\pi)^2 q} \int_0^{k_F} dk\, k \ln\left|\frac{q + 2k}{q - 2k}\right|$$

$$= -\frac{k_F \Omega_0}{8\pi^2}\left[\frac{1-\eta^2}{2\eta} \ln\left|\frac{1+\eta}{1-\eta}\right| + 1\right], \tag{17-3}$$

where $\eta = q/2k_F$. Notice the similarity of this equation to Eq. (16-31). We may now rewrite the band-structure energy of Eq. (17-2) as

$$E_{bs} = \sum_q S^*(\mathbf{q})S(\mathbf{q})F(q), \tag{17-4}$$

with $F(q)$, the **energy wave-number characteristic**, given by

$$F(q) = -\frac{mk_F \Omega_0 w_q w_q^0}{4\pi^2 \hbar^2}\left[\frac{1-\eta^2}{2\eta} \ln\left|\frac{1+\eta}{1-\eta}\right| + 1\right]. \tag{17-5}$$

If we rewrite this result in terms of w_q^0 and the dielectric function, from Eq. (16-31), we obtain

$$F(q) = -\frac{\Omega_0 q^2 (w_q^0)^2}{8\pi e^2}\left[\frac{\varepsilon(q) - 1}{\varepsilon(q)}\right], \tag{17-6}$$

and w_q^0 may be taken from Eq. (16-7) by setting κ equal to zero:

$$w_q^0 = \frac{-4\pi Z e^2 \cos q r_c}{\Omega_0 q^2}. \tag{17-7}$$

Use of Eq. (17-6) and the dielectric function of Eq. (16-31) corresponds to the accurate screening calculation. Use of Eq. (17-6) with Eq. (16-30) corresponds to the use of the Fermi-Thomas approximation throughout; this is presumed to be less accurate, though apparently it has not been tested except in Problem 17-1.

The energy of the crystal is then given by three terms. The first is the free-electron energy, which depends only upon the electron density (and the number of electrons per ion, since it is usually written as the energy per ion); the free-electron energy is given by the first three terms in Eq. (15-16). Second is the electrostatic energy, from Eq. (15-6), which depends upon the positions of the ions because the geometrical factor α does. Third is the band-structure energy, which is given in Eq. (17-4). It is best calculated from Eq. (17-4) with the unscreened form factors from Eq. (17-7) and the dielectric constant from Eq. (16-31). This corresponds to the use of form factors that are slightly more accurate than those used in studying the electronic properties; the extra accuracy is justified for the total energy calculation since it is just as easy.

We did not include the band-structure energy in discussing the bond length and compressibility. This is justified by our expansion of the energy in the pseudo-potential, as in Eq. (17-1). We included the zero-order and first-order terms in Chapter 15 but neglected the second-order terms. Now we have added the second-order terms but discard the higher-order terms. The expansion is essentially an expansion in w_q/E_F for the q that have nonvanishing structure factors; that is, for the lattice wave numbers. For the lattice wave numbers in the metals, the ratio is typically 0.1, so it is not implausible. For aluminum, for example, the electrostatic energy, free-electron energy, and band-structure energy are about -87, $+48$, and -3 eV per ion, respectively (Harrison, 1966a, p. 200), so the expansion makes sense. In general, the use of terms only to second order has proved quite satisfactory for calculating changes in energy under rearrangement of the ions at constant volume. We will return at the end of Section 17-B to a case where second-order theory was not adequate.

17-B   The Effective Interaction Between Ions, and Higher-Order Terms

For most purposes the band-structure energy is conveniently calculated by performing the sum over wave number space directly as indicated in Eq. (17-4). We shall use that procedure when we discuss phonons. There is, however, conceptual value in making a transformation back into position space where we may see that the band-structure energy can be written as a contribution to an effective interaction between ions. We rewrite the band-structure energy of Eq. (17-4), inserting the explicit form of the structure factors from Eq. (16-5):

$$E_{bs} = \frac{1}{N_a^2} \sum_{q, i, j} F(q)e^{-i\mathbf{q} \cdot (\mathbf{r}_i - \mathbf{r}_j)}. \tag{17-8}$$

It is remarkable that if we define an *indirect interaction*,

$$V_{ind}(r) = \frac{2\Omega_0}{(2\pi)^3} \int F(q)e^{-i\mathbf{q}\cdot\mathbf{r}}\, d^3q$$

$$= \frac{\Omega_0}{\pi^2} \int_0^\infty dq q^2 F(q) \sin qr/qr, \tag{17-9}$$

where the last step utilized the spherical symmetry of $F(q)$, the total band-structure energy for the N_a ions can be written

$$N_a E_{bs} = \frac{1}{2} \sum_{i,j}' V_{ind}(\mathbf{r}_i - \mathbf{r}_j) + \frac{\Omega}{(2\pi)^3} \int F(q)\, d^3q, \tag{17-10}$$

with the prime indicating that the term $i = j$ is to be omitted. The final term, like the free-electron energy, depends only upon the total volume, so that it does not change when the ions are rearranged at constant volume. For such rearrangements, the electrostatic energy changes according to $\frac{1}{2}\sum' Z^2 e^2/|\mathbf{r}_i - \mathbf{r}_j|$, so we may add $Z^2 e^2/r$ to V_{ind} to obtain the *effective interaction between ions*, which contains all of the terms in the energy that change under such rearrangements:

$$V(r) = V_{ind}(r) + Z^2 e^2/r. \tag{17-11}$$

Notice that though the energy is written as a central force interaction of two bodies, there are additional terms depending upon volume, so that the system is not in equilibrium under the influence of these interactions alone. Thus the Cauchy relations discussed in Chapter 13 are not expected to be obeyed, and indeed they are not.

Given $F(q)$, it is a simple task to evaluate $V(r)$. Such a calculation for aluminum gives the interaction shown in Fig. 17-2. The most striking aspect of this interaction is the oscillatory tail that remains at large distances. This arises directly from the logarithmic singularity in $F(q)$ at $q = 2k_F$. We may obtain the asymptotic form by integrating Eq. (17-9) by parts twice, keeping the most divergent term, giving

$$V(r) \sim \frac{9\pi Z^2 w_{2k_F}^2}{E_F} \frac{\cos 2k_F r}{(2k_F r)^3}. \tag{17-12}$$

For a discussion of how this is done, see Harrison (1970, p. 442).

There is no question but that the oscillations in the interaction are real; they occur also in the charge distribution when an impurity is present and are obtained for that case if accurate screening theory is used. They are called *Friedel oscillations* (Friedel, 1952, appendix). There is some question as to whether important phase shifts occur in the oscillations because of higher-order terms in the pseudo-potential. These shifts would be in the form of a constant δ, called the phase shift, added to the argument of the cosine, $2k_F r$ (see Heine and Weaire, 1970, and Harrison, 1973b). Surprisingly, the direct effect of these Friedel oscillations has not turned out to be strong enough that one can tell experimentally if there are shifts;

theoretical estimates, based on the formula of Harrison (1973b), and atomic phase shifts from Harrison (1970 p. 186) give phase shifts of some 78° for the oscillations, which would drastically change their positions. The question of the phase shifts remains open.

The effective interaction between ions can give the result of carrying out the full calculation for a number of properties quite directly. For example, the nearest neighbors usually lie near a minimum in the interaction energy. (See Fig. 17-2.) A little algebra shows that the effective interaction between ions also gives the interaction between vacancies. Thus we can directly read off the binding energy of divacancies from Fig. 17-2 as about 0.05 electron volts. The slope of the interaction there is quite large, so the result may not be accurate, and it does not include effects of distortion. However, it gives the answer directly, and carrying out the calculation in wave number space does not eliminate these uncertainties; it should give the same answer.

The fact that the nearest-neighbor distance lies outside the minimum suggests

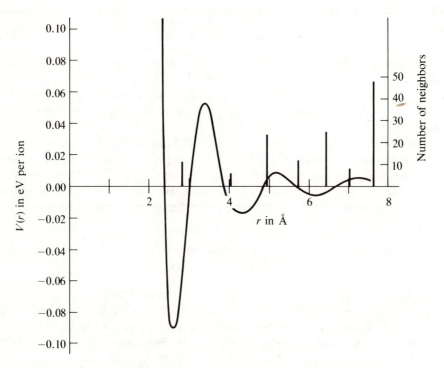

FIGURE 17-2

The effective interaction between ions in aluminum. Also shown is the distribution of neighbors as a function of distance in the face-centered cubic structure. [After Harrison, 1964.]

that the force between nearest neighbors is attractive and therefore that the neighbors nearest to a vacancy should relax outward. This also turns out to be a sensitive point; other pseudopotential theories as accurate as that which led to Fig. 17-2 give the minimum outside the nearest neighbor distance. It is good to realize that this aspect of the results is uncertain even with accurate pseudopotentials; with the cruder model used here there is additional uncertainty. The pseudopotential theory of the vibrational spectra appears to be better, and we turn to that in the next section.

Let us first look a little further into how higher-order terms in the pseudopotential affect properties. The counterpart of the energy wave-number characteristic given in Eq. 17-5, for third-order terms, has been evaluated (Lloyd and Sholl, 1968; Brovman, Kagan, and Holas, 1971). It contains three powers of the form factor, and a double sum over a q and a q' is required in the evaluation. An interesting test for the importance of such terms was noted several years ago by E. G. Brovman, who pointed out that there is a set of three frequencies which must satisfy a particular relation in the hexagonal close-packed structure in any second-order theory. The relation is analogous to the Cauchy relations discussed in Chapter 13, but unlike the Cauchy relations, it does not depend upon the ions being in equilibrium under the influence of central forces alone. The relation noted by Brovman is in fact reasonably well satisfied in magnesium and zinc, but Roy et al. (1973) found that the relation was seriously violated in beryllium. This suggests that the elements in the beryllium row of the periodic table do not obey the simple theory. To correct the error it was necessary to go to higher order in the perturbation theory, as did Bertoni et al. (1973), who included third-order terms in the calculation of the vibration spectrum of beryllium (without adjustable parameters), and indeed the correct behavior of these modes was obtained. A similar application of third-order theory to indium (Garrett and Swihart, 1976) gave poor results. Apparently, second-order theory may be adequate for all simple metals but beryllium and lithium, and third-order theory should suffice for those.

Attention has been given also to the influence of the higher-order terms on the effective interaction between ions (Harrison, 1973b). These terms modify the two-body interaction, including giving the phase shifts discussed earlier in this section. In addition, third-order terms give intrinsic three-body interactions, depending upon the coordinates of three ions. Fourth-order terms give four-body interactions, and so on. Harrison succeeded in obtaining the asymptotic form for these many-body interactions, of which the two-body asymptotic form, from Eq. (17-12), is a special case. The form is not very complicated, but to date it has not proved very useful. There was some indication (Harrison, 1973b) that many-body terms may be the determining factor in the crystal structures, as we indicated at the beginning of this chapter. Structure determination, as we indicated there, would seem to be one of the likeliest natural properties to study with the second-order pseudopotential theory, but second-order theory does not appear to contain the determining factors.

17-C   The Phonon Spectrum

Evaluation of Structure Factors

In all properties studied with pseudopotential theory, the first step is the evaluation of the structure factors. For simplicity, let us consider a metallic crystal with a single ion per primitive cell—either a body-centered or face-centered cubic structure. We must specify the ion positions in the presence of a lattice vibration, as we did in Section 9-D for covalent solids. There, however, we were able to work with the linear force equations and could give displacements in complex form. Here the energy must be computed, and that requires terms quadratic in the displacements. It is easier to keep everything straight if we specify displacements as real. For a lattice vibration of wave number $\mathbf{k}$, we write the displacement of the ion with equilibrium position $\mathbf{r}_i$ as

$$\delta\mathbf{r}_i = \mathbf{u}e^{i\mathbf{k}\cdot\mathbf{r}_i} + \mathbf{u}^*e^{-i\mathbf{k}\cdot\mathbf{r}_i}; \qquad (17\text{-}13)$$

$\mathbf{u}$ is a vector amplitude and depends upon time as $e^{-i\omega t}$, but we do not need that time-dependence at present. We may immediately write the structure factor and expand it in the displacements:

$$S(\mathbf{q}) = \frac{1}{N_a}\sum_i e^{-i\mathbf{q}\cdot(\mathbf{r}_i+\delta\mathbf{r}_i)}$$

$$= \frac{1}{N_a}\sum_i e^{-i\mathbf{q}\cdot\mathbf{r}_i} - \frac{i}{N_a}\sum_i \mathbf{q}\cdot\delta\mathbf{r}_i e^{-i\mathbf{q}\cdot\mathbf{r}_i} - \frac{1}{2N_a}\sum_i (\mathbf{q}\cdot\delta\mathbf{r}_i)^2 e^{-i\mathbf{q}\cdot\mathbf{r}_i} + \cdots$$

$$(17\text{-}14)$$

The first term is exactly the structure factor for the undistorted crystal, giving a value of 1 for the lattice wave numbers and 0 for other wave numbers. We may immediately see that the structure factor is corrected at the lattice wave number. Notice that the second term vanishes for lattice wave numbers because $e^{-i\mathbf{q}\cdot\mathbf{r}_i}$ is the same at every site, though $\delta\mathbf{r}_i$ oscillates from one site to another, averaging to zero. The third term, on the other hand, does not vanish; $e^{-i\mathbf{q}\cdot\mathbf{r}_i}$ is again the same at every site, but $(\mathbf{q}\cdot\delta\mathbf{r}_i)^2$ can be seen to average to $2(\mathbf{q}\cdot\mathbf{u})(\mathbf{q}\cdot\mathbf{u}^*)$ by squaring Eq. (17-13) and summing over i. Let us then indicate the lattice wave numbers of the perfect crystal by $\mathbf{q}_0$; we have found that, to second order in $\mathbf{u}$,

$$S(\mathbf{q}_0) = 1 - (\mathbf{q}_0\cdot\mathbf{u})(\mathbf{q}_0\cdot\mathbf{u}^*). \qquad (17\text{-}15)$$

The reduction of diffraction intensity by lattice vibrations is a familiar effect (see, for example, James, 1950, or discussions of the Debye-Waller factor in solid state texts); the factor by which the diffraction intensity is reduced is called the **Debye-Waller factor** when a thermal average over the vibrational amplitudes is made. We have obtained only the terms to second order in that amplitude.

Let us next look at the first-order term in Eq. (17-14), in which displacements obtained from Eq. (17-13) are substituted:

$$S_{\text{first}}(\mathbf{q}) = \frac{-i}{N_a} \sum_i (\mathbf{q} \cdot \delta\mathbf{r}_i)e^{-i\mathbf{q}\cdot\mathbf{r}_i}$$

$$= -\frac{i\mathbf{q}\cdot\mathbf{u}}{N_a} \sum_i e^{-i(\mathbf{q}-\mathbf{k})\cdot\mathbf{r}_i} - \frac{i\mathbf{q}\cdot\mathbf{u}^*}{N_a} \sum_i e^{-i(\mathbf{q}+\mathbf{k})\cdot\mathbf{r}_i}. \qquad (17\text{-}16)$$

These sums also can be evaluated immediately. They are of the same form as those for the perfect crystal, but with $\mathbf{q}$ replaced by $\mathbf{q} \pm \mathbf{k}$. Thus the first will be N_a if $\mathbf{q} - \mathbf{k}$ is a lattice wave number, and zero otherwise; the second will be N_a if $\mathbf{q} + \mathbf{k}$ is a lattice wave number, and zero otherwise. The lattice distortion has given nonzero structure factors at "satellites" to each of the lattice wave number, as indicated in Fig. 17-3; they lie at wave numbers $\mathbf{q}_0 \pm \mathbf{k}$ and have structure factors $-i(\mathbf{q}_0 \pm \mathbf{k}) \cdot \mathbf{u}$.

These are all the structure factors we need for obtaining the energy to second order in the displacements. The energy is obtained by summing $F(q)S^*(\mathbf{q})S(\mathbf{q})$ over wave numbers. At the satellite points, the first-order structure factors we have obtained give terms of second order in u, and higher-order terms can be neglected. At the lattice wave numbers, $S^*(\mathbf{q})S(\mathbf{q}) = 1 - 2(\mathbf{q} \cdot \mathbf{u})(\mathbf{q} \cdot \mathbf{u}^*)$, plus higher-order terms, as seen from Eq. (17-15), and any terms of higher order than those given in Eq. (17-15) can be discarded.

These satellite points will give rise to electron scattering, just as did the structure factors that arose from defects in the crystal. The satellite matrix elements $-i\mathbf{q} \cdot \mathbf{u}w_{\mathbf{q}}$ and $-i\mathbf{q} \cdot \mathbf{u}^*w_{\mathbf{q}}$ are sometimes called the "electron–phonon interaction"; we shall return to them. It may be desirable first to discuss the vibration spectrum itself.

FIGURE 17-3

A lattice vibration of wave number $\mathbf{k}$ reduces the structure factor at the lattice wave numbers (solid dots), but introduces a new structure factor at satellite points differing from the lattice wave numbers by $\mathbf{k}$.

Energy of Distortion

We have obtained the structure factors that will enable us to calculate the potential energy of distortion or elastic energy of the crystal, which is due to a periodic distortion given by Eq. (17-13). We may also take the time derivative of the displacements, noting that $d\mathbf{u}/dt = -i\omega\mathbf{u}$, and obtain the average kinetic energy (or the kinetic energy per ion), by summing over i. We obtain

$$E_{kin} = M\omega^2\mathbf{u} \cdot \mathbf{u}^*, \tag{17-17}$$

where M is the mass of the ion. We shall next equate the kinetic and potential energies and solve for the frequency. (This is a short-cut giving the same answer as a solution of the dynamical equations.)

We immediately have the band-structure energy from the structure factors that were obtained. It is convenient to sum the lattice wave numbers and the satellites together in the form

$$E_{bs} = \sum_{\mathbf{q}_0} [|(\mathbf{q}_0 + \mathbf{k}) \cdot \mathbf{u}|^2 F(\mathbf{q}_0 + \mathbf{k}) + |(\mathbf{q}_0 - \mathbf{k}) \cdot \mathbf{u}|^2 F(\mathbf{q}_0 - \mathbf{k}) - 2|\mathbf{q}_0 \cdot \mathbf{u}|^2 F(\mathbf{q}_0)].$$
$$\tag{17-18}$$

In this sum the final term is ill defined for $q_0 = 0$ and should be taken to vanish. The form is somewhat more complicated (Harrison, 1966a, p. 244) if there is more than one ion per primitive cell. The evaluation of such a sum is straightforward once we are given a form for $F(q)$, such as that in Eq. (17-6). The convergence is rather good since the three terms in Eq. (17-18) at each lattice wave number tend to cancel, and $F(q)$ drops as q^{-4} at large q. (This is a reason for *not* using the Fermi-Thomas approximation, which would lead to q^{-2}.) For propagation in symmetry directions, the waves will be purely longitudinal and transverse, so the direction of $\mathbf{u}$ is known, and there is only one parameter, u, which factors out of the sums, and the solution for ω^2 is direct. If one were to seek modes with $\mathbf{k}$ in an arbitrary direction, there would be six sums giving terms proportional to u_x^2, u_y^2, u_z^2, $u_y u_z$, $u_z u_x$, and $u_x u_y$. This exactly corresponds to the calculation of the modes for three coupled harmonic oscillators, and can also be solved directly (see Harrison, 1966a, Chapter 7), but here we will stay with the simplest case.

Although the free-electron energy does not change under the distortion, the electrostatic energy does. There are various ways of addressing that problem (Harrison, 1966a) but one rather direct way is to carry out the calculation in wave number space; that is, for the electrostatic energy, we simply undo the transformation that took us from band-structure energy in Eq. (17-8) to an indirect interaction in Eq. (17-10). The necessary Fourier transform of the Coulomb interaction was evaluated earlier as a step in obtaining Eq. (16-7), and is

$$\int e^{-i\mathbf{q} \cdot \mathbf{r}} \frac{e^2}{r} d^3r = \frac{4\pi e^2}{q^2}. \tag{17-19}$$

It is convenient to multiply this by a convergence factor $e^{-q^2/4\eta}$ and to let η go to infinity at the end (Harrison, 1966a). Then it is clear that the final term in

Eq. (17-10) does not enter the calculation and the electrostatic energy is obtained in the form

$$E_{electro} = \sum_{q} S^*(q)S(q)2\pi Z^2 e^2 e^{-q^2/4\eta}/(q^2\Omega_0). \qquad (17\text{-}20)$$

Without the convergence factor the convergence of the sum at large q is not so rapid as for the band-structure energy, but owing to the cancellation between satellites and lattice wave numbers, the sum still converges satisfactorily. It is best done (Harrison, 1966a, p. 247ff) by summing along lines of wave numbers parallel to k, for which analytic forms are easily obtained, and then the sum over these lines converges very rapidly, with only a few lines usually being required. We may then use Eq. (17-17) to write

$$\omega^2 = (E_{bs} + E_{electro})/(Muu^*). \qquad (17\text{-}21)$$

The Spectra

There have been a very large number of applications of this theory to the vibration spectra (see Heine and Weaire, 1970, for a review) since the earliest studies (Harrison, 1964, and Sham, 1965). Perhaps the most relevant here is the calculation by Ashcroft (1968) for the alkali metals; that calculation makes use of the empty-core model of the pseudopotential, described here. The results of his calculation, along with experimental points, are shown in Fig. 17-4 for potassium; similar results were obtained for sodium.

The agreement with experiment for potassium and sodium is spectacular, but it should be noted that for the alkali metals, the band-structure contribution to the energy is typically (at least for the high-frequency modes) only a few percent of the electrostatic energy. Since the electrostatic energy does not depend upon the pseudopotential, the results are not sensitive to the pseudopotential. The comparison tests the formulation of the vibrational problem but not the details of the model. With increasing Z, there is an increased cancellation between the electrostatic and band-structure terms in the energy. In aluminum, the band-structure term cancels about half of the electrostatic energy, and the cancellation is still larger in lead. For aluminum, it has frequently been found that when the pseudopotentials are determined from first principles or from a property without close relation to the vibration spectrum, errors in the frequencies of the order of a factor of two (but usually less) may be obtained. The adjustment of a single parameter (such as r_c) is generally sufficient to bring the agreement to within a few percent for most modes. This makes it difficult to test particular theories, since choices may have been made, such as approximations to the exchange potential, and an approximation may have been selected which gives good results for a particular material without appearing to make an adjustment. For lead, the cancellation is so large that it is not easy to obtain a fit with a single parameter.

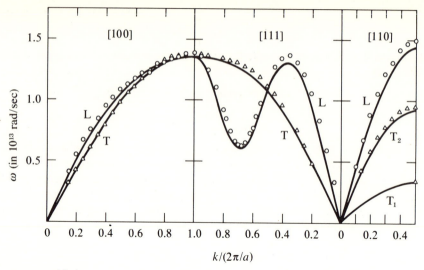

FIGURE 17-4

The vibration spectrum for potassium calculated by Ashcroft (1968), who used the empty-core pseudopotential from Eq. (15-13) and $r_c = 1.13$ Å (rather than the value of 1.20 Å from Table 16-1). Experimental points are from Cowley, Woods, and Dolling (1966).

The Bohm-Staver Formula

Two aspects of the spectra should be considered before going on to the electron–phonon interaction. First is the behavior at long wavelengths, $q \sim 0$. The contributions of satellites of $\mathbf{q}_0 = 0$ to the band-structure energy and to the electrostatic energy both diverge, but because we have done the screening correctly, the divergences will cancel. The structure factors at these satellites are $\pm i\mathbf{k} \cdot \mathbf{u}$; they vanish for transverse waves and are $\pm iku$ for longitudinal waves. For the longitudinal wave, we may substitute into Eq. (17-21), using Eqs. (17-6) and (17-18) for the band-structure energy and Eq. (17-20) for the electrostatic energy. Keeping only the two satellites to $\mathbf{q}_0 = 0$, we obtain, after a little algebra,

$$\omega^2 = \frac{4\pi Z^2 e^2}{M\Omega_0}\{1 - \cos^2(kr_c)[\varepsilon(k) - 1]/\varepsilon(k)\}. \tag{17-22}$$

By dropping the contributions from other $\mathbf{q}_0$ and their satellites, we effectively dropped the structure-dependent terms, replacing the ionic background by a jelly, but we have retained an average effect from the repulsive term by keeping $r_c \neq 0$. The right side can be expanded in k, by using Eq. (16-31) for $\varepsilon(q)$, to obtain

$$\omega^2 = \frac{4\pi Z^2 e^2 k^2}{M\Omega_0 \kappa^2}[1 + (\kappa r_c)^2/2]. \tag{17-23}$$

Again, κ is the Fermi-Thomas screening parameter, given by Eq. (16-29), though we have not used the Fermi-Thomas approximation in the calculation. As it should be, ω is proportional to k; the slope is the speed of sound. The value obtained with $r_c = 0$ is called the ***Bohm-Staver speed of sound*** (Bohm and Staver, 1951). The result is qualitatively correct and of the correct order of magnitude, but the terms from other satellites are important. For example, the values for the speed of sound calculated when the $(\kappa r_c)^2/2$ term is neglected in Eq. (17-23)—the equation then gives the Bohm-Staver values—the values found when keeping that term, and the experimental values are all given in Table 17-1 for sodium, magnesium, and aluminum; clearly the full summation is needed. The Bohm-Staver formula does reasonably well for the alkali metals, but this is because of cancellation.

For pure transverse waves, the satellites of $\mathbf{q}_0 = 0$ do not enter, and at long wavelengths (small k), only components of the pseudopotential corresponding to large q enter. It is possible either to calculate the speed of sound by expanding these terms for small k, as we have done here, or to compute the shear constants by distorting the lattice slightly and recomputing the band-structure and electrostatic energies. Using the shear constants thus computed and elasticity theory must give the same value for the speed of sound. The result of such calculations is as would be expected from our discussion of the vibration spectra; in the alkali metals, the shear constants are dominated by the electrostatic energy, and reasonably good values are obtained even if the band-structure contributions are ignored altogether (Mott and Jones, 1936, p. 149). The cancellation of the electrostatic terms by the band-structure energy increases with increasing Z and is quite essential in the polyvalent metals.

Kohn Anomalies

A second set of features that should be mentioned in connection with the vibration spectra are the ***Kohn anomalies*** (Kohn, 1959). We noted that $\varepsilon(q)$ had an infinite derivative with respect to q at $q = 2k_F$. It follows immediately from Eqs. (17-6) and (17-22) that $d\omega/dk$ must also be infinite when $k = 2k_F$. In fact,

TABLE 17-1
The speed of sound (in units of 10^5 cm/sec) calculated
.by using the Bohm-Staver formula, by including the
effect of the empty core (Eq. 17-23), and
from experiment.

Metal	Bohm-Staver	Eq. (17-23)	Experiment
Na	3.0	4.2	3.0
Mg	6.2	8.5	4.6
Al	9.1	12.2	5.1

SOURCE of experimental data: Weast (1975).

going back to Eq. (17-18), we see that $d\omega/dk$ will be infinite (positively or negatively) whenever any of the satellite points crosses $2k_F$. Such irregularities in the dispersion curves are called Kohn anomalies. They tend to be quite weak since the singularity is a logarithmic one and in the alkali metals where the band-structure term is small in any case they have not been observed. However, in the polyvalent metals, particularly lead, they are quite conspicuous and are rather well accounted for by pseudopotential theory (Harrison, 1966a, p. 254ff).

17-D   The Electron–Phonon Interaction and the Electron–Phonon Coupling Constant

Once we have obtained the dispersion curves for the metal, we may proceed to other properties just as we did with the covalent solids. In particular, we may quantize the vibrations as was done for the covalent solids and obtain the appropriate contribution to the specific heat. We shall not repeat that analysis now for the simple metals but shall wish to use the customary terminology by referring to the vibrations as phonons.

Electron Scattering

Let us return now to the electron–phonon interaction itself and to the effect of such interactions on the electronic properties. We have used $\mathbf{k}$ to specify wave numbers of both phonons and electrons, and it will be convenient to change to $\mathbf{Q}$ for the wave number of the phonon so that it is not confused with the electron wave numbers $\mathbf{k}$. We have seen that a phonon with a wave number $\mathbf{Q}$ introduces matrix elements between states that differ in wave number by $\mathbf{Q}$ and in fact between states that differ by $\mathbf{q}_0 \pm \mathbf{Q}$ for all $\mathbf{q}_0$. We shall see that these matrix elements produce a scattering probability proportional to the square of the phonon amplitude for transitions between states differing in wave number by $\mathbf{q}_0 \pm \mathbf{Q}$. It is customary to refer to those scattering processes for $\mathbf{q}_0 = 0$ as *normal scattering*; scattering processes for $\mathbf{q}_0 \neq 0$ are called *umklapp scattering*. At small Q the distinction is sharp, but when Q is large it is not. In our discussion of vibration spectra we noted that the satellites of $\mathbf{q}_0 = 0$ give nonzero structure factors only for longitudinal modes, so that for long-wavelength phonons only the longitudinal modes contribute to normal scattering. The matrix element takes the value $-i\mathbf{Q} \cdot \mathbf{u}w_Q$ exactly as if the phonon had produced a potential equal to $-w_Q$ times the local dilatation. Thus $-w_Q$ can naturally be called the longitudinal *electron–phonon interaction*. Recall that it is $2E_F/3$ for small Q. Unfortunately, the same term is used to describe other parameters, also. Since the change in wave number of the electron is the same as that of the phonon absorbed or emitted in the process, we may think of the scattering as conserving momentum, and can associate a momentum of $\hbar Q$ with the phonon.

Umklapp scattering can occur from transverse as well as from longitudinal modes, and if we are to retain conservation of momentum, we must say that the lattice as a whole takes up momentum equal to $\hbar\mathbf{q}_0$. Umklapp scattering tends to be important to the resistivity, since it allows large electron-scattering angles with low-frequency (small-Q) phonons. These qualitative statements are intended only to introduce some of the terminology and to relate it to this formulation of the problem. Any attempt at a quantitative calculation of a property such as the temperature-dependent resistivity of a metal requires an extensive calculation of the frequencies and forms of the modes and a careful organization of the scattering processes that can occur. We have given the basis for such an organization but shall not carry it further.

The Dimensionless Coupling Constant

A calculation that should be described is the one that leads to a familiar dimensionless coupling constant λ, which is of importance in the theory of superconductivity (see McMillan, 1968, or Chakraborty, Pickett, and Allen, 1976). We want to find the *total* scattering rate for electrons at high temperatures, in order to relate it to other properties. We use the first form given in Eq. (16-21), but with the factor $(1 - \cos\theta)$ removed. We have seen that the matrix elements $\langle\mathbf{k} + \mathbf{q}|W|\mathbf{k}\rangle$ are proportional to the amplitude of the mode u. Furthermore, the energy in each mode, proportional to u^2, will be equal to kT at high temperatures so that the integral is proportional to kT. All of the other effects that enter the calculation are obtainable from the pseudopotential and the form and frequency of the vibrational modes. Thus it is convenient to write the total scattering time τ_t for the electron as

$$\frac{1}{\tau_t} = \frac{2\pi kT}{\hbar}\lambda; \tag{17-24}$$

λ is the *dimensionless electron–phonon coupling constant* characteristic of the metal. We may write it in terms of the matrix elements for this problem as

$$\lambda = \frac{1}{2kT}\left(\frac{dN}{dE}\right)\frac{\Omega}{A_{FS}}\int |\langle\mathbf{k} + \mathbf{q}|W|\mathbf{k}\rangle|^2 \, dA_{FS}. \tag{17-25}$$

It is an interesting problem to see what parameters of the material the coupling constant depends upon and what general magnitude it should have. The analysis that relies on the forms obtained here for the electron–phonon interaction and the vibrational frequencies (See Problem 17-2) leads to the result that

$$\lambda \sim \frac{1}{E_F^2 k_F^2}\int_0^{2k_F} dq\, q w_q^2. \tag{17-26}$$

A numerical factor of order unity has been omitted; this is appropriate in view of the many approximations entering the analysis, such as the neglect of umklapp, but the lack of dependence upon features such as the mass of the ions is correct and important.

TABLE 17-2
Empirically determined
values of the dimensionless
coupling constant.

2	3	4
Be 0.23		
Mg —	Al 0.38	
Zn 0.38	Ga 0.40	
Cd 0.38	In 0.69	Sn 0.60
Hg 1.00	Tl 0.71	Pb 1.12

SOURCE: Values from Mc-Millan (1968). Other determinations show considerable variation from these values.

Eq. (17-26) would suggest λ to be of order 0.01 if we take w_q as being of order 0.1 E_F and note that the factor q suppresses the contribution at small q where w_q is of order E_F. The experimental estimates of λ listed in Table 17-2 indicate, however, that it can be of order unity. This difference presumably comes from umklapp, which, particularly in the polyvalent metals, does not have the suppression of contributions where w_q is large.

On the scale of validity of Eq. (17-26), the distinction between total scattering rate $1/\tau_t$ and the rate entering conductivity is not important, and this provides an estimate of the high-temperature resistivity, proportional, as is well known, to temperature.

Enhanced Specific Heat and Superconductivity

Another effect of the electron–phonon interaction is a shift in the velocity of the electrons at the Fermi surface, in some ways analogous to the polaron effect in ionic crystals. Because of the "wake" of lattice distortion that accompanies the electron, its velocity is reduced, as it turns out, by a factor $(1 + \lambda)^{-1}$. (For a discussion of this effect, and references, see Quinn, 1960, p. 58, or Harrison, 1970, p. 418ff.) The reduction in velocity corresponds to a decrease in dE/dk at the Fermi surface and, therefore, to an increase in the density of states by the same factor. We noted in Chapter 15 that the electronic specific heat is proportional to the density of states, so we may expect an enhancement of the experimental specific heat by the same factor, $1 + \lambda$. There can also be a modification from the free-electron value of the electronic specific heat because of distortion of the energy bands themselves by the finite pseudopotential. McMillan sought to take this into account also in determining the experimental values of λ given in Table 17-2.

A third, and most important, role played by the coupling constant is in the theory of superconductivity. It is well known that this phenomenon arises from the electron–phonon interaction. The temperature of the transition is higher the stronger the electron–phonon coupling is, with a dependence given theoretically by

$$T_c = \Theta e^{-1/(\lambda - \mu*)} \tag{17-27}$$

for weakly coupled superconductors (that is, λ is not large). For derivation of this, see Morel and Anderson (1962) and McMillan (1968). The parameter Θ is the Debye temperature, $\Theta = \hbar \omega_D / k$, based on the Debye frequency (see Section 9-C) or some other representative phonon frequency, and $\mu*$ is an electron–electron interaction coupling constant, typically about 0.1. Superconductivity occurs only if λ exceeds $\mu*$. Estimates of λ for the alkali metals seem to confirm that the values are less than a tenth, indicating that superconductivity is not to be expected.

17-E   Surfaces and Liquids

The study of surfaces does not follow as directly from the theory of the bulk for the simple metals as it did for the covalent and ionic solids. In the LCAO description of the latter two, one simply terminates the lattice at a plane and terminates the LCAO wave function at the same plane. Such a termination of a free-electron plane-wave is not adequate.

The Jellium Model

One of the earliest treatments of a metal surface was based upon a jellium model (Bardeen, 1936). If the electron gas terminated abruptly at the surface of the jellium there would be no net potential to contain the electrons in the metal. Therefore the electron gas extends beyond the metal, giving a dipole layer, as illustrated in Fig. 17-5. Bardeen attempted the self-consistent calculation of the resulting potential. It should be mentioned that the Fermi-Thomas approximation is not adequate for this task and was not used by Bardeen; it is not difficult to see that it would predict the Fermi energy to be at the vacuum level, corresponding to a vanishing work function.

More recently, Lang and Kohn (1970) treated the jellium model with modern methods and modern computers. In particular, they included exchange terms in the free-electron approximation, as we have discussed. In this context it is interesting that the potential well that holds the electrons in the metal comes predominantly from the exchange potential; the electrostatic potential itself gives only

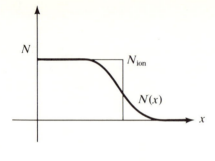

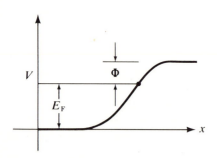

FIGURE 17-5

The jellium model of a metal surface. The ion density N_{ion} terminates abruptly at the surface but the electron density extends beyond it. The net charge density, proportional to $N_{ion} - N$, gives a dipole layer and a potential, V, which holds the electrons in the metal. The minimum energy required to remove an electron from the metal is the work function, Φ. The more complete calculations indicate that the work function arises from the exchange energy rather than from this kind of electrostatic dipole.

a tiny step, and not necessarily in the right direction. Friedel oscillations in the charge density at the surface give opposing dipole layers, cancelling the surface dipole.

Before discussing Lang and Kohn's results on the interface between a metal and a vacuum, let us examine the much simpler problem of the interface between two metals or between a solid and a liquid metal. This is simpler because the same kinds of approximations are applicable on both sides of the interface. It is clear that in equilibrium, the energy of the highest filled states is the same throughout the system; we used this fact in the approximate theory of screening illustrated in Fig. 16-11. To the extent that the system is describable by a set of one-electron energies, the ground state is obtained by filling them, independent of location, to a cut-off energy, the Fermi energy. It also follows from Poisson's equation that, far from the interface, where there are no electric fields, the electron density must be uniform and the same as in the infinite metal. Thus in the free-electron approximation there will be a potential *step*, a difference in the energy of an electron at rest ($\mathbf{k} = 0$) on the two sides, given by the difference in free-electron Fermi kinetic energy E_F on the two sides. Using Eq. (15-4), this gives

$$V_1 - V_2 = \frac{\hbar 2}{2m}(3\pi^2)^{2/3}[N_2^{2/3} - N_1^{2/3}], \qquad (17\text{-}28)$$

where $V_1 - V_2$ is electron potential energy on side one minus that on side two, and N_1 and N_2 are the electron densities on the two sides. Notice that we have not assumed that the differences are small. Notice also that the result applies only to differences far from the junction. Corrections could be made, by applying perturbation theory, for the shift in the energy of the $\mathbf{k} = 0$ state due to the W_q for the lattice wave numbers, and these corrections do modify the size of the step by modifying the energy spread of the occupied region of the band; of course they do not modify the matching at the Fermi energy. The corrections are small in metals, but we shall see that they cannot be neglected at interfaces between metals and semiconductor or between semiconductors.

The Full Jellium-Model Calculation

Lang and Kohn (1970) focused on the total surface energy—that is, the energy required to create the surface—and compared this with the surface tension of the liquid metal. In the jellium model, the only parameter that enters the calculation is the electron density, but Lang and Kohn also included estimates of the electrostatic energy, as in the free-electron approximation used here. The energy they obtained is the dashed line in Fig. 17-6. It is seen to work quite well for the alkali metals, but for the polyvalent metals, which have higher electron densities, the theory fails badly. Lang and Kohn then introduced the effects of the pseudopotential, using the empty-core model. This brought the values into good agreement with experiment.

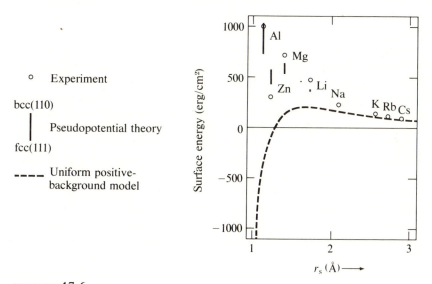

FIGURE 17-6

Theoretical values of surface energy, obtained by Lang and Kohn (1970), as a function of r_s, where r_s is related to the electron density by $N = (4\pi r_s^3/3)^{-1}$. The key is shown at the left. Calculations based on pseudopotential theory were made for bcc and fcc structures, with results given as top and bottom, respectively, of a vertical line at the electron density for each metal. For Na, K, Rb, and Cs, the lines would almost be contained within the experimental points and were not drawn.

The calculations as performed by Lang and Kohn required major computational effort, and meaningful simplified approaches appear not to have been developed yet. Simpler calculations may be possible and one could expect the electrostatic contributions to vary as Z^2/d^3 and the electronic contributions to vary as r_s^{-4} (r_s is defined in Fig. 17-6.) Even in the absence of such studies, there are a number of things which can be learned, particularly since the calculations are direct applications of well-defined and well-understood approximations. The surface energy appears to be understandable in terms of the contributions to the energy that have been discussed, and with the same approximations. No fundamental difficulties arise in regions where the electron density drops to zero, and there are correspondingly large density gradients.

Lang and Kohn (1971) also calculated the work function itself, correctly finding sizable variations from one face to another. The results agreed with existing data to within 5 to 10 percent.

The success of this approach encouraged Lang and Williams (1976) to extend the theory to the adsorption of atoms on the surfaces of simple metals. The problem is of interest in its own right and the necessary concept of a " resonance " will be important in later discussions. For the metal, they again used a jellium model with an electron density appropriate to aluminum. The sharp level associated with the free atom becomes mixed with the free-electron-like levels of the metal when the atom adsorbs to the metal surface so that it becomes a part of the energy band of the metal. However, the added density of electron states (per unit energy) is localized to an energy range near the energy of the state of the isolated atom. This localization in energy is called a **resonance** and will be discussed in some detail in Chapter 20 in connection with the d states in transition metals. As the atom is brought to the surface, the atomic state is thought of as a very sharp peak (a delta function) in the density of states, which broadens in energy as the atom overlaps the surface more and more. If it were brought into the interior of the metal, the peak might be so broad that we could forget that it was a peak and include the state as part of the free-electron gas. At the observed spacing, however, the peak has not broadened that far, as is shown in Fig. 17-7.

That figure gives the results found by Lang and Williams (1976) for three different added atoms. The lithium $2s$ level lies well above the average of the $3s$ and $3p$ levels of aluminum that give rise to the metallic energy bands. (This might be surmised semiquantitatively from the plot of the atomic energy levels in Fig. 1-8. One expects the resonance from the last occupied lithium state to lie at high energy, and this is seen in Fig. 17-7. Thus, when the electrons are placed in the energy levels for the *system* (that is, the metal plus the added atom), only a small fraction of the lithium resonance is " occupied." The charge density on the lithium is small and to a good approximation we may say that the lithium atom has given its electron to the metal and sits as a Li^+ ion outside the surface.

In a similar way, the chlorine $3p$ level lies low in comparison to the aluminum levels and, as indicated in the figure, we may think of the chlorine resonance as almost completely occupied. It is a Cl^- ion outside the metal surface. For silicon, the situation is quite different, as is shown in the figure. Although the $3s$ level of

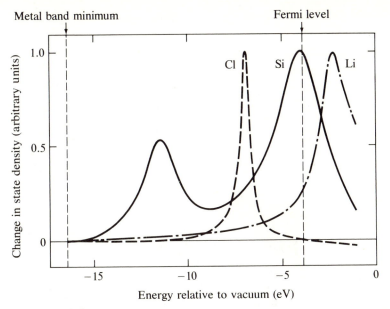

FIGURE 17-7

The change in the density of electron states per unit energy, as a function of energy, when three different types of atoms are adsorbed to an aluminum surface. [After Lang and Williams, 1976.]

the silicon lies low, the 3p level is near the Fermi energy and the p resonance is half occupied. In terms of bonding and antibonding combinations of silicon and metal states, only those combinations that tend to bond are occupied. This gives an asymmetric charge distribution around the silicon and some localization of charge (some bond charge) between the atom and the metal. This is illustrated in Fig. 17-8, where the three cases can be compared.

The three cases span the range of simple adsorptions on simple-metal surfaces. There are also exceptional cases, such as oxygen, which seems to have a tendency to burrow into the metal, rather than sit outside the surface (Yu, Miller, Chye, Spicer, Lang, and Williams, 1976). Similarly, nitrogen adsorbed on titanium actually goes *below* the top layer of titanium atoms instead of sitting on top (Shih et al., 1976). The upper two atomic layers may then be thought of as a single molecular layer of TiN. Although ordinarily one expects an adsorbed atom to remain outside the surface (Strozier, Jepsen, and Jona, 1975), there will certainly be other cases where such complex surfaces are formed.

Liquid Metals

Going from a solid to a liquid requires a considerably smaller generalization of the theory than going to a surface. The ions in a liquid are still arranged in a rather close-packed way, with a decrease in density comparable to that obtained by shaking up marbles from a close-packed arrangement in a box. The free-electron

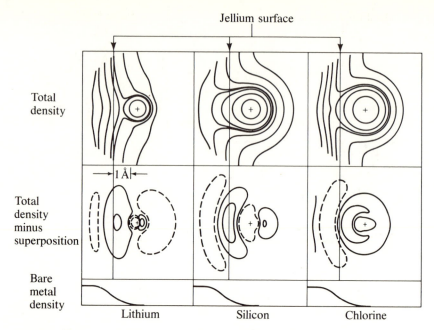

Jellium surface

FIGURE 17-8

Contours of constant electron density for adsorbed atoms on aluminum. The straight line is the surface of the jellium. The center row shows contours of total density minus the superposition of free-atom and bare metal densities. The bottom row is the profile of the bare metal density. [After Lang and Williams, 1976.]

theory with which we began is just as applicable to the liquid structure as to the crystal, and the pseudopotential can be introduced as a perturbation, just as it was in the solid. (For a review, see Faber, 1969, p. 282.)

Without the periodic array, there are no longer lattice wave numbers but a distributed structure factor $S(\mathbf{q})$. The phase will vary in a complicated way, but an average measurable $S^*(\mathbf{q})S(\mathbf{q})$ exists and is spherically symmetric. This is just what is needed for a calculation of the resistivity of the liquid metals. The first such calculation using pseudopotentials (Harrison, 1963b) followed an earlier and conceptually similar calculation by Ziman (1961). It involved the direct substitution of $S^*(\mathbf{q})S(\mathbf{q})$, obtained by X-ray diffraction experiments on the liquid, into Eq. (16-23). Subsequently, it became clear that a theoretical form for $S^*(\mathbf{q})S(\mathbf{q})$ given by Percus and Yevick (1958) and Percus (1962) was more convenient, and probably as accurate as the experiment for the resistivity calculation. This approach was used by Ashcroft and Lekner (1966) for an extensive study of the resistivity of all the simple liquid metals. The form due to Percus and Yevick depends only upon two parameters, a hard-sphere diameter and a packing fraction; these lead to a simple form in terms of elementary functions; Ashcroft and Lekner discuss the choice of parameters. This form is presumably just as appropriate for other elemental liquids.

PROBLEM 17-1   *The band-structure energy*

The observed elastic shear constant c_{44} of aluminum is 2.8×10^{11} erg/cm^3, whereas the electrostatic contribution is 14.8×10^{11} erg/cm^3 (Harrison, 1966a, p. 179; the value there was based upon an effective charge 7.9 percent larger than the 3.0 appropriate here). The band-structure energy is an estimate of the difference, -12.0×10^{11} erg/cm^2. Even if we sum all terms, the result is approximate because of the neglect of terms of higher order than two and the use of an approximate pseudopotential. We obtain the effect of the nearest lattice wave numbers here.

Aluminum has a face-centered cubic structure with cube edge $a = 4.04$ Å. The wave number lattice, therefore, is body-centered cubic, and contains shortest lattice wave numbers of the form $[111](2\pi/a)$, with all eight combinations of ± 1. A distortion ε_4 has been applied to the wave number lattice and modifies these eight lattice wave numbers as indicated in Fig. 17-9.

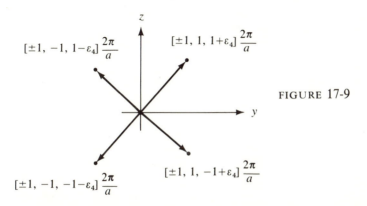

FIGURE 17-9

Obtain the energy change per ion for $\varepsilon_4 = 0.1$ (relative to the energy for $\varepsilon_4 = 0$), and thus the change in energy per unit volume for this distortion. Use the energy wave-number characteristic of Eq. (17-6), with the unscreened pseudopotential of Eq. (17-7), and the Fermi-Thomas approximation to $\varepsilon(q)$—so that $[\varepsilon(q) - 1]/\varepsilon(q) = \kappa^2/(\kappa^2 + q^2)$—and parameters from the Solid State Table.

Noticing that the elastic energy density is $(c_{44}/2)\varepsilon_4^2$, you may obtain this band-structure contribution to c_{44}, to be compared with the experimental -12.0×10^{11} erg/cm^3.

PROBLEM 17-2   *The electron–phonon coupling constant*

Carry out an approximate evaluation of the dimensionless coupling constant λ given in Eq. (17-25). This may be done by neglecting umklapp scattering and treating the vibrations as pure longitudinal and transverse waves; then only longitudinal modes contribute and their structure factors are $-iqu$. The kinetic energy per ion of each mode was given but must be multiplied by N_a before equating it to $kT/2$. This gives u^2 for each mode, and we must multiply the integral by two, since each scattering can be made by either of two modes (corresponding to each mode providing two satellites). At this stage, the result depends upon the phonon frequencies, but we may make the Debye approximation (Chapter 9), using the Bohm-Staver expression for the speed of sound and the formula for the screening constant which enters it—Eq. (16-29). This should lead to Eq. (17-26) and an appropriate numerical constant. One could readily carry out the numerical integral for any material, using the empty-core pseudopotential, but it is probably not significantly better because of the neglect of umklapp scattering.

CHAPTER 18

Pseudopotential Theory
of Covalent Bonding

SUMMARY

By equating selected band energies from the nearly-free-electron bands with those from the LCAO bands, interatomic LCAO matrix elements are deduced. The pseudopotential can then be added to the nearly-free-electron picture to generate realistic bands, as in metals. However, for covalent solids, the perturbation expansion used in simple metals becomes invalid. We must instead take the pseudopotential as the large quantity. Assuming that the [111] Fourier component of the pseudopotential, which is the largest matrix element, dominates all others leads to a semiquantitative theory of the bonding properties of the covalent systems. This theory allows us to identify the even and odd parts of that pseudopotential with the covalent and polar energies of the LCAO theory, and provides another rationalization for the d^{-2}-dependence of the covalent energy. As we deform the covalent structure into an ionic structure, we see that the covalent energy ceases to contribute to the bonding properties in this approximation.

The pseudopotential formulation also provides a simple way to estimate the relative band positions at interfaces between two semiconductors. The estimate is in reasonable accord with the results of the LCAO calculations for heterojunctions; moreover, it provides an approach for analysis of junctions between metals and semiconductors.

In Chapter 2, it was mentioned that there is a strong resemblance between bands obtained from nearest-neighbor interactions in the LCAO approximation and the bands obtained from nearly-free-electron theory. In fact, formulae for the interatomic matrix elements based upon that similarity were used to estimate properties of covalent and ionic solids in the chapters that followed. Now that a full description of the nearly-free-electron bands has been given (Chapter 16), the calculation of the coefficients in those formulae can be completed; in Chapter 2 the results were merely quoted.

The resemblance between LCAO bands and nearly-free-electron bands is an assertion that *at the observed interatomic distances in solids both the metallic and the atomic descriptions are tenable.* This would certainly not be true at very large spacing, where the electronic states are truly atomic and the free-electron description has no relevance. It is also not true at sufficiently high density, where the LCAO description has no relevance. The assertion applies only near the observed spacing.

However, near the observed spacing, both descriptions are tenable and it is interesting to seek within the metallic (or pseudopotential) description for the qualitative distinction between metallic and covalent solids noted in the LCAO description in Fig. 2-3. We saw there that as the interatomic matrix elements were increased in magnitude a qualitative change in the electronic structure occurred as the system went from metallic to insulating. We shall see that the same change occurs in the context of pseudopotential theory; when the pseudopotential becomes sufficiently large, the nearly-free-electron metal becomes insulating. Beyond that point, we should think of the pseudopotential as the dominant quantity, not the kinetic energy as in the simple metals. We shall see that this is directly analogous to the finding in the LCAO context that though the splitting $\varepsilon_p - \varepsilon_s$ is the dominant quantity when the interatomic spacing is large enough that the system is metallic, the interatomic matrix element producing the bonding–antibonding splitting becomes the dominant quantity in covalent solids.

The remarkable conclusion of this argument is that though pseudopotentials can be used to describe semiconductors as well as metals, the pseudopotential perturbation theory which is the essence of the theory of metals is completely inappropriate in semiconductors. Pseudopotential perturbation theory is an expansion in which the ratio W/E_F of the pseudopotential to the kinetic energy is treated as small, whereas for covalent solids just the reverse quantity, E_F/W, should be treated as small. The distinction becomes *un*important if we diagonalize the Hamiltonian matrix to obtain the bands since, for that, we do not need to know which terms are large. Thus the distinction was not essential to the first use of pseudopotentials in solids by Phillips and Kleinman (1959) nor in the more recent application of the Empirical Pseudopotential Method used by M. L. Cohen and co-workers. Only in approximate theories, which are the principal subject of this text, must one put terms in the proper order.

18-A   The Prediction of Interatomic Matrix Elements

Let us now complete the derivation of formulae for the interatomic matrix elements, which was described in Section 2-D, by equating band energies obtained from LCAO theory and those obtained from nearly-free-electron bands. This analysis follows a study by Froyen and Harrison (1979). The band energies obtained from nearest-neighbor LCAO theory at symmetry points were given in

Eqs. (6-1) through (6-6), in terms of interatomic matrix elements defined in Eq. (3-26). We shall utilize the bands at points of highest symmetry. These are the s-like levels at Γ with energy $\varepsilon_s \pm 4V_{ss\sigma}$, the p-like levels at Γ with energy $\varepsilon_p \pm (4V_{pp\sigma} + 8V_{pp\pi})/3$, and the two valence-band levels at X. We do not use the conduction-band levels at X, which are in poor agreement with the true bands. (This was shown in Fig. 3-8.) When we set the first six levels mentioned equal to their nearly-free-electron counterparts, the six equations may be solved to obtain the four interatomic matrix elements and the term values ε_s and ε_p.

We saw in detail in Section 16-D, and in Fig. 16-6 in particular, how nearly-free-electron bands are constructed. The diamond structure has the translational symmetry of the face-centered cubic structure, so the wave number lattice, the Brillouin Zone, and therefore the nearly-free-electron bands for the diamond structure are identical to the face-centered cubic nearly-free-electron bands and are those shown in Fig. 3-8,c. The bands that are of concern now are redrawn in Fig. 18-1,b. The lowest energy at X, relative to the lowest energy at Γ, is $(\hbar^2/2m) \times (2\pi/a)^2 = (3\pi^2/8)(\hbar^2/md^2)$, as shown; clearly it is to be identified with the lower level X_1 from Fig. 18-1,a. The other valence-band level at X and the four values at Γ are also easily identified. (The levels Γ_2' and Γ_{25}' have the same energy in the nearly-free-electron picture.) The six equations are easily solved to give the four coefficients $\eta_{ss\sigma} = -9\pi^2/64$, $\eta_{sp\sigma} = 3\sqrt{15}\,\pi^2/64$, $\eta_{pp\pi} = 21\pi^2/64$, $\eta_{pp\sigma} = -3\pi^2/32$, and an sp-splitting of $\varepsilon_p - \varepsilon_s = (3\pi^2/4)(\hbar^2/md^2)$.

Use of these parameters, including the sp-splitting, would give a vanishing energy gap at Γ (the bands of gray tin are close to this situation—see Fig. 6-1) and identical forms for the bands of C, Si, Ge, and Sn. However, the sp-splitting does not vary with bond length nearly as strongly as this equation indicates, and use of the atomic term values from the Solid State Table does give a gap and appropriate trends among the four band structures, as indicated in Fig. 18-2. This then is essentially the scheme used throughout our discussion of covalent and ionic solids, except that we used adjusted rather than predicted values of $\eta_{ll'm}$ from Table 2-1.

There is a peculiarity in using atomic term values from the Solid State Table and using a value of $\eta_{sp\sigma}$ from the solution of the six equations, because in the solutions, $\varepsilon_p - \varepsilon_s$ and $\eta_{sp\sigma}$ depend upon each other. Froyen and Harrison (1979) chose to avoid that difficulty by evaluating $\eta_{sp\sigma}$ from the requirement that the effective mass at Γ_1 be equal to the free-electron mass. This gave a value of $\eta_{sp\sigma} = 9\pi^2/32\sqrt{1 - 16/3\pi^2} = 1.88$ rather than the value 1.79 obtained in the direct solution. Use of $\eta_{sp\sigma} = 1.88$, which makes no assumption about the splitting $\varepsilon_p - \varepsilon_s$, therefore seems a better procedure, and it is the evaluation used in Froyen's calculation of the bands given in Fig. 18-2. The values obtained in this way were listed in Table 2-1 for the simple cubic and the tetrahedral structures.

The agreement between the theoretical η parameters found recently and the empirically adjusted values published a few years ago was shown in Table 2-1. It was truly remarkable that such a simple calculation could reproduce not only the form of the matrix elements but also the detailed numerical coefficients that had been obtained empirically. The differences are large enough, however, that the adjusted values give significantly better results for the properties of semiconductors.

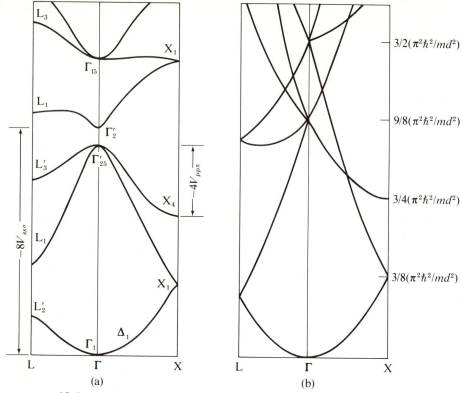

FIGURE 18-1

Energy bands for germanium. Part (a) shows the LCAO bands fitted by Chadi and Cohen (1975), based on nearest-neighbor matrix elements. Part (b) shows the nearly-free-electron bands. Two energy differences are shown in part (a) which can be matched immediately with part (b) to obtain predictions for the corresponding interatomic matrix elements. [After Froyen and Harrison, 1979.]

Froyen and Harrison (1979) also used this procedure to obtain $\eta_{ll'm}$ for a nearest-neighbor fit to the bands for a face-centered cubic structure. The four values (in comparison to the tetrahedral values) were -0.62 (-1.39), 2.33 (1.88), 2.47 (3.24), and 0 (-0.93). These are major differences, but the nearest-neighbor LCAO theory of close-packed structures is of little use in any case. The tetrahedral parameters seem to be adequate for covalent and ionic systems but not to be relevant to the simple metals.

18-B   The Jones Zone Gap

We turn now to the effects of a finite pseudopotential. Let us think specifically of silicon and begin with a free-electron gas of four electrons per ion, using the extended-zone representation, with energy equal to $\hbar^2 k^2/2m$ for all **k**, rather than

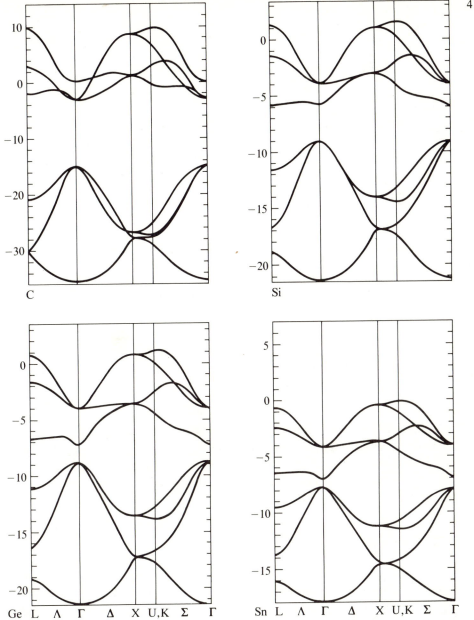

FIGURE 18-2

Nearest-neighbor LCAO bands for the homopolar semiconductors, found by using
interatomic parameters predicted as in Fig. 18-1 (and listed in Table 2-1) and term values
from the Solid State Table. Energies are in electron volts. Notice that the vertical scale
is reduced for the carbon bands. [After Froyen and Harrison, 1979.]

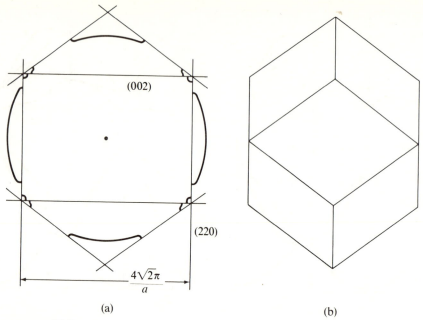

(a)

(b)

FIGURE 18-3

Part (a) shows a free-electron Fermi sphere for silicon cut by various Bragg reflection planes, reducing the area of free Fermi surface. Part (b) shows the Jones Zone, made up of (220) Bragg planes, into which all of the silicon Fermi surface has disappeared. The view is along a $[1\bar{1}0]$ direction in both parts.

the reduced-zone scheme of Fig. 18-1,b. Imagine slowly introducing the pseudo-potential. At first, diffraction occurs at each Bragg plane, and as the pseudopotential becomes stronger, the Fermi sphere becomes distorted and some of the Fermi surface disappears into the Bragg planes, as we illustrated for aluminum in Fig. 16-9. This is redrawn for silicon in Fig. 18-3.

The Jones Zone

The Bragg planes are specified in Fig. 18-3 in terms of the lattice wave number giving rise to them. For example, there are lattice wave numbers $[220]2\pi/a$, with a the cube edge of Fig. 3-1. The plane bisecting this vector is called the (220) plane. The (111) planes, which were shown in Fig. 16-9, are omitted here for clarity. The diagonal lines in part (a) are actually edges made by the intersections of planes of the (202) or (022) type with various combinations of signs of components.

We know that when the pseudopotential is at full strength, all of the Fermi surface must disappear, since none is present in the semiconductor. We can, in fact, see from the figure that what must happen is that it disappears into the slanted and vertical planes of Fig. 18-3,a; the horizontal planes as well as the omitted (111) planes are nonessential. Indeed, the vertical and slanted planes are among the twelve (220) Bragg planes that make up the **Jones Zone** (Mott and

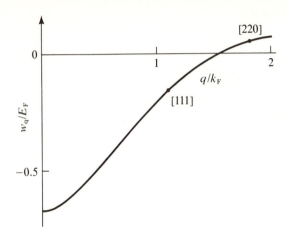

FIGURE 18-4

The ratio (w_q/E_F) of pseudo-potential form factor to free-electron Fermi energy for silicon, showing that the direct Jones-Zone diffraction [220] should be weak.

Jones, 1936; 1958, p. 159), which has long been thought to dominate the electronic structure of the covalent solids. A sketch of that zone is shown in Fig. 18-3,b. The volume contained within it is just sufficient to contain the four electrons per atom, and it would be natural to assume that all Bragg planes other than those bounding the Jones Zone are unimportant and could be neglected in a semiquantitative theory.

This view runs into difficulties that have only recently been completely resolved. The principal one is that the pseudopotential form factor happens to be very small for this particular diffraction. In Fig. 18-4 is sketched the pseudopotential form factor for silicon obtained from the Solid State Table; the form factor that gives the [220] diffraction is indicated. Because it lies so close to the crossing, it is small and the diffraction is not expected to be strong. Heine and Jones (1969) noted, however, that a second-order diffraction can take an electron across the Jones Zone; this could be a virtual diffraction by a lattice wave number of $[111]2\pi/a$ followed by a virtual diffraction by $[11\bar{1}]2\pi/a$. (Virtual diffraction is an expression used to describe terms in perturbation theory; it can be helpful but is not essential to the analysis here.) This second-order diffraction would involve the large matrix elements associated with the $[111]2\pi/a$ lattice wave number indicated in Fig. 18-4, and Heine and Jones correctly indicated that these are the dominant matrix elements.

The Larger Hamiltonian Matrix

If we wish to study a state at the face of the Jones Zone, we must consider not only the plane wave with wave number at that face, say $k_{110} = [110]2\pi/a$, and that at the opposite face, $k_{\bar{1}\bar{1}0}$, but also the two states through which they are coupled, k_{001} and $k_{00\bar{1}}$. Notice that each of these states differs from the others by a lattice wave number, so that if the free-electron bands were plotted in the reduced-zone scheme, they would all be at the same point, the point $[001]2\pi/a$, which is at the center of one of the square faces of the Brillouin Zone, for example, the point X in

Fig. 3-6. To calculate the energies of these states, we write a state as a linear combination of the four waves and construct the Hamiltonian matrix, as we did when we calculated the energy bands in Chapter 6. That matrix is rather easily written down. (We shall consider the matrix elements of greatest importance in more detail later.) The Hamiltonian matrix is

$$
H = \begin{vmatrix}
\hbar^2 k_{\bar{1}10}^2/2m & W_{220} & W_{\bar{1}11}^* & W_{\bar{1}11}^* \\
\\
W_{220} & \hbar^2 k_{\bar{1}\bar{1}0}^2/2m & W_{111} & W_{111} \\
\\
W_{111} & W_{\bar{1}11}^* & \hbar^2 k_{001}^2/2m & 0 \\
\\
W_{111} & W_{\bar{1}11}^* & 0 & \hbar^2 k_{00\bar{1}}^2/2m
\end{vmatrix}
\tag{18-1}
$$

It can be diagonalized analytically (Harrison, 1976a), and shows that the two pairs of energies, appearing on the diagonal in Eq. (18-1), are split into four levels. This is illustrated between the two vertical lines to the left of Fig. 18-5. These may be matched with the four lowest levels at X in the band structure of silicon. The energies of these levels, measured from the band minimum, are direct predictions rather than fits, so the agreement is impressive. The biggest discrepancy—the splitting of the lower levels—is directly understandable since the states $k_{1\bar{1}0}$ and $k_{\bar{1}10}$ are coupled also through the same set k_{001} and $k_{00\bar{1}}$, which lowers the second level and restores the degeneracy.

The splitting of the upper two levels *has* given the band gap between the valence and conduction bands, as hoped. This was explicitly the splitting at X, since that point, $[001]2\pi/a$, is the point in the zone differing by a lattice wave number from the four plane waves included in the analysis. The calculation may be readily extended to neighboring points in the Brillouin Zone. In particular, we may move along the line from X to Γ (see Fig. 18-5) by adding a small multiple of $-[001]2\pi/a$ to each of the four wave numbers. For the two wave numbers at the Jones Zone, $\pm[110]2\pi/a$, this simply moves them along the Jones Zone face (the planes bisecting $\pm[220]2\pi/a$) and both diagonal energies increase quadratically with the added wave number. (This is seen explicitly in the free-electron bands of Fig. 3-8,c.) It is directly the reason why the lowest conduction band and highest valence bands rise along parallel quadratic curves when the wave number varies from X to Γ; indeed, the variation of these parallel bands along Γ to L, as well as along X to Γ, traces out essentially constant band gaps between conduction and valence bands as we move over the surface of the Jones Zone. We mentioned the phenomenon of parallel bands in discussing the observed bands and the origin of the strong optical absorption peaks E_2, in Chapter 4, though it was not clear in the LCAO context why those bands should be parallel. A closely related parallel-band absorption gives rise to the principal interband peaks in the optical absorption of polyvalent metals (Harrison, 1966b) but is not nearly so important there. In the covalent solids, the Jones Zone gap should be identified with the principal optical absorption peak previously identified with LCAO interatomic matrix elements. Thus it allows a direct relation between the parameters associated with the LCAO and with the pseudopotential theories. It is best, however, to simplify the pseudopotential analysis still further before making that identification.

The Simplest Approximation

The dominant effect here, the separation of the valence bands from the conduction bands, has arisen from the pseudopotential, and from the W_{111} matrix element in particular, suggesting that a simple theory might be obtainable by neglecting all matrix elements except W_{111}. Thus we approximate the Hamiltonian of Eq. (18-1) by setting W_{220} equal to zero and neglecting the difference between $\hbar^2 k_{110}^2/2m$ and $\hbar^2 k_{001}^2/2m$. That matrix is even easier to diagonalize, and leads to two eigenvalues unshifted at the energy corresponding to the average of the diagonal elements (all of which are the same when the kinetic energy differences are ignored) and single eigenvalues at $\pm 2|W_{111}|$ measured from the same average. Thus the approximate value of the splitting between the conduction band and the valence band becomes $2W_{111}$, where we henceforth use W_{111} to indicate the magnitude of the matrix element, ignoring the phase. In a sample calculation for silicon this turns out to be about 4.1 eV rather than the value 3.2 eV from the diagonalization of Eq. (18-1) (Harrison, 1976a). The error is not large and we can correct it if we wish to. The important point is that we have neglected the kinetic energy difference, and correcting it corresponds to making an expansion in $(\hbar^2 k_{100}^2/2m)/W_{111}$, whereas just the reverse expansion, W/E_F, was appropriate in the simple metals. We have changed the expansion parameter, just as required in the LCAO theory, and as we indicated would be necessary at the beginning of this chapter.

The finding of a band gap at the (220) face equal to $2W_{111}$ is remarkable. A similar solution for the states arising from two coupled states at the (111) Bragg planes immediately gives values raised and lowered by W_{111} and also gives a gap of $2W_{111}$. These are the points L of the Brillouin Zone, and we see that the lower states L_2' and L_1 are indeed split by an amount close to that of the upper X_1-X_4 splitting. The X_1-X_4 splitting has arisen from quite different pseudopotential matrix elements, the W_{111}, than one would have at first guessed. Furthermore, the splitting is of a nonperturbative form, $2W_{111}$, rather the perturbative $2W_{111}^2/(\hbar^2 k_{100}^2/2m)$.

18-C   Covalent and Polar Contributions

Let us now look at the matrix element W_{111} in more detail. It is the only matrix element we are retaining. It is desirable at the same time we examine it to extend the theory to polar semiconductors. We return to Eq. (16-3) for the unscreened matrix element of the pseudopotential. We may in fact write it for the screened matrix element, since the Fermi-Thomas screening parameter depends only upon k_F, and therefore on bondlength, and is independent of polarity. For gallium arsenide, for example, we let $\mathbf{r}_i$ run over all gallium atoms (and denote the gallium pseudopotential by w^c), and include the arsenic atom displaced from each gallium atom by τ as a separate term (denoting the arsenic pseudopotential by w^a). Eq. (16-3) becomes

$$\langle \mathbf{k} + \mathbf{q} | W | \mathbf{k} \rangle$$

$$= \frac{1}{\Omega} \sum_i e^{-i\mathbf{q}\cdot\mathbf{r}_i} \left[\int e^{-i\mathbf{q}\cdot(\mathbf{r}-\mathbf{r}_i)} w^c(\mathbf{r} - \mathbf{r}_i)\, d^3r + e^{-i\mathbf{q}\cdot\tau} \int e^{-i\mathbf{q}\cdot(\mathbf{r}-\mathbf{r}_i-\tau)} w^a_q(\mathbf{r} - \mathbf{r}_i - \tau)\, d^3r \right]$$

$$= \tfrac{1}{2}[w^c_q + e^{-i\mathbf{q}\cdot\tau} w^a_q]. \qquad (18\text{-}2)$$

In obtaining the final form, we used the fact that $\mathbf{q}$ is a lattice wave number for the gallium lattice, to take the sum of $e^{-i\mathbf{q}\cdot\mathbf{r}_i}$ over the $N_a/2$ gallium atoms as $N_a/2$, and changed dummy variables in the two integrals to write them as pseudopotential form factors. The position, τ, is $[111]a/4$ and the dot product of τ and a wavenumber of the $[111]2\pi/a$ type gives $\pm\pi/2$, depending upon the signs of the components (for example, $[1\bar{1}\bar{1}]$ or $[111]$). Thus the second term in Eq. (18-2) is 90° out of phase with the first, and the magnitude is

$$W_{111} = |\langle \mathbf{k} + \mathbf{q} | W | \mathbf{k} \rangle| = \tfrac{1}{2}(w^{c2}_q + w^{a2}_q)^{1/2}, \qquad (18\text{-}3)$$

with $q = 3^{1/2}2\pi/a$. For homopolar materials in particular, $W_{111} = w_q/2^{1/2}$, giving the values used above for silicon.

What has been accomplished is a very simple relation between the pseudopotential and the important gap in the band structure. What is more, we have provided such a simple representation of the band structure that we may use it to calculate other properties of the semiconductor, just as we did with the LCAO theory once we had made the Bond Orbital Approximation.

The Relation to LCAO Theory

The first such application is the completion of the identification of the parameters of the two theories that we mentioned earlier. The separation of the parallel bands, which has been given here, for pseudopotential theory, by $2W_{111}$, was written for LCAO theory in Chapter 4 as $2(V_2^2 + V_3^2)^{1/2}$, with V_2 and V_3 written in terms of matrix elements of the Hamiltonian between atomic p states.

(There the relation was made in terms of the splitting at Γ rather than X, since the corresponding formulae are simpler.) The first comparison we make is between the LCAO values and the empty-core pseudopotential. We shall find only qualitative correspondence between the values because of errors in the empty-core model, which become large here. We shall then go on to consider other properties, using pseudopotential matrix elements obtained without resort to the empty-core model.

For the purposes of relating the parameters, we may rewrite Eq. (18-3) in the form

$$W_{111} = [1/8(w_q^c + w_q^a)^2 + 1/8(w_q^c - w_q^a)^2]^{1/2}. \tag{18-4}$$

Then we may make the identification

$$V_2 \leftrightarrow \left(\frac{w_q^c + w_q^a}{2}\right)\Big/\sqrt{2};$$

$$V_3 \leftrightarrow \left(\frac{w_q^c - w_q^a}{2}\right)\Big/\sqrt{2}. \tag{18-5}$$

Making first a comparison of the covalent energy, notice that in homopolar semiconductors, W_{111} becomes simply $w_{111}/2^{1/2}$. The various geometrical factors in the empty-core pseudopotential may be directly evaluated. Then, the pseudopotential matrix element becomes

$$W_{111} = \frac{0.2444e^2k_F \cos(1.108k_F r_c)}{1 + 1.0371e^2m/(\hbar^2 k_F)}. \tag{18-6}$$

The values for the homopolar semiconductors are listed, along with V_2, in Table 18-1.

TABLE 18-1
Pseudopotential matrix elements for the homopolar semiconductors compared with V_2, all in eV. Values of k_F for the diamond structure appear in the Solid State Table.

	V_2 (Eq. 4-16)	W_{111} (Eq. 18-6)	W_{111} (EPM)
C	6.94	2.41	7.8
Si	2.98	1.35	2.0
Ge	2.76	1.43	2.2
Sn	2.10	1.28	2.4

SOURCE of values for the Empirical Pseudopotential Method (EPM): Cohen and Bergstresser (1966).

The trends are the same, but the magnitudes differ by a factor of about two. This discrepancy arises from the inaccuracy of the empty-core model in fitting the pseudopotential in this range, as can be seen by considering the third column in Table 18-1, where Empirical Pseudopotential Method matrix elements are listed directly. These are in fact taken from the same calculated pseudopotential which was fitted to the empty-core model to obtain the values in the second column. A look at Fig. 16-1, where an accurate pseudopotential is plotted along with the empty-core fit indicates that the region near $q/k_F = 1.108$ (corresponding to the [111] matrix element) is where the fractional discrepancy is the largest; thus the large discrepancy here does not argue against the use of the empty-core potential in metals, and the comparison between the first and third columns in Table 18-1 confirms the relation between the LCAO and pseudopotential theories, the pseudopotential model based on a single matrix element, in particular.

We may also compare the polar energies. This is particularly simple in the context of the empty-core model. Then, in gallium arsenide, for example, we may expect the unscreened pseudopotential (both Z and r_c) for both gallium and arsenic to be the same as in each pure material. The screening depends only upon the Fermi wave number, and since the bond length does not change appreciably in an isoelectronic series, the entire denominator $\Omega_0(q^2 + \kappa^2)$ should be the same for all form factors that enter in a series such as Ge, GaAs, ZnSe, and CuBr. We may therefore very easily estimate the polarity of any compound in a series by taking the form factor, w_q^c or w_q^a in Eq. (18-5), as proportional to $Z \cos(1.108\, k_F r_c)$, with k_F taking a value appropriate to the homopolar semiconductor. The results are given in Table 18-2. Again, the trends are the same but discrepancies are rather large. In this case, the error may not be principally in the empty-core model itself. The value of the pseudopotential difference indicated in Eq. (18-5), obtained directly from the empirical pseudopotential calculations (Cohen and Bergstresser, 1966) give values of 2.55 eV and 2.73 eV for GaAs and ZnSe, respectively, to be compared with values of V_3 of 1.51 eV and 3.08 eV, respectively.

TABLE 18-2

Polarities predicted from the empty-core pseudopotential and the relations given at Eq. 18-5. Values from LCAO theory (Table 4-1) are given in parentheses.

AlP	
0.41 (0.51)	
GaAs	ZnSe
0.36 (0.48)	0.52 (0.75)
InSb	CdTe
0.31 (0.52)	0.52 (0.78)

The identification of the covalent energy with a pseudopotential has relevance to the d^{-2}-dependence that has recurred throughout our studies. The pseudo-potential form factors, when divided by the Fermi energy and plotted against q/k_F, as in Fig. 18-3, are almost a universal curve, approaching $-2/3$ at small q and crossing the axis near $q/k_F = 1.6$ or 1.7 in most systems. To the extent that this curve is universal, the value at $q/k_F = 1.108$, and therefore, W_{111} would be a universal constant times $\hbar^2/(md^2)$, as we have found to be true. The case for such a dependence under pressure, or the case for the polar energy, cannot be made as strongly. Indeed, the experimental support for d^{-2}-dependence is much weaker in these cases also. Notice that this rationalization for the d^2-dependence is completely independent of the derivation given in Section 18-A.

18-D   Susceptibility

Let us now turn to a few other informative applications of the pseudopotential model to properties (discussed extensively in Harrison, 1976a). For that purpose it will be most convenient to take the pseudopotential parameters from the LCAO values for V_2 and V_3, using Eq. (18-5). We see from Table 18-1 that for V_2 this is roughly equivalent to using Empirical Pseudopotential Model parameters. Let us look first at the dielectric susceptibility, which was so important in the development of the LCAO theory. For this, the most convenient form for the susceptibility is Eq. (4-5), which we rewrite in simple form,

$$\chi = (2Ne^2\hbar^2/m^2) \sum_a \langle b|p_x|a\rangle\langle a|p_x|b\rangle/(\varepsilon_a - \varepsilon_b)^3. \tag{18-7}$$

This is to be averaged over all occupied states $|b\rangle$ (notice that the leading factor includes the density of electrons from a sum over occupied states) and then summed over all unoccupied states $|a\rangle$. We must consider the states more carefully. (The details of this analysis appear in Harrison, 1976a). The upper valence-band state at the Brillouin Zone face, X_4 in Fig. 18-5, is a simple sine function, $(2/\Omega)^{1/2} \sin \mathbf{k}_{111} \cdot \mathbf{r}$ for homopolar semiconductors. The lower conduction-band states, the higher of the two X_1 states in Fig. 18-5, is a simple cosine function, $(2/\Omega)^{1/2} \cos \mathbf{k}_{111} \cdot \mathbf{r}$, except for a correction of order $(\hbar^2 k_{100}^2/2m)/4W_{111}$, which we may neglect. For polar semiconductors, both sine and cosine are shifted by the same phase. Thus if we take the upper valence-band state at the zone face to be the typical bond state and the corresponding lowest conduction-band state as the dominant antibonding state—as in the optical absorption, only empty states at the *same* point in the Brillouin Zone, here X, have nonvanishing matrix elements and contribute to Eq. (18-7)—the matrix element becomes

$$\langle b|p_x|a\rangle = \gamma_p \hbar \mathbf{k}_{110x}, \tag{18-8}$$

the same for homopolar or heteropolar semiconductors. The term $\mathbf{k}_{110x}$ is the x-component of $\mathbf{k}_{110}$. We have introduced a scale factor of γ_p in the matrix

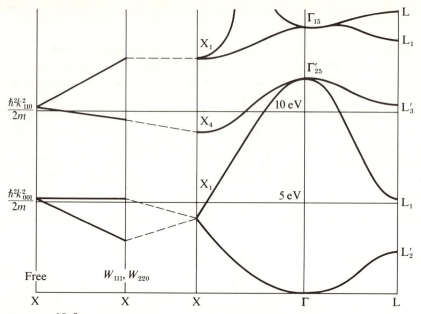

FIGURE 18-5

At the right are shown the energy bands of silicon (after Herman, Kortum, and Kuglin, 1966). At the far left are the four levels at X from the free-electron approximation. The left panel shows their splitting due to the pseudopotential, based on essentially the same values as deduced from Table 16-1. The dashed lines identify the resulting levels with the corresponding ones in the true band structure.

element, as we introduced γ in the LCAO theory. In the simplest model, the scale factor would be unity, but a number of approximations have been made which we hope to absorb in this parameter. The energy denominator is the gap energy, $2(V_2^2 + V_3^2)^{1/2}$. The matrix elements and band gap may be directly substituted into Eq. (18-7) to obtain

$$\chi = \frac{Ne^2\gamma_p^2\hbar^4 k_{110}^2}{12m^2(V_2^2 + V_3^2)^{3/2}} \tag{18-9}$$

This form does not resemble very closely the form of Eq. (4-26), which was obtained from LCAO theory. It can be brought into the same form, however, by multiplying by the square of V_2 and then dividing by the square of the form $2.16\hbar^2/md^2$, from (Eq. 4-16), for V_2^2. After a little algebra, this leads to

$$\chi = \frac{Ne^2d^2(3.2\gamma_p^2)V_2^2}{12(V_2^2 + V_3^2)^{3/2}}. \tag{18-10}$$

Thus the theories become equivalent if $3.2\gamma_p^2 = \gamma^2$.

A remarkable feature of the result is that these two models for susceptibility—one based on the simple local orbital theory and the other based on the simple plane-wave theory given here—*can be simultaneously valid only if the band gaps scale with* d^{-2}. This involves the relationship between two crude theories, and is closely related to the identification of the LCAO and nearly-free-electron theories which we made use of in Section 18-A.

We do not attach great significance to the numerical relationship that γ_p is approximately half γ. A number of corrections are absorbed in each coefficient, but it is gratifying that both γ and γ_p are of order unity.

18-E   Bonding Properties

We turn next to the total energy, and in particular to the structure-dependent part, which we discussed in terms of LCAO theory in Chapter 9. (Details, again, appear in Harrison, 1976a.) We saw in Fig. 18-5 that the pseudopotential raised the conduction-band edge by $2W_{111}$ and, correspondingly, lowered one of the deep valence bands. The counting of levels is not immediately easy here, but becomes clearest if we look at the complete bands to the right in that figure, and focus in particular on their values at X. We see that the two lowest bands at X have been lowered, but the upper two are unshifted. (The point X_4 is doubly degenerate; it represents two bands.) Again taking the states at the point X as typical, we see that the effect of the pseudopotential has been to lower two of the four bands by $2W_{111}$, or to reduce the energy per electron by W_{111}. This corresponds to the LCAO result of a lowering per electron of V_2, but here it comes from the lowering of two bands by twice as much. It is also the hybrid covalent energy of Eq. (3-6), rather than V_2, which is appropriate to the total energy in the LCAO theory.

Angular Rigidity

In addition to making an identification between pseudopotential and LCAO theories, let us make one application, namely, the calculation of the stability of the diamond structure. To be specific, imagine expanding the crystal along the z-axis by a factor $(1 + \varepsilon)$, and contracting it in the two lateral dimensions by $(1 + \varepsilon)^{-1/2}$. Then the wave number lattice is contracted in the z-direction by a factor $(1 + \varepsilon)^{-1}$ and expanded in the lateral directions by $(1 + \varepsilon)^{1/2}$. This increases the length of each [111] wave number by a factor of $(1 + \varepsilon^2/2)$ to order ε^2, and we may estimate the resulting change in W_{111} as $(k \, \partial W_{111}/\partial k)\varepsilon^2/2 = (k \, \partial w_q/\partial k)\varepsilon^2/2^{3/2}$ for a homopolar solid. A new parameter has entered, $\partial w_q/\partial k$, which may be estimated from the empty-core pseudopotential or, still better, from the accurate pseudopotentials of Animalu and Heine (1965). Using these accurate pseudopotentials, we find that $k \, \partial w_q/\partial k = 0.56E_F$ at the appropriate $q = 1.11k_F$ for silicon, germanium, and tin, to within one percent. The fact that the single coefficient 0.56 suffices again reflects a universality of the normalized form factors when plotted as in Fig. 18-4, and from this universality, the variation of V_2 as d^{-2} follows. This value of $k \, \partial w_a/\partial k$ yields an energy per electron of $0.20E_F \varepsilon^2$. We may relate this energy per electron to the elastic constants by multiplying it by the electron density and equating the result to the energy density in terms of the elastic constants for the same distortion, $(3/4)(c_{11} - c_{12})\varepsilon^2$. This immediately gives estimates of $c_{11} - c_{12}$ for diamond, silicon, germanium, and tin of 8.7, 1.1, 0.87, and 0.44 respectively (in units of 10^{12} erg/cm^3). These are in remarkable agreement with the experimental values for diamond, silicon, and germanium of 9.5, 1.0, 0.80. (We have no value for Sn.) As the LCAO theory does, this predicts a variation from material to material with d^{-5}, which is well satisfied by experiment. It is interesting that in the pseudopotential model only a single covalent energy enters the optical and the elastic properties, but since the scaling with d^{-2} is built in, this only requires the coincidence of one parameter and we should not attach too much significance to it.

In order to treat polar semiconductors, we must make some assumption as to how the odd part of the pseudopotential, V_3, varies with distortion. The assumption that it is independent of shear, which was used in the LCAO theory, gives a polar value different from the homopolar value by a factor $V_2/(V_2^2 + V_3^2)^{1/2} = \alpha_c$, or values for the isoelectronic series of Ge, GaAs, and ZnSe of 0.87, 0.81, and 0.74, respectively, compared to the experimental values of 0.80, 0.65, and 0.32. The trend is right, though it is not quantitatively very accurate. To estimate α_c, we used the empty-core polarities from Table 18-2. The agreement is better if LCAO values are used but not significantly so.

A Zero-Order View of Rigidity

One might question our use of $-W_{111}$ per electron as the total energy, particularly in light of our observation that the inclusion of all components of the pseudopotential in second-order perturbation theory leads to an unstable lattice (Harrison, 1976a). A consideration of the other terms in the energy brought an interesting point to light that fits well with the central theme of this chapter, which has been the necessity of treating the pseudopotential as the dominant quantity. In simple-metal theory, the zero-order energy was the kinetic energy of the free electrons, and it did not change when the lattice was distorted at constant volume. In applying the theory to the covalent solids, one should carry out the summation over the Jones Zone rather than over a sphere, yet the spirit of the approach using the Jones Zone is consistent with the replacement of the zone by a sphere. The difference comes only when we shear the lattice. Then the distortion deforms the zone and the corresponding sphere becomes an ellipsoid. The corresponding change in the zero-order energy is absent in the metal, where the Bragg planes move through the Fermi sphere like a sieve, but is very important in the covalent solid. The average kinetic energy $\langle \hbar^2 k^2/2m \rangle$ in the ellipsoid can be calculated exactly, and it is given by $(\hbar^2 k_F^2/10m)[2 + 2\varepsilon + (1 + \varepsilon)^{-2}]$; the change in energy is obtained immediately as $3\hbar^2 k_F^2 \varepsilon^2/10m$. Notice that the value is independent of the magnitude of the pseudopotential matrix element as long as it is sufficiently large to contain the electrons within the zone. This term contributes 3.2×10^{12} erg/cm^3 to $c_{11} - c_{12}$ in silicon, a value three times the experimental value. The pseudopotential terms, which led to the instability of the covalent structure when it was treated as a metal, must cancel this contribution. The presence of this very large term, of zero order in the pseudopotential and independent of its magnitude, again reflects the inappropriateness of the pseudopotential as an expansion parameter in the covalent solids.

The Charge Distribution

One other property of the covalent solids should also be mentioned in the context of pseudopotential theory: the electronic charge distribution. (Again, the details appear in Harrison, 1976a.) We may estimate the charge distribution, again by discarding all terms in the pseudopotential except the [111] components that we have considered dominant in determining the bonding properties. The result is a surprise in that it does not lead to a charge density with its peak at the bond center as observed experimentally; it leads to a density with a saddle point at the bond center. In contrast, treating the electrons as a simple metal, ignoring the nonperturbative effects of W_{111} but keeping all components in perturbation theory, *did* lead (Harrison, 1966a, p. 224) to a good description of the charge density; the result is shown in Fig. 18-6. The meaning is then clear: the maximum in charge density at the bond center that one intuitively associates with bond

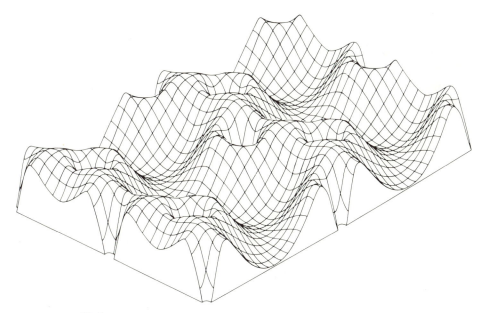

FIGURE 18-6

A projection showing the electron density on a (110) plane in silicon, calculated as if silicon were a simple metal. [From Harrison, 1966a.]

charges is in some sense independent of the effects that produce stability of the structure. This requires careful distinctions, because a second aspect of the charge distribution is also frequently associated with bond charges: there is X-ray diffraction at the lattice wave number [222], though this diffraction would be forbidden if the charge density could be written as a sum of spherically symmetric densities centered at the atoms: this follows from the vanishing of $S(\mathbf{q})$ for this wave number (see, for example, Kittel, 1967, p. 73). This diffraction cannot be explained by the charge distribution given in Fig. 18-6, since that distribution can be written rigorously as a sum of spherically symmetric charge densities. There are corrections to this distribution that permit the diffraction, because of the nonperturbative effects of W_{111}, which we have discussed here, and because states fill a Jones Zone rather than a sphere in wave number space; we may think of both of these as being essential to structural stability. Contributions permitting diffraction also come from higher-order terms in pseudopotential perturbation theory; these seem not to be essential to structural stability. It has not been established what mix of these contributions is important to each property. Full pseudopotential theory would certainly seem to be capable of describing semiconductor properties and charge densities (for example, see Chelikowsky and Cohen, 1976a); what are uncertain are the roles of different contributions. A recent study of the transverse charge e_T^* in terms of pseudopotentials (Vogl, 1977) suggests that the more elementary theory can be carried considerably further.

18-F   Ionic Bonding

Finally we wish to consider ionic solids briefly in terms of pseudopotential theory, again lending support to the procedures which were used in the LCAO theory. Certainly the central difference in the two systems is the difference in structure, and structure factors are the relevant aspects; in those materials that occur in either ionic or covalent structures, we would expect to use the same form factors for both kinds of structures. This can be most conveniently studied by using a procedure described by Schiferl (1974). Noticing that both the zincblende and the rocksalt structures consist of interlocking face-centered cubic lattices of each of the two atomic species making up the compound, we simply displace one sublattice with respect to the other to convert the zincblende lattice to a rocksalt lattice. This is illustrated in Fig. 18-7. In this process the lattice wave numbers do not change, and we neglect any changes in the form factors, but the phase factors do change. We may use the form for the matrix element given in Eq. (18-2), namely, $W_{111} = (w_q^c + e^{-i\mathbf{q}\cdot\boldsymbol{\tau}}w_q^a)$, with $\mathbf{q} = [111]2\pi/a$. In the zincblende lattice, $\mathbf{q}\cdot\boldsymbol{\tau}$ is $\pi/2$, so that, as we have indicated, the two species of atoms in the semiconductor scatter 90° out of phase, and writing the even and odd parts of the pseudopotential as V_2 and V_3 as in Eq. (18-5), we have $W_{111} = (V_2^2 + V_3^2)^{1/2}$.

We found earlier in this chapter that this is the dominant matrix element that opens up the gap at the Jones Zone, eliminates the Fermi surface, and dominates the bonding properties. As we slowly displace the second sublattice, as in

FIGURE 18-7

Conversion of a zincblende lattice to a rocksalt lattice by displacement of one sublattice. The view is along a [100] direction. We may think of the atoms ⬤ in the plane of the figure, the atoms ◯ at $a/4$ above that plane and atoms ◌ at $a/4$ below. The displacements indicated are in the plane of the figure.

Fig. 18-7, the gap at the Jones Zone remains open, and as $\mathbf{q} \cdot \boldsymbol{\tau}$ approaches π for the rocksalt structure, W_{111} approaches simply V_3. The even part of the pseudo-potential, corresponding to the covalent energy V_2 or to the interatomic matrix elements from neighboring hybrids, has not changed in the process, but it no longer contributes to the band gap that dominates the properties. In terms of orbitals, this corresponds to the fact that once each atom has six nearest neighbors, there is no longer a unitary transformation that can produce hybrids to bond with all of the neighbors. In the rocksalt structure, only the polar energy can contribute to the bonding in the simplest form of the theory.

The identification of the band gap in ionic crystals with pseudopotentials suggests one other property that may be attributable to ionic crystals. The general insensitivity of w_q/E_F to material or structure gives a rationalization to the observation that band gaps in ionic solids and even inert-gas solids vary as d^{-2}. However, the point cannot be made as strongly for ionic solids as it can for covalent solids.

18-G Interfaces and Heterojunctions

In Section 10-F, the description of the interface between two semiconductors came quite naturally in terms of LCAO theory. That is probably the best way to treat interfaces, but the pseudopotential method also provides a correct way to proceed, and it is of some interest to make the comparison, particularly since that is presumably the way that semiconductor–metal interfaces should be formulated. Pseudopotential treatments of semiconductor interfaces have in fact been made by Chelikowsky and Cohen (1976c), Frensley and Kroemer (1976), and Baraff et al. (1977); the treatment we follow here is that of Harrison (1977b).

The Step in the Average Pseudopotential

We may imagine superimposing pseudopotentials as in Eq. (16-1) but with different types on the two sides of the system. If we are in fact willing to use the screened pseudopotentials from Eq. (16-7) that we have used everywhere else, the Hamiltonian will be completely determined. Notice, however, that this is the first time in the treatment of covalent solids that the entire form factor is used, including the small q values that depend sensitively upon use of metallic screening. The difference in the pseudopotential, averaged over volume on the two sides of the junction between two solids, can be calculated by making a Fourier transformation of the total pseudopotential (Harrison, 1977), but the result in fact must (and does) come out exactly as given in Eq. (17-28), since pseudopotentials have been used that correspond to metallic screening.

The result is not rigorous in this case, and charge redistribution near the interface can shift the values, though not the values for two metals. The point is that a redistribution of charge, giving a dipole layer, may raise all the levels on one side with respect to those on the other, but as long as it does not raise the valence-band edge above the conduction-band edge on the other side, all levels in the system below a certain energy (or range of energies) will be occupied, and all those above will be empty, as required in the ground state. The neglect of such shifts can only be justified after the fact by finding that the results agree with LCAO theory, with experiment, or with self-consistent calculations. We shall return to the matter of justification shortly.

The result, Eq. (17-28), gives the difference between average pseudopotentials, $\langle \mathbf{k} | W | \mathbf{k} \rangle$ (which is an arbitrary constant in most pseudopotential calculations), on the two sides. To obtain the relative band energies on the two sides, it is necessary in principle to carry out band calculations on the two sides which incorporate this difference, though we shall be able to make use of existing band calculations. The most direct result is a value for the valence-band minimum, Γ_1, which has energy $\langle \mathbf{k} | W | \mathbf{k} \rangle$ corrected for the effects of the various W_q. We may make this estimate with perturbation theory, including only the effects of the eight W_{111} components that dominate the covalent solid. The effect of the other components can readily be estimated; it is of the order of 0.1 eV, and its variation from material to material can be neglected. The resulting value for the valence-band minimum (measured from a free-electron Fermi energy) can be written down immediately as

$$E_{\text{v-min}} = -\frac{\hbar^2}{2m}(3\pi^2 N)^{2/3} - \frac{8W_{111}^2}{\hbar^2(3\pi/2d)^2/2m}. \tag{18-11}$$

These values could be subtracted to obtain the difference between valence-band minima on the two sides, but it is the difference between valence-band *maxima* which is of central interest and which we treated in Section 10-F. We therefore estimate a valence-band maximum as $E_{\text{v-min}}$ plus the band width (taken from Chelikowsky and Cohen, 1976). Only differences for different systems are meaningful, so we add -7.60 eV to every E_v so estimated to bring the value for

germanium into register with the LCAO values from Table 10-1. This gives values
of

$$E_v = E_{v\text{-min}} + \text{Band width} - 7.60 \text{ eV}, \qquad (18\text{-}12)$$

which can be compared directly with the values in Table 10-1.

Comparison with LCAO Theory

We make two comparisons with the LCAO estimates in Table 18-3, one for the
homopolar systems (showing the dependence upon metallicity) and one for the
heteropolar semiconductors isoelectronic with germanium (showing dependence
upon polarity). We see that the trend with polarity is rather well described; it
comes largely from the variation of the final term in Eq. (18-11). The degree of
consistency is in fact somewhat remarkable in view of the small differences in
comparison to the large terms that enter the calculation. It is interesting that the
values are also reasonably consistent (see Harrison, 1977) with the pseudopotential estimates for these three materials made by Frensley and Kroemer (1976), who
matched the value of the pseudopotential at interstitial positions rather than using
screened pseudopotentials, as have been used here.

We obtain the same direction of the trend with metallicity as in LCAO theory,
but considerably overestimate it. (In contrast, Frensley and Kroemer found the
opposite trend.) The discrepancy between the trend shown in Table 18-3 and that
of Frensley and Kroemer seems to be due to dipole layers that arise when the

TABLE 18-3
Comparison of pseudopotential and LCAO values in homopolar systems and
heteropolar semiconductors (all values in eV).

Material	$\hbar^2(3\pi^2 N)^{2/3}/2m$	W_{111}	Band width	$-E_v$ (Eq. 18-12)	$-E_v$ (LCAO)
	VARIATIONS WITH METALLICITY				
Si	12.47	−2.47	12.5	10.77	9.50
Ge	11.57	−2.13	12.6	9.12	9.12
Sn	8.79	−1.83	11.3	7.57	8.04
	VARIATIONS WITH POLARITY				
Ge	11.57	−2.13	12.6	9.12	9.12
GaAs	11.48	−2.12	12.55	9.08	9.53
ZnSe	11.48	−2.49	12.25	10.35	10.58

SOURCE of values: W_{111} and band width, from Chelikowsky and Cohen (1976b). The zero of energy was
chosen to bring the Ge values of $-E_v$ (Eq. 18-12) and $-E_v$ (LCAO) into accord.

electron density differs greatly between the two sides; these require a self-consistent pseudopotential calculation, and ours *is* approximately self-consistent. However, the LCAO results are expected to be more reliable than the pseudo-potential results obtained here. For this reason, the approach used here for covalent solids and in Section 17-E for metals could be quite misleading if applied directly to metal–semiconductor junctions.

The Semiconductor–Metal Junction

It is easy to see qualitatively what must happen at a semiconductor–metal junction. A metallic state (whether occupied or not) having an energy in the semiconductor gap must have an exponentially decaying amplitude in the semi-conductor, with a decay length, in analogy with Eq. (9-16), depending on the energy difference between that state and the band edge. Thus, occupied metallic states near the conduction-band edge would produce electron density far into the semiconductor and produce a dipole layer that tends to increase the conduction-band energy and raise the conduction-band edge in comparison to the metallic Fermi energy. Similarly, an empty metallic state near the valence-band maximum will tend to lower the energy of the semiconductor states in comparison to the metallic Fermi energy. Thus a self-consistent solution of the problem will tend to give dipole moments, though a first estimate of the effect indicated it to be small. R. Sokel carried out the calculation and found that the dipole produced does not diverge as the Fermi energy approaches a band edge, as one might guess (see the appendix to Harrison, 1977b). For a symmetric treatment of conduction and valence bands in the semiconductor and a free-electron metal, Sokel found that the dipole contributes a step δE in the energy, given by

$$\frac{\delta E}{(E_c - E_v)} = \frac{e^2}{4\pi\varepsilon_1}\left(\frac{2m^*}{\hbar^2(E_c - E_v)}\right)^{1/2}\left[\left(\frac{E_c - E_F}{E_c - E_v}\right)^{1/2} - \left(\frac{E_F - E_v}{E_c - E_v}\right)^{1/2}\right]. \quad (18\text{-}12)$$

Here $m^* = m_c = m_v$, and E_F is the final Fermi energy, including the step δE; that is, the expression tells what contribution to the final step has arisen from the dipole; ε_1 is the dielectric constant, Eq. (18-12) gives a maximum shift of the order of 0.02 eV in GaAs.

This suggests that the dipole effects may not be very important in fixing the metallic Fermi energy with respect to the conduction-band edges and would argue against the idea that the "metal-induced surface states" postulated by Louie and Cohen (1975) are important in pinning the Fermi level. It is probable that the observed pinning arises from atomic rearrangements at the interface, such as those discussed in Section 10-C. We should note finally that all of the discussion in this section has related to discontinuities at the interface, not the band bending dis-cussed in Section 10-C, which is also present in heterojunction systems as well as in metal–semiconductor contacts.

PROBLEM 18-1 *Optical mode frequencies*

(a) Write out the energy W_{111} explicitly in terms of the V_2 and V_3 of Eq. (18-5) for arbitrary vectors τ between the two atoms per primitive cell.

(b) Expand this energy to second order in the deviation **u** of this τ from the value for the zincblende structure, and average over directions of that deviation. Since the magnitude of W_{111} can be taken to be the lowering in energy per electron, this directly gives the increase in elastic energy per electron and therefore we may equate the elastic energy per atom pair to the kinetic energy per atom pair, $2\,(M/2)(\omega u/2)^2$, and solve for the frequency. Compare with the experimental values of TO(0) from Table 9-1 for germanium and gallium arsenide. Clearly, we seriously overestimate the effect of polarity.

(c) Similarly, evaluate the magnitude of W_{111} for τ differing by **u** from the value for the rocksalt (sodium chloride) structure, to second order in the magnitude of **u**. Assuming $V_2 \ll V_3$, does this effect tend to stabilize or destabilize the rocksalt lattice against this distortion?

PROBLEM 18-2 *Metal–semiconductor contact*

We calculated the position of the valence-band maximum, $E_{v\text{-min}}$ + band width, with respect to an artificial metallic Fermi energy in Section 18-G. This may be used to give an estimate of the position of the Fermi energy of a metal contact with respect to the valence-band maximum. Give this estimate for all of the semiconductors listed in Table 18-3, and compare with the empirical estimate of one third the minimum band gap (Table 10-1). One feature of the result is experimentally correct: the position is quite insensitive to the work function of the metal. In addition, the trends are mostly correct, though the values are not accurate.

CHAPTER 19

Transition–Metal Compounds

SUMMARY

The d states of transition metals have energies comparable to the valence s states, but because they have greater angular momentum, d electrons do not range as far from the nucleus; thus their behavior is intermediate between that of valence electrons and core electrons. In some cases, the intra-atomic Coulomb matrix elements dominate the inter-atomic matrix elements that produced energy bands in the other systems discussed earlier in the book. When this is so, the electrons should be considered "local" rather than "itinerant." The transition-metal monoxides have local electrons, and the multiplet electronic structure that follows from viewing monoxides in this manner describes conduction properties and the electronic spectra of these systems. In contrast, compounds that have a perovskite structure and that contain transition elements to the left in the transition series have well developed d bands. They can be well described by an LCAO fit; three additional interatomic matrix elements are required to describe the coupling between the d states and the oxygen s and p states. These vary with internuclear distance as $d^{-7/2}$.

Bonding properties of the perovskites can best be studied by treating the interatomic matrix elements in perturbation theory in terms of a bonding unit that consists of the transition-metal d states and six neighboring sp hybrids (on the oxygens in $SrTiO_3$, for example). Second-order theory describes both the ion softening for this case and a contribution to the bonding energy. However, fourth-order theory is required to calculate angular forces. This fourth-order contribution, called the chemical grip, explains the angular forces in alkali halides, as well as the angular forces in the perovskites. It is, however, the electrostatic energy that is the dominant factor in the stability of the perovskite structure.

Systems with partially filled *d* shells comprise some of the most important technological solids. They are stable and strong and show a very wide diversity, for reasons that will be discussed. We shall nevertheless find it possible to cover them in a relatively limited space. It will be seen that many of the bonding properties are determined by *s* and *p* electrons, with bonding of just the same type discussed in the preceding chapters; to that extent, we need only to add the effects of the *d* electrons here. Also, many of the novel magnetic properties are associated with magnetic moments localized at the transition-metal ions. This is a specialized topic that will not be undertaken in detail, though some features will be mentioned.

We begin with a general discussion of the nature of *d* states in solids and a survey of implications with respect to properties. This discussion covers transition metals as well as transition-metal compounds. We then turn to specific systems; compounds are discussed first, in Chapter 19, since they are somewhat simpler to understand. Transition metals themselves are discussed in Chapter 20.

19-A   *d* States in Solids

Elements of the transition series have *d* states of energy comparable to the valence *s* states, so that some of the ten *d* states in the corresponding shell are filled; these elements appear to the left in the Solid State Table. Similarly the *f*-shell metals have *f* states with energy comparable to the valence *s* state and appear to the right in the table. We shall specifically discuss transition metals, with occasional reference to *f*-shell metals, but for all, the principal ideas are the same. (The similarities are discussed in Elliot, 1972; Mackintosh, 1977; Freeman and Koehling, 1974; and in Skriver, Andersen, and Johansson, 1978.) We noted in Chapter 1 that atomic *d* states are confined more closely to the nucleus than are *s* states of the same energy, so that in solids containing transition metals, the overlap interaction arising from the *s* electrons tends to hold the atoms far enough apart that the *d* electrons do not strongly overlap the neighboring atoms. In this sense, they are very much like core electrons, though their energies are comparable to those of the valence electrons.

The limitation of the overlap between neighboring *d* states introduces the possibility of behavior that is qualitatively different from that describable in terms of energy bands. There are two kinds of competing energies in solids containing transition metals—the matrix elements between *d* states and states on adjacent atoms, and an energy U equal to the Coulomb energy of interaction between two *d* electrons on the same atom. In the nontransition solids we used the one-electron approximation, and the energy U became a part of the periodic potential and was never considered explicitly. This approach led us correctly to a description of bands of electron states and electrons propagating through the crystal. If, however, the coupling between adjacent levels is small compared to U, a better starting approximation is to say that each atom has the same number of electrons and that an activation energy is required to remove an electron from one atom, to place it on a second, even if the orbitals on the two atoms are the same. In the latter approximation, a description in terms of *local electrons* is appropriate, while in the former a description in terms of *itinerant electrons* is appropriate, "itinerant" referring to the motion of electrons in band states. The monoxides are of the localized type, and we discuss them first. Many perovskite-structure solids have itinerant d states, and we discuss them second, to be followed by a survey of other systems. In both the local and itinerant descriptions, the d states are introduced as atomic d orbitals and the difference will come in how we use them. It has been stated that the atomic d states are much like core states, but there are important differences that we should enumerate before addressing specific systems.

First, though localized, the d states extend much further from the nucleus than typical core electrons. Thus, each added d state, corresponding to an added electron and a compensating nuclear charge, deepens the potential well at the atom. (As can be seen from the treatment of pseudopotentials in Appendix D, the repulsive term in the pseudopotential arising from d electrons does not affect the s and p electrons.) The deepened potential lowers the energy of the s electrons more than the p electrons since, roughly speaking, the s electrons vibrate through the core but the p electrons orbit around it. The lowered s-state energy increases the magnitude of V_1 and the promotion energy and suppresses tetrahedral covalent bonding in the sense we have discussed; there are very few such tetrahedral solids containing transition metals, as will be seen in Section 19-D.

Second, the additional potential well tends to contract the s states, reducing the nearest-neighbor spacing in comparison to the nontransition-metal counterpart; the well has the same effect in metals, as one can see by consideration of the concepts leading to Eq. (15-16). It follows that the transition metals and their compounds are more compact, more tightly bound, and more rigid than their nontransition-metal counterparts. The effect initially increases to the right through the series as more d electrons are added, but then decreases as the increased nuclear charge reduces the size of the d orbitals. We shall see this in particular for the transition metals, in Chapter 20.

Third, the d shells are only partially filled and, because of Hund's rule (Chapter 1), they tend to be entirely, or predominantly, of the same spin. Thus the magnetic moments of the d electrons reinforce each other, and the partly filled shell behaves as a small magnet. (In the core, or for a full d shell, the net moment is zero.) This is the origin of the remarkable magnetic properties of transition metals and transition-metal compounds. If the coupling between the moments on different atoms were negligible, the moments would be randomly oriented with no net magnetic moment for the specimen. Then, an applied field would tend to align them, giving a large paramagnetic susceptibility.

There is, however, always coupling between different moments. The magnetic interaction is very small, but there is a nonmagnetic interaction through the valence electrons, arising ultimately from exchange, which can tend to favor parallel or antiparallel alignment of neighboring spins. For example, the **Ruderman-Kittel interaction** in simple metals (Ruderman and Kittel, 1954) couples moments through an electron gas very much as atoms are coupled through the effective interaction between ions, as was shown in Fig. 17-2. The interaction shows the same Friedel oscillations corresponding to alternate favoring of parallel and anti-parallel spin. When the coupling between them is significant the moments will have a ferromagnetic ordering if they align parallel and an antiferromagnetic or ferrimagnetic ordering otherwise. The magnetic properties of such coupled moments are quite well understood (see, for example, Herring, 1966) but, as was indicated, this is not a subject we shall undertake in detail.

Fourth, the difference in localization between *d* electrons and *s* electrons becomes less for larger quantum numbers (the difference is less for the 4*d* series and especially so for the 5*d* series, in comparison to the 3*d* series) and is also less far to the left in each series. Both trends are expected intuitively from consideration of the potentials and the orbits and they will be apparent in the electronic structure. In titanium, at the left end of the transition series, the 3*d* states will be much more like *s* and *p* valence states and will contribute to the bonding. Even at the far left of the 5*f* series, the *f* levels will behave as valence states and will form bands in metallic thorium (Freeman and Koelling, 1974). These two trends—a decrease in the extent of localization as we move left in each series and as we move down in the periodic table—constitute the two most important trends in the transition series. They will show up in different ways in different systems, but have the same origin.

We have seen, particularly in the discussion of covalent crystals in terms of pseudopotentials, the importance of recognizing which matrix elements or effects are dominant and which should be treated as corrections afterward. This is also true in transition-metal systems, and different effects are dominant in different transition-metal systems; thus the correct ordering of terms is of foremost importance. For many transition-metal systems, we find that band calculations, particularly those by L. F. Mattheiss, provide an invaluable guide to electronic structure. Mattheiss uses the ***Augmented Plane Wave method*** (APW method), which is analogous to the OPW method discussed in Appendix D.

19-B   Monoxides: Multiplet *d* States

The first systems we consider are simple ionic solids like those discussed in Chapters 13 and 14, but the systems here have additional electrons in *d* states. The simplest such systems are the transition-metal oxides, TiO, VO—but the next metal, Cr, apparently does not oxidize as CrO—MnO, FeO, CoO, and NiO; all of these form the rocksalt structure. Most of these transition-metal oxides are insulators, but we shall see that two show metallic conductivity. Their bonding properties may be understood in terms of the overlap interaction and the electrostatic energy discussed in Chapters 13 and 14, so we shall focus here on new features of their electronic structure that arise from *d* states. Let us follow the treatment given by Koiller and Falicov (1974), who followed a similar approach by Feinleib and Adler (1968), and who based their work on APW band calculations made for all of these materials by Mattheiss (1972a). A recent study of a similar series, SmS, SmSe, SmTe, in which *f*-shell electrons behave as *d* states, has been made by Batlogg, Kaldis, Schlegel, and Wachter (1976).

The Electronic Structure

Transition metals are numbered $D3$, $D4$, $D5$, ..., Dn in the Solid State Table, n being the number of electrons beyond the last inert-gas shell (or beyond the filled f shell in the $5d$ series). The transition-metal oxides contain oxygen ions O^{2-} which have taken on two electrons from the metal. This is not contradicted by the fact that there will be softening of the oxygen ion; in incorporating the coupling that gives that softening, the p states on the oxygen will be reduced still further in energy and there will be no crossing of levels. The qualitative behavior of the monoxide is correctly described in terms of oxygen ions. The remaining $n - 2$ electrons per metallic atom (two for Ti, three for V, and so on) can be considered initially to be in d states on the metallic ion. This corresponds to an electron configuration of argon plus $(3d)^{n-2}$.

Setting aside the atomiclike d states for the moment, the electronic structure of the transition metal monoxide is that of an ionic crystal, like CaO, illustrated for MnO in Fig. 19-1. The valence bands are essentially oxygen $2p$ bands and the conduction bands are largely metallic $4s$ bands. The $4s$ bands are actually the beginning of an indefinite series of overlapping and mixed conduction bands, but it is convenient to extract a simple s band from this set (Mattheiss, 1972a). The center of gravity and width of that band are indicated by E_s and W_s, respectively, in Fig. 19-1. Koiller and Falicov noted that the band minimum lies below E_s by approximately $3W_s/4$ (see Kittel, 1971, Advanced Topic F).

Let us now turn to the d electrons in the monoxide series. Following Hund's rule, for $n - 2$ values up through 5, all electrons have the same spin. Mn^{2+} has one electron in each of the five d orbitals, all with the same spin. Beginning with iron, the added electron is of opposite spin and the total spin decreases by one electron at each step. The total orbital momentum and the total angular momentum quantum number J of the ground state follow rules of atomic physics that we need not consider in detail. For Mn^{2+}, illustrated in Fig. 19-1, with one electron in each d orbital, the total orbital momentum is zero (this state of zero orbital angular momentum is denoted by an S) and the spin is 5/2 in the ground state. (There are six orientations of spin $S_z = -5/2, -3/2, -1/2, 1/2, 3/2$, and 5/2, so the state is written ^6S.)

The energy required to shift one of the d electrons to a metallic s state in the free ion can be obtained directly from spectroscopic tables (see the standard tables of atomic energy levels in Moore, 1949, 1952). The energies are listed for the ions in the $3d$ series in Table 19-1. Energy would also be required to transfer an s electron in the ground state of the ion to a d state. Thus, for these systems, we cannot define meaningful s-state and d-state energies unless we include also the Coulomb interaction, U. Furthermore, the energy required to transfer an electron to an s state depends upon whether the spin of the s electron is parallel or antiparallel to the total spin of the d electrons, as indicated in Table 19-1; we may use the average of the two when describing a band state.

In order to identify this transfer energy with energies in the solid, we may imagine bringing the metallic ions together with the O^{2-} ions to form the solid, so

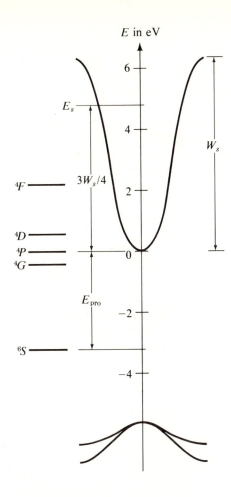

E in eV

FIGURE 19-1

The electronic structure of MnO contains valence bands that are principally of oxygen $2p$ character and a conduction band that is principally of manganese $4s$ character. The energy required to remove an electron from a Mn d state and to place it in the conduction band is E_{pro}. See the text for further description of the figure. [After Koiller and Falicov, 1974.]

TABLE 19-1

Electronic structure of the transition-metal monoxides. The metal-ion configuration has a formula $(Ar)(3d)^{n-2}$. Excitation energies (in eV) are to a configuration $(Ar)(3d)^{n-3}4s$. If ΔE (Eq. 19-2) is negative, metallic conductivity is expected.

	Ion:	Ti^{2+}	V^{2+}	Cr^{2+}	Mn^{2+}	Fe^{2+}	Co^{2+}	Ni^{2+}
	Column, Dn:	$D4$	$D5$	$D6$	$D7$	$D8$	$D9$	$D10$
Excited states	Configuration:	^3D	^4F	—	^6D	^5G	^6D	^5F
Excited states	Energy, $E_{s-d}^{\parallel}$:	4.7	5.4	—	7.8	7.9	5.7	6.7
Excited states	Configuration:	^1D	^2F	—	^4D	^3G	^4D	^3F
Excited states	Energy, $E_{s-d}^{\ddagger}$:	5.1	6.1	—	8.9	8.7	6.9	7.6
	Band width, W_s:	7.2	6.8	—	6.4	6.3	6.1	5.9
	ΔE (Eq. 19-2):	−0.6	0.5	—	3.4	3.4	1.6	2.6

NOTE: The superscripts $\parallel$ and $\ddagger$ denote s-electron spin relative to the d states; W_s was estimated by Koiller and Falicov (1974) from the band calculation of Mattheiss (1972a).

that the s levels are broadened into bands. Since the principal coupling of the metallic s state is with the neighboring oxygen p states, we would expect, as in the alkali halides discussed in Chapter 14, for the minimum of the band to remain near the s-state energy for the ion, with the conduction band spreading upward in energy, and with the more local d level remaining fixed relative to the band minimum. Koiller and Falicov instead assumed that the band broadens out and has a constant center of gravity, as if the principal coupling were directly between the metallic s states. The different conclusions result from the assumption that different parameters are dominant in the system; the choice made by Koiller and Falicov seems to give a good account of conducting properties, so we will follow their treatment. However, we must question their choice of parameters.

Koiller and Falicov identified the promotion energy in the ion (spin-averaged) as the sum of the promotion energy in the solid and the energy from the band minimum to the center of gravity of the band; it is $E_{pro}^{ion} = E_{pro} + 3/4 \, W_s$, and is illustrated in Fig. 19-1. The energy denoted by the atomic term value ^6S functions as a full valence band; optical absorption of a photon of energy E_{pro} may take an electron from this state to the conduction-band minimum.

The Optical Spectra

Let us now illustrate the optical spectra by considering MnO. The description we have obtained so far is valid for this but quite incomplete. First, there are other states of the $(3d)^5$ configuration of Mn^{2+}. In particular, the states with total spin of 3/2 rather than 5/2 have, by Hund's rule, higher energy, and there are four such states of varying orbital angular momentum designated ^4F, ^4D, ^4P, and ^4G in Fig. 19-1 (the superscript 4 refers to the four possible z-components of spin, $S_z = -3/2, -1/2, 1/2,$ and $3/2$). These also represent excited states of the system, all of which correspond to an empty conduction band.

Similarly, if an electron is taken from a manganese ion and put in the conduction band, leaving a Mn^{3+} ion behind, there are a number of different states in which that ion could be left. These all have higher energy than the ^5D term assumed in determining the position of the ^6S level. More energy is required for excitation, so these would appear as levels below the ^6S level in Fig. 19-1. The set of possible states of the system is quite intricate, but the picture is basically equivalent to a CaO ionic crystal, with additional atomic d electrons that can be put in the conduction band. There are also d states that behave as empty states in the sense that a transition to them can be made from the ^6S state but that do not behave as empty states with respect to the valence band, since they cannot be reached by transferring an electron from the valence band. They are excitations of the system but do not really belong in a one-electron diagram. Finally, there are states that can be reached by transferring an electron from the valence band, to make a Mn^+ ion, which can be in various orbital states. To understand the optical properties, one must organize these possible states and consider the possible transitions. The fact that there is no simple one-electron set of states that describes the properties is an inconvenience. Figure 19-1 gives the bands and the states of

Mn^{2+}. We need to consider also the states of Mn^+ and Mn^{3+} to describe the other excitations of the system.

There is a second complication that we have omitted by describing the states of the free ion. Although the ^6S state of Mn has spherical orbital symmetry (one electron in every *d* orbital), other orbital states, such as the ^5D ground state of Fe^{2+} (the ^5G state listed in Table 19-1 is an excited state), are not, so that the energy will be different depending upon how the ion is oriented with respect to the crystal axes. It is well known that a potential with cubic, rather than spherical, symmetry will split states of *d* symmetry into two sets of two and three states each. This is called ***crystal-field splitting***. Correspondingly, the ^5D ground state of Fe^{2+} is split by about one electron volt, and this triply degenerate state would be the lowest or ground state. Thus in the energy diagram for iron, the counterpart of what is shown in Fig. 19-1 for manganese would show an additional splitting of the ground-state level (^5D for iron rather than the ^6S shown for manganese), which would give additional structure. (This splitting also modifies the magnetic properties by quenching the orbital momentum; notice that though the crystal-field-split state is made up of states of *d* symmetry, they are combined to give no net angular momentum. The state is said to be ***quenched***). We will not give further details of the optical spectra; diagrams like Fig. 19-1 for each monoxide are given by Koiller and Falicov (1974). The important point is that it appears to be essentially correct to describe their electronic structure in terms of atomiclike *d* states (with crystal-field splitting) added to a simple CaO ionic solid.

Metallic versus Insulating Behavior

Let us now follow Koiller and Falicov in discussing the nature of the ground state in the monoxides. We saw in Table 19-1 that in all cases, energy is required to promote a *d* electron in the free ion to a free-ion *s* state. In the solid, however, the *s* state is broadened into a band, as illustrated in Fig. 19-1, so some of that energy can be gained back by placing the corresponding electron in a state at the bottom of the *s* band rather than in an ionic state having an energy expectation value at the center of gravity of the *s* band. Indeed, if more energy is gained in this way than was lost in promoting the electron to an *s* state, the starting $(3d)^{n-2}$ state must have been unstable: electrons will pour into the *s* band. The *s* band width is sufficiently large that it is always favorable to fill the lower states with electrons of both spins (as was true in the simple metals), so the system will show metallic conductivity.

Koiller and Falicov used this behavior as a basis of a criterion for whether metallic or insulating behavior is to be expected. They estimated the promotion energy per ion as the atomic promotion energy from Table 19-1, weighted by the multiplicity of the parallel and antiparallel spin configurations; for example, for NiO the promotion energy is

$$\Delta E_{pro} = \tfrac{5}{8}E[3d^74s(^5F)] + \tfrac{3}{8}E[3d^74s(^3F)] - E[3d^8(^3F)]. \qquad (19\text{-}1)$$

They estimated the energy gain in going from free-ion states to band states as the value obtained from an LCAO calculation for s bands in a face-centered cubic structure. The resulting gain of $3W_s/4$ is 4.4 eV for NiO. Thus if the difference,

$$\Delta E = \Delta E_{\text{pro}} - \tfrac{3}{4}W_s, \tag{19-2}$$

is negative, the ground state can be expected to be metallic. The values obtained by Koiller and Falicov are given in Table 19-1. They predict TiO to be metallic and indeed it is. VO shows a small positive ΔE and is not experimentally simple. (For a discussion, see Mott, 1974, p. 234.) VO occurs with a high concentration (15 percent) of vacancies and shows complex conducting and magnetic properties. The other monoxides, with large positive ΔE values, are all good insulators.

In all of this discussion of monoxides, it has been adequate to treat the d states as sharp atomic states—the only banding has occurred for the s states. One consequence of this view is that the spin magnetic moment on each ion is free to orient itself or to form antiferromagnetic states, and the magnetic properties are dramatically affected (Wilson, 1972; Goodenough, 1971; Adler, 1968). It is not difficult to see how this situation might be quite different in other compounds by returning to the crystal-field splitting mentioned earlier. We will describe the effect in terms of the MnO electronic structure of Fig. 19-1, though the crystal-field splitting is not large enough in that system to produce effects of interest.

It would be possible in principle for the crystal-field splitting to be sufficiently large that, for example, the lowest level derived from the ^4G states of Mn^{2+} shown in Fig. 19-1 would drop below the unsplit ^6S state and the ground state would have spin 3/2 rather than 5/2. This would represent a qualitative change in the state of the system; it would mean that differences in orbital energy dominated the exchange energy that led to Hund's rule, and we might expect that more energy would be gained by reducing the spin further until finally the net spin were zero. We could describe the electronic structure in terms of doubly occupied bands, and the one-electron picture used for nontransition-metal systems would again become valid. This occurs in many of the perovskite-structure systems we consider next.

19-C Perovskite Structures; d Bands

We turn now to a very important class of materials that have the formula ABC_3, with the C frequently oxygen. Strontium titanate is a familiar example and one we shall use for illustrative purposes. Titanium is in the $D4$ column of the Solid State Table, having four electrons beyond its argonlike core. Strontium has two electrons outside its kryptonlike core, so we may think of the six valence electrons as having been transferred to the three oxygen atoms to form a simple ionic system. As we shall see, however, the titanium d states form the lowest conduction band and are important in the bonding properties as well.

The Perovskite Structure

Strontium titanate forms in the perovskite structure illustrated in Fig. 19-2, in which each transition-metal ion is surrounded by six oxygen ions. There are also perovskites in which no elements are transition-metal elements; for example, $NaMgF_3$. This is also the structure of the tungsten bronzes such as Na_xWO_3 in which x can be varied and even reduced to zero. (In the corresponding structure of WO_3, there is a noncubic distortion, but the topology of the structure is not modified.) The structure of WO_3 (with no Na) is so open that it suggests the existence of some kind of covalent effects, though there is not the tetrahedral coordination that arises in *sp* bonding. It has thus been natural to think of these systems in terms of *sd* hybrids, and initially the idea looks quite promising. However, as with the fluorites (tetrahedrally surrounded fluorine ions) discussed at the end of Section 13-A, any reasonable attempt to construct a bonding unit leads to the individual ion's being that unit, and we will see in the end that an ionic bonding unit does make sense. Use of *sd* hybrids as bonding units may have been a case of forcing a concept onto a system it did not fit.

WO_3
ionic
not
coval

The Perovskite Electronic Structure

The electronic structure of $SrTiO_3$ is now very well understood. Accurate APW band calculations for this compound, and for a series of related compounds that we shall discuss, were carried out by Mattheiss (1972b). Experimental studies of these compounds continue to be made; see, for example, Pertosa and Michel-Calendini (1978).

Mattheiss interpreted the bands in terms of an LCAO description and we shall follow that interpretation. His results for wave numbers in a [100] direction are shown in Fig. 19-3, for four of the compounds he treated. Let us consider the $SrTiO_3$ bands; the Sr core states are not of interest, but we shall include the oxygen *s* states, which lie some ten electron volts below. We refer to the oxygen *p* states as the valence bands; they are completely occupied. The transition-metal *d*

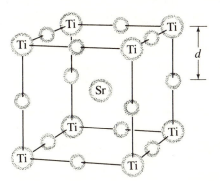

FIGURE 19-2

The perovskite structure of $SrTiO_3$ may be understood as a simple cubic lattice of titanium ions with oxygen ions at the center of every cube edge. The Sr ions reside at the cube centers. Na_xWO_3 has the same structure but with variable numbers of the cube centers occupied.

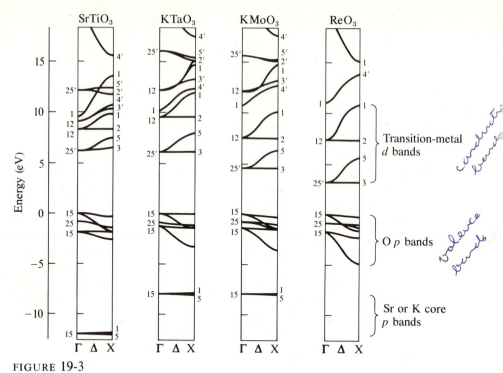

FIGURE 19-3

The energy bands along a [100] direction for four perovskite-structure compounds as obtained by Mattheiss (1972b), and his interpretation of the atomic origin of the bands. The oxygen 2s bands lie at about $-16\,eV$, as would be suggested by the atomic sp-splitting.

bands, which are empty in $SrTiO_3$, are called the conduction bands. We shall not be interested in bands originating from the Ti s state (Γ_1–X_1) nor the Sr d bands (upper Γ_{12} and Γ'_{25}).

The second material shown in Fig. 19-3 is $KTaO_3$, which has the same structure as $SrTiO_3$; Ta is located in column $D5$ of the Solid State Table; it has one more outer electron than Ti, and K has one less than Sr, so again the valence bands are full and the conduction bands empty. Similarly, WO_3 which was discussed earlier, is insulating, with full valence bands. In the third compound, $KMoO_3$, there is an additional outer electron, since Mo is an element from column $D6$, and it enters the conduction band to form a good metal. The same is true of ReO_3 which, Re being in the $D7$ column, has the same number of outer electrons and also forms a good metal with one conduction electron per atom in a d-like band. This is rather characteristic of the perovskite compounds incorporating transition metals far to the left in the periodic table. (See the survey of more than fifty perovskite oxides by Goodenough, 1971, Vol. 5, p. 145.) One can count electrons to see if they are insulating or conducting, as we have done here. There are, however, counterparts of the ten-electron compounds discussed in Chapter 13. In these, for example, in $PbTiO_3$, two electrons per primitive cell remain in Tl or Pb s bands, which lie below the conduction d bands. When they are metallic, it is very clear that the electrons are in bands rather than in multiplet states such as arose in the monoxides. Even in the insulating systems, carriers may be introduced by doping, or by varying x in $K_x MoO_3$, for example.

When the counting of electrons leads to many d electrons per ion, a multiplet structure may form. $KMnF_3$ is such a case (as are $KFeF_3$, $KCoF_3$, and $KCuF_3$). The K and Mn together contribute eight electrons; three are contributed to the flourines, leaving five electrons in a manganese multiplet. These order to form a ferromagnetic structure, and the system is insulating. If the electrons were in bands, the odd number of five electrons per primitive cell would have guaranteed that the system would be metallic. In such multiplet systems, substituting, for example, Fe atoms for Mn atoms does not put electrons in bands, but simply puts different multiplets on some ions.

There can also be intermediate situations. An antiferromagnetic insulator can be understood as an ordered array of multiplets, but an empty state produced optically can migrate; it might be more easily understood as intersite tunnelling. An antiferromagnetic state can also be understood in terms of splitting of band states by exchange; one can then describe the motion of a hole as being complicated by electron correlations. We shall not discuss these intermediate systems further (for extensive discussion and classification of the behavior, see Goodenough, 1971) but consider only cases such as those shown in Fig. 19-2, in which the electronic structure clearly has a banded nature.

The LCAO Description

Mattheiss fitted his calculated bands with LCAO bands, much as we did for tetrahedral solids in Chapter 6. We used the parameters for these tetrahedral bands to adjust the coefficients giving the matrix elements in Table 2-1, and then used those coefficients in all of the studies of *sp*-bonded materials. In a similar way we shall use the results of Mattheiss's fit for the perovskites to adjust the coeffecients in the expressions for *sd* and *pd* matrix elements, which will be derived in Chapter 20. These are listed in the Solid State Table and will be used wherever needed.

Mattheiss used a basis of fourteen atomic orbitals per primitive cell for his LCAO analysis; these were the five transition-metal *d* states and the nine oxygen *s* and *p* states for the three oxygens. At first, one might be inclined to include the oxygen *s* states as core states and therefore drop them from the analysis, as we did for SiO_2, but Mattheiss's APW calculation of the bands indicated that the *s* bands had appreciable width (0.66 eV in $SrTiO_3$) and important effects upon the other bands. It is important, therefore, to include them, and we shall even find that it simplifies the calculation of bonding properties. The APW calculation allowed this insight into the electronic structure, which would not have been available earlier.

The only nearest-neighbor matrix elements between the fourteen orbitals are those between the transition-metal *d* states and the neighboring oxygen *s* and *p* orbitals; this gives three independent parameters, as we shall see. In order to obtain a good fit to the bands, Mattheiss introduced a series of small second-neighbor matrix elements, different diagonal energies for the σ- and π-oriented p states, and overlaps (as in Appendix B). For our purposes it will be sufficient to give a considerably simplified analysis, and to include as parameters only the atomic energies for the three states of the basis, the three nearest-neighbor matrix elements, and one second-neighbor matrix element between oxygen p states. This is largely a conceptual and algebraic simplification; the computational complexity depends principally on the size of the basis set, and we will use the same set he did. We obtain our parameters from Mattheiss's fit, except for our second-neighbor matrix element, which we fit independently.

Clearly, with fourteen basis states, we must make use of symmetry. This is most directly and completely done with group theory (see, for example, Tinkham, 1964). Let us consider only states with $\mathbf{k}$ in a [001] direction (z-direction). The symmetry analysis will be given for the benefit of the reader who is familiar with group theory; others may skip to the results, since the analysis is not necessary for an understanding of the bands. The symbols may be regarded as simply a way of labelling the various bands.

The symmetry group of the wave vector for [001] propagation is the group of the square in the xy plane. This is the group Δ, of eight elements, with five irreducible representations. The five d states reduce to $\Delta_1 + \Delta_2 + \Delta_2' + \Delta_5$ in this symmetry. There are three oxygens per primitive cell and, of the three linear combinations of the s states, two transform as Δ_1 and one as Δ_2. Similarly, we may reduce the nine p states as indicated in Table 19-2, where we give also the

TABLE 19-2
The symmetry of orbitals entering states for $\mathbf{k}$ parallel to the z-axis in the perovskite structure.

Atomic orbital	Angular Form	Representation
Oxygen s states	Isotropic	$2\Delta_1, \Delta_2$
Oxygen p states	$\dfrac{z}{r}$	$2\Delta_1, \Delta_2$
	$\dfrac{x}{r}, \dfrac{y}{r}$	$3\Delta_5$
Transition-metal d states	$\dfrac{\sqrt{3}}{2}\left(\dfrac{z^2}{r^2} - \dfrac{1}{3}\right)$	Δ_1
	$\dfrac{x^2 - y^2}{2r^2}$	Δ_2
	$\dfrac{xy}{r^2}$	Δ_2'
	$\dfrac{zx}{r^2}, \dfrac{yz}{r^2}$	Δ_5

angular dependence of the atomic orbitals, expressed as in Chapter 1. The way the different combinations of oxygen states transform under the rotations and reflections that leave the crystal and the wave number invariant can be seen from the forms of the corresponding symmetries for the atomic *d* states.

Each energy-band state will have one of the symmetries indicated to the right in Table 19-2 and will contain only orbitals of the same symmetry. That was the point of the analysis, namely, to reduce the number of orbitals which need to be considered at the same time. In particular, we see that there is one set of *d* states—Δ'_2—that does not mix with any others. Since we ignore any matrix elements between neighboring *d* states (since they are second-neighbor ions), the bands will be completely flat. Consideration of the conduction band of this symmetry given by Mattheiss indicates that this is a very good approximation (see Fig. 19-3; in the notation there, the Δ'_2 is labelled by the X_3 end-point). These are represented as pure *d* states in this approximation.

Interatomic Matrix Elements

All other bands depend upon interatomic matrix elements; therefore we need to generalize the *s* and *p* matrix elements that were given in Section 2-D. Again taking spherical coordinates along the internuclear distance, we define interatomic matrix elements $V_{sd\sigma}$, $V_{pd\sigma}$, and so on. One example is illustrated in part (a) of Fig. 19-4. We may then write the angular forms given in Table 19-2 in terms of $Y_l^m(\theta, \varphi)$ to obtain representations of the simple matrix elements shown in the remainder of Fig. 19-4. For more complicated orientations we will need the form of the angular transformations, but these do not enter here. Therefore we postpone their tabulation till Table 20-1. Here we need only the three matrix elements $V_{pd\sigma}$, $V_{pd\pi}$, and $V_{sd\sigma}$, which are illustrated in the parts (a), (b), and (c) of Fig. 19-4, and the one second-neighbor matrix element $E_{x, x} = \frac{1}{2}(V_{pp\sigma}^{2nd} + V_{pp\pi}^{2nd})$ that will be included. We also need values for ε_d, ε_s, and ε_p; values are given in Table 19-3 and are obtained from the Solid State Table. The corresponding parameters obtained by Mattheiss are also given in Table 19-3; we shall, in fact, use his values in our plots of the bands. (Comparison of bands based on the Solid State Table with those of Mattheiss would be somewhat circular, since in compiling this portion of the Solid State Table, Mattheiss's results were relied on extensively, as will be seen.)

The Hamiltonian Matrix

We turn first to the bands transforming as Δ_1. They can contain only the first *d* state listed in Table 19-2, but can contain one *s* state and one *p* state from each of the oxygen ions. The orbitals that enter are shown in Fig. 19-5, where two types of oxygen ions numbered 1 and 3 are shown; the third type, which would be numbered 2, lies on top of or below the transition-metal ion. For this symmetry the coefficients of the oxygen ions numbered 1 and 2 but lying in the same plane of

TABLE 19-3

Parameters for cubic perovskites, in eV.

	SrTiO$_3$	KTaO$_3$	KMoO$_3$	ReO$_3$
OBTAINED FROM THE SOLID STATE TABLE				
d(Å)	1.95	1.99	1.96	1.87
ε_s	-29.14	-29.14	-29.14	-29.14
ε_p	-14.13	-14.13	-14.13	-14.13
ε_d	-11.04	-9.57	-11.56	-12.35
$V_{sd\sigma}$	-2.61	-3.35	-2.99	-3.53
$V_{pd\sigma}$	-2.43	-3.12	-2.78	-3.29
$V_{pd\pi}$	1.13	1.45	1.29	1.53
$E_{x,x}{}^*$	1.22	1.17	1.21	1.70
FROM MATTHEISS (1972)				
ε_d†	-10.22	-9.99	-10.14	-12.66
$V_{sd\sigma}$	-2.56	-3.41	-3.00	-3.53
$V_{pd\sigma}$	-2.25	-3.06	-2.77	-3.54
$V_{pd\pi}$	$+1.14$	1.39	1.25	1.62
$E_{x,x}$	0.159	0.097	0.117	0.129

* From nearest-neighbor formula, not appropriate here.
† The average of Mattheiss's values for $\varepsilon_d - \varepsilon_p$ were added to the ε_p value given in the upper part of the table.

constant z must be the same. The phase factors giving the phase difference between corresponding ions in different planes of constant z are indicated to the left. It is not too difficult to construct the Hamiltonian matrix as we did for other systems. By using states $|s_{1,2}\rangle$ equal to $(|s_1\rangle + |s_2\rangle)\sqrt{2}/2$ and $|p_{1,2}\rangle$ equal to $(|p_{z1}\rangle + |p_{z2}\rangle)\sqrt{2}/2$ rather than the four individual states, we reduce it to a five-by-five matrix given by

$$
H = \begin{array}{c} \\ s_3 \\ p_3 \\ s_{1,2} \\ p_{1,2} \\ d \end{array}
\begin{array}{ccccc}
s_3 & p_3 & s_{1,2} & p_{1,2} & d \\
\left| \begin{array}{ccccc}
\varepsilon_s & 0 & 0 & 0 & 2V_{sd\sigma}\cos kd \\
0 & \varepsilon_p & 0 & 2\sqrt{2}\,E_{x,x}\cos kd & 2iV_{pd\sigma}\sin kd \\
0 & 0 & \varepsilon_s & 0 & -\sqrt{2}\,V_{sd\sigma} \\
0 & 2\sqrt{2}\,E_{x,x}\cos kd & 0 & \varepsilon_p & 0 \\
2V_{sd\sigma}\cos kd & -2iV_{pd\sigma}\sin kd & -\sqrt{2}\,V_{sd\sigma} & 0 & \varepsilon_d
\end{array} \right|
\end{array}
$$

$$(19\text{-}3)$$

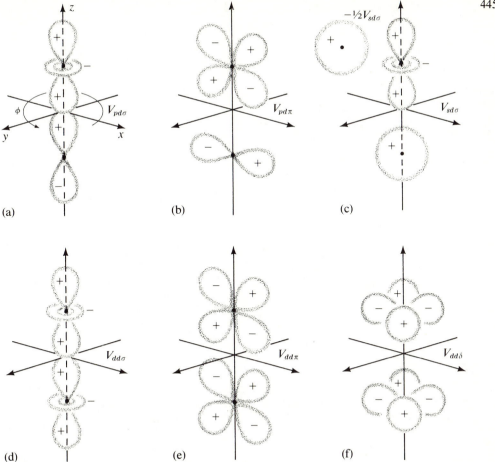

(a)   (b)   (c)

(d)   (e)   (f)

FIGURE 19-4

Matrix elements between atomic orbitals can be defined in terms of spherical harmonics based on an axis along the internuclear separation, as indicated at (a) in the upper left. There, the lower state is an $m = 0$ p orbital and the upper state is an $m = 0$ d orbital, which has angular dependence as $3^{1/2}[(z^2/r^2) - 1/3]/2$. For other d orbitals $(m \neq 0)$, it is easy to visualize other cartesian forms, such as the symmetry zx/r^2 shown at (e) or the xy/r^2 shown at (f), which lead to the same values for the matrix elements as do the spherical harmonics. There is only one independent sd matrix element; for it, the different cases indicated at (c) in the upper right can be related by algebraic manipulation or the transformations described in Eq. (19-21). The two independent pd matrix elements are shown at (a) and (b) and the three independent dd matrix elements are shown at (d), (e), and (f). Signs of the wave functions are chosen such that all but $V_{pd\pi}$ and $V_{dd\pi}$ are expected to be negative.

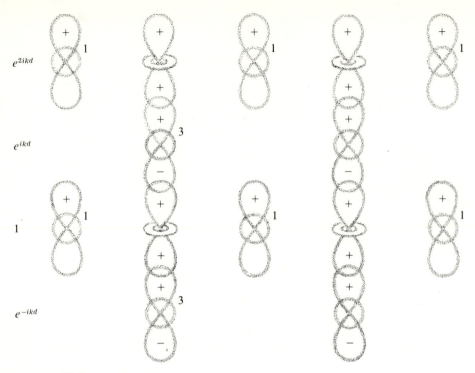

e^{2ikd}

e^{ikd}

e^{-ikd}

FIGURE 19-5

The orbitals entering the Δ_1 states propagating in a z-direction (upward in the figure), and the phase factors entering for each plane. Oxygen atoms, each with an s and a p orbital shown, are numbered as type 1 or 3; those which would be numbered 2 are displaced by d from the plane of the figure.

In constructing this matrix we may, for example, focus on a d state. It is coupled to two $|s_3\rangle$ states with equal matrix elements but with phases differing by e^{ikd} and e^{-ikd}, and from Fig. 19-4, we see that the matrix elements are $V_{sd\sigma}$, which leads to the $2V_{sd\sigma} \cos kd$ of Eq. (19-3). The d state is also coupled to two states $|p_3\rangle$, but with matrix elements having signs opposite each other, which leads to $2iV_{pd\sigma} \sin kd$ in Eq. (19-3). Furthermore, it is coupled to four atomic states of the $|s_1\rangle$ or $|s_2\rangle$ type with a matrix element $-V_{sd\sigma}/2$, but each contains a factor $\sqrt{2}/2$ in the $|s_{1,2}\rangle$ state. We neglect coupling of the s states except with their neighboring d states, so the other elements involving the s states are zero. Finally, each $|p_3\rangle$ state is coupled to four $|p_1\rangle$ or $|p_2\rangle$ orbitals, but each enters $|p_{1,2}\rangle$ with a factor $\sqrt{2}/2$. This completes the evaluation of the matrix elements.

Solution of the Matrix

The eigenvalues of Eq. (19-3) are quite easy to obtain at $\Gamma(k = 0)$ and at $X(k = \pi/(2d))$. At Γ the states $|p_3\rangle$ and $|p_{1,2}\rangle$ are coupled only with each other and are obtained by a solution of the corresponding two-by-two matrix,

$$\Gamma_{15} = \varepsilon_p \pm 4\sqrt{2}\, E_{x,\,x}. \tag{19-4}$$

The secular determinant for the remaining three-by-three matrix may be obtained immediately and solved to give

$$\left. \begin{aligned} \Gamma_1 &= \varepsilon_s; \\[2mm] \Gamma_{12} &= \frac{\varepsilon_s + \varepsilon_d}{2} \pm \sqrt{\left(\frac{\varepsilon_d - \varepsilon_s}{2}\right)^2 + 6V_{sd\sigma}^2}. \end{aligned} \right\} \tag{19-5}$$

The notation for the states at Γ corresponds with that given in Fig. 19-3.

The solutions at X are even simpler, since there the $|s_3\rangle$ state and the $|p_{1,2}\rangle$ states become uncoupled from the others and take energies

$$\left. \begin{aligned} X_1 &= \varepsilon_s, \\[1mm] X_4' &= \varepsilon_p, \end{aligned} \right\} \tag{19-6}$$

and the remaining three-by-three matrix gives a secular equation

$$(\varepsilon_p - E)(\varepsilon_s - E)(\varepsilon_d - E) - 2(\varepsilon_p - E)V_{sd\sigma}^2 - 4(\varepsilon_s - E)V_{pd\sigma}^2 = 0, \tag{19-7}$$

which may be solved numerically to give three X_1 values, -6.85, -16.78, and -29.85 eV for $SrTiO_3$ with the parameters of Table 19-3. A convenient way to do this is, for example, to divide through by $(\varepsilon_s - E)(\varepsilon_d - E)$ and take E to the right to obtain

$$E = \varepsilon_p - [2(\varepsilon_p - E)V_{sd\sigma}^2 + 4(\varepsilon_s - E)V_{pd\sigma}^2]/[(\varepsilon_s - E)(\varepsilon_d - E)],$$

which can be iterated, starting with $E = \varepsilon_p$.

The construction of the Hamiltonian matrix for states of symmetry Δ_5 is very similar to that for Δ_1 and is described in Problem 19-1. The eigenvalues obtained at Γ are

$$\left. \begin{aligned} \Gamma_{25}' &= \varepsilon_d; \\[2mm] \Gamma_{25} &= \varepsilon_p; \\[2mm] \Gamma_{15} &= \varepsilon_p \pm 4\sqrt{2}\,E_{x,\,x}; \end{aligned} \right\} \tag{19-8}$$

and at X they are

$$\left. \begin{aligned} X_5' &= \varepsilon_p \pm 4E_{x,\,x}; \\[2mm] X_5 &= \frac{\varepsilon_d + \varepsilon_p}{2} \pm \sqrt{\left(\frac{\varepsilon_d - \varepsilon_p}{2}\right)^2 + 4V_{pd\pi}^2}. \end{aligned} \right\} \tag{19-9}$$

There is also one band of symmetry Δ_2' with energy ε_d, independent of k.

The states of symmetry Δ_2 have a simple feature: the condition that the wave function become negative under reflection in the plane $x = y$ rules out all orbitals on the oxygen atom numbered 3 and therefore all coupling between orbitals of different phases. The solutions become independent of k. We consider the orbitals in a particular plane of constant z, as shown in Fig. 19-6. Symmetry requires that the coefficients on the $|s_1\rangle$ and $|s_2\rangle$ states be equal and opposite. The same is true of the $|p_1\rangle$ and $|p_2\rangle$ states (not shown in Fig. 19-6). Thus we have an $|s_{1,2}\rangle$ state coupled with a d state an an uncoupled $|p_{1,2}\rangle$ band, both of symmetry Δ_2, and we obtain final energies

$$\Delta_2 = \varepsilon_p;$$

$$\Delta_2 = \frac{\varepsilon_s + \varepsilon_d}{2} \pm \sqrt{\left(\frac{\varepsilon_d - \varepsilon_s}{2}\right)^2 + 6V_{sd\sigma}^2}.$$

(19-10)

Notice that the last two values are identical to the Γ_{12} levels of Eq. (19-5) at that point and become X_2 at the point X. The first terminates at Γ_{25} and at X_3'.

Using the values for the parameters given in Table 19-3, we have a complete description of the bands and could readily extend it to wave numbers in other

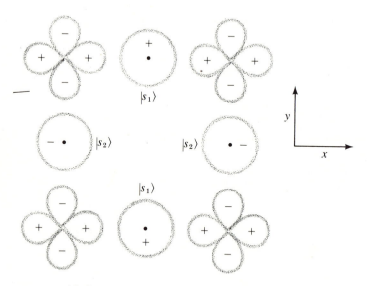

FIGURE 19-6

The s and d orbitals in a plane of constant phase that enter the Δ_2 states. They are not coupled to orbitals in other planes by the parameters we have kept. Nor are they coupled to the p_z states that are in this plane but oriented perpendicular to it. The matrix element between one of the s states shown and its neighboring d states is $-\sqrt{3}\,V_{sd\sigma}/2$.

directions. We compare the fit we obtain along $\Gamma\Delta X$ with the more accurate fit obtained by Mattheiss, in Fig. 19-7. His fit over many lines in the Brillouin Zone is included in the figure, but only the panel on the left in part (b) is relevant for comparison with the simplified treatment used here. Notice also that his calculation gave energies relative to the maximum in the valence band, and we have taken the absolute energy by matching his average *p*-state energy to our value from the Solid State Table.

The slight differences between our set of bands and his arise from our neglect of a number of matrix elements, the overlap of orbitals, and the crystal-field splitting of the *p* state. These corrections would seem to be quite negligible on the scale of the uncertainties. The electronic structure is well explained by Mattheiss, and we find, moreover, that a good description can be obtained even with a considerable reduction in the number of parameters. There are low-lying oxygen *s* states and valence bands that are principally oxygen *p* states. The conduction bands are transition-metal *d* states, and the states of the strontium (or potassium) do not have an important effect. The intra-atomic electron–electron interaction that dominated the electronic structure of the monoxides by forcing a multiplet structure is not important here except insofar as it determines the parameters of the bands. We shall discuss those parameters shortly.

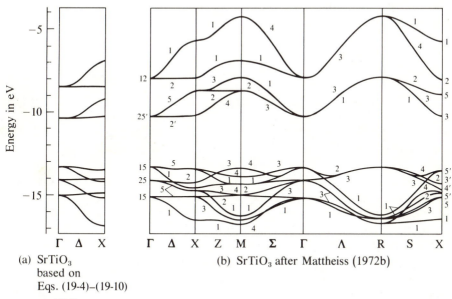

(a) $SrTiO_3$
based on
Eqs. (19-4)–(19-10)

(b) $SrTiO_3$ after Mattheiss (1972b)

FIGURE 19-7

The valence and conduction bands in $SrTiO_3$. In the left panel are those obtained in the text, using Mattheiss's parameters from Table 19-3 and Eqs. (19-4) through (19-10). The rest are directly from Mattheiss (1972b). The two $\Gamma\Delta X$ panels can be directly compared. Notice that the symmetries of the bands at both Γ and X are given by Mattheiss's bands. The oxygen *s* bands lie at approximately -29 eV on this scale.

The Role of the Body-Centered Ion

A very striking feature of this discussion of electronic structure is that the K ion or the Sr ion has no other function than that of providing electrons to the system, which may occupy valence-band p states or conduction-band d states, but the potassium or strontium conduction-band states are so much higher in energy that they are of no importance. This has the consequence that, for example, there should be very little difference between $SrTiO_3$ and $BaTiO_3$, or in fact between KWO_3 and ReO_3 (except for slight differences in the d states on W and Re). This is largely the case, though we shall see that the positively charged ion at the cube center (K^+, for example) can affect the stability of the structure just through its contribution to the Madelung energy.

This may come as somewhat of a surprise since we can directly estimate matrix elements between the oxygen $p\pi$ states and the Sr s states toward which they are directed from the Solid State Table, obtaining 3.69 eV and a much larger matrix element with the Sr p states. These states *are* higher in energy, but are still at negative energy, and the matrix elements are much too large to be neglected in comparison to an energy difference of the order of 10 eV. The point would seem to be that, as in the LCAO description of a simple metal, the matrix elements are so large and numerous that the LCAO description becomes irrelevant and the many excited states become merged with the simple free-electron energy (or ionized states of the ion in question). Any effect they have should be absorbed in the parameters and not included by introducing additional LCAO states.

Parameters

The parameters used in plotting the first panel in Fig. 19-7 were obtained from Mattheiss's APW calculation, after adjustment to fit the observed gap. These parameters were listed in Table 19-3. The results have been used, in conjunction with various theoretical expectations, to generate the general set of parameters given in the Solid State Table; the parameters in the Solid State Table give other values also listed in Table 19-3. We shall give the arguments which led to them.

First, consider the d-state energy, ε_d. The only serious discrepancy between Mattheiss's calculation and the experimental optical spectra was that Mattheiss's calculation appeared to overestimate the band gap by about three electron volts. Since this gap is dominated by the energy $\varepsilon_d - \varepsilon_p$, the discrepancy suggested an overestimate of this difference. In fact, his calculated bands were positioned roughly in accord with the splitting predicted by term values of Herman and Skillman (1963). Finally, the same overestimate applies to the Herman-Skillman term values in comparison to Hartree-Fock term values. This suggested, then, that values for ε_d should be taken from Hartree-Fock calculations, and those are what appear in the Solid State Table and therefore also in Table 19-3. C. Calandra has suggested independently (unpublished) from consideration of transition metals themselves that Hartree-Fock parameters give better relative energies of d-like and sp-like bands.

Consideration of Hartree-Fock term values from Fischer (1972), as discussed in Appendix A, indicates that Hartree-Fock values for valence s- and p-state energies are quite similar to the Herman-Skillman values given in the Solid State Table. Thus a more systematic treatment would result from use of the Hartree-Fock values throughout; they are more appropriate for the transition metals and it would make little difference which set of values is used for other systems. The reasons for retaining Herman-Skillman values here are largely historical; it is also possible that use of Hartree-Fock values would increase discrepancies with existing band calculations since, as indicated in Appendix A, the approximations almost universally used in solids are the same as those used in the Herman-Skillman calculation. For s and p states the differences are small in any case.

Next, we consider the sd and pd matrix elements. The analysis based upon transition-metal pseudopotential theory in Section 20-E predicts the form

$$V_{ldm} = \eta_{ldm} \frac{\hbar^2}{m} \frac{r_d^{3/2}}{d^{7/2}}. \tag{19-11}$$

The parameter r_d is listed in the Solid State Table for each transition element. The theory used in Section 20-E does not give useful values for the three universal constants $\eta_{sd\sigma}$, $\eta_{pd\sigma}$, and $\eta_{pd\pi}$, but Muffin-Tin Orbital theory (Section 20-D) suggests that the latter two should be related by $\eta_{pd\sigma} = -3^{1/2}\eta_{pd\pi}$. Consideration of the values obtained by the fit to Mattheiss's bands suggests

$$\eta_{pd\sigma} = -2.17\eta_{pd\pi} \tag{19-12}$$

is a better relation. We choose the three constants as an average of the values obtained by fitting the four perovskites, and these coefficients,

$$\left. \begin{aligned} \eta_{sd\sigma} &= -3.16, \\[6pt] \eta_{pd\sigma} &= -2.95, \\[6pt] \eta_{pd\pi} &= 1.36, \end{aligned} \right\} \tag{19-13}$$

are listed in the Solid State Table. The values for the matrix elements obtained from the formula at Eq. (19-11) and the parameters in the Solid State Table are shown in Table 19-3. The extent of agreement with the values obtained from the fit to the bands is impressive confirmation of Eq. (19-11). Notice that the coefficients at Eq. (19-13) would have been essentially the same had they been obtained from a fit to one perovskite, say $SrTiO_3$. Then the predictions for the other three could have been regarded as tests of that formula, since none of the parameters used (except d) depended upon experimental information about the compounds in question. The agreement reflects an accurate prediction of the variation from material to material.

The second-neighbor matrix element is somewhat problematic. The corresponding internuclear distance is 40 percent larger than the nearest-neighbor distance, and to be systematic, we should set it equal to zero as we did in treating other systems, for example, as we did for the alkali halides. This would in fact cause no problem in the treatment of dielectric and bonding properties for these systems, since we see that $E_{x,x}$ does not enter those calculations; it is a coupling between occupied oxygen p states, as V_1 coupled occupied states in tetrahedral solids, and these properties are dominated by coupling between occupied and empty states. On the other hand, the valence bands themselves are strongly effected by $E_{x,x}$, so in order to make a comparison, we have included it. An estimate from the nearest-neighbor formulae from the Solid State Table (which presumes vanishing second-neighbor coupling) is given in the upper part of Table 19-3 and is clearly inappropriate. Therefore, in plotting Fig. 19-7,a, the $E_{x,x}$ value given in the lower part of that table was used; it gives a reasonable fit to the bands and is a number we will not have occasion to use again.

19-D   Other Compounds

The rocksalt structure and the perovskite structure are the only ones we shall discuss in detail here, but we should make a brief survey of other systems. For a more complete account see, for example, Goodenough (1971) and for the bands themselves, see Calais (1977).

The monoxides discussed in Section 19-B form the rocksalt structure for just the reasons that the alkali halides do, that is, because of electrostatic considerations in conjunction with ionic radii. As in the nontransition-metal solids, there are many other compounds that do not have equal numbers of oppositely charged ions and that will therefore minimize their electrostatic energy by forming more complex structures such as those listed in Table 13-1; indeed, a number of those structures were labelled by a transition-metal compound that forms the corresponding structure. An important example is the case of the rutile structure, that of TiO_2. The transition-metal dioxides alone contain examples that are metal, semiconducting, or insulating, and which are ferromagnetic, antiferromagnetic, or nonmagnetic. A recent study of three of the metallic oxides, RuO_2, OsO_2, and IrO_2, has been made by Graebner, Greiner, and Ryden (1976); likewise, a detailed study of the electronic band structure has been made by Mattheiss (1976) along much the same lines as his study of the perovskites, which was discussed in the preceding section.

Rutile, and many other transition-metal compound structures, are characterized by dense packing and high coordination numbers (numbers of nearest neighbors). Their bonding properties are those of ionic solids, and many of the structures have been rationalized in terms of ionic radii and Madelung energies.

Recall, however, that the structure of CsCl has also been rationalized in this way and that the rationalization fails upon closer inspection. Although this aspect of the theory of transition-metal compounds was not studied sufficiently to judge the validity of the sphere-packing arguments while writing this book, notice, as symptomatic of limited reliability, the need to introduce different ionic radii for different charge states, and so on. We choose to bypass the question of structure for the most part, as we did for other systems. With a few exceptions, to be enumerated shortly, we can regard all transition-metal compounds as being essentially ionic. We can then introduce the chemical grip, which describes angular forces in ionic solids, and see how it affects the stability of open structures such as the perovskite structure. We shall also see that the polar covalent solids could have been understood by expanding around the polar limit, rather than the covalent limit, using the chemical grip.

Let us look first for transition-metal compounds that are truly covalent in the sense of tetrahedral structures and two-electron bonds, which we discussed earlier. There are only a few examples. NbN and TaN both form in the wurtzite structure. We presume that bond orbitals of sp^3 hybrids must be present to stabilize the structure; this requires three electrons from each transition-metal ion. Both ions are found in column $D5$ of the Solid State Table, so we anticipate that the remaining two electrons would form a ^3F multiplet (as in the ground state of Ti^{2+}). Thus the effects of the d state are simply added onto an otherwise simple covalent system, just as they were added to a simple ionic system in the monoxides. MnS, MnSe, and MnTe also form a wurtzite structure and presumably may be understood in just the same way. This class of compounds is apparently too small to have been studied extensively.

On the other hand, the transition-metal counterparts of the mixed tetrahedral solids are numerous and important. Zircon ($ZrSiO_4$) for example, has a structure like SiO_2 and its bonding should be understood in the same way. Since zirconium is an element from column $D4$ of the Solid State Table, there are just enough electrons to replace the silicon and, as in $SrTiO_3$, there are no occupied d states in the ground state but there are conduction bands of the character of zirconium d states. There are also some transition-metal arsenates and phosphates with the same structure. There are intermediate situations where the essence of the bonding is not so clear. Garnets have the formula $R_3^{2+}R_2^{3+}(SiO_4)_3$ where the metallic atoms R may or may not be transition metals, and where in some cases the Si is replaced by a transition-metal atom such as iron. We would anticipate that since iron is located in column $D8$, far to the right in the series, it would have residual electrons in multiplet states. Indeed, the garnets are one of the most important of the magnetic transition-metal compounds. To understand the bonding of these, one must see if an appropriate bonding unit can be constructed, or whether these must be considered ionic solids. This has not yet been undertaken. A similar situation exists with respect to the spinels, of formula $R_2^{3+}R^{2+}O_4$. As with the other compounds, these structures have been rationalized with sphere-packing arguments that will not be evaluated in this text.

Calculation of Properties

In all of these compounds, even the tetrahedral ones, a possible starting point for the calculation of properties is an ionic electronic structure with the effects of interatomic matrix elements treated in perturbation theory. As we have indicated, and as will be seen in detail in the next section, it is even possible to treat the polar covalent nontransition-metal solids in this way. Thus we should be able to calculate properties of the transition-metal compounds just as we did for the simple ionic compounds.

All of the needed interatomic matrix elements are listed in the Solid State Table, and a comparison of values in Table 19-3 indicates that very good predictions can be made of the variations of these from material to material. L. F. Mattheiss has noted (unpublished) that the same LCAO parameters he obtained for $SrTiO_3$ give a reasonable account of TiO and TiO_2, and A. C. Switendick has found the same to be true of a number of other transition-metal compounds (also unpublished). Thus there is every reason to believe that the theoretical forms and adjusted η_{ldm} given in the Solid State Table will be adequate for studying properties, though few studies have so far been made.

There is more uncertainty about the diagonal matrix elements ε_s, ε_p, and ε_d, particularly the difference between ε_d and the other two. The Hartree-Fock values listed in Table 19-3 gave a good account for the perovskites, but it seems unlikely that the values of ε_d with magnitudes larger than the ε_p for oxygen (that is, the ε_d values that occur to the right in the transition series) will prove appropriate. There may well be important shifts in the d-state energy because of changes in occupation of orbitals from one system to another; these are effects of the intra-atomic Coulomb interaction U that was discussed in Section 19-A. More experience will be required to learn how to select the ε_d parameter with confidence.

This same intra-atomic energy also produces the multiplet structures that were discussed in Section 19-B, which required a very different conception of electronic structure in order to understand the conducting and optical properties. However, there may be many other properties for which the distinction between the multiplet and band conceptions can be ignored. The point is that we calculate corrections to atomic orbitals in perturbation theory, and then, in the formation of multiplets, as in the formation of bands, the relative energies change (and, for example, so do optical properties), but to a good approximation the total energy and total charge distribution do not change. As in the simple solids, this covers most of the bonding properties.

A perovskite-structure system, $KNiF_3$, which does form a multiplet structure and is an antiferromagnetic insulator, was treated by Mattheiss (1972b) in the same way other perovskites were treated: he obtained the parameters for this system also, using an LCAO fit. The interatomic matrix elements he obtained, $V_{sd\sigma} = -1.15$ eV, $V_{pd\sigma} = -1.05$ eV, and $V_{pd\pi} = 0.51$ eV, accord well with the values $V_{sd\sigma} = -1.26$ eV, $V_{pd\sigma} = -1.17$ eV, and $V_{pd\pi} = 0.55$ eV obtained from the Solid State Table when $d = 2.01$ Å, as given by Mattheiss (1972b). From the Solid State Table we may establish the zero of energy, using the ε_p for fluorine of -16.99 eV, but from the Solid State Table we obtain $\varepsilon_d - \varepsilon_p = -1.05$ eV, whereas Mattheiss found $\varepsilon_d - \varepsilon_p = 8.27$ eV. The discrepancy reflects the difficulty in selecting values for the diagonal elements, mentioned earlier; these are influenced by the intra-atomic Coulomb interaction U; the best solution at present may be to use atomic spectra, as were used for the monoxides in Section 19-B; a general scheme is not now available for establishing values of ε_d for transition-metal systems to the right in the transition series.

Metallic Compounds

Before going on to a detailed description of the bonding properties of the perovskites, two important classes of materials should be mentioned that might be included either with the transition-metal compounds or with the transition metals. First are the transition-metal carbides or nitrides, which form in the rocksalt structure; NbC is an example. They are called *refractory compounds* because of their high binding energies and resulting high melting points. The energy bands for a number of these have been calculated; a recent calculation for NbC, with references to other work, has been given by Schwarz (1977). A simple view of the electronic structure, given by Weber (1973) and by Gale and Pettifor (1977), is consistent with these bands. In this view, two of the transition-metal d states—the last two in Eq. (1-21) if coordinate axes are taken along the crystal cube axes— couple strongly with the p states of the carbon much as in the perovskites, and if the transition metal is from column $D4$, the system can be insulating, as many of the perovskites are. Niobium is from column $D5$ and therefore NbC will be metallic, with one electron in bands arising from the other d states—the first three in Eq. (1-21).

A second class of compounds that are good metals are called the *A15 compounds* (the terminology is based upon the crystal structure); these contain long chains of closely spaced transition-metal atoms; Nb_3Sn is an example. The band structure for some of these has also been calculated (Mattheiss, 1975). A simplified view of the band structure was suggested by Labbé and Friedel (1966) on the basis of these chains. It leads to an independent band for each type of d state, and to a simple description of the superconducting properties. This class of materials is particularly important since it includes superconductors of the highest known critical temperature.

19-E   The Perovskite Ghost

We have seen that in essentially all transition-metal compounds, a possible start-ing point for a calculation of the electronic structure is an ionic description. In the perovskites, however, it may be determined from Table 19-3 that perturbation theory runs into trouble. We may, for example, calculate the effective charge Z^* on oxygen, starting with a closed shell O^{2-}. The perturbation of each oxygen $p\sigma$

state by a neighboring $d\sigma$ state transfers $[V_{pd\sigma}/(\varepsilon_d - \varepsilon_p)]^2$ electrons from the oxygen to the transition-metal neighbor; this is 0.33 for $SrTiO_3$. The transfer occurs for both Ti neighbors and for both spins, reducing the charge by 1.32 electrons. Another 0.68 electrons are transfered from the $p\pi$ states and another 0.33 from the s states, leaving the oxygen positive. This unreasonable conclusion indicates that we must treat the coupling between atoms more carefully.

Use of sp Hybrids

The parameters of Table 19-3 suggest a way to achieve a more careful treatment of the coupling. We have seen that the matrix elements $V_{sd\sigma}$ and $V_{pd\sigma}$ are very nearly equal. This means that if we construct two sp hybrids on the oxygen, $(|s\rangle \pm |p\rangle)/2^{1/2}$, each will have a very large matrix element coupling one hybrid to the d state on one side and a negligible matrix element coupling that hybrid to the d state on the other. Thus we might imagine making the corresponding unitary transformation on the states, obtaining bonding and antibonding combinations of the d states and the neighboring sp hybrids. The antibonding state in particular has almost no coupling with the other states, since its dominant d term has negligible matrix elements with the neighboring oxygen hybrids, and its p terms are sufficiently small that though the intra-atomic matrix elements are large, that is, $\langle h'|H|h\rangle = (\varepsilon_p - \varepsilon_s)/2 = 7.5$ eV, their *effect* is not large. Thus these antibonding states, which are labelled d bands in Fig. 19-3, can be rather easily and accurately obtained. We also saw that the $V_{pd\pi}$ matrix elements were considerably smaller (by a factor 0.46) than $V_{sd\sigma}$ and $V_{pd\sigma}$, and since the coupling enters the properties squared (by a factor 0.21), in a first treatment we might ignore their contribution to the bonding properties; they could be added later. In $SrTiO_3$ and $KTaO_3$, the antibonding states are empty, but they contain one electron per transition-metal atom in the other two compounds of Table 19-3.

This information is in fact all we need. We began with atomic orbitals for the oxygen and the transition metal. If they were all occupied, we would have full oxygen shells and ten electrons on the transition metal. The same is true after the unitary transformation. Then we empty the six π-oriented d states, leaving the oxygen levels unaltered since we neglect $V_{pd\pi}$ here. Finally, we empty the remaining d states and their admixed (in an antibonding relationship) sp hybrids from the oxygen. These states are like "ghosts" of the d states; they dominate the bonding properties and are called *perovskite ghosts*. The average energy of the ghosts and all oxygen s and p states remains unaltered at the average of the atomic energies, since the transformation by which they were constructed is unitary. Thus an increase in the energy of the ghosts is identical to the lowering of the total energy of all the occupied states, so the bonding energies can be calculated by considering the ghost alone. Furthermore, the real charge densities can be obtained from those corresponding to full shells by correcting for the charge density of the ghost, by the same unitarity argument.

To calculate the properties in terms of the perovskite ghosts, we will need a covalent energy, equal to the matrix element between a σ-oriented d state and a neighboring sp hybrid,

$$W_2 = -(V_{pd\sigma} + V_{sd\sigma})\sqrt{2}/2, \tag{19-14}$$

and a polar energy defined in terms of the difference,

$$W_3 = \tfrac{1}{2}[\varepsilon_d - (\varepsilon_s + \varepsilon_p)/2]. \tag{19-15}$$

These are listed for the four materials we consider in Table 19-4.

The admixture of each oxygen hybrid state in the ghost corresponds to $W_2^2/(2W_3)^2$—this is the fraction of each electron that resides on each of the six neighboring oxygens; from lowest-order perturbation theory, it is 0.09 for $SrTiO_3$. This is small enough to justify the use of perturbation theory, though with this simple system it is not difficult to do better. We shall see that the coupling between the d state and the sp hybrids raises the ghost energy 3 eV above the d-state energy. The matrix element between the ghost and a neighboring oxygen hybrid (or bonding state) is $W_1 W_2/(2W_3)$, with W_1 equal to half the oxygen sp splitting, giving a shift in energy of the ghost, of $W_1^2 W_2^2/(2W_3)^3$ or 0.44 eV in $SrTiO_3$. We neglect this shift in treating each ghost like a bonding unit, that is, as independent. This is analogous to the use of the Bond Orbital Approximation in the covalent solids, and is the only approximation we shall make here.

Using the ghost in calculations of properties is slightly tricky because a d state that is σ-oriented with respect to one oxygen will also have matrix elements with other oxygens. These matrix elements will be sorted out in the next section, but here, we shall extract two interesting results which are obtainable if we neglect these smaller matrix elements.

TABLE 19-4
Matrix elements associated with the σ-oriented d states and oxygen sp hybrids that make up the perovskite ghost.

Matrix element	$SrTiO_3$	$KTaO_3$	$KMoO_3$	ReO_3
W_2	3.40	4.57	4.08	5.00
W_3	5.71	5.82	5.75	4.49
Z^*	1.65	1.38	1.50	0.76
δZ_π^*	-0.68	-0.90	-0.65	—

NOTE: The intra-atomic matrix element coupling two oxygen hybrids, $W_1 = (\varepsilon_p - \varepsilon_s)/2$, is 7.5 eV; Z^* is the effective charge of oxygen, as reduced from 2 by the perovskite ghost. There is an additional transfer of δZ_π^* from the π-oriented states.

The Effective Charge

The ghost removes charge from each oxygen neighbor, thus modifying its effective charge. The modification is most simply calculated by returning to one oxygen hybrid and calculating the charge that is transferred to the transition-metal d state to which it is most strongly coupled—$[W_2/(2W_3)]^2$ electrons of each spin. The same transfer occurs from the other hybrid, giving an effective charge for oxygen of

$$Z^* = 2 - W_2^2/W_3^2. \tag{19-16}$$

These effective charges are listed in Table 19-4. These numbers include only the effect of the ghost; there is also transfer from the π-oriented p states. That transfer is also readily calculated in perturbation theory for the four π electrons that are coupled with their two neighbors,

$$\delta Z_\pi^* = -8[V_{pd\pi}/(\varepsilon_d - \varepsilon_p)]^2 \qquad \text{for the insulating case.} \tag{19-17}$$

In the metallic case, one of the d states is occupied, reducing $Z^* + \delta Z_\pi^*$ by a factor of 5/6. These shifts reduce the oxygen charge to nearly 1, as seen in Table 19-4, and as in the mixed tetrahedral solids. The particular case of ReO_3 gives a value for δZ_π^* that is greater than 7, and the perturbation theory is not meaningful. These calculations neglect any transfer to Sr (or K). This is presumably appropriate since, as we have indicated, the corresponding states seem not to be important to the electronic structure—they are simply absorbed in the vacuum (or ionized) states. Thus Z^* for $BaTiO_3$ should be the same as that for $SrTiO_3$.

Calculation of the meaningful physical charge e_T^* requires use of the dependence of the matrix elements upon bond length, the $d^{-7/2}$ dependence given in Eq. (19-11). The value of e_T^* is estimated with this form for $SrTiO_3$ in Problem 19-2.

It would also be possible to estimate the susceptibility in terms of these parameters as we did for the simple ionic solids in Section 14-B. However, the rather large coupling in comparison to the band gap noted in our discussion of Z^* here would suggest that, as in strongly coupled simple ionic crystals, the results might not be very accurate.

Energy Levels

Finally, it is of interest to relate the levels obtained in terms of this simple bonding unit of transition-metal d states and six oxygen hybrids to the bands we discussed earlier. Three of the d states have no σ matrix elements with the hybrids and we are dropping π matrix elements; these three correspond to the triply degenerate state Γ_{25}' at the energy ε_d. The other two d states are equally shifted upward. Again, it is easier to equate twice this shift to six times the shift $W_2^2/(2W_3)$ of the oxygen hybrids. This leads to the energy of the ghost states,

$$\varepsilon_{gh} = \varepsilon_d + 3W_2^2/(2W_3). \qquad (19\text{-}18)$$

For $SrTiO_3$ this gives two levels 3.04 eV above the d energy, or at -7.18 eV, which we see from Fig. 19-7 is roughly in accord with the two bands terminating at Γ_{12}. The corresponding oxygen hybrid levels and these d states are like the bond orbitals in the covalent solids. Their energies determine the bonding properties, though in the real solid, these sharp levels are broadened out into bands by other matrix elements. In particular, the matrix element W_1 splits the hybrid level into two sets of bands separated by nearly 15 eV.

19-F   The Chemical Grip

We turn finally to the total energy in the transition-metal compounds. As in the ionic solids, we expect the radial forces and cohesion to be dominated by the overlap interaction between ions and the Madelung energy. The overlap interaction will not be reconsidered, and discussion of the Madelung energy will be postponed to Section 19-G. We turn to the angle-dependent energy, the chemical grip, which we mentioned in the discussion of ionic crystals but did not treat in detail. The *chemical grip* is the effect of *inter*atomic matrix elements between the upper valence states and the lower conduction states. It should not be confused with the coupling between occupied d states and empty s states on the *same* atom, which led to ion distortion in the noble-metal halides (Section 8-B). A consideration of the bands is required in each system to see which matrix elements are important.

We wish to proceed by perturbation theory, but we shall see in a moment that angular forces do not arise in second-order perturbation theory, and we must extend the result, Eq. (1-14), to fourth order in the perturbation. Such derivations are very tricky. Let us state the problem explicitly, give the result, and then outline a derivation that leads to the correct answer. We are interested in two sets of orbitals: the first set (we shall call it the lower set) is made up of, for example, the p states on the oxygen ions in $SrTiO_3$, with diagonal matrix elements $-W_3$, with no off-diagonal matrix elements between them. We indicate them with indices α or β. The second set of orbitals (the upper set), for example, d states on the Ti, has diagonal matrix elements $+W_3$, which we index γ or δ, and which are coupled to the lower set of orbitals. The eigenvalue equations, corresponding to Eq. (1-10) or (1-26), take the form

$$\left.\begin{array}{l} -W_3 u_\alpha + \sum_\gamma W_{\alpha\gamma} u_\gamma = E u_\alpha; \\[2mm] \sum_\beta W_{\delta\beta} u_\beta + W_3 u_\delta = E u_\delta. \end{array}\right\} \qquad (19\text{-}19)$$

The sum of the n_1 energies of the lower set can then be evaluated with the formula

$$\sum_i E_i = -n_1 W_3 - \frac{1}{2W_3} \sum_{\alpha\gamma} W_{\alpha\gamma} W_{\gamma\alpha} + \frac{1}{(2W_3)^3} \sum_{\alpha\beta\gamma\delta} W_{\alpha\gamma} W_{\gamma\beta} W_{\beta\delta} W_{\delta\alpha}. \tag{19-20}$$

One way to derive this is to eliminate the coefficients for the upper set between Eqs. (19-19). This leads to an eigenvalue equation for eigenvalues E^2 of the matrix $\delta_{\alpha\beta} W_3^2 + \Sigma_\gamma W_{\alpha\gamma} W_{\gamma\beta}$. The sum of eigenvalues (sum of E^2) over this lower set is exactly equal to the trace of this matrix. Similarly, the sum of E^4 over the set is exactly equal to the trace of the squared matrix. One can also write these sums as sums over $E_i = -W_3 + \Delta_i$, noticing that Δ_i has second-order and fourth-order terms, and solve for the sum of Eq. (19-20). It is also readily confirmed that this is correct to fourth order for the special case of only two coupled states by expanding the exact solution, $\pm\sqrt{W_{12}^2 + W_3^2}$, in W_{12}.

Second-Order Terms

We first show that the second-order term in Eq. (19-20) does not lead to angular forces. We focus on a particular titanium ion. For each of the d states, γ, we must sum over the neighboring oxygen ions, but these are independent sums. We may have initially constructed our d states as listed in Table 19-2, with respect to some laboratory coordinate system, but we can reexpress them in terms of d states defined with respect to a coordinate system with z-axis in the direction of the oxygen ion being considered. The details are not crucial here but will be when we go to fourth order, so the point should be stated carefully. (More detail is given by Rose, 1957, or Weissbluth, 1978.)

Let us consider two coordinate systems (with the same origin) with angles θ_1, φ_1 measured with respect to the first and angles θ_2, φ_2 measured with respect to the second. The two coordinate systems are related by three Euler angles, a, b, and c. These do not need to be specified in detail here, except to note that b is the angle between the two z-axes. It is convenient to express the atomic states in terms of spherical harmonics as in Eq. (1-19) at this stage. Then a spherical harmonic $Y_l(\theta_2, \varphi_2)$ in the second set of axes can be written as a sum of spherical harmonics in the first:

$$Y_l^m(\theta_2, \varphi_2) = \sum_{m'} D_{mm'}^l(a, b, c) Y_l^{m'}(\theta_1, \varphi_1). \tag{19-21}$$

The set of coefficients $D_{mm'}$ is a unitary matrix (since within each set the spherical harmonics are orthogonal to each other). Notice, for example, that $l = 2$ states are expanded only in $l = 2$ states.

Let us now focus on the sum over γ in the second-order term of Eq. (19-20). We take a coordinate system with coordinates θ_2 and φ_2, in which the states γ are expressed; these states can be numbered m, with m taking $2l + 1$ values. Let the state $|\alpha\rangle$ lie on a particular oxygen neighbor, specifically, let it be either σ-oriented or π-oriented with respect to the titanium–oxygen separation. (Of most interest will be σ-oriented sp hybrids.) In general, it could have matrix elements with each of the $2l + 1$ states $|m\rangle$, but if we were to expand each of these states in states $Y_l^m(\theta_1, \varphi_1)$ defined in terms of a coordinate system with z-axis along the titanium–oxygen separation, only the state $m' = 0$ would have a nonvanishing matrix element with the σ-oriented p state, and the determination also becomes simple for the π-oriented state. Let us write

$$W_{\alpha\gamma} = \langle \alpha | H | m \rangle = \sum_{m'} \langle \alpha | H | m' \rangle D_{mm'}^l(a, b, c), \tag{19-22}$$

and

$$W_{\gamma\alpha} = \sum_{m''} D_{m''m'}^{*l}(a, b, c) \langle m'' | H | \alpha \rangle. \tag{19-23}$$

Then, in $\Sigma_\gamma W_{\alpha\gamma} W_{\gamma\alpha}$, the sum over γ is a sum over m, and because of unitarity,

$$\sum_m D_{m''m}^{*l}(a, b, c) D_{mm'}^l(a, b, c) = \delta_{m''m'}. \tag{19-24}$$

Thus

$$\sum_\gamma W_{\alpha\gamma} W_{\gamma\alpha} = \sum_{m'} \langle \alpha | H | m' \rangle \langle m' | H | \alpha \rangle, \tag{19-25}$$

with the m' states constructed with respect to a z-axis along the interion distance. This completes the proof for the second-order terms, since the result for each oxygen does not depend upon the starting coordinate system nor, therefore, on the orientations of any of the other oxygens. The sum in Eq. (19-25) is, in fact, simply W_2 for a σ-oriented oxygen sp state and $V_{pd\pi}$ for a π-oriented oxygen p state. In second order, the titanium "looks" spherically symmetric to the oxygen, there are no angular forces, and this term could be included with the overlap interaction. The result depended only upon summing over the entire shell with the same W_3 appropriate to each state.

Fourth-Order Terms

For those fourth-order terms where both the state $|\alpha\rangle$ and $|\beta\rangle$ lie on the same oxygen neighbor, we can make the same argument (once for the sum over γ and once for the sum over δ), and there are again no angular forces. However, for terms in which $|\alpha\rangle$ and $|\beta\rangle$ are on different oxygen atoms, the fourth-order sum does not simplify. In the sum $\Sigma_\gamma W_{\alpha\gamma} W_{\gamma\beta}$, we may choose our coordinate system along the vector to the α atom and only a single matrix element $W_{\alpha\gamma}$ will contribute, but we are left with a factor $D_{mm'}^l(a, b, c)$, where b is the angle between the two oxygens. The m and m' are the quantum numbers for those titanium d states with which each oxygen orbital is coupled ($m = 0$ for a σ-oriented oxygen state and $m = \pm 1$ for a π-oriented oxygen state).

The simplest case, and the only one that will be carried out completely here, is for terms in which both the states $|\alpha\rangle$ and $|\beta\rangle$ are σ oriented. Then only the d states for m and $m' = 0$ have nonvanishing matrix elements. (The $m = 0$ state, with respect to the z-axis, is the state labelled Δ_1 in Table 19-2. The states labelled Δ_2 and Δ_2' are linear combinations of states of $m = \pm 2$, and the states labelled Δ_5 are linear combinations of states of $m = \pm 1$.) Then for two oxygens separated by an angle b, the contribution to the fourth-order term of Eq. (19-20) becomes

$$\frac{1}{(2W_3)^3} \sum_{\gamma\delta} W_{\alpha\gamma} W_{\gamma\beta} W_{\beta\delta} W_{\delta\alpha} = \frac{W_2^4}{(2W_3)^3} D_{00}^2(b)^* D_{00}^2(b), \qquad (19\text{-}26)$$

and the coefficient D_{00}^2 is a particularly simple one:

$$D_{00}^2(b) = \frac{3}{2} \cos^2 b - \frac{1}{2}. \qquad (19\text{-}27)$$

In just the same way, we could consider the case where $|\alpha\rangle$ was σ oriented but $|\beta\rangle$ was π oriented. The more general coefficient is almost as simple,

$$|D_{0m}^l(a, b, c)| = \left(\frac{4\pi}{2l+1}\right)^{1/2} Y_l^m(b, c) \qquad (19\text{-}28)$$

of which Eq. (19-27) is a special case. But notice that the product of matrix elements in this case is $W_2^2 V_{pd\pi}^2$, and since $V_{pd\pi} = 0.34 W_2$ for $SrTiO_3$, the contribution is only 11 percent as large. In this first treatment, we include only the σ-oriented p states. The argument for doing this is even stronger in sp-bonded materials where the ratio of matrix elements that are $pp\pi$ to those that are $pp\sigma$ is 0.25. In Table 20-1, we shall see how the complete set of matrix elements can be included.

Let us then isolate the angular terms in the energy from Eq. (19-20). We call all of these the chemical grip, but here we shall include only the σ-oriented contributions from among them. At each ion (here, a titanium) we construct vectors to each neighboring ion and write the angle between a pair as $\theta_{\alpha\beta}$ (which we called b before). Each pair enters twice, so we may write the result in the form

$$E_{\text{grip}} = \frac{2W_2^4}{|2W_3|^3} \sum_{\alpha>\beta} P_l(\cos\theta_{\alpha\beta})^2 \qquad (19\text{-}29)$$

where $P_l(\cos\theta)$ is the Legendre polynomial, given by $D_{00}^2(\theta)$ in Eq. (19-27) for the special case of d states. We shall wish to consider it also for p states: $P_1(\cos\theta) = \cos\theta$.

We have written absolute values on the W_3 to emphasize the fact that this expression is positive definite. It was based upon the coupling of two sets of states for which the fourth-order term was found to reduce the average separation in energy. (In contrast to second-order terms, the levels attract each other rather than repel each other.) Thus, when only the lower levels are occupied, and since the average energy of *all* levels is not changed by the unitary transformation that gives the solution, this term always raises the total energy.

Application to the Alkali Halides

Before making application of this formula to the perovskites, let us make a brief application to ionic crystals—we summarized the results of this application in Chapter 13—and to simple tetrahedral solids. In the alkali halides, we focus upon the occupied p states in the halogen ion and calculate the chemical grip associated with interaction of the halogen ion with the alkali s states. These are the same couplings that were included in the calculation of ion softening in Section 14-C. The coupling W_2 of Eq. (19-29) becomes the matrix element $V_{sp\sigma} = 1.84\ h^2/(md^2)$, and $2W_3$ is to be identified with the $E_g = 9.1\ h^2/(md^2)$ used in Table 14-2. Then (Eq. 19-29) becomes

$$E_{\text{grip}} = \frac{0.030\hbar^2}{md^2} \sum_{\alpha < \beta} \cos^2 \theta_{\alpha\beta}, \tag{19-30}$$

for alkali halides.

We now consider two independent shear distortions in the alkali halides—the same distortions we considered for the tetrahedral solids in Chapter 8. That, associated with c_{12}, corresponds to extension in the y-direction and contraction in the x-direction (see, for example, Fig. 8-1), but no change in angle between neighbors nearest to a halogen; thus Eq. (19-30) does not lead to angular contributions to c_{12}. The distortion ε_4 associated with c_{44}, on the other hand, corresponds (see, for example, Fig. 8-5) to the sliding of planes of constant z over each other in the y-direction. Thus, each neighbor in the z-direction from a halide changes its angle with the two neighbors in the y-direction by $\delta\theta_{\alpha\beta} = \pm\varepsilon_4$ radians.

Notice first that in NaCl, with the six neighbors lying in cube directions, the $\cos\theta_{\alpha\beta}$ for alkali atoms separated by 90° are zero, so any change in angle raises the energy, stiffening the lattice against that distortion; $\cos\theta_{\alpha\beta} = \delta\cos(\pi/2 \pm \varepsilon_4) = \delta\sin(\pm\varepsilon_4) \approx \pm\varepsilon_4$. In the distortion we are considering, there is no change in angle for those separated by 180°. Thus the change in E_{grip} for each halogen ion is, from Eq. (19-30),

$$\delta E_{\text{grip}} = \frac{0.12\hbar^2}{md^2} \varepsilon_4^2. \tag{19-31}$$

TABLE 19-5
Deviations $c_{44} - c_{12}$ predicted from Eq. (19-32). The values in parentheses are from direct subtraction of experimental values from Table 13-5. All values are in units of 10^{11} erg/cm^3.

	Li	Na	K	Rb
F	0.45	0.22	0.11	0.08
	(1.8)	(0.37)	(−0.21)	(−0.47)
Cl	0.13	0.08	0.05	0.04
	(0.18)	(0.02)	(−0.03)	(−0.14)
Br	0.09	0.06	0.04	0.03
	(0.06)	(−0.07)	(−0.04)	(−0.10)
I	0.06	0.04	0.03	0.02
	—	(−0.16)	(−0.08)	(−0.08)

We may multiply this by the density of halogen ions, $4/(2d)^3$ to obtain the energy density, equal to $(\delta c_{44}/2)\varepsilon_4^2$, with δc_{44} the angular contribution to c_{44} from the chemical grip:

$$\delta c_{44} = \frac{0.12 h^2}{m d^5} = \frac{14.6 \times 10^{11}}{d^5} \frac{\text{erg-Å}^5}{cm^3}. \tag{19-32}$$

This gives a deviation from the Cauchy relations.

We compare in Table 19-5 the values predicted from Eq. (19-32) with the value $c_{44} - c_{12}$ obtained from the experimental elastic constants (see Table 13-5). The comparison was discussed in Chapter 13. The agreement is only semiquantitative. Notice first that Eq. (19-32) can only give positive contributions to δc_{44} whereas, experimentally, the heavier alkalis have negative deviations. This means simply that there must be other contributions to angular forces. For example, ion A induces a dipole in ion B, which exerts a force on ion C; this is an angular force contributing to deviations from the Cauchy relations. Nonetheless, general magnitudes and the trends from material to material in the compounds involving Li and Na seem to be rather well given, particularly in view of uncertainty in the experimental numbers, which is reflected in their lack of smooth behavior. It is tempting to attribute the deviations among the compounds of K and Rb to the role of the unoccupied d states, since these two elements immediately precede the transition-metal series. We could incorporate the effect of the chemical grip arising from these states in the same manner as we will treat the effect of the empty d states in the perovskites, where it will be seen that this d-state grip does contribute negatively to c_{44}, as the data in Table 19-5 requires. However, we have not sought parameters to see if it is in quantitative agreement. One should also examine the effects of the polarizable ion mentioned here, but this has not yet been done either.

Application to Tetrahedral Solids

Let us also make an application of the chemical grip to polar covalent solids. This corresponds to beginning in the limit of unit polarity, $V_3/(V_2^2 + V_3^2)^{1/2} = 1$, and introducing the covalent energy as a perturbation. We could do this just as we did for the alkali halides, treating the grip associated with p states on the nonmetallic atom and perhaps also that associated with the ghost p states on the metallic one. However, it will be of more interest to proceed somewhat schematically in order to make identification with the theory given in Chapter 8. Let us think of the grip associated with p states on the nonmetallic atom and with s states, or sp hybrids, on the metallic atom, but write the interatomic matrix element V_2. (We also do not distinguish between V_2 and V_2^h.) This replaces W_2 in Eq. (19-29). The gap $2W_3$ can be written $2(V_2^2 + V_3^2)^{1/2}$ rather than $2V_3$, which differs only to higher order in V_2. Then Eq. (19-29) becomes

$$E_{\text{grip}} = \frac{2V_2^4}{8(V_2^2 + V_3^2)^{3/2}} \sum_{\alpha > \beta} \cos^2 \theta_{\alpha\beta}. \tag{19-33}$$

This may be expanded for small deviations $\delta_{\alpha\beta}$ from the tetrahedral angle (the cosine of which is $-1/3$). Linear terms cancel when summed, and we obtain

$$E_{\text{grip}} = \frac{7}{36} \frac{V_2^4}{(V_2^2 + V_3^2)^{3/2}} \sum_{\alpha > \beta} \delta_{\alpha\beta}^2. \tag{19-34}$$

This formula, however, has the same form—$(1/2)C_1\sum_{\alpha > \beta} \delta_{\alpha\beta}^2$—as the valence force field introduced in Eq. (8-18), except that there we associated contributions with angles both at the metallic and nonmetallic atoms. We must therefore divide this expression by two before equating it to $(1/2)C_1\sum_{\alpha > \beta} \delta_{\alpha\beta}^2$ if we wish to identify it with the C_1 given in Eq. (8-22); doing this, we rewrite Eq. (8-22) as

$$C_1 = 2\lambda V_2^h \alpha_c^{h3}/3 = \frac{2\lambda V_2^{h4}}{3(V_2^{h2} + V_3^{h2})^{3/2}}. \tag{19-35}$$

By comparison of Eqs. (19-34) and (19-35) we see that the calculation based upon the chemical grip corresponds to the form we obtained from consideration of bond orbitals for tetrahedral solids with λ equal to 7/24. In particular, we obtain the $V_2 \alpha_c^3$ dependence suggested in Eq. (8-15).

Because we have not distinguished between the different matrix elements that enter, the precise value of λ is not so relevant. Indeed, we indicated that the form based upon simple hybrid matrix elements, Eq. (19-35), is only an approximation to the results R. Sokel obtained in the full analysis. Even the identical forms obtained have been partly forced by the use of $(V_2^2 + V_3^2)^{3/2}$ in the denominator rather than simply V_3^3. Furthermore, the two calculations are really just different mathematical approximations to the same formulation of the total energy, so they *should* agree if both approximations are sound. It is gratifying, however, to obtain the V_2^4-dependence of Eq. (19-35), which did not appear in the Bond Orbital Approximation, but was required only in the more complete treatment by Sokel. Furthermore, it is gratifying that in the limit of high polarity, the chemical grip can also describe tetrahedral systems.

It is interesting to notice here that the only large discrepancies we found when calculating rigidities in the tetrahedral solids in Chapter 8 occurred in the noble-metal halides. It is tempting to associate these discrepancies with the d states on the noble metals and one might at first think that these could be described in terms of the chemical grip. The difficulty is that the d states are full, and so (in the sense of an ionic crystal) are the neighboring halide s and p states. Thus the lowering of one cancels against the raising of the other, and this is not an example of the chemical grip. Instead, the important terms in this case more approximately reflect ion distortion and were treated in Chapter 8. In all such analyses of the total energy, *the important first step is to decide which atomic orbitals are necessary to describe the electronic structure and to note which are occupied*; then one may be able to apply perturbation theory to calculate shifts in the total energy.

Application to the Perovskites

We return finally to the chemical grip based upon the d state in the perovskite structure. We combine Eqs. (19-27) and (19-29) and evaluate the expression for small deviations $\delta_{\alpha\beta}$ of $\theta_{\alpha\beta}$ from 90°;

$$E_{\text{grip}} = -\frac{3W_2^4}{8W_3^3} \sum_{\alpha>\beta} \delta_{\alpha\beta}^2. \qquad (19\text{-}36)$$

(A contribution twice this large for deviations $\delta_{\alpha\beta}$ from 180° will not enter here.)

Let us consider the distortion ε_4 (illustrated for another structure in Fig. 8-5) and the corresponding elastic constant c_{44}. This distortion changes four of the angles at each transition-metal ion by ε_4. The energy given in Eq. (19-36) may be divided by the volume per transition-metal ion $(2d)^3$ to obtain the energy density, equal to $(c_{44}/2)\varepsilon_4^2$, and the corresponding contribution to c_{44} may be evaluated. Using values from Tables 19-3 and 19-4, we obtain -0.6×10^{11} erg/cm^3 for SrTiO$_3$ and -1.7×10^{11} erg/cm^3 for KTaO$_3$. The chemical grip tends to reduce the angular rigidity of the cubic perovskites, just as it did for the heavy-alkali halides. To understand the stability of the cubic structure, we must return to the electrostatic energy. We shall do that in the following section, but discuss the results here.

The electrostatic contribution to the elastic constant for SrTiO$_3$ is estimated by considering point charges of $2e$ at Sr sites, $4e$ at Ti sites, and $-2e$ at oxygen sites. We will find that this approach leads to a contribution to c_{44} of 24.7×10^{11} erg/cm^3 for SrTiO$_3$ and 26.2×10^{11} erg/cm^3 for KTaO$_3$ (also 27.8 and 34.7 for KMoO$_3$ and ReO$_3$, respectively, in the same units). These values are considerably higher than the observed values near 5×10^{11} erg/cm^3. (Specific experimental values were not found for these but were estimated instead from similar materials.) The estimate made here of the reduction due to the chemical grip is not great enough, either. There may be some reduction due to ion softening, as suggested by the values for high-valence ionic compounds listed in Table 13-5. In addition, with a fourth-order term, a modification of W_2 by a factor of two would be sufficient to bring the values into agreement. It seems likely that the principal physical mechanisms have been given correctly but that the numerical estimates are quite inaccurate.

Structural Stability

In particular, two trends in stability appear to be understandable in terms of these results. The series $SrTiO_3$, $KTaO_3$, and WO_3 has a constant number of outer electrons, and we can see in Table 19-6 that the electrostatic angular rigidity constant increases by 20 percent through the series. Nonetheless, it is found experimentally that WO_3 is unstable against ε_4-like distortion. This can be understood in terms of an increasingly strong chemical grip; the estimate given two paragraphs earlier indicates growth by a factor of three already in going from $SrTiO_3$ to $KTaO_3$. Of course a more quantitative test should be carried out for WO_3 to confirm this conclusion.

The second trend concerns the addition of sodium (or other alkali metals) to WO_3, which restores the cubic symmetry. We see in Table 19-6 that even adding a sodium for every tungsten only decreases the electrostatic rigidity constant by 3 percent. It would appear to be the weakening of the destabilizing chemical grip that causes the return to cubic symmetry. The grip arises from the coupling between occupied oxygen states and empty transition-metal d states. By filling some of the d states, we reduce its effect, roughly in proportion to the fraction of

TABLE 19-6
Electrostatic rigidity constant α for the perovskite structure.

	$Z = 0$	$Z = 1$	$Z = 2$	$Z = 3$
$Q = 0$	-0.593	1.075	5.29	6.68
$Q = 1$	10.95	14.00	14.24	11.68
$Q = 2$	29.39	28.48	24.76	18.24

NOTE: The contribution to c_{44} is $\alpha e^2/a^4$, with a the cube edge; Z is the charge at the body-centered position (Sr in $SrTiO_3$); Q is the charge at the edge-centered position (O in $SrTiO_3$); and $6 - Z$ is the charge at the cube corner (Ti).

the states filled. Since the d states can accommodate ten electrons per primitive cell, the added electron reduces the strength of the grip some 10 percent, presumably enough that the electrostatic tendency toward a cubic structure dominates.

19-G   The Electrostatic Stability of Perovskites

The calculation of the angular stability of perovskite structures provides a good opportunity to see in detail how the electrostatic energies discussed in connection with both simple metals and ionic crystals are calculated. In Eq. (13-1) we gave the electrostatic energy for a crystal with two ions per primitive cell. Now let us rewrite the result for a general set of ions, each of charge Z_i, making up a crystal of N_c primitive cells:

$$E_{\text{electro}} = \frac{1}{2N_c} \sum_{i, j} Z_i Z_j e^2 / r_{ij}. \tag{19-37}$$

For $SrTiO_4$, there would be one Sr ion with $Z_i = 2$, one Ti ion with $Z_i = 4$, and three oxygens with $Z_i = -2$ in each primitive cell. We may use a trick to evaluate the change in this energy with distortion. We write the electrostatic potential due to the collections of ions as $\varphi(\mathbf{r}) = \Sigma_{\mathbf{q}'} \varphi_{\mathbf{q}'} e^{i\mathbf{q}' \cdot \mathbf{r}}$. The Fourier coefficients can readily be written explicitly by multiplying both sides by $e^{-i\mathbf{q} \cdot \mathbf{r}}$ and integrating

$$\varphi_{\mathbf{q}} = \frac{e}{\Omega} \sum_i \int \frac{e^{-i\mathbf{q} \cdot \mathbf{r}} Z_i}{|\mathbf{r} - \mathbf{r}_i|} d^3 r$$

$$= \frac{4\pi e}{q^2 \Omega_c} \frac{1}{N_c} \sum_i Z_i e^{-i\mathbf{q} \cdot \mathbf{r}_i}. \tag{19-38}$$

The sum over i will be identically zero unless $\mathbf{q}$ is a lattice wave number, in which case the sum over each cell is identical:

$$S_c(\mathbf{q}) = \sum_{i \text{ in cell}} Z_i e^{-i\mathbf{q} \cdot \mathbf{r}_i}. \tag{19-39}$$

We wish to obtain the electrostatic energy by multiplying each charge by the potential at that charge, dividing by two, and summing over all charges. The potential $\varphi(\mathbf{r})$ at each charge is infinite, owing to the charge itself, so the result would diverge. However, if we ask only for the change in electrostatic energy under distortion, these divergences cancel out and the correct result is obtained by ignoring them. This can be demonstrated using the Ewald-Fuchs method (see, for example, Harrison, 1966a, Chapter 5.) Thus we write the electrostatic energy per cell as

$$E_{electro} = 1/2 \frac{4\pi e^2}{\Omega_c} \sum_{\mathbf{q}} \frac{S_c^*(\mathbf{q})S_c(\mathbf{q})}{q^2} \tag{19-40}$$

Strictly speaking, a convergence factor of $e^{-q^2/4\eta}$ is required in the sum, but we shall perform the calculation in such a way that it does not affect the result (Harrison, 1966a). We may directly compute changes in this under any distortion of interest, and apply it to the special case of the perovskite structures. (It should be mentioned in passing that the results we obtain could be written in terms of the corresponding expressions for the face-centered cubic, body-centered cubic, and simple cubic structures, and expressions for the simple cubic structure could be written in terms of the face-centered cubic and rocksalt expressions, by suitable additions in wave number and real space; the approach is much the same as that by which the value for the diamond structure was obtained (Harrison, 1966a). However, it is a very tricky calculation.)

The lattice wave numbers for the perovskite structure, shown in Fig. 19-2, are given by

$$\mathbf{q} = (2\pi/a)[n_x, n_y, n_z], \tag{19-41}$$

where a is the cube edge, $a = 2d$, and the n_x, n_y, and n_z are integers. We write the charge on the oxygens as $Z_i = -2$, that on the Sr as Z, and that on the Ti as $6 - Z$, so that the structure factor can be evaluated for other perovskites; notice that we have maintained charge neutrality in the cell though this is not done in the simple metals. Then,

$$S_c(\mathbf{q}) = 6 - 2[(-1)^{n_x} + (-1)^{n_y} + (-1)^{n_z}] + Z[(-1)^{n_x + n_y + n_z} - 1]. \tag{19-42}$$

Further, $\Omega_c = a^3$ and $q^2 = (2\pi/a)^2(n_x^2 + n_y^2 + n_z^2)$.

Now, imagine a shear distortion such that an atom at $[x, y, z]$ is displaced in the y-direction by εx. (Symmetry appears to rule out internal displacements for this case.) The lattice wave numbers (Eqs. 16-11 and 16-16) may be recomputed for the distorted structure and are

$$\mathbf{q} = (2\pi/a)[n_x, n_y - n_x\varepsilon, n_z]. \tag{19-43}$$

Then $1/q^2$ can be expanded in ε, keeping the second-order term

$$\delta\left(\frac{1}{q^2}\right) = \left(\frac{2\pi}{a}\right)^{-2}\left(\frac{4n_x^2 n_y^2}{n^6} - \frac{n_y^2}{n^4}\right)\varepsilon^2, \tag{19-44}$$

with $n^2 = n_x^2 + n_y^2 + n_z^2$. The value of $S_c(\mathbf{q})$ is unaffected, so the second-order term in Eq. (19-40) can be directly evaluated. The change in energy in each cell under this distortion can be written in terms of the elastic constant, $\delta E = 1/2 c_{44} \varepsilon^2 a^3$, so that the sum gives us the electrostatic contribution to the shear constant:

$$ c_{44} = \frac{e^2}{\pi a^4} \sum_{\mathbf{n}} S_c^2 \left(\frac{2\pi \mathbf{n}}{a} \right) n_y^2 \left(\frac{4 n_x^2}{n^6} - \frac{1}{n^4} \right). \tag{19-45} $$

We perform the sum over n_z first, using the identity

$$ \sum_{n_z = -\infty}^{\infty} \frac{1}{a^2 + b^2 n_z^2} = \frac{\pi}{ab} \coth(\pi a/b) \tag{19-46} $$

(see Harrison, 1966a). In this sum, a^2 is $n_x^2 + n_y^2$ (and should not be confused with the cube edge a). The structure factor S_c is seen to alternate between even and odd n_z, so two sums over n_z are required, one with $b = 1$ and S_c^2 for odd n_z, and one with $b = 2$ and S_c^2 (even n_z) $- S_c^2$ (odd n_z) replacing the S_c^2. An expression for the sum over n^{-4} is obtained from Eq. (19-46) by differentiating with respect to a; an expression for the sum over n^{-6} is obtained from that by another differentiation with respect to a.

This is one of the few calculations in this text that is inconvenient unless one has access to a computer with greater versatility than a Hewlett-Packard HP25 has. However, the sum over n_z is easily programmed, as is the remaining sum over n_x and n_y. The convergence is better than one might at first think from looking at Eq. (19-45). The leading terms drop only as n^{-2} as we sum along n_y for small n_x; however, the convergence is rather good if we sum over n_x first (a few hundred values); then summing n_y to four appears to give results well within 1 percent (though summing to much larger n_y leads to errors).

A convenient way to present the results is in terms of a dimensionless rigidity constant α chosen such that $c_{44} = \alpha e^2 / a^4$, with $a = 2d$ the cube edge. (See Fig. 19-2.) This class of systems is sufficiently important that we list values for various parameters in Table 19-6. In terms of the ions shown in Fig. 19-2, the charge on the Ti is $6 - Z$, that on the Sr is Z, and that on the O is Q. Thus, for the four perovskites we have been discussing, $Q = 2$. Also, $Z = 2$ for $SrTiO_4$, $Z = 1$ for $KTaO_3$ and $KMoO_3$, and $Z = 0$ for ReO_3. The corresponding values of c_{44} were listed in Section 19-F.

As we indicated in the preceding section, these values are considerably larger than the experimental values, and part of the reduction comes from the chemical grip. A part must also come from ion softening, which reduces the effective charges on the ions, though this was not conspicuous in the alkali-halide elastic constants. We noted that the full integral charge enters the cohesive energy, since it contained the energy required to bring together widely spaced ions. Under distortion, use of the full charge assumes that the electronic charge moves rigidly with the ions; for example, O^{2-} moves as a unit. This becomes particularly questionable

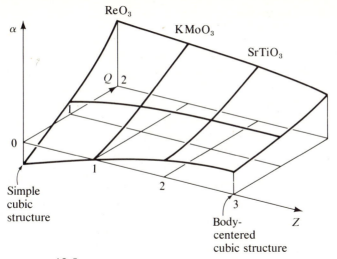

FIGURE 19-8

A fishnet plot of the electrostatic rigidity constant α for perovskite structures as a function of the charge Z on the body-centered ion and the charge Q on the edge-centered ion; the cube-corner ion would have charge $6 - Z$. Three compounds being discussed here are indicated, as well as the simple cubic and body-centered cubic structure, which can also be described this way.

when it corresponds to large charges on the ions Ti^{4+} or Re^{6+}. (Re is located in column $D7$ of the Solid State Table, but the additional electron is assumed to be in a conduction band concentrated on the Re ion.) We may well expect this to be the dominant correction.

Values for the rigidity constant are also of interest for other values of the parameters. By adding Na to WO_3, the value of Z can be continuously increased from zero to one, so a plot for nonintegral Z, as in Fig. 19-8, can be of interest. The charge Q also may be taken different from two, corresponding to deviations from neutrality, or, more realistically, ions in a compensating background electron gas. Thus $Z = 3$, $Q = 0$ corresponds to a body-centered cubic simple metal (trivalent) with a value in agreement with that given much earlier (Fuchs, 1936). The charges $Z = 0$, $Q = 0$ corresponds to a simple-cubic structure (hexavalent), and the negative value indicates that it is electrostatically unstable.

19-H The Electron–Phonon Interaction

It was noted in Section 19-D that in A15 compounds, the metallic bands may be thought of as being based entirely upon orbitals of the same symmetry. In such a one-band system or in a system we so approximate, with nearest-neighbor interactions, the electron–phonon interaction can be written in a very simple and informative way, following an analysis given by Barišić, Labbé, and Friedel (1970). We will also see that it leads (within a scale factor of order unity) to the pseudopotential form for coupling to long wavelength modes in the appropriate limit, suggesting it has rather general validity as an approximation.

At the same time we should recognize that the restriction to coupling within a single band may be very misleading. It would suggest, for example, that a dimensionless electron–phonon coupling constant λ, discussed in Section 17-D, could be computed for each set of bands with an equation such as Eq. (17-25) and that the constant could be very different for different bands. In fact, recent determinations of $\lambda(\mathbf{k})$ for niobium by Crabtree et al. (1979) show only rather minor variations over the Fermi surface, though band properties—particularly the electron velocity—vary widely. (See also Butler and Williams, 1978, for a related discussion of thermal conductivity, with a similar conclusion, for a variety of transition metals.) The approximation which leads us to significant errors is our assumption, at Eq. (19-50), that we need include only changes in the matrix element that result from changes in radial distance. Angular terms can readily be included and will couple different bands, contributing to the relative constancy of $\lambda(\mathbf{k})$ obtained in detailed calculations. We shall nevertheless proceed here with the simpler theory. This is the appropriate approach to the electron–phonon interaction in the LCAO context, and the generalization to include other terms is immediate.

In the one-orbital approximation, the electronic states in the undistorted crystal can be written in the LCAO form

$$|\mathbf{k}\rangle = N_a^{-1/2} \sum_j e^{i\mathbf{k} \cdot \mathbf{r}_j} |j\rangle, \tag{19-47}$$

with an energy given, in analogy with Eq. (2-5), by

$$E(\mathbf{k}) = \varepsilon_d + \sum_j H_{ij} e^{i\mathbf{k} \cdot \mathbf{d}_j}. \tag{19-48}$$

The H_{ij} are, of course, interatomic matrix elements between some state i and its neighbor a distance $\mathbf{d}_j$ away. It will be useful to note also that the electron velocity (Eq. 2-9) is given by

$$\mathbf{v_k} = \frac{1}{\hbar} \frac{dE(\mathbf{k})}{d\mathbf{k}} = \frac{i}{\hbar} \sum_j H_{ij} \mathbf{d}_j e^{i\mathbf{k} \cdot \mathbf{d}_j}. \tag{19-49}$$

If we deform the lattice, as for a lattice vibration, then so long as the H_{ij} did not change, Eq. (19-47) would still describe an eigenstate (with the same values of $\mathbf{k} \cdot \mathbf{r}_j$) with the same energy, and there would be no coupling between different states. It is therefore the *change* in the H_{ij} that produces the electron–phonon interaction in this representation. The only change we include is the change with internuclear distance,

$$\frac{\partial H_{ij}}{\partial d} = \frac{\eta H_{ij}}{d}. \tag{19-50}$$

We shall see that for d bands, $\eta = -5$ is appropriate. We have included only nearest neighbors, so the magnitude of $\mathbf{d}_j$ has been written as d.

Let us then introduce a lattice distortion

$$\delta\mathbf{r}_j = \mathbf{u}e^{i\mathbf{Q}\cdot\mathbf{r}_j}. \tag{19-51}$$

This distortion does not introduce matrix elements with other orbitals, so the only interaction is with other states based upon the same orbitals. (This would no longer be true if we took a more general change in H_{ij} than Eq. (19-50).) We wish to evaluate the matrix element between two such states, to first order in $\mathbf{u}$. Thus we write

$$H_{ij} \rightarrow H_{ij} + \frac{\eta H_{ij}}{d^2}\mathbf{d}_j \cdot (\delta\mathbf{r}_j - \delta\mathbf{r}_i) = H_{ij} + \frac{\eta H_{ij}}{d^2}\mathbf{u} \cdot \mathbf{d}(e^{i\mathbf{Q}\cdot\mathbf{r}_j} - e^{i\mathbf{Q}\cdot\mathbf{r}_i}). \tag{19-52}$$

Only the second term gives matrix elements between states of different wave number, $\mathbf{k}$ and $\mathbf{k}'$, and we have

$$\langle\mathbf{k}'|H|\mathbf{k}\rangle = \sum_{i,j}\frac{\eta H_{ij}}{N_a d^2}\mathbf{u} \cdot \mathbf{d}(e^{i\mathbf{Q}\cdot\mathbf{r}_j} - e^{i\mathbf{Q}\cdot\mathbf{r}_i})e^{-i\mathbf{k}'\cdot\mathbf{r}_i}e^{i\mathbf{k}\cdot\mathbf{r}_j}. \tag{19-53}$$

To evaluate the sum over the first term in parentheses, we write $e^{-i\mathbf{k}'\cdot\mathbf{r}_i}$ as $e^{-i\mathbf{k}'\cdot\mathbf{r}_j}e^{i\mathbf{k}'\cdot\mathbf{d}_j}$ and sum over $\mathbf{r}_j$, holding $\mathbf{d}_j$ fixed. The sum is identically zero unless $\mathbf{k} + \mathbf{Q} - \mathbf{k}'$ is zero or is a lattice wave number. In this representation, there is no distinction between normal and umklapp scattering. If $\mathbf{k}' = \mathbf{k} + \mathbf{Q}$, the sum is N_a and the remaining sum has become a sum over $\mathbf{d}_j$. Similarly, the sum over the second term in parentheses leads to a factor $N_a e^{-i\mathbf{k}'\cdot\mathbf{d}_j}$, and Eq. (19-53) becomes

$$\langle\mathbf{k}'|H|\mathbf{k}\rangle = \frac{\eta}{d^2}\mathbf{u} \cdot \sum_j H_{ij}\mathbf{d}_j(e^{i\mathbf{k}'\cdot\mathbf{d}_j} - e^{-i\mathbf{k}\cdot\mathbf{d}_j}). \tag{19-54}$$

We may now make use of the Eq. (19-49) to write this result in terms of the velocities,

$$\langle\mathbf{k}'|H|\mathbf{k}\rangle = +\frac{\eta\hbar}{d^2}i\mathbf{u} \cdot (\mathbf{v}_\mathbf{k} - \mathbf{v}_{\mathbf{k}'}), \tag{19-55}$$

which is essentially the form of the result found by Varma and Weber (1977). A very important feature of this expression noted by Varma and Weber is that the coupling is large where the electron velocities are large and therefore the densities of states are small. The matrix element enters squared for many properties and enters the density of states only linearly, so that the important parts of the electronic structure may be just the opposite of what one would guess if judging the electronic structure in terms of the density of states alone.

Finally, it is interesting to relate the Varma-Weber result to that obtained from pseudopotential theory in Chapter 17. In pseudopotential theory, $\mathbf{v_k} - \mathbf{v_{k'}}$ becomes $h(\mathbf{k} - \mathbf{k'})/m = -h\mathbf{Q}/m$, for normal scattering. Then the electron–phonon interaction becomes

$$\langle \mathbf{k'} | H | \mathbf{k} \rangle \rightarrow -i\mathbf{Q} \cdot \mathbf{u}\eta h^2/(md^2). \tag{19-56}$$

We saw in Section 17-D that pseudopotential theory led to $-i\mathbf{Q} \cdot \mathbf{u}w_Q$ for this matrix element, and we saw that w_Q approached $-2E_F/3$ at long wavelengths.

Remarkably enough, we find the same dependence upon d^{-2} even for d states where the interatomic matrix elements vary as d^{-5}. Furthermore, with η negative, we obtain the same sign and general order of magnitude for the electron–phonon interaction. Thus we gain some confidence in using this form, or generalizations of it, in a variety of circumstances.

PROBLEM 19-1   *Perovskite bands*

The energy bands of symmetry Δ_5 (propagation along the z axis) in $SrTiO_3$ are doubly degenerate. They may be calculated much as the Δ_1 bands were obtained in this chapter. One set may be constructed from the d states of symmetry zx/r^2 and the three oxygen p_x states. All other orbitals are ruled out by symmetry, such as the condition that the wave function becomes negative under reflection in the yz plane. (Only this set of four orbitals need be considered; the second set of identical bands is obtained with orbitals yz/r^2 and p_y.) Construct the appropriate figure, corresponding to Fig. 19-5, for the orbitals zx/r^2 and the oxygen p_x states. Construct the four-by-four Hamiltonian matrix in analogy with Eq. (19-3) and verify the solutions at Γ and X given in Eqs. (19-8) and (19-9).

PROBLEM 19-2   *Transverse charge in the perovskites*

In Section 19-E we found contributions to the effective charge Z^* of the oxygen due to the matrix element W_2 between an sp hybrid and a d state and due to the matrix element $V_{pd\pi}$. For an oxygen displacement perpendicular to the Ti—O axis, the predicted transverse charge is simply this Z^*. However, for displacements along the Ti—O axis there are contributions from the transfer of charge between ions just as there were in the alkali halides, for which we obtained the transverse charge in Section 14-C.

Evaluate these contributions and the total transverse charge for oxygen displacements along the Ti—O axis in $SrTiO_3$ by using the relation

$$\frac{d}{W_2}\frac{\partial W_2}{\partial d} = \frac{d}{V_{pd\pi}}\frac{\partial V_{pd\pi}}{\partial d} = \eta$$

with $\eta = -7/2$, which follows directly from Eq. (19-11).

Indeed, very large couplings do occur in $SrTiO_3$ (see Bäuerle, 1974, and references contained therein), but a quantitative experimental estimate of the transverse charge is not available for comparison with this theoretical estimate.

PROBLEM 19-3   *The chemical grip*

Use Eq. (19-30) to estimate any contribution of the chemical grip to the $k = 0$ optical mode frequency in NaCl.

CHAPTER 20

Transition Metals

SUMMARY

Energy bands for the transition metals are constructed, using a minimal basis set of atomic orbitals. The eleven parameters required are reduced to two, the d-band width W_d and its position E_d relative to the s-band minimum, using Muffin-Tin Orbital theory. Relations giving W_d and all interatomic matrix elements in terms of a d-state radius r_d and the internuclear distance are listed in the Solid State Table, along with values of r_d and E_d for all of the transition elements; this makes possible elementary calculations of the bands for any transition metal, at any atomic volume.

The nature of the Fermi surfaces arising from such bands is illustrated for body-centered cubic chromium; the nesting surfaces that give rise to antiferromagnetism in this metal are evident. The density of states is also shown for this system and for a close-packed structure, and provides the basis for understanding the variation of properties with the electron-to-atom ratio. This density of states is then approximated by the Friedel model, the sum of a free-electron density of states and a uniform d-band density of states $10/W_d$. The occupation of only the lower part of the d bands gives the high cohesive energy at the center of each transition-metal series; the variation of the band width with the inverse fifth power of the internuclear distance causes the high density of the same metals.

The electronic structure is reformulated in terms of free electrons and a d resonance in order to relate the band width W_d to the resonance width Γ, and is then reformulated again in terms of transition-metal pseudopotential theory, in which the hybridization between the free-electron states and the d state is treated in perturbation theory. The pseudopotential theory provides both a definition of the d-state radius and a derivation of all interatomic matrix elements and the free-electron effective mass in terms of it. Thus it provides all of the parameters for the LCAO theory, as well as a means of direct calculation of many properties, as was possible in the simple metals.

Ferromagnetism in the transition metals is treated first by using the Friedel model of the density of states to derive a condition for ferromagnetism. That condition is satisfied only for cobalt and nickel; failure to predict ferromagnetism in manganese and iron is attributed to deviations from the Friedel model. Second, the formation of local moments on individual ions is treated in terms of d-state resonances, allowing a description of ferromagnetic metals at temperatures above the Curie temperature as well as of antiferromagnetic metals.

In this chapter we turn finally to transition metals themselves. We discussed at the beginning of Chapter 19 the general effects that partially filled d shells and f shells have upon metals. We mentioned in particular the two distinct types of behavior that are associated with localized electrons (as in the monoxides) and itinerant electrons (as in $SrTiO_3$). The local view is appropriate to the rare earths and the heavier actinides. In these cases, the f shells are directly added to an otherwise simple metal and the properties follow rather directly from the approaches used in the preceding few chapters.

In all of the transition metals—the d shell series to the left in the Solid State Table—the itinerant picture is appropriate and the effects of the corresponding d bands require further attention. It should be mentioned at the outset that though the itinerant picture is appropriate for all of the transition metals, there is a complication analogous to the multiplet states that will be discussed in Section 20-F. This is the formation of local moments, and it is well described—as excitons were—as a correction to the band picture.

20-A   The Bands

The energy bands have been calculated for the $3d$ transition-metal series by Mattheiss (1964) and are shown in Fig. 20-1. Calculations for the $4d$ series have been made by Morruzi, Janak, and Williams (1978). Obviously, there is a difficulty in studying trends through the series because of the appearance of different crystal structures. (The structures are indicated above each panel; hcp is hexagonal close-packed; fcc is face-centered cubic; and bcc is body-centered cubic.) The difficulty is avoided by looking at the set V, Cr, Fe (Ti is also bcc at high temperature), and the set Co, Ni, and Cu, each of which has its own structure. In each case, there is a set of five bands that we associate with the d states, crossed by a free-electron-like band. We shall see that it is indeed convenient to think of these as d resonances in a free-electron gas for the purposes of calculating some properties, though the form of the bands is easily understood in terms of LCAO theory.

It is apparent in Figure 20-1 that there is a decrease in the energy of the d bands relative to the level Γ_1, as we move through the series, and a slight narrowing of the d bands. The shift is qualitatively in accord with the shift in relative energies of the atomic states through the series shown in Fig. 1-8. The extent to which the bands are filled—the Fermi energy is indicated by a dashed line in Fig. 20-1—increases to the right also. In the next element after copper—zinc—the d bands

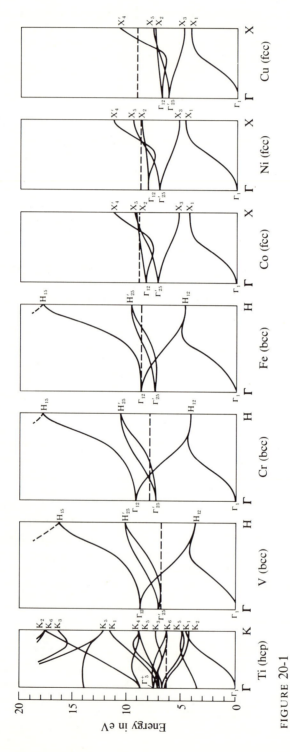

FIGURE 20-1

Energy bands for the transition-metal series Ti, V, Cr, (Mn, with a complex structure, is omitted), Fe, Co, Ni, and Cu, as a function of wave number along a symmetry line in the appropriate Brillouin Zone. For the face-centered cubic (fcc) and body-centered cubic (bcc) structures, the symmetry line is in a [100] direction; for the hexagonal close-packed (hcp) structure, it is parallel to a nearest-neighbor distance in the basal plane. Dashed lines indicate estimated bands. [After Mattheiss, 1964.]

drop below Γ_1 and rapidly become corelike as we continue to increase the atomic number.

The bands have long been interpreted as LCAO d bands, crossed by and hybridized with a free-electron-like band (Saffren, 1960; Hodges and Ehrenreich, 1965; and Mueller, 1967). By including some eleven parameters (pseudopotential matrix elements, interatomic matrix elements, orthogonality corrections, and hybridization parameters), it is possible to reproduce the known bands very accurately. We shall also make an LCAO analysis of the bands but shall take advantage of recent theoretical developments to reduce the number of independent parameters to two for each metal, each of which can be obtained for any metal from the Solid State Table. These two parameters will also provide the basis for understanding a variety of properties of the transition metals.

The LCAO Formulation

An LCAO description of the electronic structure requires at least the minimal basis set (all orbitals that may be occupied in the ground state of the atom) of five d states per atom and the s state. Consideration of the bands from Fig. 20-1 indicates that in fact the highest-energy states shown (for examples, H_{15} and X_4') have p-like symmetry, and we shall not reproduce this with our minimal set, but the bands at this energy are unoccupied in any case and it will be of little consequence. For constructing bands in solids, the angular forms for the d states in terms of cartesian coordinates, shown in Eq. (1-21), are most convenient. Here we shall carry out the calculation explicitly for chromium, in the body-centered cubic structure; it is carried out for the face-centered cubic structure in Problem 20-1. (A similar treatment of the hexagonal close-packed structure has been given recently by Bertoni, Bisi, and Manghi, 1978.) In the cubic structures, the axes upon which the angular forms are based are taken along the directions of cube edges.

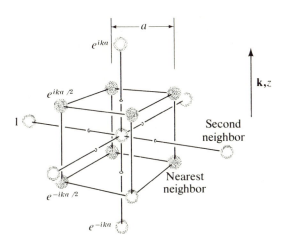

FIGURE 20-2

The body-centered cubic structure. The central atom sits at the center of a cube formed by its eight nearest neighbors, shaded to distinguish them, though every atom and its environment (in the extended crystal) is identical. The six second neighbors lie a distance 15 percent further away. We construct a Bloch sum with wave number in the z-direction, giving phase factors shown for atoms in each plane of constant z.

We shall also limit our calculation to wave numbers along such a direction, the z-axis.

The body-centered cubic structure of chromium is the same as the cesium chloride structure illustrated in Fig. 2-1,a, with both Cs and Cl ions replaced by Cr atoms; it is redrawn in Fig. 20-2, with the z-axis being considered vertical. For each of the six orbitals ($\alpha = 1, 2, \ldots, 6$) on each atom, we construct a Bloch sum over the N_a atoms in the system, in analogy with Eq. (3-19):

$$\chi_\alpha(\mathbf{k}) = \sum_i e^{i\mathbf{k}\cdot\mathbf{r}_i}|i, \alpha\rangle/N_a^{1/2}. \tag{20-1}$$

The phase factors for a few atomic planes (planes of constant z) are shown in the figure. We must construct a Hamiltonian matrix based upon these six Bloch states for each wave number $\mathbf{k}$. The corresponding matrix elements are

$$\begin{aligned}
H_{\alpha\beta}(\mathbf{k}) &= \sum_{i,j} e^{i\mathbf{k}\cdot(\mathbf{r}_i-\mathbf{r}_j)}\langle j, \alpha|H|i, \beta\rangle/N_a \\
&= \sum_i e^{i\mathbf{k}\cdot\mathbf{r}_i}\langle 0, \alpha|H|i, \beta\rangle,
\end{aligned} \tag{20-2}$$

where we have obtained the final form by noting that in this structure, with one atom per primitive cell, the result of a sum over all neighbors i to each j gives the same result; thus it is evaluated for $j = 0$ at $\mathbf{r}_j = 0$ and multiplied by N_a. The bands are obtained from solution of the six simultaneous equations,

$$\sum_\beta H_{\alpha\beta}(\mathbf{k})u_\beta = E_k u_\alpha. \tag{20-3}$$

The Interatomic Matrix Elements

We need finally the atomic matrix elements $\langle 0, \alpha|H|i, \beta\rangle$, the form of which has been given by Slater and Koster (1954). The vector $\mathbf{r}_i$ is written $(l\hat{\mathbf{x}} + m\hat{\mathbf{y}} + n\hat{\mathbf{z}})d$, with $\hat{\mathbf{x}}, \hat{\mathbf{y}}$, and $\hat{\mathbf{z}}$ unit vectors along the cube axes; that is, l, m, and n are the direction cosines of the vector from the left state to the right state. Then the matrix elements are written as an E, and the states represented by their angular form are written as subscripts, the first for the left state (α), and the second for the right state (β); for example, the symbol $E_{xy, yz}$ represents $\langle 0, \alpha|H|i, \beta\rangle$, with α of symmetry xy and β of symmetry yz. It depends on the vector $\mathbf{r}_i - \mathbf{r}_0 = \mathbf{r}_i$ and therefore upon l, m, and n. Slater and Koster have given these matrix elements in terms of the values for $V_{dd\sigma}$, $V_{dd\pi}$, and so on, which we illustrated in Fig. 19-4. These depend upon d, and estimates for them are given in the Solid State Table. We shall return to the origin of these expressions later. The expressions used by Slater and Koster, which they derived from the transformation given at Eq. (19-21), are listed in Table 20-1. We have reproduced the full table, including matrix elements involving s and p states, for completeness. We now have everything required to obtain the Hamiltonian matrix and, therefore, the bands.

TABLE 20-1 481

The Slater and Koster (1954) tables of interatomic matrix elements as functions of the direction cosines, l, m, and n, of the vector from the left state to the right state. Other matrix elements are found by permuting indices. General formulae for these expressions and explicit expressions involving f and g orbitals have been given recently by Sharma (1979).

$E_{s,s} = \quad V_{ss\sigma}$

$E_{s,x} = \quad l V_{sp\sigma}$

$E_{x,x} = \quad l^2 V_{pp\sigma} + (1 - l^2) V_{pp\pi}$

$E_{x,y} = \quad lm V_{pp\sigma} - lm V_{pp\pi}$

$E_{x,z} = \quad ln V_{pp\sigma} - ln V_{pp\pi}$

$E_{s,xy} = \quad 3^{1/2} lm V_{sd\sigma}$

$E_{s,x^2-y^2} = \quad \frac{1}{2} 3^{1/2} (l^2 - m^2) V_{sd\sigma}$

$E_{s,3z^2-r^2} = \quad [n^2 - \frac{1}{2}(l^2 + m^2)] V_{sd\sigma}$

$E_{x,xy} = \quad 3^{1/2} l^2 m V_{pd\sigma} + m(1 - 2l^2) V_{pd\pi}$

$E_{x,yz} = \quad 3^{1/2} lmn V_{pd\sigma} - 2lmn V_{pd\pi}$

$E_{x,zx} = \quad 3^{1/2} l^2 n V_{pd\sigma} + n(1 - 2l^2) V_{pd\pi}$

$E_{x,x^2-y^2} = \quad \frac{1}{2} 3^{1/2} l(l^2 - m^2) V_{pd\sigma} + l(1 - l^2 + m^2) V_{pd\pi}$

$E_{y,x^2-y^2} = \quad \frac{1}{2} 3^{1/2} m(l^2 - m^2) V_{pd\sigma} - m(1 + l^2 - m^2) V_{pd\pi}$

$E_{z,x^2-y^2} = \quad \frac{1}{2} 3^{1/2} n(l^2 - m^2) V_{pd\sigma} - n(l^2 - m^2) V_{pd\pi}$

$E_{x,3z^2-r^2} = \quad l[n^2 - \frac{1}{2}(l^2 + m^2)] V_{pd\sigma} - 3^{1/2} ln^2 V_{pd\pi}$

$E_{y,3z^2-r^2} = \quad m[n^2 - \frac{1}{2}(l^2 + m^2)] V_{pd\sigma} - 3^{1/2} mn^2 V_{pd\pi}$

$E_{z,3z^2-r^2} = \quad n[n^2 - \frac{1}{2}(l^2 + m^2)] V_{pd\sigma} + 3^{1/2} n(l^2 + m^2) V_{pd\pi}$

$E_{xy,xy} = \quad 3l^2 m^2 V_{dd\sigma} + (l^2 + m^2 - 4l^2 m^2) V_{dd\pi} + (n^2 + l^2 m^2) V_{dd\delta}$

$E_{xy,yz} = \quad 3lm^2 n V_{dd\sigma} + ln(1 - 4m^2) V_{dd\pi} + ln(m^2 - 1) V_{dd\delta}$

$E_{xy,zx} = \quad 3l^2 mn V_{dd\sigma} + mn(1 - 4l^2) V_{dd\pi} + mn(l^2 - 1) V_{dd\delta}$

$E_{xy,x^2-y^2} = \quad \frac{3}{2} lm(l^2 - m^2) V_{dd\sigma} + 2lm(m^2 - l^2) V_{dd\pi} + \frac{1}{2} lm(l^2 - m^2) V_{dd\delta}$

$E_{yz,x^2-y^2} = \quad \frac{3}{2} mn(l^2 - m^2) V_{dd\sigma} - mn[1 + 2(l^2 - m^2)] V_{dd\pi} + mn[1 + \frac{1}{2}(l^2 - m^2)] V_{dd\delta}$

$E_{zx,x^2-y^2} = \quad \frac{3}{2} nl(l^2 - m^2) V_{dd\sigma} + nl[1 - 2(l^2 - m^2)] V_{dd\pi} - nl[1 - \frac{1}{2}(l^2 - m^2)] V_{dd\delta}$

$E_{xy,3z^2-r^2} = \quad 3^{1/2} lm[n^2 - \frac{1}{2}(l^2 + m^2)] V_{dd\sigma} - 3^{1/2} 2lmn^2 V_{dd\pi} + \frac{1}{2} 3^{1/2} lm(1 + n^2) V_{dd\delta}$

$E_{yz,3z^2-r^2} = \quad 3^{1/2} mn[n^2 - \frac{1}{2}(l^2 + m^2)] V_{dd\sigma} + 3^{1/2} mn(l^2 + m^2 - n^2) V_{dd\pi} - \frac{1}{2} 3^{1/2} mn(l^2 + m^2) V_{dd\delta}$

$E_{zx,3z^2-r^2} = \quad 3^{1/2} ln[n^2 - \frac{1}{2}(l^2 + m^2)] V_{dd\sigma} + 3^{1/2} ln(l^2 + m^2 - n^2) V_{dd\pi} - \frac{1}{2} 3^{1/2} ln(l^2 + m^2) V_{dd\delta}$

$E_{x^2-y^2,x^2-y^2} = \quad \frac{3}{4}(l^2 - m^2)^2 V_{dd\sigma} + [l^2 + m^2 - (l^2 - m^2)^2] V_{dd\pi} + [n^2 + \frac{1}{4}(l^2 - m^2)^2] V_{dd\delta}$

$E_{x^2-y^2,3z^2-r^2} = \quad \frac{1}{2} 3^{1/2} (l^2 - m^2)[n^2 - \frac{1}{2}(l^2 + m^2)] V_{dd\sigma} + 3^{1/2} n^2 (m^2 - l^2) V_{dd\pi}$
$$+ \frac{1}{4} 3^{1/2} (1 + n^2)(l^2 - m^2) V_{dd\delta}$$

$E_{3z^2-r^2,3z^2-r^2} = \quad [n^2 - \frac{1}{2}(l^2 + m^2)]^2 V_{dd\sigma} + 3n^2(l^2 + m^2) V_{dd\pi} + \frac{3}{4}(l^2 + m^2)^2 V_{dd\delta}$

SOURCE: Slater and Koster (1954).

The Hamiltonian Matrix

When $\mathbf{k}$ is in a special direction such as along the z-axis, cancellation among terms in the sum in Eq. (20-2) will cause many of the $H_{\alpha\beta}$ values to vanish. Which ones vanish can be learned readily by symmetry, using group theory, or by direct evaluation. It is found in fact that there are matrix elements between Bloch sums based upon the s state and upon the d state of symmetry $3z^2 - r^2$, but all other $H_{\alpha\beta}$ for $\beta \neq \alpha$ vanish. For these others, Eq. (20-3) becomes a set of independent equations $E_k = H_{\alpha\alpha}$, and we evaluate them directly.

Our first impulse might be to include only interatomic matrix elements between each atom and the eight nearest neighbors it has in the body-centered cubic structure. We shall see, however, that this gives a very poor representation of the bands. The difficulty is special to the body-centered cubic structure, in which the second neighbors are only 14 percent more distant than the nearest. A good set of bands will be obtained by including the six second-nearest neighbors, using the same formulae from the Solid State Table for the interatomic matrix elements. For the face-centered cubic structure (Problem 20-1) and hexagonal close-packed structure, the twelve nearest neighbors suffice.

For the eight nearest neighbors in the body-centered cubic structure (Fig. 20-2), the direction cosines l, m, and n entering Table 20-1 are all $3^{-1/2}$, all combinations of plus and minus being used: $3^{-1/2}(111)$, $3^{-1/2}(\bar{1}11)$, $3^{-1/2}(1\bar{1}1)$, $3^{-1/2}(11\bar{1})$, $3^{-1/2}(1\bar{1}\bar{1})$, $3^{-1/2}(\bar{1}1\bar{1})$, $3^{-1/2}(\bar{1}\bar{1}1)$, $3^{-1/2}(\bar{1}\bar{1}\bar{1})$. Those with positive direction cosines n (in the positive z direction) have phase factors $e^{ika/2}$; those with negative n have phase factors $e^{-ika/2}$. Let us then make the evaluation explicitly for states of symmetry zx. The interatomic matrix element $E_{zx,\,zx}$ does not appear in Table 20-1 but it is clear that interchanging z and y in the subscripts (in the orbitals) simply interchanges the direction cosines n and m (describing z- and y-components of $\mathbf{r}_i$). Thus

$$E_{zx,\,zx} = 3l^2n^2V_{dd\sigma} + (l^2 + n^2 - 4l^2n^2)V_{dd\pi} + (m^2 + l^2n^2)V_{dd\delta}. \quad (20\text{-}4)$$

The matrix element $V_{dd\delta}$ is always quite small and, as indicated in the Solid State Table, we take $\eta_{dd\delta} = 0$, and drop the final term in Eq. (20-4). This step was taken to reduce the number of free parameters before theoretical ratios between them had been established. In a subsequent analysis it would probably be better to retain all the free parameters. For the particular case of Eq. (20-4), the interatomic matrix element is the same for all nearest neighbors, and is $1/3V_{dd\sigma} + 2/9V_{dd\pi}$. (They are ordinarily different for different neighbors.) Half have phase $e^{ika/2}$ and half $e^{-ika/2}$. Summing over the eight neighbors, we obtain the nearest-neighbor contribution to $H_{\alpha\alpha}(\mathbf{k})$ for this case, of $(8/3V_{dd\sigma} + 16/9V_{dd\pi})\cos ka/2$. Similarly, the second neighbor matrix elements with direction cosines (100), (010), (001), $(\bar{1}00)$, $(0\bar{1}0)$, and $(00\bar{1})$, and phase factors 1, 1, e^{ika}, 1, 1, and e^{-ika}, respectively (see Fig. 20-2) contribute $2V_{dd\pi}^{2nd}(1 + \cos ka)$. The energy for this band, with $\mathbf{k}$ along the z-axis, is obtained by adding the two contributions to the intra-atomic term ε_d and is displayed in Table 20-2 as E_k^5. The superscript 5 represents the symmetry of the state, Λ_5 in the notation of group theory, and may here be taken simply as a band label. The Bloch state of angular form zy gives an identical energy; symmetry requires the two bands to have the same energy for **k** in this direction.

The evaluation for the state of symmetry xy is similar but gives a different expression since, for example, the second neighbor at $(00a)$, which has a phase factor e^{ika}, now is coupled with a matrix element of $V_{dd\sigma}$ rather than $V_{dd\pi}$. Other second-neighbor couplings are also interchanged. The result is given in Table 20-2 as the first E_k^2. Similarly, the energy of the Bloch sum based upon the d states of symmetry $x^2 - y^2$ gives the band listed as the second E_k^2 in Table 20-2.

Finally we turn to the two coupled Bloch sums based upon the s state and the d state of symmetry $3z^2 - r^2$. The diagonal energy for the d state Bloch sum is $\varepsilon_d + 16/3V_{dd\pi} \cos ka/2 + 2V_{dd\sigma}^{2nd}(\cos ka + 1/2)$. That for the s state Bloch sum is $\varepsilon_s + 8V_{ss\sigma} \cos ka/2 + 2V_{ss\sigma}^{2nd}(2 + \cos ka)$. The s-like band is very poorly described by this LCAO description for exactly the reasons discussed for the simple metals at the beginning of Chapter 15, and we lose little more by simplifying its representation by taking $\varepsilon_d = \varepsilon_s$ and $V_{ss\sigma}^{2nd} = 0$ and adjusting $V_{ss\sigma}$ to fit the known bands at **k** = 0. The result is given as E_k^s in Table 20-2. There, it is written as V_{ss} rather than $V_{ss\sigma}$ since the treatment of the s band is artificial enough that we expect little relation to the $V_{ss\sigma}$ defined earlier for solids not involving transition metals. Of course a better fit could be made by using the two additional parameters. The matrix element between the two Bloch sums is $2V_{sd\sigma}^{2nd}(\cos ka - 1)$, shown as E_k^{sd} in Table 20-2. The inadequacy of the s-state expansion for this broad band suggests the use of a free-electron description of it, as in the simple metals. This then becomes the fit used by Saffren (1960) and others, discussed at the beginning of the chapter. We shall return to that description in Section 20-E. Here the use of s states has the advantage of giving an estimate of the coupling from an independent

TABLE 20-2
Energy bands along a cube edge for the body-centered cubic structure, based on first-neighbor and second-neighbor interactions. Matrix elements $V_{dd\delta}$, $V_{dd\delta}^{2nd}$, $V_{ss\sigma}^{2nd}$, and $\varepsilon_d - \varepsilon_s$ have all been set equal to zero. All energies are measured relative to ε_d.

$$E_k^s = 8V_{ss} \cos ka/2$$

$$E_k^d = 16/3V_{dd\pi} \cos ka/2 + 2V_{dd\sigma}^{2nd}(\cos ka + 1/2)$$

$$E_k^{sd} = -4V_{sd\sigma}^{2nd} \sin^2 ka/2 = 2V_{sd\sigma}^{2nd}(\cos ka - 1)$$

$$E_k^2 = (8/3V_{dd\sigma} + 16/9V_{dd\pi})\cos \frac{ka}{2} + 4V_{dd\pi}^{2nd}$$

$$E_k^2 = 16/3V_{dd\pi} \cos \frac{ka}{2} + 3V_{dd\sigma}^{2nd}$$

$$E_k^5 = (8/3V_{dd\sigma} + 16/9V_{dd\pi})\cos \frac{ka}{2} + 2V_{dd\pi}^{2nd}(1 + \cos ka)$$

source, the energy bands of the perovskites. The solution of the two simultaneous equations, Eq. (20-3), for these bands immediately gives

$$E_k = \frac{E_k^d + E_k^s}{2} \pm \sqrt{\left(\frac{E_k^d - E_k^s}{2}\right)^2 + E_k^{sd\,2}}. \tag{20-5}$$

Choice of Parameters

We see clearly why so many parameters have been required in fitting the transition-metal bands. Our formulation was based initially upon the parameters $V_{dd\sigma}$, $V_{dd\pi}$, $V_{dd\sigma}^{2nd}$, $V_{dd\pi}^{2nd}$, $V_{sd\sigma}^{2nd}$, $V_{ss\sigma}$, $V_{ss\sigma}^{2nd}$, $V_{dd\delta}$, $V_{dd\delta}^{2nd}$, and $\varepsilon_d - \varepsilon_s$. (The parameter $V_{sd\sigma}$ would also enter for other directions of **k**.) In principle, all eleven of these would need to be fitted for each metal. However, the similar form of the bands for all body-centered cubic structures in Fig. 20-1 suggests correctly that there are really only two important independent parameters: the average energy E_d of the d band relative to the bottom of the s band and the width W_d of the d band. We shall use Muffin-Tin Orbital theory (explained in Section 20-D) and pseudopotential theory to relate all of the necessary parameters to these two; the numbers in the Solid State Table that are necessary for the calculation for each metal have been determined in this way.

The four dd matrix elements $V_{dd\sigma}$, $V_{dd\pi}$, $V_{dd\sigma}^{2nd}$, and $V_{dd\pi}^{2nd}$ are all obtained from

$$V_{ddm} = \eta_{ddm} \frac{\hbar^2 r_d^3}{md^5}, \tag{20-6}$$

where r_d is a length that is characteristic of each transition metal and is listed in the Solid State Table; it is 0.90 Å for Cr. The two coefficients $\eta_{dd\sigma}$ and $\eta_{dd\pi}$ are found theoretically in Section 20-E to be -14.3 and 9.6, respectively, but were adjusted to -16.2 and 8.75 to fit values for $\Gamma'_{25} - H'_{25}$ and $\Gamma_{12} - H_{12}$ read from Mattheiss's bands for chromium. These adjusted values appear in the Solid State Table. The matrix element $V_{sd\sigma}^{2nd}$ can similarly be found from the Solid State Table; its use in the calculated bands indicated it to be too small, so a value twice as large was used in this calculation. Finally, notice from Table 20-2 that the bottom of the s-like band lies at $8V_{ss}$ relative to the energy ε_d for the body-centered cubic structure. (It lies at $12V_{ss}$ for the face-centered cubic structure.) Thus $E_d = -8V_{ss}$ for chromium, or, in terms of the parameter k_d given for each transition metal in the Solid State Table (and the relation between E_d and k_d given there),

$$V_{ss} = -\frac{\hbar^2 k_d^2}{16m}\left(1 + \frac{5r_d^3}{\pi r_0^3}\right). \tag{20-7}$$

All matrix elements are readily obtained; notice that $r_0 = 1.42$ Å for chromium, that $d^3 = 3^{1/2}\pi r_0^3$ for nearest neighbors in a body-centered cubic structure, and that the second neighbors have d larger by a factor $3^{-1/2}2$. We obtain $V_{dd\sigma} = -0.93$ eV, $V_{dd\pi} = 0.50$ eV, $V_{dd\sigma}^{2nd} = -0.45$ eV, $V_{dd\pi}^{2nd} = 0.24$ eV, and $V_{ss} = -1.02$ eV.

The LCAO Bands

By using these values and the expressions in Table 20-2, the bands shown in Fig. 20-3,b are obtained; the bands found by Mattheiss are shown on the right, in panel (c). (The result of setting the second-neighbor matrix elements equal to zero is shown in Fig. 20-3,a. Clearly the second-neighbor matrix elements are essential to the bands and presumably to other properties of the metal.) Slight adjustments in four parameters have been made to improve the fit $(\eta_{dd\sigma}, \eta_{dd\pi}, V_{sd\sigma}, \text{and } E_d)$, but in view of the complexity of the bands, the correspondence may be regarded as remarkably good. Furthermore, we would use the corresponding adjusted parameters to obtain bands without further adjustment for any other body-centered cubic system, and in fact for any other structure or for any volume.

For different structures, one must redo the analysis that led to Table 20-2. That

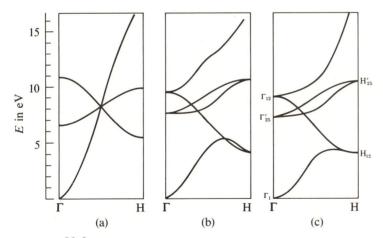

FIGURE 20-3

In the central panel, part (b), are shown the LCAO bands for body-centered cubic chromium, with four parameters adjusted slightly to accord with the bands for chromium from Fig. 20-1, which are reproduced to the right, in part (c). The corresponding adjusted values appear in the Solid State Table, except for $V_{sd\sigma}^{2nd}$, which was taken to be twice the value obtained from the Solid State Table. (The matrix element $V_{sd\sigma}$ for nearest neighbors does not enter.) In part (a) are shown the bands obtained if all second-neighbor interactions are dropped. [Part (c), after Mattheiss, 1964.]

is done for face-centered cubic metals in Problem 20-1, a somewhat simpler calculation, since only nearest neighbors enter. The resulting bands for nickel are shown in Fig. 20-4,a. (For these bands, the $V_{sd\sigma}$ value from the Solid State Table is used directly.) Since all parameters are taken from the Solid State Table, panel (a) in the figure is not a fit to Mattheiss's bands but is an independent band calculation. In this context, the agreement with Mattheiss's bands for nickel, shown in panel (b) of Fig. 20-4, is very impressive. The biggest discrepancy would have been eliminated by use of a larger value of E_d than is obtained from the k_d listed in the Solid State Table, a value obtained by scaling from the chromium value for E_d.

The Band Width

In the body-centered cubic structure it is natural to define the band width W_d as the difference between the levels H_{12} and H'_{25}. (See Fig. 20-3.) These are at the wave number $ka = 2\pi$ in Table 20-2. Making the evaluation there and subtracting leads to

$$W_d = -\frac{8}{3} V_{dd\sigma} + \frac{32}{9} V_{dd\pi} - 3V_{dd\sigma}^{2nd} + 4V_{dd\pi}^{2nd}. \tag{20-8}$$

We may combine this with Eq. (20-6) and the adjusted coefficients to obtain

$$W_d = 115.h^2 r_d^3/(md^5) = 6.83h^2 r_d^3/(mr_0^5), \tag{20-9}$$

where in the first form, d is the nearest neighbor distance in the body-centered cubic structure. The second form is written in terms of the atomic sphere radius and can be used to define a d-band width, for other structures as well, in terms of the atomic parameter r_d and the atomic sphere radius.

In studying the properties of the transition metals it is most convenient to

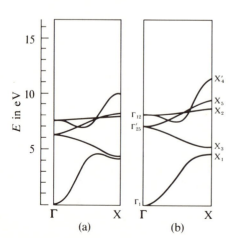

FIGURE 20-4

Nearest-neighbor LCAO bands for nickel in the face-centered cubic structure are shown in panel (a), with all parameters being predicted from the Solid State Table; the calculation is carried out in Problem 20-1. The bands for nickel from Fig. 20-1 are shown for comparison in panel (b). Panel (a) is an independent calculation rather than a fit to the bands in (b). [Part (b), after Mattheiss, 1964.]

describe the electronic structure in terms of the two parameters W_d and E_d. The value given by Eq. (20-9) depended upon an adjustment of the parameters η_{ddm} to fit the chromium bands of Fig. 20-1, so Fig. 20-3 does not present a real test of these values. A more general test is made by obtaining W_d values from such a fit to $H'_{25} - H_{12}$ for the other body-centered cubic metals from the $3d$ series and comparing them with values obtained from Eq. (20-9) and the Solid State Table. Similarly, for face-centered cubic metals, a W_d value is taken as the energy difference between X_1 and X_5. (See Fig. 20-1 for this comparison.) An empirical E_d was obtained from band energies read from Table 20-1, and by using $E_d = (\frac{2}{5}\Gamma_{12} + \frac{3}{5}\Gamma'_{25}) - \Gamma_1$; they are compared with values obtained from Eq. (20-7) and the Solid State Table. The comparison, made in Table 20-3, indicates that the trends are correctly given and that the magnitudes are approximately correct, perhaps in agreement on the scale of the accuracy of the band structures displayed in Fig. 20-1. A similar comparison can be made for the $4d$ series and the predictions of Pettifor (1977a). Pettifor's values for E_d are smaller than those from the Solid State Table, but his W_d values are in reasonable agreement. There is advantage in a single systematic set for all metals, so the scaled set has been used in the Solid State Table rather than individual fits.

TABLE 20-3

Comparison of the d-band width W_d and the d-band energy E_d obtained by fitting the bands of Fig. 20-1 with values of W_d and E_d from the Solid State Table (abbreviated as SST). The latter were obtained by using parameters from Andersen and Jepsen (1977) and values fitted to chromium. All energies are in eV. Also shown is the atomic sphere radius r_0.

Element	$r_0(\text{Å})$	W_d Fitted	W_d SST	E_d Fitted	E_d SST
Sc	1.81	—	5.13	—	7.05
Ti	1.61	—	6.08	—	7.76
V	1.49	6.56	6.77	7.71	8.13
Cr	1.42	6.85	6.56	8.18	8.01
Mn	1.43	—	5.60	—	7.91
Fe	1.41	5.25	4.82	7.90	7.64
Co	1.39	5.05	4.35	7.56	7.36
Ni	1.38	4.62	3.78	7.47	6.91
Cu	1.41	3.47	2.80	6.46	5.90

SOURCE: SST values were obtained by using parameters from Andersen and Jepsen (1977) and chromium-fitted values.

NOTE: The discrepancies for Cr are from different readings of the curves and are some measure of the accuracy of the values.

20-B   The Electronic Properties and Density of States

Fermi Surfaces

It is immediately apparent from Fig. 20-1 that along the one symmetry line in the Brillouin Zone shown, there are two or three bands crossing the Fermi energy for each metal and therefore quite a complex set of Fermi surfaces. These have been thoroughly studied, using the techniques discussed in connection with simple metals. It would, however, be quite inappropriate here to attempt any complete discussion of this problem. Instead, the Fermi surface of a single system, chromium, will be discussed. It is perhaps the most interesting case, and it illustrates the principal effects that enter considerations of the other systems. We shall then turn to the density of states, which dominates many of the electronic properties.

The Fermi surface of chromium has been studied by Rath and Callaway (1973), who used an LCAO approach. They obtained bands looking very much like those shown for chromium in Figs. 20-1 and 20-3, and also obtained some cross-sections of the Fermi surfaces. In Fig. 20-5 we give the sections for a (001) plane through the center of the Brillouin Zone containing the [100] line. We may

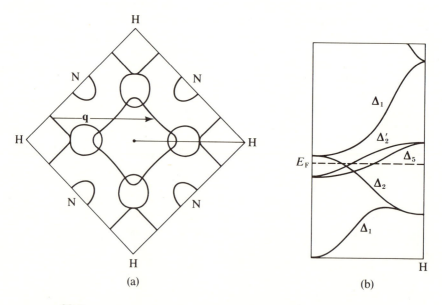

(a)   (b)

FIGURE 20-5

A (001) section of the body-centered cubic Brillouin Zone showing the Fermi surface cross-sections for chromium, as determined by Rath and Callaway (1973). The energy bands from Γ at the center to H at the right are shown for comparison; they are taken from Fig. 20-1. The surfaces separated by $\mathbf{q}$ produce antiferromagnetic order. [After Rath and Callaway, 1973.]

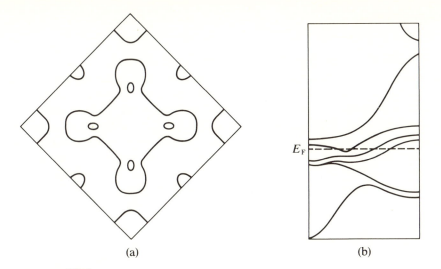

FIGURE 20-6

A schematic construction of Fermi surface sections in a plane slightly displaced from the plane of Fig. 20-5.

identify the intersections of the Fermi surface with the [100] line by comparing the diagram at (a) with the bands from Fig. 20-1, shown at (b) in Fig. 20-5 for convenience.

Starting from Γ, the first crossing is associated with the band labelled Δ_2 on the right. That band is decreasing with increasing k, so it is occupied to the right of the crossing and empty to the left. Correspondingly, the entire oval piece of Fermi surface is "electron" surface occupied within and empty without; we shall return later to the fact that it intersects another piece of Fermi surface. The second crossing is with the band Δ_2', which increases in energy with increase in k and correspondingly, the diamond-shaped segment at Γ is also electron surface. The third crossing is of the doubly-degenerate bands labelled Δ_5. One band corresponds to the outer edge of the oval surface and the other to the beginning of the "hole" surface (having empty states within), which is diamond-shaped and centered at the point H in the extended-zone scheme. They touch because the bands are degenerate along this line.

The resulting picture is somewhat confusing because of the degeneracies that are characteristic of symmetry lines or symmetry planes. If we were to construct the bands in a plane displaced from this central (001) plane, the degeneracies would all be lifted and it would be possible to see the real connectivity of the Fermi surface; this is illustrated in Fig. 20-6. We see that the large electron section centered at Γ, sometimes called the "jack," is in the shape of a regular octahedron with ball-shaped enlargements at the corners; the jack is in the fourth band (or fourth Brillouin Zone). There are also some small fifth-band sections of electron surface within it. The octahedral hole surface centered at H does not have enlargements at the corners. As expected, the Fermi surfaces of molybdenum and tungsten, also in the $D6$ column and body-centered cubic, are similar.

A treatment of transport properties in terms of this surface is no more complicated in principle than that in the polyvalent metals, but there is not the simple free-electron extended-zone scheme that made that case tractable. Friedel oscillations arise from the discontinuity in state occupation at each of these surfaces, just as they did from the Fermi sphere. When in fact there are rather flat surfaces, as on the octahedra in Fig. 20-6, these oscillations become quite strong and directional. A related effect can occur when two rather flat surfaces are parallel, as in the electron and hole octahedra, in which the system spontaneously develops an oscillatory spin density with a wave number determined by the difference in wave number between the two surfaces, the vector $\mathbf{q}$ indicated in Fig. 20-5. This generally accepted explanation of the antiferromagnetism of chromium, based upon "nesting" of the Fermi surfaces, was first proposed by Lomer (1962).

Density of States

In addition to producing complex Fermi surfaces, the many bands can also clearly produce significant variations in the density of states near the Fermi energy. This feature is important in a number of properties. The density of states for general energies may be directly calculated, using the technique discussed in Section 2-E, and in fact, Rath and Callaway did that for chromium. The result, displayed in Fig. 20-7, shows a division into two peaks characteristic of the body-centered cubic structure. The density of states for the hexagonal close-packed structure and for the face-centered cubic structure are more uniform than for the body-centered cubic structure. This is illustrated in Fig. 20-8,a by the density of states for the hexagonal close-packed metal osmium, given by Jepsen, Andersen, and Mackintosh (1975). They used a "Linearized Muffin-Tin Orbital" method due to Andersen (1975), which is closely related to the method described in Section 20-D.

Figure 20-8,a shows considerable structure which, except on the fine scale, is real and not statistical error. Notice that it is not very similar to the body-centered cubic curve, shown in Fig. 20-7. On the other hand, comparison of the density of states obtained for the other hexagonal close-packed transition metals obtained in the same study shows a strong resemblance between them. This might at first suggest the existence of a universal density of states curve for hexagonal close-packed metals (another for face-centered cubic metals and another for body-centered cubic metals). Closer inspection of the curves indicates that the energy at which various peaks and valleys occur varies significantly from system to system, but that these structures occur at the same fraction of filling; the meaning of this can be indicated with reference to Fig. 20-8,a. Notice that the deep valley in the density of states, occurring near 8 eV, occurs at an energy below which there are just four electrons per atom; this corresponds to a four on the right-hand scale for the integrated density curve which threads through the density of states curve. Just such a deep dip occurs in every hexagonal close-packed density of states at the same four electrons per atom (though at different energies) and the same correspondence tends to occur for the other features. This then means that the density of states per atom may be a reasonably universal function of the electron-to-atom ratio *within a given structure*, even though there are appreciable shifts in the bands with atomic number, indicated by the trends shown in Fig. 20-1. In fact, metals with the same number of electrons per atom tend to have the same structure, so that a property such as the density of states at the Fermi energy can be expected to be a rather universal function of electron-to-atom ratio, a fact which has long been known experimentally. It should not, however, be taken as evidence that the density-of-states curves $n(E)$ are independent of structure, nor even that $n(E)$ is rigid within one structure. It only indicates that $n(E_F)$ versus n, the total number of electrons per atom, is a universal curve for one structure. We may illustrate the result in terms of the specific heat.

We indicated in our discussion of simple metals that the electronic specific heat at low temperatures is linear in temperature and proportional to the density of states at the Fermi energy. The density of states so obtained from the specific heat is shown in Fig. 20-9, and the correlation is apparent. Notice in particular the low value at four electrons per atom for Ti, Zr, and Hf, all of which occur in the hexagonal close-packed structure; this corresponds to the dip we noted in the density of states in Fig. 20-8,a. Similarly the minimum at six electrons per atom corresponds to the minimum in the body-centered cubic density of states of Fig. 20-7, at the Fermi energy for chromium.

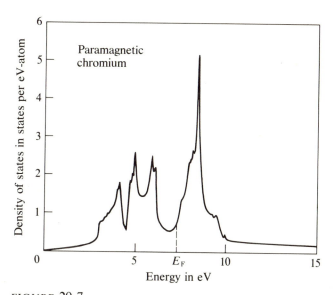

FIGURE 20-7

The total density of states for chromium, calculated by Rath and Callaway (1973), using the technique discussed in Section 2-E. The Fermi energy for chromium, at six electrons per atom, is indicated. [After Rath and Callaway, 1973.]

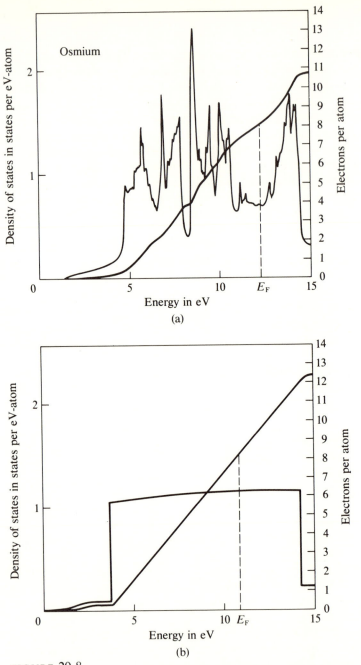

FIGURE 20-8

The density of states in hexagonal close-packed osmium, with Fermi energy indicated by the dashed line. The monotonically rising line is the integral over the density of states, with scale given to the right. The curves in part (a) are after Jepsen, Andersen, and Mackintosh (1975). Those in part (b) are from the Friedel model with parameters taken from the Solid State Table.

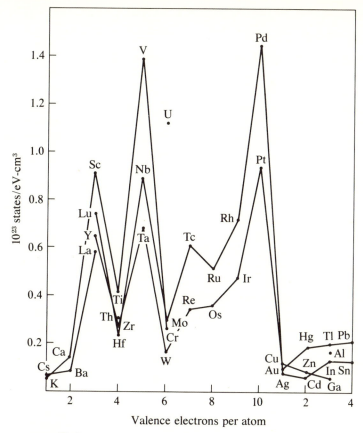

FIGURE 20-9

The density of states determined from the experimental electronic specific heat. [After Gladstone, Jensen, and Schrieffer, 1969.]

It is possible experimentally to trace out the curves for a single structure for intermediate concentrations of electrons by making alloys. A study, for example, of body-centered cubic alloys by Cheng, Gupta, van Reuth, and Beck (1962) showed a sharp peak at concentrations between six and seven electrons per atom. This peak is presumably associated with the highest-energy peak in the body-centered cubic density of states, Fig. 20-7.

The picture for the many electronic properties that are associated with the density of states is therefore quite clear. For each structure, a plot could be given of the density of states as a function of the electron-to-atom ratio, for alloys as well as pure materials. Such calculations have been carried out for the 4d transition series (the Zr row) by Pettifor (1977a) but he displayed only values for the observed structures, giving a curve of the type shown in Fig. 20-9, in fact in quite good agreement with that experimental figure. Pettifor (1977b) also discussed electronic properties such as the electron–phonon coupling constant λ, but we shall limit the discussion to treatment of the bonding properties.

20-C   Cohesion, Bond Length, and Compressibility

The procedure required to calculate the total energy follows from the discussion of the preceding two sections. The energy bands should be calculated, the density of states obtained, and the Fermi level selected to give the proper number of electrons. This should be done self-consistently, so that the potential is determined from the occupied states. The energy is obtained by making the appropriate integration over the density of states, and self-energy and ion–ion corrections are made as described in Chapter 7, to obtain the total energy. If this is done as a function of volume, it yields predictions of equilibrium density, cohesive energy, and compressibility. This would seem an absolutely formidable task, and yet it has been carried through in detail by Moruzzi, Janak, and Williams (1977) for the 3d and 4d transition-metal series. A volume compiling the intermediate results of this program has also been prepared (1978). Figure 20-10 summarizes the final results and compares them with experiment.

The remarkable agreement indicates that the one-electron approximation is capable of a very complete and adequate description of these systems if carefully carried through. Even the irregularities in the cohesion near the center of the two series, which arise from spin polarizations associated with Hund's rule, are rather well given. The only significant discrepancies are in the bulk modulus of the strongly magnetic metals at the center of the iron series.

The Friedel Model

Here we shall redo the principal parts of an equivalent calculation, using the comparatively crude approximations suggested by the electronic structure as it has been described in this chapter. A major approximation in this scheme is the replacement of the rather intricate density of states arising from d-like bands, shown in Fig. 20-8,a for osmium, by a constant density of states, shown in Fig. 20-8,b; this approximation was made by Friedel (1969). Even with the hexagonal close-packing of osmium, a considerable amount of structure is being discarded, and one might question the approximation. However, when we recognize that this approximation corresponds simply to smoothing out the ripples in the integrated density of states, shown in Fig. 20-8,a, this does not look so serious. The total energy comes from a similar integral and the grosser properties should be rather well given.

Friedel's model, then, is of a density of states containing two contributions. First is the density of d-like states, centered at an energy E_d and distributed over a band width W_d; the parameters E_d and W_d that we have used in this chapter fit well with such a description, and the corresponding density of d-like states is

$$n_d(E) = 10/W_d \text{ per atom} \qquad \text{for } E_d - \frac{1}{2} W_d < E < E_d + \frac{1}{2} W_d;$$

$$n_d(E) = 0 \qquad\qquad\qquad \text{otherwise,} \tag{20-10}$$

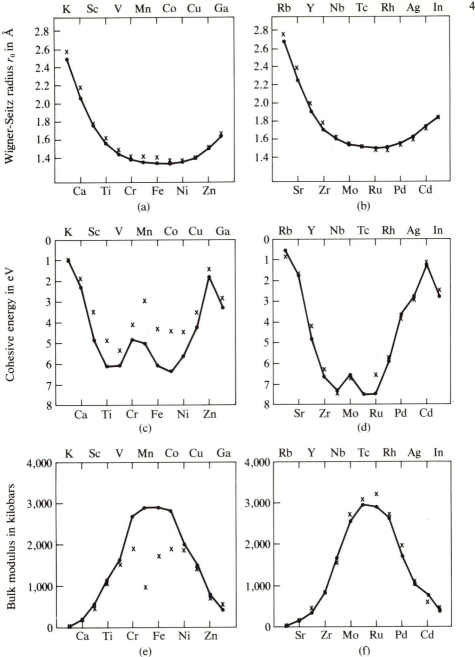

FIGURE 20-10

Results from Morruzi, Williams, and Janak (1977) for cohesive properties versus atomic number. Parts (a) and (b) show equilibrium nuclear separation in terms of the Wigner-Seitz (or atomic sphere) radius r_0. (The volume per atom is $4\pi r_0^3/3$.) Parts (c) and (d) show cohesive energy in eV per atom. Parts (e) and (f) show bulk moduli in kilobars. The atomic number increases in steps of one from 19 to 31 in (a), (c), and (e) and from 37 to 49 in (b), (d), and (f). Measured values, at low temperature where available, are indicated by crosses. [After Morruzi, Williams, and Janak, 1977.]

for the ten d states (including the two spin states) per atom. There is, in addition, a density of states arising from the s electrons. We have indicated that the corresponding bands are rather free-electron-like, so it is more appropriate, and simpler, to represent the corresponding density of states with a free-electron density (rather than an s-band density, though it is customary to use the subscript s), which may be written

$$n_s(E) = \frac{2}{3\pi}\left(\frac{2m_s r_0^2}{\hbar^2}\right)^{3/2} E^{1/2}. \tag{20-11}$$

The effective mass m_s will be calculated in Section 20-E; it is given by $m/[1 + 5r_d^3/(\pi r_0^3)]$. Adding these two contributions, using parameters from the Solid State Table, gives the density of states for osmium shown in Fig. 20-8,b. It is important to notice that the curve in Fig. 20-8,b is a predicted curve based upon the parameters from the Solid State Table, not a fit to the curve given in part (a) of the figure. Thus, since we have parameters for all transition metals, we can as easily construct an approximate density of states for all other transition metals. This approximation should enable us to understand the principal trends from material to material and, we may hope, also the volume-dependent properties. We follow Friedel's earlier analysis in doing this and are aided by a much more detailed analysis of the same questions by Pettifor (1977a,b), who used Atomic Sphere Approximation calculations for the $4d$ metals.

The Free Electrons

We look first at the role of the s-like, or free, electrons. With only these electrons being considered, the theory essentially becomes that of a simple metal and we might expect an energy, as a function of r_0, given by the simple-metal expression of Eq. (15-16), but now with an effective mass m_s. The difficulty comes in the use of an empty-core pseudopotential. In transition metals, the d electrons are far enough from the nucleus that the potential does not approach the assumed $-e^2/r$ until distances that are much larger than the core radii that would be appropriate. Thus, if the atomic s-state energy (6.76 eV for Cr) and ion charge of order unity are used, the approach described in Problem 16-2 gives core radii of order 0.6 Å. Consideration of the chromium $3d$ wave function (which will be shown in Fig. 20-13) indicates that the d-state charge density is still large at 0.6 Å. Thus the potential $-e^2/r$ for $r > r_c$ (and zero for $r < r_c$) is so artificial that, though it correctly gives the s-state energy, it does not give a correct volume dependence, nor does it give any other property correctly. One adjusted parameter has only accounted for one experimental number.

The failure to give correct volume dependence can be rectified by taking a more complicated pseudopotential, since the pseudopotential method itself, as described in Appendix D, is a general and rigorous method. However, the more complicated pseudopotential involves more parameters, and it is not clear that much is gained. Pettifor (1977a), for example, took an empty-core radius about twice our 0.6 Å, but then adjusted an effective ion charge such that a potential $-Z^*e^2/r$ was used outside the core. The resulting formulae give the band minimum correctly over a considerable range of volume but involve another parameter.

The problem may not be as severe for the f-shell metals; particularly in the $4f$ series. The f electrons (in most cases $n - 3$ electrons in the Fn column for $4 \leq n \leq 17$) are so well localized that they may be treated as core electrons. The corresponding r_c values were calculated from the ionization potentials of the atoms (see Problem 16-2,c), and are listed in the Solid State Table. Then the bonding properties of the f-shell metals can be treated exactly as the simple metals (or as beginning transition series if the effects of d states are sufficiently large to make that necessary) and effects of the f-shell electrons (such as the magnetic properties discussed in Section 20-F) can be treated separately. As an example, the equilibrium spacing of the rare earths is discussed in Problem 20-2, in which any d-state effects are ignored. This is a rather crude approximation, since with three non-f electrons there is always some occupation of d-like bands.

Although we abandon the empty-core pseudopotential for the free electrons in transition metals, we do not abandon the free-electron concept, which has its band minimum E_d below the center of the d band and its free-electron density of states given by Eq. (20-11). In discussing the volume dependence of the total energy, we could introduce a volume-dependent energy phenomenologically as we did in discussing the volume-dependent energy in covalent solids, for which the detailed calculation of that energy was complicated. This phenomenological energy would be used, as Eq. (15-16) would have been used if it had been applicable to transition metals.

The d Electrons

Let us return to the d states which, for each transition metal, at a given r_0, are characterized by an average energy E_d with respect to the minimum of the free-electron band and a width W_d. The corresponding model density of states is illustrated in Fig. 20-8,b. Given W_d and E_d, we can easily compute the value of E_F that is consistent with the number of electrons present.

Exactly how this is done is seen in Fig. 20-11, where the model density of states from Fig. 20-8,b is redrawn (rotated 90°) with the free-electron and d-like densities shown separately. For a metal in a particular Dn column of the Solid State Table ($3 \leq n \leq 11$), E_F should be picked so that the integral over the occupied portions of the two densities of states shown—and given in Eqs. (20-10) and (20-11)—comes to n electrons. This is carried out in Problem 20-3 to obtain the number Z_s of free electrons per atom. The results are shown in Table 20-4. We shall see in Section 20-F that the magnetic properties suggest a value for the $3d$ series of $Z_s = 0.6$. The discrepancy is not large in view of the assumption of a uniform density of d states.

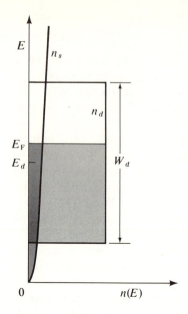

FIGURE 20-11

The approximate density of electronic states for osmium obtained with parameters from the Solid State Table, equivalent to Fig. 20-8,b. The density of d-like states, n_d, and the density of s-like states, n_s, are drawn separately and the occupation of each is shown by different shading.

TABLE 20-4

Width of d band, W_d, and its average energy, E_d, in eV; the number of free electrons, Z_s, calculated as in Problem 20-3; and the width of resonance Γ, in eV.

	Column and element								
Parameter	$D3$	$D4$	$D5$	$D6$	$D7$	$D8$	$D9$	$D10$	$D11$
	Sc	Ti	V	Cr	Mn	Fe	Co	Ni	Cu
W_d	5.13	6.08	6.77	6.56	5.60	4.82	4.35	3.78	2.80
E_d	7.05	7.76	8.13	8.01	7.91	7.64	7.36	6.91	5.90
Z_s	0.46	0.58	0.69	0.76	0.82	0.84	0.84	0.81	1.00
Γ	2.97	2.56	2.26	1.82	1.73	1.45	1.19	0.90	0.56
	Y	Zr	Nb	Mo	Tc	Ru	Rh	Pd	Ag
W_d	6.59	8.37	9.72	9.98	9.42	8.44	6.89	5.40	3.63
E_d	6.75	7.17	7.29	7.12	6.67	6.02	5.08	4.52	2.49
Z_s	0.39	0.47	0.57	0.67	0.72	0.73	0.66	0.59	1.00
Γ	3.47	2.86	2.36	1.95	1.47	1.07	0.69	0.53	0.12
	Lu	Hf	Ta	W	Re	Os	Ir	Pt	Au
W_d	7.81	9.56	11.12	11.44	11.02	10.31	8.71	7.00	5.28
E_d	8.44	9.12	9.50	9.45	8.99	8.38	7.35	6.51	5.18
Z_s	0.54	0.67	0.82	0.96	1.04	1.09	1.02	0.94	1.00
Γ	5.62	4.87	4.48	3.98	3.22	2.57	1.88	1.46	0.86

Volume Dependence

With a change in r_0, the band width varies as r_0^{-5}, as was seen in Eq. (20-9). In the LCAO theory, E_d was given by $8V_{ss}$, suggesting that it should vary as r_0^{-2}, and in fact this conclusion would not be changed by adding more-distant neighbors or p states. We use that form though the numerical calculations by Pettifor (1977a) suggest a weaker dependence.

We wish to calculate the change in total energy with a change in volume, and in particular we will find a change due to the variation in W_d that is linear in changes in r_0. For calculating that term, we may neglect any variation in Z_s with r_0, since the energy gained in transferring electrons between bands at the Fermi energy will be proportional to the square of the change in Z_s. This means also that our assumed variation of E_d as r_0^{-2} does not affect this term in the energy.

Thus the change in total energy with change in r_0 divides naturally into two parts. The first is obtained by reducing the d-band width to zero and fixing the number of electrons in it as if it were a core state. The energy of the other (non-d) electrons is essentially a simple-metal total energy though, as we indicated, there may be difficulties in treating it with the empty-core pseudopotential. The second contribution, δE_b, specifically associated with the d electrons, is the sum of the energy, relative to E_d, of the d electrons in the band of width W_d. That energy can be written down by inspection of Fig. 20-11, using Eqs. (20-10) and (20-11). The energy per atom of the Z_d electrons in the d bands is

$$\delta E_b = \int_{-W_d/2}^{-W_d/2 + Z_d W_d/10} dE(10/W_d)E = 5W_d\left[-\frac{Z_d}{10} + \left(\frac{Z_d}{10}\right)^2\right], \quad (20\text{-}12.)$$

equivalent to the form obtained by Friedel (1969). This expression has its minimum at $Z_d = 5$, corresponding to the maximum in the cohesive energy seen in Fig. 20-10. Because of the r_0^{-5}-dependence of W_d (Eq. 20-9), this energy drops rapidly with decreasing r_0;

$$\frac{\partial \delta E_b}{\partial r_0} = \frac{25W_d}{r_0}\left[\frac{Z_d}{10} - \left(\frac{Z_d}{10}\right)^2\right]; \quad (20\text{-}13)$$

and the effect is largest at the center of the transition series where the expression in square brackets is maximum. This compression of the lattice and increased cohesion at the center of the transition series make up the principal trend in the properties visible in Fig. 20-10. It has long been thought that the compression comes from the bonding energy of the half-filled d-shell, and the Friedel model illustrates the point clearly.

To make this point quantitatively requires a knowledge of the radial interactions other than those from the d states (Eq. 20-13). Assuming such terms are the same across a transition series, and fitting them to the bulk modulus for one metal, does, in combination with Eq. (20-13), predict a minimum in r_0 at the center of the series. The predicted differences in r_0 are of the right order of magnitude, but the minimum in the 4d series is predicted at molybdenum rather than ruthenium, and similar discrepancies occur in the other series. To make a quantitative prediction requires a more intricate description, such as that given by Pettifor (1977a,b).

Friedel noted that the band width W_d is rather insensitive to structure even if the details of the density of states are not, which is why smooth curves are obtained for the transition series even though the structures change. This is also the reason why the heat of fusion in transition metals is particularly small compared to the heat of sublimation; the bonding energy is quite insensitive to structure.

Duthie and Pettifor (1977) have treated d bands and s bands in the rare earths and have considered the structure dependence by calculating the density of states in detail (within the Atomic Sphere Approximation described in Section 20-D) for different structures. They indicate that they have predicted a sequence of four different structures, which occurs both for increasing pressure and for decreasing atomic number across the rare-earth series. This correlation and the ratio of core volume to atomic volume had been related by Johansson and Rosengren (1975), but Duthie and Pettifor argue that the essential feature is the number of electrons in the d bands and that this is only incidently reflected in the core volume.

20-D   Muffin-Tin Orbitals and the Atomic Sphere Approximation

Energy bands and a number of properties have been described here in terms of LCAO theory and matrix elements given by formulae such as Eq. (20-6). We turn now to the origin of those formulae and to a description of the electronic structure that proves useful for other properties. The formulae for the matrix elements will in fact be obtained from transition-metal pseudopotential theory, but the principal results can be obtained from the theory of "Muffin-Tin Orbitals," which we discuss first. Moreover, one of the central concepts of Muffin-Tin Orbital theory is necessary for using transition metal pseudopotential theory to obtain the formulae for the interatomic matrix elements. The analysis in this section and the next is somewhat analogous to the use of free-electron theory to obtain the form and estimates of the magnitudes of the matrix elements used in the LCAO theory, and here the consequences are just as rich.

The Wigner-Seitz Method

Muffin-Tin Orbital theory is in the spirit of the very early treatment of alkali metals by Wigner and Seitz (1934), who focused on a single atomic cell (those points nearer the atom being studied than any other atom) in which the potential is nearly spherically symmetric. They then replaced the cell by a sphere of equal volume, the sphere of radius r_0 that we introduced in the discussion of simple metals. This is illustrated in Fig. 20-12 for a face-centered cubic lattice. Wigner and Seitz sought the energy of the state at the bottom of the band, the state Γ_1, calculated in this spherical cell, and used it in a study of the cohesive energy. More recently, Andersen (1973) went further to obtain the average energies of bands in transition metals and the widths of those bands in terms of this sphere, an approach he called the Atomic Sphere Approximation. He has also used the atomic sphere in the construction of localized orbitals, called muffin-tin orbitals, which are analogous to the atomic orbitals in LCAO theory. These muffin-tin orbitals will give us the systematics of the matrix elements in the LCAO theory, so we treat them first. We then return to the Atomic Sphere Approximation to obtain the parameters for all transition metals.

Muffin-Tin Orbitals

Atomic states drop rapidly to zero, as illustrated in Fig. 20-13,a; the corresponding atomic potential is shown in the lower diagram there. In a crystal, there are additional potentials from the neighboring atoms as indicated by the dashed curve in the lower part of (a), and it could be inappropriate to use states obtained from potentials that continue to rise monotonically at large r. A common way of avoiding this difficulty is to approximate the potential in the solid by the spherically symmetric atomic potential within a sphere of radius r_{mt} (muffin-tin radius) and by a constant in the region between spheres. This is called a **muffin-tin potential**, owing to its resemblance to the cooking utensil of that name, which is flat except for a round cup to hold each muffin. This geometric construction is possible only if r_{mt} is sufficiently small that the spheres do not overlap; it might be the radius of a sphere inscribed in the atomic cell.

Two very important points can be made about this. First, the approximation makes sense for a range of values of r_{mt}, and so our formal results for physical

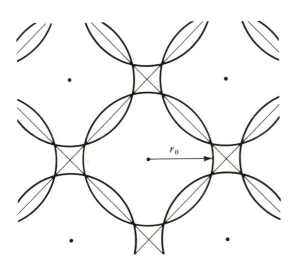

FIGURE 20-12
A (100) section of a face-centered cubic lattice with atomic cells constructed around each atom. Also shown are atomic spheres, of equal volume, constructed around each atom.

quantities should be independent of r_{mt}; this will have important consequences for the dependence of parameters upon volume. Second, experience has shown that when detailed analytical results (or even results of computer programs) continue smoothly to larger values of r_{mt} where the spheres overlap, the best accord with experiment comes when r_{mt} is taken equal to the radius r_0 of the atomic sphere (the Wigner Seitz sphere). These two points may cause a slight conceptual problem, but they will cause no mathematical difficulty. It is desirable to retain the expression r_{mt} for the present, but when we make evaluations, we take $r_{mt} = r_0$.

Andersen considered atomiclike orbitals calculated with a potential equal to the atomic potential for $r \le r_{mt}$ but constant for $r > r_{mt}$; these might have more relevance to the metal than do true atomic orbitals. The orbital at any given

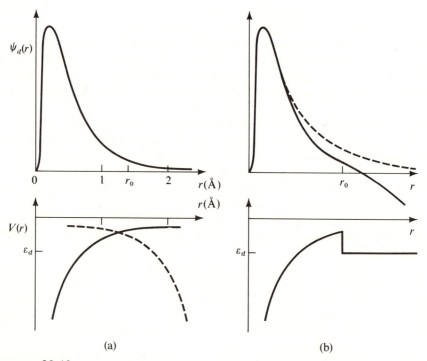

(a)   (b)

FIGURE 20-13

In part (a), the atomic d orbital $\psi_d(r)$ for chromium is shown; below is the potential $V(r)$ from which it is calculated. The radius of the atomic sphere is indicated as r_0 and the dashed curve below is the potential arising from a nearest-neighbor chromium atom. In part (b), the potential is taken as constant outside the atomic sphere with a value equal to the energy of the d state. The wave function (solid curve) computed with this potential, and at the same energy, no longer drops to zero at large distances but takes the form $Ar^2 + Br^{-3}$. A muffin-tin orbital is obtained by subtracting Ar^2 from this wave function and renormalizing; it is shown as a dashed curve.

energy can be obtained by integrating the Schroedinger equation radially from the center of the muffin-tin sphere. The general form of the solution outside the sphere is then matched at r_{mt} to the solution inside, and therefore depends upon the form inside through the value of the slope and the value of the wave function at r_{mt} obtained from the integration within the sphere.

The *form* of the wave function outside depends upon the value of the constant potential chosen outside the sphere. Andersen recognized that the results were rather insensitive to that choice, so he selected the simplest value, a value equal to the energy of the state being considered, so that the kinetic energy of the state is zero in that region. Such a potential is shown in Fig. 20-13,b. The consequences of this assumption will turn out, in this section and the next, to be extraordinary. For this case, the form of the general solution of the Schroedinger equation is simply

$$\psi_{lm}(\mathbf{r}) = [A(r/r_{mt})^l + B(r/r_{mt})^{-(l+1)}]Y_l^m(\theta, \varphi) \qquad \text{for } r \geq r_{mt}, \qquad (20\text{-}14)$$

with A and B chosen to match the value $(A + B)$ of the wave function and its slope $[lA - (l + 1)B]/r_{mt}$ at the sphere. To obtain a reasonably well-localized orbital, Andersen subtracted $A(r/r_{mt})^l$ from the solution at all r, set $r_{mt} = r_0$, and renormalized the resulting orbital, calling it a ***muffin-tin orbital***. It is illustrated in Fig. 20-13,b. What this gave was a solution outside the sphere that matched smoothly at the sphere, but was no longer a solution within the sphere.

The Form of Interorbital Matrix Elements

The basic idea of the muffin-tin theory was then to construct tight-binding combinations—as in Eq. (20-1)—of muffin-tin orbitals, and to require that the tails of all neighboring muffin-tin orbitals cancel the $-A(r/r_{mt})^l$ that had been added to the central sphere; the result is a solution of the Schroedinger equation everywhere, but the calculation has been reduced to a calculation within a single atomic sphere. The form of the condition that the $-A(r/r_{mt})^l$ be cancelled is the same as the LCAO equation, Eq. (20-3), but with the $H_{\alpha\beta}/E_k$ replaced by the coefficient in the expansion of the tails from the corresponding neighbors in spherical harmonics defined at the central cell. This formulation has provided a fast and efficient way to calculate bands and a very convenient starting point for a number of approximate analyses. The aspect of most interest here is the association of the coefficients in the expansion with the interatomic matrix elements of the LCAO theory, which allows a prediction of the relative values of different matrix elements and therefore systematizes the LCAO parameters. This identification was also made by Andersen, Klose, and Nohl (1978), as well as by Pettifor (1977a).

The evaluation of these coefficients is straightforward and will be illustrated for the case of pp matrix elements. Let us consider a muffin-tin orbital of symmetry z/r, which has been obtained by subtracting Az/r_{mt} from the solution of the Schroedinger equation. That muffin-tin orbital is $Br_{mt}^2 z/r^3$ for $r > r_{mt}$. We now consider such a muffin-tin orbital centered at a neighbor at a position $\mathbf{d} = [00d]$. Its tail may be expanded around the central atom and is given by

$$\psi_{mt} = \frac{Br_{mt}^2(\mathbf{r} - \mathbf{d}) \cdot \mathbf{d}}{|\mathbf{r} - \mathbf{d}|^3 d} = -Br_{mt}^2 \frac{1}{d^2} + \frac{2z}{d^3} + \cdots. \tag{20-15}$$

The ratio of the term proportional to z to the subtracted Az/r_{mt} is $-2Br_{mt}^3/d^3A$ and it enters the Muffin-Tin Orbital theory as the ratio of $H_{\alpha\beta}$ to the energy of the state in LCAO theory; the geometry has been chosen to correspond to a matrix element $V_{pp\sigma}$. If instead we had chosen a neighbor at $[d00]$, the expansion of the tail would have been given by Bzr_{mt}^2/d^3 and the ratio Br_{mt}^3/d^3A would have corresponded to the matrix element $V_{pp\pi}$. This contains two predictions: first, it predicts that $V_{pp\sigma} = -2V_{pp\pi}$ (we shall return to this point); second, it predicts a distance-dependence of the matrix elements between p states, of $(r_{mt}/d)^3$. It is easy to generalize this distance-dependence to matrix elements between an orbital of angular momentum quantum number l and one of quantum number l'. The corresponding analysis gives

$$V_{ll'm} = C_{ll'm} d^{-(l+l'+1)}, \tag{20-16}$$

of which the V_{ppm} given above are a special case. The coefficient $C_{ll'm}$ contains a factor $r_{mt}^{l+l'+1}$ and varies from metal to metal also, through the ratio of the value to the slope of the spherical state (of given l or l') evaluated at the muffin-tin radius. However, Andersen has argued that the result for such a matrix element should not depend upon the choice of muffin-tin radius and therefore $C_{ll'm}$ for each metal should depend only upon its three indices; in particular, it should be independent of volume. It was for this reason that we retained r_{mt} explicitly rather than set it equal to r_0, which would have suggested a dependence on volume. Andersen's argument has led to the conclusion that any apparent dependences on volume are approximately cancelled out. This is a very powerful conclusion and depends directly on the assumption of invariance under variations of r_{mt}. It means that the variation of a matrix element from one atom pair in a metal to another at fixed volume is the same as the variation for a given pair, as the separation is modified by compression. This kind of invariance was implied also by our use of the form $\eta_{ll'm}h^2/(md^2)$ for matrix elements involving only s and p states.

If we apply Eq. (20-16) directly to s and p states, we see that it predicts a d^{-1}-dependence for $V_{ss\sigma}$, a d^{-2}-dependence for $V_{sp\sigma}$, and a d^{-3}-dependence for $V_{pp\sigma}$ and $V_{pp\pi}$, while the free-electron theory gave d^{-2} for all three. This should not come as a surprise, since a decrease with distance as gradual as any of these $d^{-(l+l'+1)}$ would not give convergent results when summed over neighbors. (Notice that a sum over neighbors at large distances can be replaced by an integral over volume, $\Omega_0^{-1} \int 4\pi r^2 \, dr$, and even with a matrix element varying as $r^{-3} = d^{-3}$ in the integrand, it integrates to $\ln r$ and diverges at the upper limit.) A reformulation becomes necessary if only nearest neighbors are to be included. For s and p states, this was done by using free-electron theory to obtain nearest-neighbor matrix elements varying as d^{-2}, as if both corresponded to $l = 1/2$.

This divergence does not occur for dd matrix elements, for which Eq. (20-16) gives a decay as d^{-5}. We write that equation for d states in a form that more clearly displays the properties found here,

$$V_{ddm} = \eta_{ddm} \frac{\hbar^2}{m} \frac{r_d^3}{d^5} ; \qquad (20\text{-}17)$$

r_d has units of length, so η_{ddm} is dimensionless if V_{ddm} is to have units of energy. The ratios of the various η_{ddm} are found from Muffin-Tin Orbital theory to be independent of material, so we take them as universal constants, and r_d is a radius characteristic of each element. This form separates the dependence upon material from the dependence upon m and is the form, Eq. (20-6), used in obtaining matrix elements in Section 20-A.

Muffin-Tin Orbital theory also predicts the ratios between $\eta_{dd\sigma}$, $\eta_{dd\pi}$, and $\eta_{dd\delta}$, just as we derived the ratio of $\eta_{pp\sigma}$ to $\eta_{pp\pi}$ with this theory. However, the scale depends upon the definition of the r_d. In the next section, we shall rederive Eq. (20-17) from pseudopotential theory and make an appropriate definition of r_d. The η_{ddm} will then also be easily derived, giving the same ratios as those obtained from Muffin-Tin Orbital theory (Andersen, Klose, and Nohl, 1978).

The Muffin-Tin Orbital theory (Andersen and Jepsen, 1977) also predicts that matrix elements between states of different l (or between states on different atomic species) should scale as the geometric mean of those obtained individually, as one would expect from consideration of the pp matrix element derivation given earlier. This has important consequences with respect to alloys, for which it would predict that matrix elements between a d state on an atom of type 1 and one of type 2 would be given by

$$V_{ddm}^{(1,\,2)} = \eta_{ddm} \frac{\hbar^2 \big(r_d^{(1)} r_d^{(2)}\big)^{3/2}}{md^5} . \qquad (20\text{-}18)$$

It also *suggests* the form for the sd and pd matrix elements. We have seen that both the s and the p states have behaved as if $l = 1/2$, giving ss, sp, and pp matrix elements scaling as $\hbar^2/(md^2)$. This suggests the form

$$V_{ldm} = \eta_{ldm} \frac{\hbar^2 r_d^{3/2}}{md^{7/2}} , \qquad (20\text{-}19)$$

with universal dimensionless constants $\eta_{sd\sigma}$, $\eta_{pd\sigma}$, and $\eta_{pd\pi}$.

This form differs from the form in Eq. (20-16) obtained by Andersen and Jepsen, and at this stage is simply a speculation. We shall confirm in Section 20-E, using transition-metal pseudopotential theory, that this form is correct, and shall see how the coefficients η_{ldm} are calculated. Of the set with $l = 0$ or 1, Muffin-Tin Orbital theory gives only a prediction of the ratio $\eta_{pd\sigma}/\eta_{pd\pi}$, since it can only compare matrix elements of the same $d^{-(l+l'+1)}$-dependence; for that ratio, it gives

$-3^{1/2}$, and we use the somewhat higher value of -2.17 obtained in Chapter 19 from fitting the perovskite bands. We also saw that Muffin-Tin Orbital theory predicted $\eta_{pp\sigma} = -2\eta_{pp\pi}$ rather than the $\eta_{pp\sigma} = -4\eta_{pp\pi}$ that we obtained from fitting the semiconductor bands.

The Atomic Sphere Approximation

Let us turn now to the values of the parameters W_d and E_d, or r_d and k_d, for individual materials. We have defined W_d for the body-centered cubic structure as the energy difference between the band energies H_{12} and H'_{25}. At the corresponding wave number of $2\pi/a[100]$, the phase factors in Eq. (20-1) are exactly the negative of each other for nearest neighbors. Analysis of these states on the basis of symmetry or in terms of the approximate states we have constructed shows that the wave functions labelled H'_{25} vanish on the plane bisecting the nearest-neighbor vectors, while those labelled H_{12} have vanishing normal derivative on that plane. Thus, these two energies could be obtained from a solution of the Schroedinger equation within the atomic cell, using these boundary conditions. Andersen (1973) replaced the cell by a sphere of equal volume, as did Wigner and Seitz (1934), and applied the vanishing value and vanishing normal derivative conditions to obtain the top and the bottom of the d band respectively, and therefore, W_d. This is illustrated in Fig. 20-14. He also defined the center of each band by the condition that a muffin-tin orbital constructed at that energy be a solution inside as well as outside the sphere; that is, the ratio of the radial derivative to the value must be $-(l+1)/r_0$ at the sphere. This energy plays the same role as the ε_l in LCAO theory and is the average energy of the band. In the Atomic Sphere Approximation, all values are obtained by radial integrations in a spherical potential, integrations that were carried out for all transition metals by Andersen and Jepsen (1977). Some reduction of their parameters is necessary to obtain estimates of W_d and E_d. In particular, they defined effective masses m_l that scale the bands; that is, W_d will be proportional to $\hbar^2/(m_d r_0^2)$. Here m_d is evaluated at the observed r_0. The coefficient is approximately 25/2 (Pettifor, 1977a, Eq. 10), and we chose a value close to that, namely, $W_d = 12.39\hbar^2/(m_d r_0^2)$, to fit the chromium bands, as indicated in the preceding section. The resulting W_d for all transition metals are given in Table 20-4, and were obtained by using values for r_0 (also listed in the Solid State Table) and m_d given by Andersen and Jepsen (1977). The values of r_d obtained from these, using Eq. (20-9), are listed in the Solid State Table.

Obtaining E_d from Andersen and Jepsen is not quite so unambiguous. They list the energy difference between the center of the d band and the center of the s band, but E_d is defined as the difference between the center of the d band and the *bottom* of the s band. We must therefore add half the s-band width, proportional to $\hbar^2/(m_s r_0^2)$. The m_s values from Andersen and Jepsen (1977) have been used and the coefficient has been chosen to be 1.08, which is the average of the three values fitted to obtain the E_d values for the three body-centered cubic metals listed in Table 20-3. This gives the values listed in Table 20-4. To obtain the k_d values listed in the Solid State Table, E_d has been written to equal $\hbar^2 k_d^2/2m_s$, using the m_s values

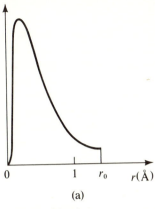 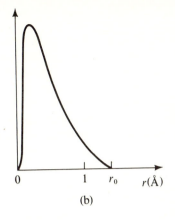

(a)   (b)

FIGURE 20-14

The minimum d-band energy is obtained in the Atomic Sphere
Approximation by using the atomic potential, as in Fig. 20-13,a, for $r < r_0$,
and choosing an energy such that the corresponding wave function has
vanishing slope at $r = r_0$, as shown in part (a). The top of the d band is
obtained by choosing an energy such that the corresponding wave
function has vanishing *value* at $r = r_0$, as shown in part (b). The center
of the band is obtained by choosing an energy such that the corresponding
integration in the muffin-tin potential, as in Fig. 20-13,b, contains no term
increasing as r^2 (that is, $A = 0$).

obtained from Eq. (20-48). These steps for obtaining both W_d and E_d have simply
reduced data, since the Atomic Sphere Approximation can give the two parameters directly if programmed to do that.

We have used the Atomic Sphere Approximation only to obtain values for W_d
and E_d. The approach can be used as the basis for more complete calculations, as
Pettifor (1977a,b) did in his detailed study of the $4d$ series.

Renormalized Atom Theory

An approach that is very closely related to the Atomic Sphere Approximation is
the ***Renormalized Atom Theory***, introduced first by Watson, Ehrenreich, and
Hodges (1970) (see also Watson and Ehrenreich, 1970, Hodges et al., 1972, and
particularly Gelatt, Ehrenreich, and Watson, 1977). The name derives from the
way the potential is constructed: a charge density for each atom is constructed on
the basis of atomic wave functions that are truncated at the Wigner-Seitz, or
atomic, sphere. The charge density from each state is then scaled up (renormalized) to make up for that density beyond the sphere which has been dropped.

This gives an approximation to the electron density; the potentials, including
free-electron exchange, derived from that density have proved to be very successful. The minimum and maximum energies for the bands are obtained from the
solution of the Schroedinger equation in a spherically symmetric potential, with a
boundary condition on the atomic sphere appropriate to bonding and antibonding states, as described above. Then the Fermi energy relative to that conduction-
band minimum is obtained through an approximation to an ingenious scheme

due to Lloyd (1967): if the only band of interest were free-electron-like, the Fermi energy would be obtained directly from the known average electron density and free-electron theory—$\hbar^2 k_F^2 / 2m$, with k_F given by (Eq. 15-4). However, the potential of each atom has given rise to d bands and, correspondingly, to extra electrons localized around each atom. The extra number of electrons, in states up to some cut-off energy, localized at an atom, can be obtained from the phase shifts (discussed in Section 20-E), using the Friedel sum rule from Eq. (20-56). The magnitude of that correction, as a function of energy, can be obtained from the potential within the atomic sphere, making it possible to adjust the Fermi energy to obtain the correct total number of electrons (n per atom in the Dn column of the transition series) without carrying out a band calculation. We accomplished the same goal approximately in Section 20-C by assuming a very simple model for the density of states in the calculation of Z_s.

Coherent Potential Approximation

Mention should perhaps also be made of a method by which the description of electronic structure can be extended to alloys of the transition metals. This method is called the **Coherent Potential Approximation** (Soven, 1967; Velicky et al., 1968; see also an account by Callaway, 1976, p. 440ff). In this method, the idea is to replace the part of the system outside of each sphere by an effective medium so that again it is possible to have a single-sphere problem. The properties of that effective medium are determined by the other spheres, however, and a self-consistency condition can be constructed such that in the effective medium there is no scattering in an average sense by the sphere in question. This allows a nonperturbative approximation to the problem (in contrast to pseudopotential approximations), which can be important when the individual potentials are strong enough to produce localized states or new bands. The coherent potential theory is sophisticated both conceptually and mathematically; it is not practical to describe it further here.

20-E d Resonances and Transition-Metal Pseudopotentials

Two other views of the d-state electronic structure should be introduced. They are related mathematical formulations but, as with the use of both LCAO and pseudopotential descriptions of covalent solids, the different approaches are relevant for answering different questions.

Scattering Theory

Imagine again that spheres are constructed around each atom of a transition metal. A muffin-tin potential, constant between spheres and spherically symmetric within, is assumed. In the context of this section, it will be best to let the spheres be nonoverlapping.

With this view of the metal it is rigorously correct to describe the electrons as free between the spheres and to consider the effect of the spheres as being that of scattering the electrons; that is, the entire effect of the spheres is absorbed in the boundary condition on the wave function at the sphere surfaces. This is very much the way we proceeded in the pseudopotential method, but when *d* bands become important, the complexity of their electronic structure must be reflected in complexity of the scattering. We may understand this effect by considering a single atom and, for convenience, it may be placed at the center of a large sphere (that is, of many atomic volumes) of radius R, with the potential constant for $r_{mt} < r < R$. The states of the system can be constructed to have the angular dependence of the spherical harmonics Y_l^m, and the complexity of the *d* states involves only the states for $l = 2$.

The solution of the corresponding radial equation is a familiar problem—we follow Schiff (1968); outside the atomic sphere, there are two independent solutions, the spherical Bessel function and Neumann function, $j_l(kr)$ and $n_l(kr)$, both corresponding to energy $\hbar^2 k^2/2m$. For conceptual purposes, they may be thought of as a sine and cosine function, respectively. The general solution outside the sphere becomes

$$R_l(r) = A[\cos \delta_l j_l(kr) - \sin \delta_l n_l(kr)]. \tag{20-20}$$

The two arbitrary constants have been written $A \cos \delta_l$ and $-A \sin \delta_l$. The asymptotic form at large r (following Schiff, 1968) is

$$R_l(r) \to (kr)^{-1} A \sin\left(kr - \frac{l\pi}{2} + \delta_l\right). \tag{20-21}$$

We fit to a vanishing boundary condition at R (it is $R_l(R) = 0$) and must fit also at the atomic sphere. Only the j_l is regular at the origin, so if the potential had the same constant value within the sphere, the correct solution would be simply j_l; that is, the "phase shift" δ_l would be zero. The effect of a scattering potential is simply to introduce a nonzero phase shift, which indeed simply shifts the phase of the sinusoidal oscillations in the asymptotic form given in Eq. (20-21).

The effect of the scattering center, for $l = 2$, is completely specified by specifying the values of δ_l as a function of energy. In this context the presence of the *d* state is called a *d* resonance and produces phase shifts, given for energy E near the resonance energy E_d by

$$\tan \delta_l(E) = \frac{\Gamma}{2(E_d - E)}, \tag{20-22}$$

with Γ a constant (see Messiah, 1965, p. 396ff; for further discussion, see Harrison, 1970, p. 197ff). Thus the *d* resonance is specified by the two numbers E_d and Γ, just as the *d* bands were specified by E_d and W_d in the previous sections. We shall in fact relate Γ and W_d, but should first notice how properties are deduced from phase shifts, particularly for the resonance specified by Eq. (20-22).

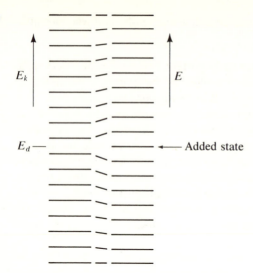

E_k

E_d —

E

— Added state

FIGURE 20-15

A schematic diagram showing the addition of an energy level, due to the introduction of a scattering resonance.

In the absence of scattering, δ_l vanishes and the outer boundary condition is satisfied by setting $kR - \pi$ equal to an integer times π (see Eq. 20-21). This gives a grid of levels in direct analogy with that in simple metals, as in Eq. (15-2). The resonance written at Eq. (20-22), however, gives a phase shift that increases from zero to π over an energy range of order Γ, which has the effect of squeezing an additional level into the sequence, as illustrated in Fig. 20-15. Levels with and without scattering match far from resonance but are shifted near the resonance, and the five additional levels (for the five m values) are the result of adding the atomic potential.

We may estimate the total shift in energy for an electron gas (in a free-electron band) due to the addition of this single atom. The outer boundary condition becomes

$$kR - \pi + \delta_l = \text{Integer} \cdot \pi. \tag{20-23}$$

The shift in energy due to the phase shift may be written

$$\delta E = (\hbar^2 k/m)\delta k = -\hbar^2 k \delta_l/(mR), \tag{20-24}$$

and the number of levels, counting spin and angular momentum components, in the range dk is

$$dn = (R/\pi)2(2l + 1)\, dk, \tag{20-25}$$

so that the total shift in band energy becomes

$$\delta E_b = \int \delta E\, dn = -2(2l + 1) \int_0^{E_F} dE\, \delta_l/\pi. \tag{20-26}$$

When the Fermi energy is very far below the resonance, so that $\tan \delta_l$ can be replaced by δ_l in Eq. (20-22), the energy shift looks very much like ordinary perturbation theory,

$$\delta E_b = 2(2l + 1) \int_0^{E_F} \frac{\Gamma/(2\pi)}{E - E_d} \, dE. \tag{20-27}$$

If we didn't make this approximation to Eq. (20-22), we could treat cases with E_F near resonance, using Eq. (20-22), with the denominator written $E - E_d + i\Gamma/2$, and with real parts taken at the end of the calculation (see Moriarty, 1972a), but Eq. (20-27) will suffice for identifying parameters.

Now let us develop the relation between Γ and the band width. The derivation is due to Heine (1967), who applies the same boundary conditions at the atomic sphere that we described for the Atomic Sphere Approximation. The bottom of the *d* band was determined by setting the gradient of the *d* state for each atom equal to zero at its atomic sphere, or

$$\cos \delta_l j_l'(kr_0) - \sin \delta_l n_l'(kr_0) = 0. \tag{20-28}$$

Similarly, the highest-lying *d*-band state was determined by setting the value equal to zero,

$$\cos \delta_l j_l(kr_0) - \sin \delta_l n_l(kr_0) = 0. \tag{20-29}$$

We may then solve for $\tan \delta_l$ in the two cases and use Eq. (20-22) to obtain the two energies. We then subtract, to obtain

$$W_d = \frac{\Gamma}{2} \left(\frac{n_l'}{j_l'} - \frac{n_l}{j_l} \right). \tag{20-30}$$

Heine then made the small-argument expansion of the Bessel functions $j_2(\sigma) \to \sigma^2/15$, $n_2(\sigma) \to -3/\sigma^3$ (see Schiff, 1968, p. 85) and evaluated the result at the wave number k_d, which gives $W_d = 225\Gamma/(4k_d^5 r_0^5)$; or, by taking W_d from the second form in Eq. (20-9), we may write

$$\Gamma = 0.1214 \hbar^2 k_d^5 r_d^3/m, \tag{20-31}$$

which is the required result. The small-argument expansion is questionable since $k_d r_0$ is typically 1.5, but perhaps it can be justified by Andersen's argument (Section 20-D) that the potential in the flat part of the muffin-tin potential is arbitrary. We may evaluate Eq. (20-31), using parameters from the Solid State Table. The results are given in Table 20-4. A comparison of the 0.56 eV for copper can be made with the value given by Moriarty (1977b) of 0.33 eV. (Moriarty called the width of the resonance W_d rather than Γ.) The resonance width Γ is seen to drop from left to right in each transition-metal series.

At first glance, it might seem that in obtaining the form for W_d given just before Eq. (20-31), we rederived the r_0^{-5}-dependence of W_d, assuming Γ and k_d were independent of r_0. In fact, as indicated in our discussion of d states in Section 20-D, k_d is expected to vary as r_0^{-1}, so Eq. (20-31) tells as that Γ should scale with r_0^{-5} as W_d does.

Transition-Metal Pseudopotentials

The expansion in $\Gamma/(E_d - E_F)$ would suggest the possibility of incorporating the effect of the resonance in perturbation theory and thus extending the pseudo-potential perturbation theory of simple metals to transition metals. This has in fact been done (Harrison, 1969) and the pseudopotential approach has been extensively developed by Moriarty (1970, 1972a,b,c), but the applications have been largely restricted to the ends of the transition series where the expansion is clearest.

Extension of pseudopotential theory to the transition metals preceded the use of the Orbital Correction Method discussed in Appendix E, but transition-metal pseudopotentials are a special case of it. In this method, the states are expanded as a linear combination of plane waves (or OPW's) plus a linear combination of atomic d states. If the potential in the metal were the same as in the atom, the atomic d states would be eigenstates in the metal and there would be no matrix elements of the Hamiltonian with other states. However, the potential *is* different by an amount we might write $\delta V(\mathbf{r})$, and there are, correspondingly, matrix elements $\langle \mathbf{k}|H|d\rangle = \langle \mathbf{k}|\delta V|d\rangle$ hybridizing the d states with the free-electron states. The full analysis (Harrison, 1969) shows that the correct perturbation differs from δV by a constant. The hybridization potential is

$$\Delta(\mathbf{r}) = \delta V - \langle d|\delta V|d\rangle. \tag{20-32}$$

Notice that adding a constant to δV does not modify $\langle \mathbf{k}|\Delta|d\rangle$ and a constant potential *should* not couple the states. Notice also that if $|\mathbf{k}\rangle$ and $|d\rangle$ were orthogonal, $\langle \mathbf{k}|\Delta|d\rangle$ would equal $\langle \mathbf{k}|\delta V|d\rangle$. The hybridization potential shifts the energy of a free-electron state,

$$E_k = \frac{\hbar^2 k^2}{2m} + \langle \mathbf{k}|W|\mathbf{k}\rangle + \sum_d \frac{\langle \mathbf{k}|\Delta(\mathbf{r})|d\rangle\langle d|\Delta(\mathbf{r})|\mathbf{k}\rangle}{E_k - E_d}. \tag{20-33}$$

Thus the effect of hybridization with the d states has been to replace the pseudo-potential by

$$W + \sum_d \frac{\Delta(\mathbf{r})|d\rangle\langle d|\Delta(\mathbf{r})}{E_k - E_d}. \tag{20-34}$$

The hybridization term can be written as a sum of individual atomic terms just as the pseudopotential was, so that it simply modifies the form factor associated with the coupling between different plane waves as illustrated in Fig. 20-16 (see Harrison, 1969, or Harrison, 1970). The diagonal matrix element—the term added to $\langle \mathbf{k} | W | \mathbf{k} \rangle$ in Eq. (20-33)—will be seen to contribute to the effective mass.

Of more interest here is a complementary application of the theory in which the hybridization potential is used to obtain a coupling between *d* states on different

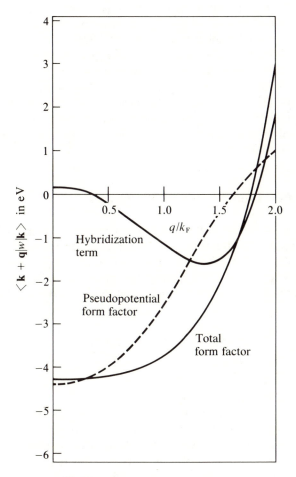

FIGURE 20-16

The pseudopotential form factor, hybridization term, and total form factor, obtained by Moriarty (1972) for copper. This includes a greatly improved treatment of exchange due to Lindgren (1971), in comparison to that used earlier by Harrison (1969, 1970) and Moriarty (1970). [After Moriarty, 1972b.]

atoms through the free-electron system, as suggested by Pettifor (1969); this will give explicit forms for the interatomic matrix elements

$$V_{ddm} = \sum_k \frac{\langle d' | \Delta | \mathbf{k} \rangle \langle \mathbf{k} | \Delta | d \rangle}{E_d - E_k}, \tag{20-35}$$

where the state $|d'\rangle$ is on a different atom from $|d\rangle$ and both states are taken to have the same angular momentum quantum number m with respect to the internuclear axis. Such a form follows quite directly from perturbation theory and was used by Moriarty (1975) to relate the V_{ddm} to the resonance width. We shall obtain much simpler forms, using ideas from Muffin-Tin Orbital theory, forms that are also obtainable from Moriarty's by taking the appropriate limit.

Andersen's assumption of a muffin-tin potential equal to the energy of the d state, which was used in the preceding section to obtain muffin-tin orbitals, is used here in two ways. (This analysis has only very recently been made; see Harrison and Froyen, 1979.) First, it is used to obtain the matrix elements $\langle \mathbf{k} | \Delta | d \rangle$ and second, it is used in writing $E_d - E_k = -\hbar^2 k^2 / 2m$ in Eq. (20-35). We begin with the evaluation of $\langle \mathbf{k} | \Delta | d \rangle$.

Evaluation of the Hybridization Matrix Element

We again take the flat portion of the muffin-tin potential equal to the energy ε_d, as we did in Fig. 20-13,b, but now it is desirable to take the muffin-tin radius r_{mt} smaller than r_0, such that the potential is continuous at r_{mt}. Fig. 20-13 has been modified to do this, in Fig. 20-17. Taking this muffin-tin potential as the potential in the metal, we immediately see that the potential in the metal minus that in the free atom is simply

$$\begin{aligned} \delta V(r) &= \varepsilon_d - V_a(r) & \text{for } r \geq r_{mt}; \\ \delta V(r) &= 0 & \text{for } r \leq r_{mt}, \end{aligned} \tag{20-36}$$

where $V_a(r)$ is the free-atom potential.

In order to write the matrix element, we select a coordinate system with z-axis along the vector from the $|d'\rangle$ atom to the $|d\rangle$ atom, and write the d state as $|d\rangle = |2, m\rangle = R_{n2}(r) Y_2^m(\theta_1, \varphi_1)$, as in Eq. (1-19). The plane wave, normalized in the volume Ω of the system, can also be written in terms of spherical harmonics centered on the same atom (Schiff, 1968, p. 119); thus

$$e^{i\mathbf{k} \cdot \mathbf{r}} = (4\pi/\Omega)^{1/2} \sum_l (2l + 1)^{1/2} j_l(kr) Y_l^0(\theta_2, \varphi_2), \tag{20-37}$$

where the angles in this case are measured from a coordinate system with the z-axis along $\mathbf{k}$. These spherical harmonics can be written in the coordinate system

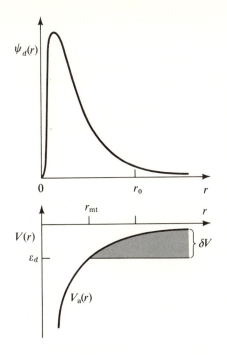

FIGURE 20-17

The atomic d state in chromium and the atomic potential, $V_a(r)$, as in Fig. 20-13. A muffin-tin potential is constructed with r_{mt} chosen to make the potential continuous, and equal to ε_d for $r \geq r_{mt}$. The difference between the muffin-tin potential and V_a is the shaded region and is the potential causing hybridization of the d state.

θ_1, φ_1, using Eqs. (19-21) and (19-28), and when the angular integration is performed, the only contributing term will be that for $l = 2$ and the m associated with the atomic d state in question,

$$\langle \mathbf{k} | \delta V | 2, m \rangle = \frac{4\pi}{\sqrt{\Omega}} Y_2^m(\theta_k, \varphi_k) \int_0^\infty j_2(kr) \, \delta V(r) R_{n2}(r) r^2 \, dr, \quad (20\text{-}38)$$

where θ_k and φ_k give the orientation of $\mathbf{k}$ with respect to the coordinate system with z-axis along the internuclear distance. To complete the evaluation of the matrix elements of the hybridization potential $\Delta(\mathbf{r})$, we also need

$$\langle 2, m | \delta V | 2, m \rangle = \int_0^\infty R_{n2}^2(r) \, \delta V r^2 \, dr, \quad (20\text{-}39)$$

and

$$\langle \mathbf{k} | 2, m \rangle = \frac{4\pi}{\sqrt{\Omega}} Y_2^m(\theta_k, \varphi_k) \int_0^\infty j_2(kr) R_{n2}(r) r^2 \, dr. \quad (20\text{-}40)$$

For purposes of computation it is very convenient to use Eq. (20-36) and the Schroedinger equation to write the matrix elements of δV in terms of matrix elements of the kinetic energy operator, and that evaluation has been carried out (Harrison and Froyen, 1979). Here we shall only make use of the form of the results that have been obtained.

The most important point is that for given $\mathbf{k}$, the integrals appearing in

Eqs. (20-38), (20-39), and (20-40) depend only on the potentials and wave functions of the free atom and not on any parameter from the metal. This remarkable result has come directly from Andersen's assumption about the form of the potential, Eq. (20-36). The evaluation is made for a well-localized d state, so that $j_2(kr)$ can be replaced by its small-argument form $k^2r^2/15$. Eq. (20-38) becomes,

$$\langle \mathbf{k} | \delta V | 2, m \rangle = \frac{4\pi}{\sqrt{\Omega}} Y_2^m(\theta_\mathbf{k}, \varphi_\mathbf{k}) \frac{k^2}{15} \int_0^\infty r^4 \, \delta V \, R_{n2}(r) \, dr. \tag{20-41}$$

We write the volume of the system as the number of atoms N_a times the atomic cell volume, $4\pi r_0^3/3$, and notice that since the matrix element has units of energy, the integral must have units of length to the three-halves power times $\hbar^2/m$. The same form obtains for the product $\langle \mathbf{k}|d\rangle\langle d|\delta V|d\rangle$, appearing in $\langle \mathbf{k}|\Delta|d\rangle$, so we may write the hybridization matrix element in the form

$$\langle \mathbf{k} | \Delta | 2, m \rangle = \frac{\hbar^2 k^2}{N_a^{1/2} m} \left(\frac{r_d}{r_0}\right)^{3/2} Y_2^m(\theta_\mathbf{k}, \varphi_\mathbf{k}), \tag{20-42}$$

where r_d is called the **d-state radius** and is a length that is characteristic of the atom. It could be obtained from integrals given here by using the tabulated atomic d orbitals. We have instead obtained it by matching results with those of the Atomic Sphere Approximation.

The small-r expansion, which led to Eq. (20-41), can readily be tested by performing the integral in Eq. (20-38), and this has been done (Harrison and Froyen, 1979). It was found that there are sizable corrections, leading to additional terms in Eq. (20-42) proportional to k^4 and higher. By fitting our r_d values to the final band width W_d given by Andersen and Jepsen (1977), we partially correct for not including these terms and retain the simpler theory based upon Eq. (20-42). Given this expression and values of r_d from the Solid State Table, we could of course make direct use of pseudopotential theory as in the simple metals, though now using the transition-metal pseudopotential of Eq. (20-34).

Interatomic Matrix Elements

Of more interest here is Eq. (20-35) for the coupling between d states on different atoms. The term E_k is a free-electron energy, measured from the flat portion of the muffin-tin potential, which, with Andersen's assumption, is taken at the d-state energy ε_d; thus the denominator in Eq. (20-35) is simply $-\hbar^2k^2/2m$. If the origin of coordinates is taken at the atom with the state $|d'\rangle$, we must shift the origin of coordinates in obtaining the second matrix element of the hybridization potential, in order to use Eq. (20-42) directly. This gives a factor $e^{i\mathbf{k}\cdot\mathbf{d}}$, where $\mathbf{d}$ is the vector distance to the second atom. Then Eq. (20-35) becomes

$$V_{ddm} = \frac{\hbar^4}{N_a m^2} \left(\frac{r_d}{r_0}\right)^3 \sum_k \frac{k^4 Y_2^{m*}(\theta_k, \varphi_k) Y_2^m(\theta_k, \varphi_k) e^{i\mathbf{k}\cdot\mathbf{d}}}{-\hbar^2k^2/2m}. \tag{20-43}$$

The sum over wave numbers is written as an integral over wave number space, with a density of states in wave number space of $\Omega/(2\pi)^3 = N_a r_0^3/(6\pi^2)$. The exponent $\mathbf{k} \cdot \mathbf{d}$ is simply $kd \cos \theta_k$, and the spherical harmonics can be written out in terms of the associated Legendre functions, $P_2^m(\cos \theta_k)e^{im\varphi_k}$ (Schiff, 1968, p. 80), and the integral over φ_k performed

$$V_{ddm} = -\frac{5\hbar^2 r_d^3 (2 - |m|)!}{6\pi^2 m(2 + |m|)!} \int e^{ikd \cos \theta_k} P_2^m(\cos \theta_k)^2 \sin \theta \, d\theta \, k^4 \, dk. \quad (20\text{-}44)$$

The evaluation of this integral is straightforward but tedious: we substitute for the associated Legendre functions, obtaining fourth-order polynomials in $\cos \theta_k$ in the integrand. The leading terms in $(P_2^m)^2$ are $9/4 \cos^4 \theta_k$, $-9 \cos^4 \theta_k$, and $9 \cos^4 \theta_k$, for $m = 0$, 1, and 2, respectively. The integral

$$\int_0^\pi e^{ikd \cos \theta_k} \cos^4 \theta_k \sin \theta_k \, d\theta_k$$

contains a term $4!2 \sin kd/(kd)^5$ plus terms that diverge less strongly at $k = 0$. We evaluate the integral over k by closing a contour in the upper half plane and, because of the k^4 in the integrand, only this term diverging as $1/k^5$ has a pole. We obtain, for the integral, $4! \pi/d^5$ multiplied by $9/4$ for $m = 0$, by -9 for $m = 1$, and by 9 for $m = 2$. This is substituted into Eq. (20-44) to obtain

$$V_{ddm} = \eta_{ddm} \frac{\hbar^2 r_d^3}{md^5}, \quad (20\text{-}45)$$

with $\eta_{dd} = -45/\pi$, $\eta_{dd\pi} = 30/\pi$, and $\eta_{dd\delta} = -15/(2\pi)$. This is an exact evaluation of Eq. (20-43) and is the desired result. It has been used with a calculation of the body-centered cubic bands just as it was in Eq. (20-8), but including $V_{dd\delta}$ as well as $V_{dd\sigma}$ and $V_{dd\pi}$, to obtain the relation, in Eq. (20-9), between W_d and r_d. In treating the bands, and in making the Solid State Table, $\eta_{dd\delta}$, which is found here to be one sixth of $\eta_{dd\sigma}$, was taken equal to zero, and $\eta_{dd\sigma}$ and $\eta_{dd\pi}$ were adjusted slightly, but such as to retain the same relation between W_d and r_d. Exactly the same ratios between the different matrix elements were obtained by Andersen, Klose, and Nohl (1978), by using the Muffin-Tin Orbital theory, and they obtained the same d^{-5}-dependence. Here we have added the relation to the r_d from pseudopotential theory.

The Free-Electron Band

The connection with pseudopotential theory will also enable us to obtain other parameters of the energy bands in terms of the d-state radius r_d. We now have an electronic structure based upon d states, which are coupled to each other according to Eq. (20-45). They are also coupled to free-electron bands, as seen in

Eq. (20-33). That equation requires more scrutiny now that we have the form of the hybridization matrix elements. Let us look first at the energy at $\mathbf{k} = 0$, $\hbar^2 k^2/2m = 0$ and, since $\langle \mathbf{k}|\Delta|d\rangle$ is proportional to k^2, the final term is zero also. The energy is simply $\langle 0|W|0\rangle$. But we have measured energies from the flat portion of the muffin tin potential, which was taken at the d-state energy, so it follows that

$$\langle 0|W|0\rangle = -E_d. \tag{20-46}$$

We also consider states for $\mathbf{k} \neq 0$. It is assumed that $\langle 0|W|0\rangle$ is independent of energy, as in the simple metals, but the final term in Eq. (20-33) gives an interesting contribution. The denominator is written as $\hbar^2 k^2/2m$, and the sum becomes

$$\sum_d \frac{\langle \mathbf{k}|\Delta|d\rangle\langle d|\Delta|\mathbf{k}\rangle}{\hbar^2 k^2/2m} = \frac{2\hbar^2 k^2 r_d^3}{N_a m r_0^3} \sum_d Y_2^m(\theta_k, \varphi_k)^* Y_2^m(\theta_k, \varphi_k). \tag{20-47}$$

The wave number is fixed in the sum, so the d states can be constructed with z-axis along $\mathbf{k}$. Only the term with $m = 0$ is nonzero, and for it the summand is $5/(4\pi)$ (Schiff, 1968, p. 80). There is one term for each of the N_a atoms, so that Eq. (20-47) becomes

$$\sum_d \frac{\langle \mathbf{k}|\Delta|d\rangle\langle d|\Delta|\mathbf{k}\rangle}{\hbar^2 k^2/2m} = \frac{5\hbar^2 k^2 r_d^3}{\pi 2 m r_0^3}.$$

When added to the first term in Eq. (20-33), the result can be written $\hbar^2 k^2/2m_s$, with the effective mass given by

$$\frac{m_s}{m} = \left(1 + \frac{5r_d^3}{\pi r_0^3}\right)^{-1}. \tag{20-48}$$

Interestingly enough, it has been possible to write the free-electron effective mass in terms of the same d-state radius r_d that determined the d-band width and determined hybridization with the free electrons. It was noted in the discussion of the Atomic Sphere Approximation that Andersen defined an effective mass for d electrons in terms of the d-band width, $W_d = 12.5\hbar^2/(m_d r_d^2)$. This can be combined with the expression for W_d in terms of r_d (Eq. 20-9), to write $m_s/m = (1 + 2.91m/m_d)^{-1}$. In the Atomic Sphere Approximation m_s, m_p, and m_d are regarded as independent quantities, but both the m_s and m_p values given by Andersen and Jepsen (1977) are rather close to the effective mass m_s obtained from $m_s/m = (1 + 2.91m/m_d)^{-1}$.

Notice that the shift in the mass m_s from the free-electron mass is a consequence of the coupling with the d states. As we now go on to treat the coupling more completely we must use $\hbar^2 k^2/(2m) + \langle 0|W|0\rangle$ for the uncoupled band; use of an m_s in that expression would count the effect of the coupling twice.

sd- and *pd*-Coupling

Finally, let us use the transition-metal pseudopotential theory to estimate matrix elements between *d* states and *s* and *p* states. These are not so useful in the transition metals themselves since the description of the electronic structure is better made in terms of *d* bands coupled to free-electron bands, $\hbar^2 k^2/(2m) + \langle 0|W|0\rangle$, rather than in terms of *d* bands coupled to *s* and *p* bands. However, the matrix elements $V_{sd\sigma}$, and so forth, directly enter the electronic structure of the transition-metal compounds, and it is desirable to obtain these matrix elements in terms of the *d*-state radius r_d. We do this by writing expressions for the bands in terms of pseudopotentials and equating them to the LCAO expressions obtained in Section 20-A.

We consider a plane wave with wave number **k** along the *z*-direction. From Eq. (20-42) it is seen to be coupled only to *d* states with $m = 0$, since $Y_0^m(0, \varphi) = [5/(4\pi)]^{1/2}$ for $m = 0$ and equals zero for $m = \pm 1$ or ± 2 (Schiff, 1968). We construct a Bloch sum (Eq. 20-1) for $m = 0$ states and sum the N_a terms—each is given by Eq. (20-42)—entering the matrix element between $|\mathbf{k}\rangle$ and that Bloch sum. We write that matrix element as

$$E_k^{sd} = \frac{\hbar^2 k^2}{m} \left(\frac{r_d}{r_0}\right)^{3/2} \left(\frac{5}{4\pi}\right)^{1/2}, \tag{20-49}$$

which should be identified with the E_k^{sd} given in Table 20-2 for the LCAO description

$$\frac{\hbar^2 k^2}{m} \left(\frac{r_d}{r_0}\right)^{3/2} \left(\frac{5}{4\pi}\right)^{1/2} = -V_{sd\sigma}^{2\mathrm{nd}} 4 \sin^2(ka/2). \tag{20-50}$$

It was a second-neighbor matrix element entering that table and it corresponded to an internuclear distance of the body-centered cubic cube edge *a*. Thus, writing $4\pi r_0^3/3 = a^3/2$ for the body-centered cubic structure and expanding the sine for small *k*, we may write

$$V_{sd\sigma}^{2\mathrm{nd}} = \eta_{sd\sigma} \frac{\hbar^2}{m} \frac{r_d^{3/2}}{a^5}, \tag{20-51}$$

with $\eta_{sd\sigma} = -(10/3)^{1/2} = -1.83$, and for this matrix element, $a = d$.

The value of $\eta_{sd\sigma}$ obtained by fitting the perovskite bands in Chapter 19 is nearly twice this, -3.16. Perhaps it is not surprising that this is the largest discrepancy that we have encountered between predicted and adjusted $\eta_{ll'm}$. It is one of the largest extrapolations in going from the small oxygen–transition-metal distance (1.95 Å for $SrTiO_3$) in the perovskite to the large second-neighbor distance in the body-centered cubic metal (2.88 Å for Cr). Furthermore, the representation of the free-electron states in the nearest-neighbor *s*-band approximation is very crude; the corresponding nearest-neighbor V_{ss} was also much different from the $V_{ss\sigma}$ obtained from the Solid State Table. Representation of the free-electron bands by coupled s and p bands gives additional terms of the form of the left side of Eq. (20-50) but with a different numerical coefficient. Thus the matrix elements $V_{pd\sigma}$ also are of the form given in Eq. (20-51), as speculated in Eq. (20-19), but the values of the coefficients depend upon how they are evaluated, and the fit to the perovskites in Chapter 19 is probably closest to the way these matrix elements will be used. Thus those are the values listed in the Solid State Table.

For analysis of the transition metals themselves, the use of free-electron bands and LCAO d states is preferable. The analysis based upon transition-metal pseudopotential theory has shown that the interatomic matrix elements between d states, the hybridization between the free-electron and d bands, and the resulting effective mass for the free-electron bands can all be written in terms of the d-state radius r_d, and values for r_d have been listed in the Solid State Table.

The transition-metal pseudopotential theory is also of interest in its own right in allowing the calculation of a number of properties directly, just as they were calculated for the simple metals. These properties include the resistivity of liquid metals (Moriarty, 1970), structural energies, and vibration spectra (Moriarty, 1972b). These were carried out without the parameterization in terms of r_d that has been described here, and were applied to noble metals and alkaline earths (Moriarty, 1972c). Little has been done in the way of exploring the consequences of the simpler theory. It can be noticed immediately that the same approach is applicable to core states, with Eq. (20-34) simply giving corrections to the simplest pseudopotential theory. Indeed, the exploration of multi-ion interactions (Harrison, 1973b) in the simple metals just preceding and just following transition series (for examples, alkali metals and noble metals) suggested that these corrections to the pseudopotential form factor may be responsible for the particular structures that these metals take.

20-F   Local Moments and Magnetism

We turn finally to the question of magnetism in metals. The ultimate mechanism producing magnetism is the same effect of exchange energy that produced Hund's rule in the atom, paramagnetism in O_2, and atomic moments in the transition-metal monoxides. Exchange lowers the energy of an electron in proportion to the number of electrons of the same spin. Thus the exchange energy is minimum if all electrons are of the same spin. Opposing this effect is the increased band energy involved in transferring electrons from the lowest band states, occupied with up and down spin, to band states of higher energy. That band energy prevents simple metals from becoming ferromagnetic, but the high density of states for transition elements illustrated in Fig. 20-11 can readily reduce the band energy required to flip a spin by a factor of ten or more, which is why magnetism is found among transition metals.

Band Ferromagnetism

A calculation of the bands of a ferromagnetic metal based on free-electron exchange is not far more complicated than that of the nonmagnetic metal and simply requires the inclusion of spin-up bands and spin-down bands, with an exchange potential for each depending upon the occupation of states of the same spin. The bands and the occupation must be determined self-consistently such that the Fermi energies are the same in each. Such calculations have been carried out and appear to describe the ferromagnetic moment adequately in Ni (Wakoh and Yamashita, 1966; Connolly, 1967). We may picture the result by thinking of the nickel bands of Fig. 20-1 as representing spin-down states and the copper bands as representing spin-up states in nickel. Then a common Fermi energy must be constructed.

Let us carry out such a calculation for ferromagnetic metals, using the simplified density of states that was illustrated in Fig. 20-11, and then consider the effect of nonuniformities in the density of states. The more difficult concept is the electronic structure of a ferromagnet at temperature high enough (above the Curie temperature) that individual moments still reside on the ions but they become disordered. We shall return to that afterward.

Notice at the outset that the competition between different electronic states we are discussing here is analogous to, but fundamentally different from, the competition discussed in the transition-metal monoxides. In atomic terms, the competition there was between different configurations, such as between the $3d^5$ and $3d^4 4s$ configuration of the Mn^{2+} ion in MnO. The formation of a broad s band tended to favor the latter. However, whatever the number of electrons residing in d states in the end, those electrons align their spins according to Hund's rule; a net moment is produced and it is best not to attempt to describe those d states in terms of bands. In the transition metal, there is a competition between states of the system of the same configuration, such as $3d^5$, but with one of the spins flipped (for a total spin of 3/2 rather than the 5/2 following from Hund's rule). In this case the formation of a d band favors the reduced spin by allowing more electrons to occupy the lower portions of the band.

We return to the simple density of states of Fig. 20-11 and again write the number of electrons in the free-electron band as Z_s. Then, for the Dn column of the periodic table, there are $Z_d = n - Z_s$ electrons in the d bands. We wrote the band energy for these Z_d electrons in Eq. (20-12), taking equal numbers in the spin-up and spin-down bands. We now let a fraction $1/2 + x$ of the electrons have spin up and a fraction $1/2 - x$ have spin down, and rewrite that band energy as

$$\delta E_b = \delta E_b^0 + 20 \left(\frac{Z_d}{10} \right)^2 W_d x^2. \tag{20-52}$$

This has a minimum at $x = 0$, where its value δE_b^0 is that given in Eq. (20-12).

We next introduce an exchange interaction $-U_x$ for each pair of d electrons of the same spin on the same atom. It is convenient here, as discussed in Appendix A, to include an exchange interaction of each electron with itself. It may easily be verified that the same total exchange energy is obtained whether we sum over band states and divide by the number of atoms or simply compute the energy for $(1/2 + x)Z_d$ electrons in each atom of spin up and $(1/2 - x)Z_d$ electrons of spin down. The exchange energy for m electrons of the same spin in one atom is $-m^2 U_x/2$, including a $U_x/2$ self-energy per electron. This leads to an exchange energy per atom of

$$E_x = -Z_d^2 U_x \left(\frac{1}{4} + x^2 \right). \tag{20-53}$$

This energy is a maximum at $x = 0$, so the nonmagetic state with $x = 0$ becomes unstable when the coefficient of x^2 in Eq. (20-53) exceeds in magnitude the coefficient in Eq. (20-52); that is, when $Z_d^2 U_x > 20 Z_d^2 W_d/100$, or when

$$U_x > W_d/5. \tag{20-54}$$

When this condition is satisfied, the electron spins will spontaneously align as Hund's rule indicates they do in atoms. If $Z_d \leq 5$, all Z_d d electrons will be of the same spin; if $Z_d \geq 5$, all holes will be of the same spin, giving a net moment of $10 - Z_d$ per ion, as illustrated in Fig. 20-18.

For the approximate, uniform density of states $10/W_d$, the formation of a moment becomes an all-or-nothing proposition, called *strong ferromagnetism*. Clearly if the density of states were nonuniform, the energy might be minimized for an intermediate shift of the bands, called *weak ferromagnetism*. We shall see that both occur.

In order to make the condition quantitative, we require a value for U_x. It can be calculated from the integrals given in Appendix A (U_x is exactly the integral given in Eq. (A-9) with atomic d states for the ψ_i and ψ_j; there it is summed over all i and j and divided by two to give the sum over pairs, E_{exch}), but we shall take it from the corresponding atomic parameters determined from experiment. The d states are much the same in the atom and in the solid, so that should be accurate. From the spectra, one can learn the energy difference between different spin states of the atoms. One learns that this simple description with a single U_x is only approximate, but nevertheless useful. We have obtained parameters by starting with the energy of the configuration $d^{n-2}s^2$, which is usually the ground state of the atom and for which the spin of the two s electrons cancel. Then the lowest energy in this configuration has all d electrons of the same spin (Hund's rule) if the shell is less than half full ($n - 2 \leq 5$). If it is more than half full the d electrons in excess of five have reversed spin. Thus the total spin S increases by steps of $1/2$ until it is $5/2$ at the $D7$ column and then drops by steps of $1/2$. In all cases, flipping a spin reduces S by one and in this simple model increases the energy by $(2S - 1)U_x$. (Notice that $2S - 1$ is the *net* number of electrons with the same spin as the one flipped.) This energy is equated to the experimentally determined energy necessary for excitation to the lowest excited state of the same configura-

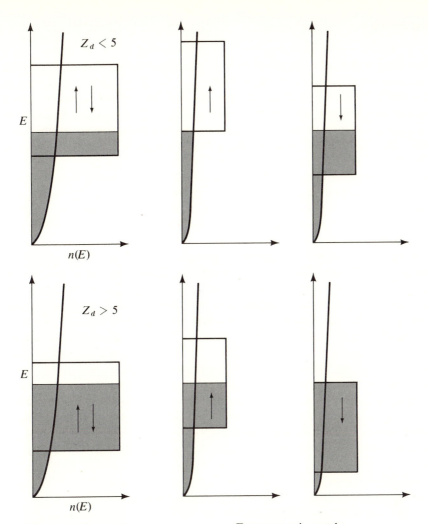

$n(E)$

Nonmagnetic metal   Ferromagnetic metal

FIGURE 20-18

On the left is a total density-of-states plot, similar to Fig. 20-11, with equal numbers of spin-up and spin-down electrons. Above, the total number of d electrons, Z_d, is less than five; below it is greater. If the system is unstable against ferromagnetism, electrons may shift, for example, to spin-down states, lowering their energy as shown in the spin-up and spin-down density-of-states plots to the right. When the shift is complete, as shown both above and below to the right, the system is said to have strong ferromagnetism.

tion, but with multiplicity (equal to $2S + 1$) decreased by one. For example, the ground state of manganese is ^6S (multiplicity 6 with the S here referring to zero orbital-angular momentum, not the spin S) and the excited state ^4G; both correspond to a configuration $3d^5 4s^2$. Their energy difference is 25,266 cm^{-1} = 3.13 eV, leading to a U_x of 0.78 eV. Values for all three transition series are given in Table 20-5. Also listed there is the critical value of U_x, suggested by Eq. (20-54): $W_d/5$.

The comparison is quite interesting. The condition for ferromagnetism is only close to being satisfied to the right of the $3d$ row, where indeed the magnetic materials occur (Pd, in the $4d$ row, might also be close). This is partly because U_x is largest in the $3d$ series, as expected, since the d states are more localized. It is also partly because the bands are narrower, in contrast to sp bands in simple solids, which are broader for the lighter elements.

Indeed, the criterion is satisfied for cobalt and nickel. For nickel this would suggest that the empty states associated with the Z_s missing d electrons per ion (nickel is in the $D10$ column) would all have the same spin, giving a magnetization of Z_s electronic magnetic moments (called **Bohr magnetons**) per ion. The observed moment in nickel (calculated from the measured saturation magnetization) is 0.60. This value of Z_s appears to be appropriate for the entire $3d$ series, though slightly higher values (0.81 for Ni from Table 20-4) were suggested by the model density of states of Fig. 20-11. The error is not large in view of the crudeness of the assumed $n_d(E)$. The other transition elements do not satisfy the criterion for ferromagnetism, though chromium, manganese, iron, and cobalt all show magnetic behavior. The discrepancy for iron and cobalt is not large considering the simplicity of the model, and we see that chromium and manganese are strongly affected by the

TABLE 20-5
Exchange interaction, U_x, in eV, for the transition metals, determined from experimental optical spectra (Moore, 1949, 1952), as indicated in the text. Also listed is the critical value of $U_x = W_d/5$, above which the uniform-density-of-states model predicts ferromagnetism; $W_d/5$ was obtained from the Solid State Table.

$2S - 1$:	1	2	3	4	3	2	1
3d series:	Ti	V	Cr	Mn	Fe	Co	Ni
U_x:	0.90	0.68	0.64	0.78	0.76	1.02	1.60
$W_d/5$:	1.22	1.35	1.31	1.12	0.96	0.87	0.76
4d series:	Zr	Nb	Mo	Tc	Ru	Rh	Pd
U_x:	0.63	0.48	0.60	—	0.88	0.81	—
$W_d/5$:	1.67	1.94	2.00	1.88	1.69	1.38	1.08
5d series:	Hf	Ta	W	Re	Os	Ir	Pt
U_x:	0.70	0.60	0.39	0.36	0.61	0.86	—
$W_d/5$:	1.91	2.22	2.29	2.20	2.06	1.74	1.40

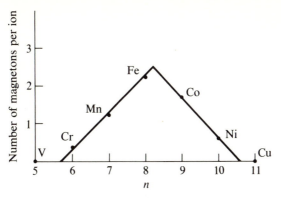

FIGURE 20-19

The Pauling-Slater curve, showing the number of electronic magnetic moments per ion as a function of column number in the periodic table. The plotted points are based on experiment (after Friedel, 1969), and the continuous lines show Friedel's prediction based upon $Z_s = 0.6$ and the involvement of only the upper-half d band.

nonuniformity of the density of states. In spite of the lack of numerical success, it is gratifying that the table suggests that the right side of the $3d$ series is the best prospect for magnetic behavior.

Notice that if the criterion for ferromagnetism *were* satisfied, the net moment would be expected to increase by one with each decrease in atomic number (up to a moment of five, after which it would decrease by one for each decrease in atomic number), as illustrated in Fig. 20-19. Indeed, this is seen to be the case for cobalt and iron, but then the moment drops for lower atomic numbers, indicating weak ferromagnetism. The resulting experimental curve is called the **Pauling-Slater curve** (Slater, 1937; Pauling, 1938), and it describes alloys as well as pure materials plotted as a function of n, the number of electrons per atom. Moments are shown for pure Cr and Mn, though they are antiferromagnetic (that is, they have antiparallel spin alignment). The origin of the antiferromagnetism in chromium was discussed in Section 20-B. Friedel (1969) explained this curve as arising from the splitting of the density of states for the body-centered cubic metals, shown here in Fig. 20-7, into two well-separated and equal peaks; most of the ferromagnetic alloys are also body-centered cubic. Thus it may be that only the upper sub-band participates, and the curve should turn down at a value of 2.5 rather than 5, in good agreement with experiment. This is consistent with the approximately constant Z_s in Table 20-4.

Local Moments

A more serious conceptual problem arises at temperatures greater than the Curie temperature, T_c, where the metal is no longer ferromagnetic. The system does *not* return to the set of normal metal bands, identical for spin up and spin down. This is known experimentally from the fact that the magnetic susceptibility corresponds to the behavior of local ionic magnets that have become disordered; this is also expected theoretically, since the formation of moments is intrinsically an intra-atomic effect with a characteristic energy U_x of the order of 1 eV, while the coupling between moments is small, of the order of kT_c, or 0.1 eV. We turn then to the formation of a magnetic moment on an individual ion. It is related to the condition for forming the ferromagnetic state described above, but now the d band, which we took to have a uniform density of state, is replaced by a d resonance such as was described in the preceding section.

The theory of the formation of local moments in metals is due to Anderson (1961); a simplified account is given by Harrison (1970, p. 480). The account that will be given here is more closely related to the actual electronic structure of real transition metals than is the discussion in either of these sources.

We begin by imagining a nonmagnetic state of, for example, an iron atom dissolved in copper or aluminum. There is a d resonance that occurs at the same energy for spin-up and spin-down electrons. The energy of that resonance is partly determined by the presence of the exchange energy from the resonant states of the same spin. We shall estimate that shift in terms of the parameter U_x and then shall develop a criterion analogous to the band criterion of Eq. (20-54).

Let us describe a resonance by a phase shift; we rewrite the phase shift given earlier in Eq. (20-22):

$$\tan \delta_l(E) = \frac{\Gamma/2}{E_d - E}. \tag{20-55}$$

It was mentioned earlier that through the resonance region, the phase shift increased by π, corresponding to the insertion of an extra state. If the Fermi energy is well above E_d, the resonance is completely occupied in the sense that the probability density for the atomic state $|d\rangle$, which we used to describe the resonance in the formulation of transition-metal pseudopotentials, is unity. Similarly, it is empty if E_d is much less than E_F. It is not difficult to show that in fact the probability density for occupation is just δ_l/π at intermediate energies also. This is a special case of the Friedel sum rule, which states that the number of excess electrons located at a scattering site is

$$n_{\text{extra}} = \frac{2}{\pi} \sum_l (2l + 1)\delta_l(E_F); \tag{20-56}$$

(Friedel, 1952; discussed in Harrison, 1970, p. 176 ff). That is, a scattering center inserted in an electron gas with Fermi energy E_F localizes n_{extra} electrons in excess of what would be in the same region if there were no scattering center. In this particular case, we are introducing a d resonance, and the resonance should localize $Z_d = n - Z_s$ electrons for a transition metal of the column Dn. Then Eq. (20-56) becomes

$$Z_d = 10\delta_2(E_F)/\pi. \tag{20-57}$$

This may be combined with Eq. (20-55) to obtain an estimate of the energy of the resonance with respect to the Fermi energy,

$$E_d - E_F = \frac{\Gamma}{2 \tan(Z_d \pi / 10)}. \tag{20-58}$$

This is consistent to within an electron volt or so of the values obtained by taking E_d from Table 20-4 and assuming Z_s free electrons and the form for $n_s(E)$ from Eq. (20-11) in obtaining E_F.

We may also use this fractional occupancy of the resonance, δ_l/π, to isolate exchange and direct contributions to the resonance energy E_d, which will enable us to establish the condition for formation of a local moment. Thus, we would say that if the phase shift for spin-up electrons is δ_l^+, the energy of the spin-up resonance contains a shift from the value E_d^0 without exchange, $E_d^+ = E_d^0 - U_x n^+$, where $n^+ = 5\delta_l^+/\pi$, and where the 5 is the orbital degeneracy $(2l + 1)$ of the d state. The corresponding expression can be written for E_d^-. There is also a contribution to E_d from the direct Coulomb interaction of each electron with electrons of both spins. In fact, U is approximately equal to U_x; it is mentioned in Appendix A that the direct and exchange terms exactly cancel for the same orbital (the self-energy term), and this is approximately true for d states of different quantum number, m. Thus we may take $U = U_x$ and combine the two contributions to obtain $E_d^+ = E_d^0 + U_x n^-$. Then, Eq. (20-55) for the spin-up state becomes

$$\tan \pi n^+/5 = \Gamma/[2(E_d^0 + n^- U_x - E_F)], \tag{20-59}$$

and for the spin-down state it becomes

$$\tan \pi n^-/5 = \Gamma/[2(E_d^0 + n^+ U_x - E_F)]. \tag{20-60}$$

We seek a self-consistent solution for the values of n^+ and n^-. Clearly one solution is always obtained with $n^+ = n^-$, that being the solution with no net moment which was assumed in deriving Eq. (20-58). We seek a second solution graphically by using each equation to plot n^+ against n^-, as in Fig. 20-20. There, the parameter $E_F - E_d = 0.36\Gamma$ was used, the value being near the 0.40Γ obtained for iron from Eq. (20-58). In part (a) U_x/Γ was taken equal to 0.3, that value being somewhat smaller than that appropriate to iron, 0.52, obtained from Tables 20-4 and 20-5. There is only one crossing of the curves and only one solution, the nonmagnetic one with $n^+ = n^-$. In part (b) U_x/Γ was taken equal to 0.7, somewhat larger than the iron value. (These illustrative parameters were chosen before the resonance widths from Table 20-4 were available.) This gave three solutions. The center one, with no magnetic moment, has a maximum energy; the other two have up or down net spin and are of minimum energy. Notice that for the solutions with net moment, $n^+ = 4.7$ and $n^- = 1.3$, or vice versa, corresponding to 6.0 electrons per ion in d states, rather than the seven of the nonmagnetic solution. A

528

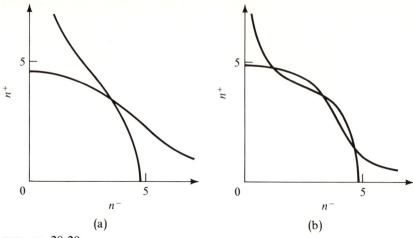

FIGURE 20-20

Plots of Eqs. (20-59) and (20-60) with intersections corresponding to self-consistent solutions. In both cases, $E_F - E_d$ was taken as 0.36Γ, consistent with iron, in the D8 column of the Solid State Table. In part (a), U_x was taken as 0.3Γ and the single solution is nonmagnetic. In part (b) a value of 0.7Γ was taken, giving a stable local moment of some 3.4 Bohr magnetons. Tables 20-4 and 20-5 give an intermediate value of $U_x = 0.52\Gamma$ for iron.

more careful study would be required to learn if such a transfer occurs in the formation of the local moment, but the Pauling-Slater curve for the pure materials (Fig. 20-19) would not suggest such a transfer. In Problem 20-4, the calculation is carried through for parameters appropriate to chromium, a slightly simpler case, to obtain a condition on U_x/Γ for formation of a local moment.

Notice that the local moment is quite analogous to the multiplet structure described for the monoxides and approached here as U_x/Γ becomes large. Notice also that there is an abrupt threshold at which the moment appears in the approximation used here. Thus, to some extent, we can sharply classify systems as being either moment-forming (or, multiplet) or simple. The sharpness of the boundary between the two disappears if we look very closely. When the moments are very weak, they fluctuate so fast as not to be recognizable as moments; when moments are nearly stable, they occur in fluctuations called *paramagnons*. Thus our description is only approximate (it is the use of the self-consistent one-electron approximation that introduces the important error; see Appendix A), but it is quite accurate far from the threshold between the magnetic and nonmagnetic states. Near the threshold, one must think of rapidly fluctuating moments that are not well described in either limit.

This description has given us a moment on an isolated ion. If there are other ions with moments, there will be coupling through the free electrons, just as the ions were coupled through the free-electrons producing the indirect interaction between ions of Eq. (17-9). This interaction between moments, the Ruderman-Kittel interaction discussed in Section 19-A (Ruderman and Kittel, 1954), also shows the Friedel oscillations that were illustrated in Fig. 17-2. Thus, at low temperatures the moments will order themselves to minimize the interaction energy. If the resulting order is ferromagnetic, with all net moments parallel, we return to two sets of energy bands, one for spin up and one for spin down, which we discussed at the beginning of this section. If the state is antiferromagnetic or disordered (the latter is called a *spin glass*), then the spin density varies from position to position and the corresponding electron states with spin depending upon position require a significant generalization beyond the band picture we have given here. However, most properties can still be studied, beginning with the local moments deduced here, and their couplings, without concern for the complexity of the underlying electron states.

PROBLEM 20-1   *Band calculation for nickel*

Carry out the LCAO energy-band calculation for nickel, in the face-centered cubic structure. Use the same set of six orbitals per atom but now neglect all but nearest-neighbor interactions (second neighbors are 40 percent more distant). The analysis is the same as for the body-centered cubic structure, except that there are now twelve nearest neighbors, in directions $2^{-1/2}(011)$, $2^{-1/2}(101)$, $2^{-1/2}(110)$, $2^{-1/2}(01\bar{1})$, $2^{-1/2}(\bar{1}01)$, $2^{-1/2}(1\bar{1}0)$, and the negative of each of these. For each band, it may be helpful to make a table of l, m, and n for each of the twelve neighbors, followed by the $E_{zx,\,zx}$ or other interatomic matrix element, and then the phase factor. They are then easily summed. Because of the cubic symmetry, the same matrix elements between Bloch sums that vanished for the body-centered cubic structure vanish here, except for the nearest-neighbor sd coupling from $E_{s,\,3z^2-r^2}$. Take all matrix elements from the Solid State Table (and take $\varepsilon_s = \varepsilon_d$).

PROBLEM 20-2   *Equilibrium density of the rare earths*

It is reasonable to consider the rare earths as simple metals, with respect to bonding properties, of charge $Z = 3$ and with core radii obtained from the ionization potential (and as given in the Solid State Table). Set $dE_{tot}/dk_F = 0$, using the appropriate equation from Chapter 15 to obtain the equilibrium k_F for cerium. Then compute the corresponding radius for the Wigner-Seitz sphere and compare it with the value given in the Solid State Table. Clearly, similar values of r_0 would be obtained for the other metals, but the fluctuations among the estimates do not correspond well with the fluctuations from experiment. In particular, the large observed values of r_0 for europium and ytterbium (both $Z = 2$ if considered as metals) are not reflected by the estimates, suggesting that the ionization energies used to obtain the r_c correspond to $Z = 3$ rather than $Z = 2$.

PROBLEM 20-3   *Number of free electrons in transition metals*

All parameters necessary to estimate the Fermi energy from the density of states to the right in Fig. 20-11 are given in the Solid State Table. Estimate the Fermi energy for chromium. The number Z_s of free electrons per atom can be written

$$Z_s = \int_0^{E_F} n_s(E)\,dE,$$

using Eq. (20-11). The number Z_d of d electrons per atom can also be written in terms of E_F, E_d, and W_d, and the sum $(Z_s + Z_d)$ can be equated to six (Cr is in column D6). Take W_d from the second form in Eq. (20-9). A convenient procedure is to solve for E_F in the expression for Z_d, substitute it into the expression for Z_s, write $Z_d = 6 - Z_s$, and iterate the expression, which converges quickly enough. Obtain the corresponding value for Z_s. Your value will differ slightly from that in Table 20-4, since those values do not have the error in rounding off that results from writing $r_d = 0.90$ in the Solid State Table.

PROBLEM 20-4   *Criterion for formation of local magnetic moments*

For chromium, of column D6, with one electron in a free-electron band (not the 0.76 indicated in Table 20-4) the d band is half full; $E_F = E_d$ in Eq. (20-58). Correspondingly, $E_d^0 + \frac{5}{2}U_x = E_F$ in Eqs. (20-59) and (20-60). Confirm that $n^+ = n^- = 5/2$ is a solution. (For this case it is convenient to take the reciprocal of both sides in Eqs. (20-59) and (20-60).)

We should like to derive a threshold ratio of U_x/Γ for the formation of a local moment. By consideration of Fig. 20-20, you may establish a condition on dn^+/dn^- at the crossing (in this case at $n^+ = n^- = 5/2$), which will indicate moment formation. Evaluate the critical U_x/Γ. Do the parameters of Tables 20-4 and 20-5 fail to predict a local moment as they did for iron with the parameters for Fig. 20-20,a?

The One-Electron Approximation

In introducing quantum mechanics in Chapter 1, we wrote a Hamiltonian for each individual electron. The true Hamiltonian of the system contains the interaction between all of the electrons, $\frac{1}{2}\Sigma_{i \neq j} e^2/|\mathbf{r}_i - \mathbf{r}_j|$ (we have included each term twice and divided by two), and a correct solution of the quantum-mechanical problem requires the evaluation of the total wave function, $\psi(\mathbf{r}_1\mathbf{r}_2 \cdots \mathbf{r}_N)$, in terms of the Hamiltonian of the entire system. It was an approximation to focus on a particular $\mathbf{r}_i$ and average the electron–electron interaction over all other $\mathbf{r}_j$ so that this term became a simple potential, $V(\mathbf{r}_i)$. Some such approximation is necessary since the solution of the full problem without it is far beyond the realm of possibility. We shall see here how we can generate the one-electron approximation as a variational solution of the full problem.

If it were really possible to write the Hamiltonian as a sum of individual one-electron Hamiltonians, we could show that a many-electron wave function could be written as a product of one-electron functions $\psi(\mathbf{r}_i)$; this is shown by direct substitution in the Schroedinger equation, Eq. (1-5):

$$\sum_i H(\mathbf{r}_i)\Pi_j\psi(\mathbf{r}_j) = E\Pi_j\psi(\mathbf{r}_j) \tag{A-1}$$

or

$$\sum_i \Pi_{j \neq i}\psi(\mathbf{r}_j)[H(\mathbf{r}_i)\psi(\mathbf{r}_i) - E\psi(\mathbf{r}_i)] = 0. \tag{A-2}$$

Dividing through by the total wave function,

$$\sum_i \frac{1}{\psi(\mathbf{r}_i)}[H(\mathbf{r}_i)\psi(\mathbf{r}_i) - E\psi(\mathbf{r}_i)] = 0. \tag{A-3}$$

Each term is a function of only the coordinates $\mathbf{r}_i$ of a single electron, so each term must be individually zero, giving the one-electron equation for each electron. If these one-electron solutions are substituted back into Eq. (A-1), we see that we have obtained a solution.

The true Hamiltonian cannot be written as a sum of one-electron Hamiltonians, but we may find the best possible one-electron-like solution by approximating the wave function by a product-wave-function and carrying out a variational calculation, just as we sought the best LCAO solutions of the one-electron problem by writing that approximate form and minimizing the energy. This simple product form for the many-electron wave function is in fact not allowed by the additional requirement that the total electron wave function be antisymmetric with respect to the interchange of any pair of electrons. The Pauli principle introduced in Chapter 1 is a direct consequence of this antisymmetry. We shall return to that complication after treating the simple product.

In the variational calculation, we must minimize

$$\langle \psi_1(\mathbf{r}_1)\psi_2(\mathbf{r}_2) \cdots \psi_N(\mathbf{r}_N) | H | \psi_1(\mathbf{r}_1) \cdots \psi_N(\mathbf{r}_N) \rangle - \sum_i \varepsilon_i \langle \psi_i(\mathbf{r}_i) | \psi_i(\mathbf{r}_i) \rangle \qquad \text{(A-4)}$$

where the Hamiltonian is given by

$$H = \sum_i \left[\frac{-\hbar^2}{2m} \frac{\partial^2}{\partial \mathbf{r}_i^2} + V_0(\mathbf{r}_i) + \tfrac{1}{2} \sum_{j \neq i} \frac{e^2}{|\mathbf{r}_i - \mathbf{r}_j|} \right], \qquad \text{(A-5)}$$

and where we have used Lagrange multiplyers ε_i for the normalization requirement for each one-electron function. The term V_0 contains all terms in the electron potential except those arising from other electrons. We may separately minimize Eq. (A-4) with respect to variations of the real part and the imaginary part of each function or, equivalently, for the function and its complex conjugate. Thus we ask that Eq. (A-4) be minimum with respect to variation in ψ_k^* and, in particular, for variations $\delta\psi_k^*$ orthogonal to ψ_k^*. Then, except for terms in the Hamiltonian containing $\mathbf{r}_i$ for $i = k$, we obtain zero. There is one such nonzero term in the kinetic energy, one for the potential, and two ($j = k$, $i = k$) for the interaction. We have

$$\int \delta\psi_k^*(\mathbf{r}_k) \left[\frac{-\hbar^2}{2m} \frac{\partial^2}{\partial \mathbf{r}_k^2} + V_0(\mathbf{r}_k) + \sum_{j \neq k} \langle \psi_j | \frac{e^2}{|\mathbf{r}_j - \mathbf{r}_k|} | \psi_j \rangle - \varepsilon_k \right] \psi_k(\mathbf{r}_k) \, d^3r_k = 0. \qquad \text{(A-6)}$$

This is satisfied, and the product solution is minimum in energy, if

$$\left[-\frac{\hbar^2}{2m} \nabla^2 + V_0(\mathbf{r}) + V_d(\mathbf{r}) \right] \psi_k(\mathbf{r}) = \varepsilon_k \psi_k(\mathbf{r}) \qquad \text{for all } k \qquad \text{(A-7)}$$

(this is called the Hartree Equation), where the direct interaction

$$V_d(\mathbf{r}_k) = \sum_{j \neq k} \int \psi_j^*(\mathbf{r}_j)\psi_j(\mathbf{r}_j) \frac{e^2}{|\mathbf{r}_j - \mathbf{r}_k|} \, d^3r_j \qquad \text{(A-8)}$$

is the potential we would obtain by using Poisson's equation and a charge density $-e\psi^*(\mathbf{r})\psi(\mathbf{r})$ for each electron. Notice that the electron potential does not include a potential from its own charge distribution ($j = k$). The best product-wave-function is that based upon the states obtained from a one-electron equation, Eq. (A-7).

Next we add the requirement that the total wave function $\Psi(\mathbf{r}_1 \cdots \mathbf{r}_N)$ be antisymmetric with respect to interchange of any two electrons, so that $\Psi(\cdots \mathbf{r}_j, \mathbf{r}_k \cdots) = -\Psi(\cdots \mathbf{r}_k, \mathbf{r}_j \cdots)$. This is accomplished by taking a wave function that is a sum of product-wave-functions with coordinates interchanged, each product having a factor $(-1)^n$ for the n permutations required to obtain that term from the starting product-wave-function.

It is not difficult to see that this gives an additional term in the energy written in Eq. (A-4), of

$$E_{\text{exch}} = \tfrac{1}{2} \sum_{i \neq j} \int \psi_i^*(\mathbf{r}_i)\psi_j^*(\mathbf{r}_j) \frac{e^2}{|\mathbf{r}_i - \mathbf{r}_j|} \psi_i(\mathbf{r}_j)\psi_j(\mathbf{r}_i)d^3r_i d^3r_j \tag{A-9}$$

for each pair of electrons of the same spin; those of opposite spin are already antisymmetric. This is the exchange energy and its inclusion leads to the **Hartree-Fock equation**; that is, Eq. (A-7) with an added term. Notice that this term allows an important simplification. We can include a term also with $i = j$ and add the same term to the direct interaction of Eq. (A-8). Both are made simpler and the two added terms exactly cancel.

The corresponding Hartree-Fock calculations have been carried out for all atoms in their ground states by Mann (1967). These could be used instead of the Herman-Skillman term values, which were plotted in Fig. 1-8. The values are quite similar, but we saw in Chapter 19 that the Hartree-Fock values are preferred for use with the transition metals. Essentially equivalent calculations were carried out by Fischer (1972), but instead of using the atomic ground state in the transition metals (alternating randomly between one and two s electrons across a series), she used two s electrons all the way across. Thus she obtained a smooth dependence upon atomic number where Mann's values showed irregularities. Since these irregularities are specific to the atom they should not, and do not, show up in the solids. Thus for studies of solids, Fischer's tables are more directly useful. They might be better (or worse) if they had systematically been done with a single s electron but that has not been done. For just this reason, irregularities in the transition series from the Herman-Skillman tables were smoothed out in Fig. 1-8. We give the relevant term values obtained from Fischer in Table A-1. The d values from this table appear in the Solid State Table and perhaps s and p values should also have been used there.

The exchange term at Eq. (A-9) considerably complicates calculations and it has generally proved appropriate in band calculations for solids to approximate it by the plausible form (discussed further in Appendix C), of

$$E_{\text{exch}} = \int d^3r E_x(n(\mathbf{r})), \tag{A-10}$$

where $E_x(n)$ is the exchange energy density for a uniform electron gas of density n. That is given (see, for example, Kittel, 1963) by

$$E_x(n) = -\frac{3}{4}e^2\left(\frac{3}{\pi}\right)^{1/3}n^{4/3}. \tag{A-11}$$

Then the variational calculation leads to just the Hartree equation; Eq. (A-7), with an exchange potential called *free-electron exchange*—or $\rho^{1/3}$ *exchange*, since in this field, $n(\mathbf{r})$ has traditionally been written $\rho(\mathbf{r})$—given by

$$V_x(\mathbf{r}) = -e^2(3/\pi)^{1/3}n^{1/3}(\mathbf{r}). \tag{A-12}$$

TABLE A-1
Hartree-Fock term values after Fischer (1972), in eV. In the upper part, ε_s values are given first for each element and ε_p values are given next; for the transition metals, ε_d values are given first and ε_s values are given second.

							He	Li		
							24.97	5.34		
							—	—		
	Be	B	C	N	O	F	Ne	Na		
	8.41	13.46	19.37	26.22	34.02	42.78	52.51	4.95		
	—	8.43	11.07	13.84	16.72	19.86	23.13	—		
	Mg	Al	Si	P	S	Cl	Ar	K	Ca	Sc
	6.88	10.70	14.79	19.22	24.01	29.19	34.75	4.01	5.32	5.72
	—	5.71	7.58	9.54	11.60	13.78	16.08	—	—	—
Cu	Zn	Ga	Ge	As	Se	Br	Kr	Rb	Sr	Y
7.72	7.96	11.55	15.15	18.91	22.86	27.00	31.37	3.75	4.85	5.34
—	—	5.67	7.33	8.98	10.68	12.43	14.26	—	—	—
Ag	Cd	In	Sn	Sb	Te	I	Xe	Cs	Ba	La
7.06	7.21	10.14	13.04	16.02	19.12	22.34	25.69	3.36	4.29	4.35
—	—	5.37	6.76	8.14	9.54	10.97	12.44	—	—	—
Au	Hg	Tl	Pb	Bi	Po	At	Rn			
6.98	7.10	9.82	12.48	15.19	17.96	20.82	23.78			
—	—	5.23	6.53	7.79	9.05	10.33	11.64			

TRANSITION METALS

	Sc	Ti	V	Cr	Mn	Fe	Co	Ni	Cu
	9.35	11.04	12.55	13.94	15.27	16.54	17.77	18.96	20.14
	5.72	6.04	6.32	6.59	6.84	7.08	7.31	7.52	7.72
	Y	Zr	Nb	Mo	Tc	Ru	Rh	Pd	Ag
	6.80	8.46	10.03	11.56	13.08	14.59	16.16	17.66	19.21
	5.34	5.68	5.95	6.19	6.39	6.58	6.75	6.91	7.06
	Lu	Hf	Ta	W	Re	Os	Ir	Pt	Au
	6.62	8.14	9.57	10.96	12.35	13.73	15.13	16.55	17.98
	5.41	5.72	5.98	6.19	6.38	6.52	6.71	6.85	6.98

A similar general approach was given by Slater (1951) and made variational by Kohn and Sham (1965). In fact, the same theory had been developed earlier as an extension of Fermi-Thomas theory by Gombas (1949). It is the basis of the one-electron approximation contemplated throughout this text, and the resulting equation,

$$-\frac{\hbar^2}{2m}\nabla^2\psi + V\psi = E\psi, \tag{A-13}$$

with

$$V = V_0 + V_d + V_x, \tag{A-14}$$

is the Schroedinger equation written in Eq. (1-5), which has provided the basis for our calculations.

Nonorthogonality
of Basis States

In our construction of LCAO bond orbitals in Chapter 3, we took the individual atomic orbitals to be orthogonal to each other. It is true that atomic orbitals on the same site are orthogonal, but atomic orbitals on adjacent sites are not; they have nonzero overlap

$$S_{\alpha\beta} = \int \psi_\alpha^*(\mathbf{r})\psi_\beta(\mathbf{r} - \mathbf{d})d^3r. \tag{B-1}$$

This does not preclude the use of atomic orbitals in a variational calculation of the electron states, but it does change the form of the resulting equations. Let us look briefly then at the effects of that nonorthogonality (Harrison and Ciraci, 1974; see also Tejeda and Shevchik, 1976).

We consider, in particular, the two hybrids $|h^a\rangle$ and $|h^c\rangle$ entering a particular bond orbital, and write

$$S = \langle h^c | h^a \rangle. \tag{B-2}$$

Estimates of S based upon real atomic $|s\rangle$ and $|p\rangle$ states give values of order 0.5, far from negligible. However, we take as an approximate bond orbital

$$|b\rangle = u^c |h_c\rangle + u^a |h_a\rangle, \tag{B-3}$$

and choose u^c and u^a to minimize the energy

$$\frac{\langle b|H|b\rangle}{\langle b|b\rangle},$$

by variation of u^c and u^a. There need be no requirement that $S = 0$. In writing the result, we define

$$M_2 = \langle h_c | H | h_a \rangle, \tag{B-4}$$

and

$$M_3 = \tfrac{1}{2} | \langle h_c | H | h_c \rangle - \langle h_a | H | h_a \rangle |, \tag{B-5}$$

and the result obtained is

$$\varepsilon_b = \frac{M_2 S - [M_2^2 S^2 + (1 - S^2)(M_2^2 + M_3^2)]^{1/2}}{(1 - S^2)}, \tag{B-6}$$

measured with respect to the average of the two hybrid energies. Notice that it reduces to $-(M_2^2 + M_3^2)^{1/2}$, as it should, for $S = 0$.

We may identify this with the form used in Chapter 3, by defining a covalent energy

$$V_2 = M_2/(1 - S^2) \tag{B-7}$$

and a polar energy

$$V_3 = M_3/(1 - S^2)^{1/2}. \tag{B-8}$$

Then the bond energy, Eq. (B-6), becomes

$$\varepsilon_b = V_2 S - (V_2^2 + V_3^2)^{1/2}, \tag{B-9}$$

and the antibonding energy becomes

$$\varepsilon_a = V_2 S + (V_2^2 + V_3^2)^{1/2}. \tag{B-10}$$

Thus, except for a shift in average energy by $V_2 S$, the results become equivalent. Since we fit V_2 and V_3 (or the full set of $V_{ll'm}$) to splittings in the known band structure, we may say that it was these values, from Eqs. (B-7) and (B-8), that we have been using. Notice that the correction for S in Eq. (B-8) would suggest a scaling up of V_3, in comparison to values obtained from term values, by 14 percent (for $S = 0.5$). There seemed to be other corrections also, and no net scaling of the term values was required.

The shift SV_2 in average energy then must be considered part of the overlap interaction. We may see that indeed it contributes only a radial interaction by relating it to an orthogonalization before bond formation. We imagine approximately orthogonalizing each atomic orbital to its neighbors in the form

$$|\psi'_\alpha\rangle = |\psi_\alpha\rangle - \tfrac{1}{2} \sum_\beta |\psi_\beta(\mathbf{r} - \mathbf{d})\rangle S_{\beta\alpha}. \tag{B-11}$$

We may readily verify that this eliminates the nonorthogonality of neighboring orbitals to first order in S:

$$\langle \psi'_\beta | \psi'_\alpha \rangle \approx S_{\beta\alpha} - 2 \cdot \tfrac{1}{2} S_{\beta\alpha} = 0. \tag{B-12}$$

(Notice again that atomic orbitals on the same atom are orthogonal. The corrections of Eq. (B-11) do introduce intra-atomic nonorthogonalities, but only to second order in S.) We may also obtain the energy of the corrected orbitals,

$$\frac{\langle \psi'_\alpha | H | \psi'_\alpha \rangle}{\langle \psi'_\alpha | \psi'_\alpha \rangle} = \frac{\varepsilon_\alpha - \frac{1}{2} \sum_\beta (S_{\alpha\beta} H_{\beta\alpha} + S_{\beta\alpha} H_{\alpha\beta}) + 1/4 \sum_\beta \varepsilon_\beta S_{\alpha\beta} S_{\beta\alpha}}{1 - 3/4 \sum_\beta S_{\alpha\beta} S_{\beta\alpha}}. \tag{B-13}$$

To first order in S, this becomes

$$\varepsilon'_\alpha = \varepsilon_\alpha - \sum_\beta S_{\alpha\beta} H_{\beta\alpha}, \tag{B-14}$$

just the counterpart of the average shifts in Eqs. (B-9) and (B-10). Because it is summed over all orbitals on the neighbors, however, it is clear that the energy shifts are not directional as might have been guessed from Eq. (B-9). Thus, these shifts are properly included in the simple radial overlap interaction. They were in fact assumed to be directional by Harrison and Ciraci (1974) and by Harrison and Phillips (1974), but that no longer appears appropriate.

Notice that $H_{\beta\alpha}$ contains kinetic energy contributions (actually, only if $S_{\beta\alpha} \neq 0$) as well as potential energy contributions. These were both included in the corresponding part of the overlap interaction and there are no additional terms to be considered.

The net result of this analysis is that nonorthogonalities shift the interatomic matrix elements, but since these were evaluated by fitting true band structures and assuming orthogonality, these shifts are automatically included. The nonorthogonalities also shift the average energies of the atomic orbitals, but these shifts were also included, though only approximately, in the overlap interaction.

The Overlap
Interaction

Let us discuss the calculation of the energy, as a function of separation, of two closed-shell atoms. It is directly applicable to rare-gas atoms and, by adding an electrostatic interaction, to ionic crystals. We also used it as one stage in the formation of covalent crystals from the isolated atoms in Chapter 7. In Chapter 7 it was stated that this energy consisted of the kinetic energy of the electrons, of electrostatic interaction between cores, of electrostatic interaction between the electrons and the cores, and of electrostatic interaction among the noncore electrons. It was also stated that this last term is the complicated one, and it was not described in detail, except to identify contributions that were included in the one-electron energies. Then, in Appendix A, an approximate method for evaluating it was described. We wish to dissect the electrostatic energies in a different way here. Let us begin by imagining an exact calculation of the total energy.

Imagine that we knew the positions of all cores and knew the exact many-electron wave function for all electrons. We could then in principle calculate the expectation value of the electronic kinetic energy. We could also compute the expectation value of the electronic charge density, $-en(\mathbf{r})$, at each point. Using that density, and the charge density from the cores, we could compute a total electrostatic energy. However, the electrostatic energy so obtained is not the correct Coulomb interaction energy for our problem. A series of corrections are required which, if made, would lead to the correct total energy.

First, in calculating the total electrostatic energy of the electron distribution, we include an interaction of each electron with its own charge distribution. That is not a physical contribution and should be subtracted out; instead, we add a cancelling contribution when we make the second correction, the electrostatic interaction associated with exchange, which we discussed in Appendix A. Exchange is the contribution to the energy arising from the antisymmetry of the wave function under permutation of electrons. It prevents two electrons from occupying the same one-electron state and also prevents two electrons of the same spin from coming too close to each other; thus it lowers the total electrostatic energy. That lowering is called the exchange energy.

We have also made an error in calculating the interaction energy of electrons by using an average potential arising from each electron as if the electron were a distributed cloud of charge. It is in fact a rapidly moving point charge that tends to avoid the other electrons. The correction of this error (which consists of errors in the way the kinetic energy is ordinarily calculated in the first step as well as errors in the Coulomb energy) is called the *correlation energy*. Adding correlation energy makes the total energy correct. However, it is much too difficult to calculate accurately, except in the very simplest of circumstances, and it forces us to use an approximate approach in any problem of interest.

We have given a complicated description of a very complicated, and in fact insoluble, problem. The resolution, from our point of view, can be found in the fact that the problem *is* insoluble: we are forced to approximate the correlation energy and it becomes appropriate to make the corresponding approximation to the other terms. This not only allows us to proceed, but makes the entire problem relatively simple. It might happen that the simplification was not accurate enough to be useful to us, and we would need either to undertake a much more difficult calculation or abandon the problem. We find the approximations made by Nikulin and Tsarev (1975), by Gordon and Kim (1972), and by Kim and Gordon (1974a,b) for rare-gas atoms, ionic solids, and generalized to covalent solids, good enough to tell us the physical origin of the principal effects, and apparently accurate enough to give reasonable estimates of the bond energy, bond length, and bulk modulus.

The conceptual basis for the approximation may be taken as the proof by Hohenberg and Kohn (1964) that the total ground-state energy of a collection of electrons in the presence of an applied potential (in this case, the collection of valence electrons in the presence of the potential due to the cores) depends only on the average density of electrons, $n(\mathbf{r})$. This is not at all obvious and in fact is untrue if we leave out the phrase "ground-state" (we saw in Chapter 7 that violation of the ground-state condition led to inaccuracies). It is also not immediately useful, since we do not know $n(\mathbf{r})$ nor would we know how to calculate the total energy if we did know $n(\mathbf{r})$. Kohn and Sham (1969) suggested an approximation which makes the problem soluble; they added total energy contributions from each volume element, computing the contribution from each element by using the formula for the exchange and correlation energy which is appropriate for a uniform electron gas. (An expansion for the uniform electron gas in terms of electron density was given by Gell-Mann and Brueckner, 1957.) For more discussion, see Harrison (1970, p. 309ff). Within this approximation, the density $n(\mathbf{r})$ itself could be obtained by a variational calculation. This leads to variational equations, Eq. (A-7), that are exactly of the form of the one-electron energy-eigenvalue equation, Eq. (1-5), which has provided the conceptual basis for our discussion; however, within that equation appear exchange potentials and correlation potentials in addition to direct Coulomb potentials and the potential due to the core. It was on the basis of a closely related approximation that Herman and Skillman calculated the atomic term values which were plotted in Fig. 1-8 and tabulated in the Solid State Table. Variations on this approximation to the electron–electron interaction (these are called by several names—Slater exchange, free-electron or $\rho^{1/3}$ exchange, self-consistent-field for exchange, Kohn-Sham exchange, and the X_α method) form the basis for almost all studies of the electronic structure and properties of solids. The approach is inevitably approximate but appears not to be the source of the largest error in even the best and most accurate calculations in solids. Approximations made at other stages ordinarily are more serious, so that we may regard the Kohn-Sham exchange as "good enough" for solids. For atoms and small molecules the rest of the problem can sometimes be done well enough that Kohn-Sham exchange becomes the feature that limits the accuracy.

Nikulin (1971) and Gordon and Kim (1972) made two additional approximations in the treatment of rare-gas interactions: they used also the free-electron formula for the electronic kinetic energy and they approximated the electron density by the superposition of free-atom densities. These two approximations are not completely plausible by themselves but seem to wed happily with the Kohn-Sham exchange. Concerning the latter, Heller, Harris, and Gelbart (1975) noted that since the Kohn-Sham calculation is a variational calculation, the *total energy* is stationary with respect to variations in $n(\mathbf{r})$; thus any first-order error in $n(\mathbf{r})$ gives only a second-order error in the total energy. This explains why the superposition of atomic densities may be sufficient, but also suggests that if we are to improve the calculation of any one term in the energy (for example, correlation energy), we must also improve the calculation of all other terms; this is because the individual terms in the energy are *not* stationary with respect to variations in $n(\mathbf{r})$. Thus Heller et al. not only showed that the superposition approximation is sensible; they also showed that it is dangerous to try to improve on it.

This last point is weakened by the use of free-electron kinetic energy rather than the accurate kinetic energy of the Kohn-Sham prescription. However, this use of free-electron kinetic energy, called the Fermi-Thomas approximation, can be seen in examples to give rather accurate values for the kinetic energy for given electron densities. The place where it becomes inaccurate is when it is used as the basis of a variational calculation. This is apparent, for example, in the case of a neutral atom. Then no term in the potential which enters the approximate theory can be nonzero except where the electron density is nonzero; the electron density, self-consistently determined, inevitably spreads too far out in comparison to an exact treatment, where each electron is influenced by the long-range attractive Coulomb potential of the rest of the atom (or the image force in the case of a solid). It is perhaps for this reason that the Fermi-Thomas approximation has fallen into disuse, but the objection does not apply if the atomic charge densities that are used are obtained from calculations using accurate kinetic energy; that is, with the electron density used by Gordon and Kim. Finally, we cannot even *make* the approximation of superimposed charge densities without an approximation such as the Fermi-Thomas approximation, since the true kinetic energy depends upon the Laplacian of the *wave function* and cannot be determined by the electron density alone.

These considerations of the Kohn-Sham exchange and of the free-electron kinetic energy do not tell us how accurate they will be. The approach is conceptually and, in comparison to other realistic options, also calculationally simple. There are plausible reasons for hoping it will be reasonably accurate, and there are good reasons to rule out many of the alternatives that first come to mind. However, only by carrying out calculations and comparing them with experimental (or accurately calculated) interactions will tell how well the approach works. Gordon and Kim did this for closed-shell configurations such as rare gases and closed-shell ions in ionic crystals. The results have appeared to be about as accurate as the corresponding interactions are known, the leading errors being in terms such as the Van der Waals interaction, which we have considered separately.

Let us then state the prescription used by Nikulin (1971) and Gordon and Kim (1972) and subsequently applied by Sokel (Harrison and Sokel, 1976) to open-shell systems. Atomic charge densities are obtained from accurate calculations or from experimental measurements; Gordon and Kim used Hartree-Fock wave functions to compute the electron density; in all cases these are spherically symmetric. The overlap interaction $V_o(d)$ is obtained by superimposing the electron densities from two atoms with internuclear separation d, leading to an electron density $n(\mathbf{r})$. The electronic kinetic energy is obtained by using the average kinetic energy per electron in a uniform electron gas of density n. That formula was derived in Section 14-A and leads to a total kinetic energy given by

$$\text{K.E.} = \frac{3h^2}{10m}(3\pi^2)^{2/3} \int n(\mathbf{r})^{5/3} \, d^3r. \tag{C-1}$$

We may also compute the electrostatic energy of this electron distribution combined with the nuclear charge from each atom. This is the most intricate part of the calculation but is a straightforward machine calculation. Next we may add the exchange energy. It is given for a free-electron gas by, for example, Kittel (1963, p. 92), and leads to a total approximate exchange energy of

$$E_{\text{exch}} = -3/4e^2\left(\frac{3}{\pi}\right)^{1/3} \int n(\mathbf{r})^{4/3} \, d^3r. \tag{C-2}$$

Different approximations for the correlation energy of a free-electron gas have been given. Gordon and Kim, and Sokel used an interpolated value between the known high-density and low-density limits. In terms of that curve the correlation energy is given by

$$E_{\text{corr}} = \int \varepsilon_{\text{corr}} n(\mathbf{r}) \, d^3r. \tag{C-3}$$

One feature of this important approach is apparent from the brief description in this Appendix: the method is very well characterized, except for the correlation energy, which is not so important in many properties. Anyone using Eqs. (C-1), (C-2), and (C-3) should be able to reproduce rather closely the curves given by Gordon and Kim or by Sokel. Frequently the approximations that enter solid state calculations are so intricate that they are never specified in the publication in which the calculations are described and would not be of interest if they were. Nevertheless, without knowledge of the approximations, it is sometimes difficult to assess a result that is characterized only as being the best the author could do.

Quantum–Mechanical Formulation of Pseudopotentials

The repulsive term in the pseudopotential was introduced in Chapter 15 as an approximate correction for the extra kinetic energy due to the presence of core states. This was in direct analogy with the way it was included in the overlap interaction. We also saw, in Appendix B, that this extra kinetic energy is directly related to the nonorthogonality of basis states. Indeed the rigorous formulation of pseudopotentials has been based upon the required orthogonality of the valence-band (or conduction-band) states to the core wave functions. Let us use that approach here. For extensive discussion of the formulation, as well as applications, see Harrison (1966a).

Imagine that a valence-electron state $|\psi(\mathbf{r})\rangle$ could be written as a smooth pseudo wave function $|\varphi(\mathbf{r})\rangle$, corrected to be orthogonal to all core states $|c\rangle$:

$$|\psi\rangle = |\varphi\rangle - \sum_c |c\rangle\langle c|\varphi\rangle. \tag{D-1}$$

No approximation is involved; we have simply made explicit the orthogonality. (We have defined a $|\varphi\rangle$ by adding a set of $|c\rangle$ to the true $|\psi\rangle$ with coefficients $\langle c|\varphi\rangle$ which we may choose.) Orthogonality may be verified by multiplying through by the complex conjugate of a core state and integrating; that is, by operation on the left with $\langle c'|$. Notice that $\langle c'|c\rangle = \delta_{c'c}$.

If $|\varphi\rangle$ is replaced by a plane wave on the right side of Eq. (D-1), this gives exactly what is called an **orthogonalized plane wave**, or OPW. The orthogonalized plane wave method of band calculation consists of expanding the true wave function in OPW's. It was invented by Herring (1940) and provides the conceptual basis of pseudopotential theory.

We substitute this form in the Schoedinger equation, Eq. (1-5), to obtain an equation for $|\varphi\rangle$:

$$H|\psi\rangle = H|\varphi\rangle - \sum_c H|c\rangle\langle c|\varphi\rangle = E|\psi\rangle$$

$$= E|\varphi\rangle - \sum_c E|c\rangle\langle c|\varphi\rangle. \tag{D-2}$$

It is customary to combine the terms from the orthogonality with the potential to define the *pseudopotential*, W:

$$W|\varphi\rangle = V|\varphi\rangle + \sum_c (E - H)|c\rangle\langle c|\varphi\rangle. \tag{D-3}$$

Then Eq. (D-2) becomes the *pseudopotential equation*,

$$-\frac{\hbar^2}{2m}\nabla^2|\varphi\rangle + W|\varphi\rangle = E|\varphi\rangle. \tag{D-4}$$

It will be useful later to notice that the core states are eigenstates of the Hamiltonian, so that $H|c\rangle$ can be replaced by $E_c|c\rangle$, but we can most clearly see the appropriateness of the empty-core model of the pseudopotential by leaving H as a kinetic and potential energy. We have seen that the pseudopotential always enters our calculations in a matrix element between plane waves, so we write such a matrix element as

$$\langle \mathbf{k}'|W|\mathbf{k}\rangle = \langle \mathbf{k}'|\left[V + \sum_c \left(E + \frac{\hbar^2\nabla^2}{2m} - V\right)|c\rangle\langle c|\right]|\mathbf{k}\rangle. \tag{D-5}$$

Now we may let the kinetic energy operator operate to the left (do two partial integrations) so that it becomes $-\hbar^2 k'^2/(2m)$.

In most calculations done in this book, the plane waves of importance were those with the Fermi wave number, and the energy E of interest was the Fermi energy, so the kinetic energy cancels E. This can be made rigorous by suitable choice of the zero of energy for each matrix element. Then we have, from Eq. (D-5),

$$W = V - \sum_c V|c\rangle\langle c|. \tag{D-6}$$

The operator $\sum_c |c\rangle\langle c|$ projects functions onto the core states. Thus in Eq. (D-1) the operator $1 - \sum_c |c\rangle\langle c|$ removed from $|\varphi\rangle$ those terms that could be expanded in $|c\rangle$. In a similar way we may think of the operation

$$W = V\left(1 - \sum_c |c\rangle\langle c|\right) \tag{D-7}$$

as removing those parts of V that can be expanded in the core states. This is not quite fair, since the integral evaluated is $\langle \mathbf{k}'|V|c\rangle$, not $\int V\psi_c d^3r$, but if we neglect the variation of $|\mathbf{k}'\rangle$ over the core the two integrals become equivalent.

Indeed, the core states do form a reasonably complete set of states within the region of the core, so that the projection written in Eq. (D-7), to a good approximation, does reduce the pseudopotential to zero in the core region, justifying the Ashcroft model which we used. An early calculation of the reduction of the potential given by Eq. (D-7) is shown in Fig. D-1.

The set of transformations made in Eqs. (D-5)–(D-7) rationalize the empty-core model. If one wishes to do better than that model, it is best to return to Eq. (D-3) and take advantage of the fact that the cores are eigenstates, to write

$$W|\varphi\rangle = V|\varphi\rangle + \sum_c (E - E_c)|c\rangle\langle c|\varphi\rangle. \tag{D-8}$$

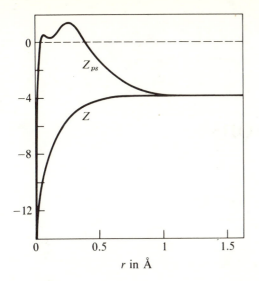

The potential and pseudopotential (for $l = 0$ states) of a Si^{4+} ion. The potential V is expressed in the form $V(r) = Z(r)e^2/r$ and V_{ps} is similarly expressed in terms of $Z_{ps}(r)$. Note V and V_{ps} both become equal to the Coulomb potential $-4e^2/r$ outside the core, which has a radius somewhat less than one Å. [After Heine, 1969, p. 7.]

An important and peculiar aspect of the pseudopotential can be learned from this form. We may show that it gives exact eigenvalues by substituting it into the pseudopotential equation, (D-4), and operating on the left with the *true* valence state $\langle\psi|$. All repulsive terms in the pseudopotential drop out since $\langle\psi|c\rangle = 0$; the Hamiltonian can be taken to operate to the left to give the true energy, which is then seen (if $\langle\psi|\varphi\rangle \neq 0$) to equal the eigenvalue E from Eq. (D-4). The peculiar point is that this would have been true no matter what constant had appeared where, in Eq. (D-8), we have $(E - E_c)$. All choices give what are called *valid pseudopotentials*. This arbitrariness is best eliminated by *optimizing* the pseudopotential in a way designed to give best convergence (see Harrison, 1966a).

There is one other point we should make concerning this formulation. The orthogonalization in Eq. (D-1) "punches a hole" in the pseudo wave function at each core. We see in fact that for each state, a fraction $\Sigma_c\langle\varphi|c\rangle\langle c|\varphi\rangle$, called the *orthogonalization hole*, is removed from the core regions. (The hole is a fraction in each cell or in the whole crystal.) This amounts to from 5 to 10 percent. When we calculate charge distributions and potentials, should we not correct for the orthogonalization holes and the associated renormalization? If we are proceeding rigorously, we should. However, notice from Eq. (D-4) that the pseudopotential has energy dependence; the repulsive term contributes an energy-dependent term

$$\langle\varphi|\frac{\partial W}{\partial E}|\varphi\rangle = \sum_c \langle\varphi|c\rangle\langle c|\varphi\rangle \tag{D-9}$$

equal to the orthogonalization hole. Shaw and Harrison (1967) have in fact shown that the energy dependence and the orthogonalization hole are directly and generally related, and it is an error to include one and not the other. Since we did not include the energy dependence of the pseudopotential (we made a local approximation to the pseudopotential), it was correct not to include these corrections to the charge distribution.

Orbital Corrections

In the LCAO description of covalent and ionic solids, a very accurate description was obtained based upon the minimal basis set of orbitals consisting of a single s state and three p states on each atom. This might seem puzzling at first glance, since it is well known that in molecular orbital calculations it is necessary to modify the atomic orbitals in going to the molecule, and in atomic terms, modifying the orbitals means admixing other atomic wave functions. (The full set of wave functions, including ionized states, constitutes a complete set.) Second thought gives an immediate answer: we did not use wave functions but fitted the matrix elements to the bands in the solid, thereby taking these modifications into account. That answer is incomplete, since we should expect that the modifications would be different under different circumstances, but we shall see in this appendix that even the variations in orbitals with changing circumstances may have been approximately included in our fitting. This finding is an incidental result of a method for refining the LCAO theory, the Orbital Correction Method, described in detail elsewhere (Harrison, 1973a).

Imagine a set of orbitals $|\alpha\rangle$ in terms of which it is wished to approximate the occupied states in the system—these are best thought of as bond orbitals rather than as individual atomic orbitals. Thus we write our approximate state

$$|\varphi_k\rangle = \sum_\alpha u_{k\alpha}|\alpha\rangle. \tag{E-1}$$

We then write the difference between this approximation and the true eigenstate $|\psi_k\rangle$ as the **orbital correction** $|\chi_k\rangle$,

$$|\psi_k\rangle = |\varphi_k\rangle + |\chi_k\rangle. \tag{E-2}$$

The idea of the method is to treat this correction as small and to obtain corrections to the energy to second order in $|\chi_k\rangle$.

In the analysis, it becomes necessary to make an expansion of $|\chi_k\rangle$ in some basis set, and plane waves have been used here. This is an important difference from the simple use of a larger number of atomic orbital basis states. The latter *appears* to converge reasonably well as one sums over excited states but converges to the wrong answer: after obtaining convergence in the sum over excited bound states, one must begin summing over positive energy (ionized) states that give additional corrections of the order of those due to bound states. For our purposes, the single set, plane waves, are much more appropriate and are a complete set. Indeed we noted at a number of points that the upper bands based upon the minimal set, and certainly the bands from excited states, are so mixed with the ionized states that they cannot be fruitfully isolated.

We shall do the simple treatment here for a single orbital. This would, for example, be appropriate for the correction of one of the atomic orbitals on oxygen in water, such as would be needed because of the nonsphericity which the protons in water cause. (Such corrections have been studied in detail in this method by Meserve, 1975.) We shall then indicate the complications that arise in extending it to solids and give the result of the full analysis.

We substitute the exact form of the state, Eq. (E-2), into the Schroedinger equation, $H|\psi\rangle = E|\psi\rangle$,

$$H|\varphi\rangle + H|\chi\rangle = E|\varphi\rangle + E|\chi\rangle. \tag{E-3}$$

We take $|\varphi\rangle$ to be normalized and operate with $\langle\varphi|$ on the left, to obtain

$$\langle\varphi|H|\varphi\rangle + \langle\varphi|H|\chi\rangle = E + E\langle\varphi|\chi\rangle. \tag{E-4}$$

We solve for E in the form

$$E = \langle\varphi|H|\varphi\rangle + \langle\varphi|H - E|\chi\rangle. \tag{E-5}$$

The first term is the expression used throughout our LCAO studies; the second is the effect of the orbital correction.

We wish to expand $|\chi\rangle$ in plane waves, $\sum_q |q\rangle\langle q|\chi\rangle$, and obtain the coefficient by multiplying Eq. (E-3) on the left by $\langle q|$. Noting that

$$\langle q|H = \frac{\hbar^2 q^2}{2m}\langle q| + \langle q|V,$$

we obtain

$$\langle q|\chi\rangle = \frac{\langle q|H - E|\varphi\rangle + \langle q|V|\chi\rangle}{E - \hbar^2 q^2/2m}. \tag{E-6}$$

We now take $\langle q|V|\chi\rangle$ to be of higher order than $|\chi\rangle$ and drop the final term in the numerator. There is actually arbitrariness in the orbital correction, just as there was arbitrariness in the pseudopotential, and we could imagine optimizing the choice to make $V|\chi\rangle$ small, just as we optimized pseudopotentials. Dropping the term $\langle q|V|\chi\rangle$ and substituting the expanded form for $|\chi\rangle$ back into Eq. (E-5), we obtain

$$E = \langle\varphi|H|\varphi\rangle + \sum_q \frac{\langle\varphi|H - E|q\rangle\langle q|H - E|\varphi\rangle}{E - \hbar^2 q^2/2m}, \tag{E-7}$$

a form very analogous to ordinary second-order perturbation theory, as written in Eq. (1-14).

In carrying out this analysis, we have imagined that $|\varphi\rangle$ was an atomic orbital or an LCAO, but it could be any approximate state which we wished to correct. It is interesting to notice that pseudopotential theory is simply a special case of the Orbital Correction Method in which the approximate state is an orthogonalized plane wave, as described following Eq. (D-1). The first term in Eq. (E-7) is just the expectation value of the energy with respect to an OPW; taking care that this be a normalized OPW, we find the expectation value to be the expectation value of the kinetic energy operator and the pseudopotential, in essentially the form of Eq. (D-8), with respect to the plane wave. Similarly, the matrix elements in the second term in Eq. (E-7) become matrix elements of the same pseudopotential. In this context, it may be noted that the term $\langle q|V|\chi\rangle$ that we discarded earlier *is* of higher order in the pseudopotential.

The transition-metal pseudopotentials that were discussed in Chapter 20 are also a special case of the Orbital Correction Method, in which we correct starting OPW states by admixture of other OPW's and also correct starting atomic d states by adding OPW's.

Let us return to the LCAO starting states appropriate to molecules and nonmetallic solids; we have expanded the orbital correction in plane waves. It would also have been permissible to expand in OPW's, since the final states must be orthogonal to the core states and the starting LCAO can have been constructed to be orthogonal. This complicates the analysis somewhat unless we ignore the nonorthogonality of different OPW's. The ultimate effect is to replace the potentials in Eq. (E-7) everywhere by pseudopotentials.

The evaluation of the orbital correction terms in Eq. (E-7) for a starting LCAO state is perfectly straightforward but becomes intricate if the potentials are handled well. Notice first that the sum in Eq. (E-7) is second order in the orbital correction and that replacing E in the matrix elements by $\langle\varphi|H|\varphi\rangle$ makes only a third-order error in the energy; therefore the replacement can be made. Second, when we let H operate on $|\varphi\rangle$ (in either matrix element) we obtain, as in any LCAO calculation, one term from the kinetic energy and one from the potential associated with the atom upon which the atomic orbital rests, which is just the atomic energy times the same atomic orbital. We then have corrections due to the deviation from that potential.

The only case which appears to have been treated in detail with this method was that of the central hydrides, HF and CH_4 in particular, by Meserve (1975). There, the starting orbitals are essentially atomic orbitals for the central atom, though in detail, Meserve obtained improved atomic orbitals by taking a spherical average of the nuclear potentials and then recalculating the atomic orbitals for the full neon configuration. The orbital corrections then arose from the replacement of the spherical average of the proton charge distributions (a globe or shell of positive charge at the C—H distance, for example) by point charges at the appropriate positions. The calculations of bond length, total energy, and vibrational frequency proved quite accurate (see also Harrison, 1973a, for values). One might at the same time question the importance of refining LCAO calculations which, in many cases, we have found to be reasonably accurate without corrections. We see, however, that in the case of the central hydrides it is just the orbital correction term that determines the geometry of the molecule. If we ignore orbital corrections, the electronic charge density becomes spherically symmetric and the protons take angles which minimize the electrostatic energy. We would find a minimum energy in water, in particular, with the protons separated at 180°. It is just the orbital correction that causes the bent configuration of water. Meserve's calculations on methane indicated that an essentially correct geometry for water would have been obtained, though he did not carry out the H_2O calculation in detail.

Finally, let us discuss the evaluation when the starting states really are linear combinations of atomic orbitals rather than individual atomic orbitals. Imagine in particular having constructed a set of bond orbitals. The corresponding molecular or crystalline states are linear combinations of these, and we might seek corrections to the resulting states. The difficulty then is that the corrections do in fact depend upon our making the proper choice of linear combinations of bond orbitals. We must in principle diagonalize the Hamiltonian matrix based upon bond orbitals, Fig. 3-4, before obtaining orbital corrections. We cannot use unitarity as we did in Chapter 3 to avoid the diagonalization of the matrix.

This would seem at first to have ruled out the development of a general theory of bonding that included the orbital corrections. However, it proved possible (Harrison, 1973a) to write the condition for diagonalization to lower order and to use that condition to evaluate the sums necessary to obtain the energy to second order in the orbital correction. This was a complex piece of analysis and will not be repeated here. Instead, only the result will be given. We first define the average value of the bond-orbital energies as

$$\bar{\lambda} = \text{Average over } \alpha \text{ of } \langle \alpha | H | \alpha \rangle. \tag{E-8}$$

Then we define a Green's function G by

$$G = \sum_{\mathbf{q}} \frac{|\mathbf{q}\rangle\langle\mathbf{q}|}{\bar{\lambda} - \hbar^2 q^2 / 2m}.$$

Then the sum over all eigenvalues λ_k, to second order in the orbital corrections, becomes

$$\sum_k \lambda_k = \sum_\alpha \langle \alpha | H | \alpha \rangle - \sum_{\alpha, \beta}' \langle \alpha | \beta \rangle \langle \beta | H - \bar{\lambda} | \alpha \rangle$$
$$+ \sum_\alpha \langle \alpha | H - \bar{\lambda} | G | H - \bar{\lambda} | \alpha \rangle + \sum_{\alpha\beta} \overline{\langle \alpha | G | \alpha \rangle} \langle \alpha | H - \bar{\lambda} | \beta \rangle \langle \beta | H - \bar{\lambda} | \alpha \rangle$$
$$- \sum_{\alpha\beta} [\langle \alpha | H - \bar{\lambda} | G | \beta \rangle + \langle \alpha | G | H - \bar{\lambda} | \beta \rangle] \langle \beta | H - \bar{\lambda} | \alpha \rangle. \tag{E-9}$$

The first term is as expected. The third term is essentially what we got when we considered an individual orbital, in Eq. (E-7). The second term arises since we did not assume that the different bond orbitals were orthogonal to each other; that is, $\langle \alpha | \beta \rangle \neq 0$. It is just the counterpart of the terms $S_{\alpha\beta} H_{\beta\alpha}$ which arose in Eq. (B-13) in our discussion of nonorthogonalities of bond orbitals in Appendix B. These we planned to incorporate with the overlap interaction in that context.

The remaining terms are new, but are of the same order in the orbital corrections as the third term. They could be evaluated first to give corrections for terms $\beta = \alpha$ and then for pairs of nearest-neighbor interactions.

To do the calculation properly, one should carry out the calculation of the potentials self-consistently. This might at first seem a formidable task, but it turns out that it also becomes straightforward, because the calculations are only to be carried to second order in the orbital corrections, and therefore the screening can be systematically carried out as it was in metals. The resulting total energy (Harrison, 1973a, Eq. 36) resembles Eq. (E-9) but has corrections, including a contribution analogous to the overlap interaction of Chapter 7.

This provides a path for attempting first-principles calculations of the total energies, without the use of the parameters from the Solid State Table. In addition, since it incorporates orbital corrections, it should be more accurate. However, there has not been a serious attempt to carry it through for any solid state system. The calculation of elastic constants by Bullett, mentioned in Chapter 7, was based on a related method, but it is questionable whether it was carried out in a useful way.

We return finally to the approach used in this text, that of fitting the parameters to observed band structures rather than taking explicit atomic orbitals. We noted in Appendix B that the nonorthogonality, represented by the second term in Eq. (E-9), could be absorbed in the appropriate interatomic matrix elements and overlap interaction so that by fitting we may say that we have already incorporated the second term in Eq. (E-9). Similarly, the third term in Eq. (E-9) can be regarded as an additional contribution to the first, a term $(H - \bar{\lambda})G(H - \bar{\lambda})$ added to the Hamiltonian and therefore can be regarded as being included in the fit. Finally, even the contributions of the last two sums corresponding to $\beta = \alpha$ can be regarded as being included in the fit. Only the final terms for $\beta \neq \alpha$ are of a different form. Thus, by using the fitting procedure that led to the solid state matrix elements, we may be including a much larger part of the problem than we had in mind. This may account for the surprising accuracy of the predictions that have frequently been encountered.

Bibliography
and Author Index

Abarenkov, I. V., and Heine, V. (1965), *Phil. Mag.*, **12** (529). 343

Adachi, E. (1969), *J. Phys. Chem. Solids*, **30** (776). 157

Adler, D. (1968), in *Solid State Physics*, vol. 21, (p. 1), Ehrenreich, H., Seitz, F., and Turnbull, D., eds., Academic Press, New York. 438

Adler, D. *See* Feinleib, J.

Aggarwal, R. L. (1967), *Bull. Am. Phys.*, **12** (100).

Aggarwal, R. L. *See* Reine, M.

Albert, J. P., Jouain, C., and Gout, C. (1977) *Phys. Rev.*, **B16** (925). 323

Allen, F. G. *See* Gobeli, G. W.

Allen, P. B. *See* Chakraborty, B.

Andersen, O. K. (1973), *Solid State Commun.*, **13** (133). 501–507, 511, 514ff

Andersen, O. K. (1975), *Phys. Rev.*, **B12** (3060). 490

Andersen, O. K., and Jepsen, O. (1977), *Physica* **91B** (317). 487, 505, 506, 516, 518

Andersen, O. K., Klose, W., and Nohl, H. (1978), *Phys. Rev.* **B17** (1209). 503, 505, 517

Andersen, O. K. *See* Jepsen, O.; Skriver, H. L.

Anderson, P. W. (1961), *Phys. Rev.*, **124** (41). 526ff

Anderson, P. W. (1975), *Phys. Rev. Letters*, **34** (953). 246, 252

Anderson, P. W. *See* Morel, P.

Anderson, R. L. (1960), *Proceedings of the International Conference on Semiconductors, Prague, 1960* (Czech. Acad. Sci., Prague) (563). 255

Andreatch, P., Jr. *See* McSkimin, H. J.

Andrews, D. H., and Kokes, R. J. (1965), *Fundamental Chemistry*, 2nd ed., John Wiley and Sons, New York. 127

Animalu, A. O. E. (1966), *Proc. Roy. Soc. A*, **294** (376). 343, 362

Animalu, A. O. E., and Heine, V. (1965), *Phil. Mag.*, **12** (1249). 343, 362, 381, 421

Appapillai, M., and Williams, A. R. (1973), *J. Phys.*, **F3** (759). 343

Appelbaum, J. A., Baraff, G. E., and Hamann, D. R. (1975), *Phys. Rev.*, **B12** (5749). 239

Appelbaum, J. A. *See* Baraff, G. E.

Appleman, D. E. *See* Skinner, B. J.

Arlinghaus, F. J. *See* Gay, J. G.

Ashcroft, N. W. (1966), *Physics Letters*, **23** (48). 352, 363

Ashcroft, N. W. (1968), *J. Phys.*, **C1** (232). 393f

Ashcroft, N. W., and Langreth, D. C. (1967), *Phys. Rev.*, **155** (682). 344, 355, 362

Ashcroft, N. W., and Lekner, J. (1966), *Phys. Rev.*, **145** (83). 404

Aspnes, D. E. *See* Van Vechten, J. A.

Austin, B. J., Heine, V., and Sham, L. J. (1962), *Phys. Rev.*, **127** (276). 343

Aven, M. *See* Hite, G. E.

Babalola, I. A. *See* Chye, P. W.

Bagus, P. S. *See* Gilbert, T. L.

Bailey, S. M. *See* Wagman, D. D.

Bailly, F., and Manca, P. (1972), in *Chemical Bonds in Solids*, vol. I (28). Sirota, N. N., ed., Plenum, New York. 135

Baker, W. M. *See* Czyzak, S. J.

Baldereschi, A. (1973), *Phys. Rev.*, **B7** (5212). 81, 182

Baldini, G. (1962), *Phys. Rev.*, **128** (1562). 295, 296

Balkanski, M. *See* Kunc, K.

Baraff, G. A., Applelbaum, J. A., and Hamann, D. R. (1977), *J. Vac. Sci. and Tech.* **14** (999). 425

Baraff, G. E. *See* Applelbaum, J. A.

Bardeen, J. (1936), *Phys. Rev.*, **49** (653). 399

Bardeen, J. (1947), *Phys. Rev.*, **71** (717). 244

Barišić, S., Labbé, J., and Friedel, J. (1970) *Phys. Rev. Letters*, **25** (919). 471

Barker, A. S., Jr. (1968), *Phys. Rev.*, **165** (917). 115

Bateman, T. B. *See* McSkimin, H. J.

Batlogg, B., Kaldis, E., Schlegel, A., and Wachter, P. (1976), *Phys. Rev.*, **B14** (5503). 433

Bäuerle, D. (1974), *Physica Status Solidi*, **(b)63** (177). 475

Beck, P. A. *See* Cheng, C. H.

Bell, R. J., Bird, N. F., and Dean, P. (1968), *J. Phys.*, **C1** (299). 278f, 282, 286

Bennett, A. J., and Roth, L. M. (1971a), *Phys. Rev.*, **B4** (2686). 263

Bennett, A. J., and Roth, L. M. (1971b), *J. Phys. Chem. Solids*, **32** (1251). 263

Benoit à la Guillaume, C., Salvan, F., and Voos, M. (1970), *Proceedings of the Tenth International Conference on the Physics of Semiconductors* (Cambridge, Mass.), Keller, S. P., Hensel, J. C., and Stern, F., eds. (516). 164

Bergstresser, T. K. *See* Cohen, M. L.

Berlincourt, D., Jaffe, H., and Shiozawa, L. R. (1963), *Phys. Rev.*, **129** (1009). 196

Bernardes, N. (1958), *Phys. Rev.*, **112** (1534). 293, 294

Bertoni, C. M., Bisi, O., and Manghi, F. (1978), *Phys. Rev.*, **B17** (3750). 479

Bertoni, C. M., Bortolani, V., Calandra, C., and Nizzoli, F. (1973), *Phys. Rev. Letters*, **31** (1466). 344, 389

Beveridge, D. L. *See* Pople, J. A.

Bharatiya, N. R. *See* Gilat, G.

Biegelsen, D. K. (1975), *Phys. Rev.*, **B12** (2427). 116

Bieniewski, T. M., and Czyzak, S. J. (1963), *J. Opt. Soc. Am.*, **53** (496). 115

Bir, G. L., Pikus, G. E., Suslina, L. G., and Fedorov, D. L. (1970), *Fiz. Tverd. Tela*, **12** (1187) [*Sov. Phys.-Solid State (English Transl.)*, **12** (926).] 157

Bird, N. F. *See* Bell, R. J.

Bisi, O. *See* Bertoni, C. M.

Bloembergen, N. *See* Chang, R. K.; Wynne, J. J.; Yablonovitch, E. C.

Blyholder, G., and Coulson, C. A. (1968). *Theor. Chem. Acta*, **10** (316). 46

Bohm, D., and Staver, T. (1951), *Phys. Rev.*, **84** (836). 395

Bond, W. L. *See* McSkimin, H. J.

Born, M., and Huang, K. (1954), *Dynamical Theory of Crystal Lattices*, Clarendon Press, Oxford. 204, 311

Bortolani, V. *See* Bertoni, C. M.

Boyd. G. D. *See* McFee, J. H.

Braunstein, R., and Ockman, N. (1964), *Final Report*, Contract No. NONR-4128100, ARPA Order No. 306–62, Washington, Office of Naval Research. 122

Breckenridge, R. A. *See* Huang, C.

Brockhouse, B. N., and Dasannacharya, B. A. (1963), *Solid State Commun.*, **1** (205). 208

Brodsky, M. H. *See* Burstein, E.

Brooks, H. (1963), *Trans. Met. Soc. AIME*, **227** (546). 356

Brovman, E. G., Kagan, Yu., and Holas, A., (1970), *Fiz. Tverd. Tela* **12** (1001) [*Soviet Phys.-Solid State (English Trans.)*, **12** (786).] 344

Brovman, E. G. Kagan, Yu., and Holas, A. (1971), *Zh. Eksperim. i Teor. Fiz.*, **61** (737) [*Soviet Phys. JETP (English Transl.)*, **34** (394), (1972).] 344, 389

Brown, F. C. (1967), *Physics of Solids*, Benjamin, New York. 304, 326

Brown, F. C. *See* Kanazawa, K. K.

Brown, R. N. *See* Pidgeon, C. R.

Brueckner, K. A. *See* Gell-Mann, M.

Bullett, D. W. (1975), *J. Phys.*, **C8** (3108). 186, 550

Burstein, E., Brodsky, M. H., and Lucovsky, G. (1967), *Int. J. Quant. Chem.*, **IS** (759). 115

Burstein, E. *See* Lucovsky, G.

Butler, W. H., and Williams, R. K. (1978), *Phys. Rev.* **B12** (6483). 472

Byer, R. *See* Choy, M. M.

Calais, J.-L. (1977), *Advances in Physics*, **26** (847). 542

Calandra, C. See Bertoni, C. M.

Callaway, J. (1976), Quantum Theory of the Solid State, Academic Press, New York. 508

Callaway, J., and Hughes, A. J. (1967) Phys. Rev., **156** (860). 87

Callaway, J. See Rath, J.

Carabatos, C., Hennion, B., Kunc, K., Moussa, F., and Schwab, C. (1971), Phys. Rev. Letters, **26** (770). 208

Carabatos, C. See Hennion, B.; Prevot, B.

Cardona, M. (1963), J. Phys. Chem. Solids, **24** (1543). 158

Cardona, M. (1965), J. Phys. Chem. Solids, **26** (1351). 158

Cardona, M. (1972), in Atomic Structure and Properties of Solids, p. 514, Burstein, E., ed., Academic Press, New York. 116

Cardona, M., Gudat, W., Sonntag, B., and Yu, P. Y. (1970), in Proceedings of the Tenth International Conference on the Physics of Semiconductors (Cambridge, Mass.), Keller, S.P., Hensel, J. C., and Stern, F., eds. 102

Cardona, M., McElroy, P., Pollak, F. H., and Shaklee, K. L. (1966), Solid State Commun., **4** (319). 101

Cardona, M., and Pollak, F. H. (1972), in Physics of Optoelectronic Materials, Albers, W. A., Jr., ed., Plenum, New York. 106

Cardona, M., Shaklee, K. L., and Pollak, F. H. (1967), Phys. Rev., **154** (696). 161

Cardona, M. See Gavini, A. A.; Higginbotham, C. W.; Sanchez, C.; Shileika, A. Yu.; Van Vechten, J. A.; Yu, P. Y.

Castner, T., and Känzig, W. (1957), J. Phys. Chem. Solids, **3** (178). 326

Cavallini, M., Dondi, M. G., Scoles, G., and Valbusa, U. (quoted by Gordon and Kim) (1972). 293

Chadi, D. J. (1977), Phys. Rev., **B16** (790). 161

Chadi, D. J. (1978), Phys. Rev. Letters, **41** (1062; [1332-E]). 236f

Chadi, D. J., and Cohen, M. L. (1973a), Phys. Rev., **B7** (692) 81

Chadi, D. J., and Cohen, M. L. (1973b) Phys. Rev., **B8** (5747). 81, 182ff

Chadi, D. J., and Cohen, M. L. (1975), Phys. Stat. Sol. (b)**68** (405). 49, 53f, 74ff, 104, 109, 142, 143, 410

Chadi, D. J. and Martin, R. M. (1976), Solid State Commun., **19** (643). 54, 182ff, 193, 196

Chadi, D. J., White, R. M., and Harrison, W. A. (1975), Phys. Rev. Letters, **35** (1372). 131ff

Chadi, D. J. See Martin, R. M.

Chakraborty, B., Pickett, W. E., and Allen, P. B.

(1976), Phys. Rev., **B14** (3227). 397

Chang, Chin-An. See Sakaki, H.

Chang, L. L. See Sakaki, H.

Chang, R. K., Ducuing, J., and Bloembergen, N. (1965), Phys. Rev. Letters, **15** (415). 122

Chelikowsky, J. R., and Cohen, M. L. (1976a), Phys. Rev. Letters, **36** (229). 424

Chelikowsky, J. R., and Cohen, M. L. (1976b), Phys. Rev., **B14** (556). 100, 101, 106, 139ff, 426f

Chelikowsky, J. R., and Cohen, M. L. (1976c), Phys. Rev., **B13** (826). 425

Chelikowsky, J. R., and Schlüter, M. A. (1977), Phys. Rev., **B15** (4020). 257, 263f, 267–272

Chelikowsky, J. R. See Cohen, M. L.; Louie, S. G.

Chemla, D., Kupecek, P., Schwartz, C., Schwab, C., and Goltzene, A. (1971), J. Quant. Electr., **QE-7** (126). 122

Chen, A. B. (1976), Bull. Am. Phys. Soc., **22** (446). 56

Cheng, C. H., Gupta, K. P., van Reuth, E., and Beck, P. A. (1962), Phys. Rev., **126** (2030). 493

Choy, M. M., Byer, R. L., and Ciraci, S. (1975), J. Quant. Electr., **QE-11** (40). 119, 122, 124

Chun, H. U. See Klein, G.

Chye, P. W., Babalola, I. A., Sukegawa, T., and Spicer, W. E. (1975), Phys. Rev. Letters, **35** (1602). 246

Chye, P. W., Sukegawa, T., Babalola, I. A., Sunami, H., Gregory, P., and Spicer, W. E. (1977), Phys. Rev., **15** (2118). 105

Chye, P. W. See Yu, K. Y.

Ciraci, S. See Choy, M. M.; Harrison, W. A.

Cohen, M. H., and Heine, V. (1961), Phys. Rev., **122** (1821). 343

Cohen, M. L., and Bergstresser, T. K. (1966), Phys. Rev., **141** (789). 417f

Cohen, M. L. and Heine, V. (1970), in Solid State Physics, vol. 24, Ehrenreich, H., Seitz, F., and Turnbull, D., eds., Academic Press, New York. 344, 362

Cohen, M. L., 139ff. See Chadi, D. J.; Chelikowsky, J. R.; de Alvarez, C. V.; Joannopoulos, J. D.; Larsen, P. K.; Louie, S. G.

Collins, A. T., Lightowlers, E. C., and Dean, P. J. (1967), Phys. Rev., **158** (833). 115

Collins, T. C. See Reynolds, D. C.

Connolly, J. W. D. (1967), Phys. Rev., **159** (415). 521

Connolly, J. W. D. See Johnson, K. H.

Cook, E. L. See Strehlow, W. H.

Corey, A. J. See Tsay, Y. F.

Coulson, C. A., Rèdei, L. R., and Stocker, D. (1962), *Proc. Roy. Soc. A*, **270** (357). 43, 60, 144

Coulson, C. A. (1970), in *Physical Chemistry, an Advanced Treatise*, vol. 5, Eyring, H., Henderson, D., and Jost, W., eds., Academic Press, New York. 22, 65

Coulson, C. A. *See* Blyholder, G.

Cowley, E. R. (1971), *J. Phys.*, **C4** (988). 312

Cowley, R. A., Woods, A. D. B., and Dolling, G. (1966), *Phys. Rev.*, **150** (487). 394

Cowley, R. A. *See* Warren, J. L.

Crabtree, G. W., Dye, D. H., Karim, D. P., Koelling, D. D., and Ketterson, J. B. (1979), *Phys. Rev. Letters*, **42** (390). 472

Cracknell, A. P. (1969), *Adv. Phys.*, **18** (681). 369

Crane, R. C. *See* Czyzak, S. J.

Czyzak, S. J., Baker, W. M., Crane, R. C., and Howe, J. B. (1957), *J. Opt. Soc. Am.*, **47** (240). 115

Czyzak, S. J. *See* Bieniewski, T. M.

Dasannacharya, B. A. *See* Brockhouse, B. N.; Roy, A. P.

de Alvarez, C. V., Cohen, M. L., Kohn, S. E., Petroff, Y., and Shen, Y. R., (1974), *Phys. Rev.*, **B10** (5175). 130

Dean, P. J. *See* Bell, R. J.; Collins, A. T.; Yarnell, J. L.

Debye, P. (1912), *Ann. Physik*, **39** (789). 216–218

Decarpigny, J. N., and Lannoo, M. (1976), *Phys. Rev.*, **B14** (538). 97

Decarpigny, J. N. *See* Lannoo, M.

Dennis, R. B. *See* Summers, C. J.

Dick, B. G., Jr., and Overhauser, A. W. (1958), *Phys. Rev.*, **112** (90). 212

Dickey, D. H., Johnson, E. J., and Larsen, D. M. (1967), *Phys. Rev. Letters*, **18** (599). 157

Dimmock, J. O. *See* Stillman, G. E.; Wheeler, R. G.

Dingle, R., Wiegmann, W., and Henry, C. H. (1974), *Phys. Rev. Letters*, **33** (827). 253, 255

DiStefano, T. H., and Eastman, D. E. (1971a), *Phys. Rev. Letters*, **27** (1560). 271

DiStefano, T. H., and Eastman, D. E. (1971b), *Solid State Commun.*, **9** (2259). 270

Dolling, G. (1963), *Inelastic Scattering of Neutrons in Solids and Liquids*, vol. II (37), International Atomic Energy Agency, Vienna. 208

Dolling, G., and Waugh, J. L. T., (1965), in *Lattice Dynamics* p. 19, Wallis, R. F., ed., Pergamon Press, Oxford. 207, 208, 217

Dolling, G. *See* Cowley, R. A.; Warren, J. L.

Dondi, M. G. *See* Cavallini, M.

Dresselhaus, G., and Dresselhaus, M. S. (1967), *Phys. Rev.*, **160** (649). 149

Dresselhaus, M. S. *See* Dresselhaus, G.

Ducuing, J. *See* Chang, R. K.

Duke, A. R. *See* Lubinsky, A. R.

Duthie, J. C. and Pettifor, D. G. (1977), *Phys. Rev. Letters*, **38** (564). 500

Dye, D. H. *See* Crabtree, G. W.

Eastman, D. E. *See* DiStefano, T. H.; Grobman, W. D.

Ehrenreich, H. (1967), in *The Optical Properties of Solids*, p. 106, Tauc, J., ed., Academic Press, New York. 102

Ehrenreich, H. *See* Esposito, E.; Gelatt, C. D. Jr.; Hodges, L.; Philipp, H. R.; Velický, B.; Watson, R. E.

Einstein, A. (1907), *Ann. Physik*, **22** (180; 800). 217

Einstein, A. (1911), *Ann. Physik*, **34** (170). 217

Elliot, R. G., (1972), ed., *Magnetic Properties of Rare Earth Metals*, Plenum, New York. 431

Ellis, D. E. *See* Painter, G. S.

Era, K. *See* Langer, D. W.

Esaki, L. *See* Sakaki, H.

Eschrig, H. *See* Paasch, G.

Eschrig, H. *See* Hafner, J.

Esposito, E., Ehrenreich, H., and Gelatt, C. D., Jr. (1978), *Phys. Rev.*, **B18** (3913).

Euwema, R. N. *See* Langer, D. W.

Evans, W. H. *See* Wagman, D. D.

Ewald, A. W. *See* Lindquist, R. E.

Faber, T. E. (1969), in *The Physics of Metals*, p. 282, Ziman, J. M., ed., Cambridge University Press, New York. 404

Falicov, L. M. *See* Koiller, B.

Fan, H. Y. *See* Nahory, R. E.

Farnsworth, H. E. *See* Schlier, R. E.

Farrell, H. H. *See* Larsen, P. K.

Fedorov, D. L. *See* Bir, G. L.

Feinleib, J., and Adler, D. (1968), *Phys. Rev. Letters*, **21** (1010). 433

Felber, B. *See* Paasch, G.

Feldkamp, L. A., Steinman, D. K., Vagelatos, N., and King, J. S., and Venkatamaran, G. (1971), *J. Phys. Chem. Solids*, **32** (1573). 208

Fischer, C. F. (1972), *Atomic Data*, **4** (301). 451, 533f

Fisher, T. E. (1965), *Phys. Rev.*, **139** (A1228). 254

Fisher, T. E. (1966), *Phys. Rev.*, **142** (519). 254

Flytzanis, Ch. (1970), *Phys. Rev. Letters*, **31A** (273). 119

Flytzanis, Ch. (1975), in *Quantum Electronics: A Treatise*, vol. I, p. 9, Rabin, H., and Tang,

C. L., eds., Academic Press, New York. 119, 124

Flytzanis, Ch. See Yablonovitch, E. C.

Fowler, W. B. See Yip, K. L.

Frank, F. C., and Turnbull, D. (1956), *Phys. Rev.*, **104** (617). 129

Freeman, A. J., and Koelling, D. D. (1974), in *The Actinides: Electronic Structure and Related Properties*, Freeman, A. J., and Darby, J. R., eds., Academic Press, New York. 431, 433

Freeman, A. J. See Rath, J.

Freeouf, J. L. See Grobman, W. D.

Frensley, W. R., and Kroemer, H. (1976), *J. Vac. Sci. Technol.*, **13** (810). 255, 425, 427f

Friedel, J. (1952), *Phil. Mag.*, **43** (153). 387, 526

Friedel, J. (1969), in *The Physics of Metals*, Ziman, J. M., ed., Cambridge University Press, New York. 494ff, 525

Friedel, J. (1978), *Journal de Physique*, **39** (651, 671). 90, 273

Friedel, J. See Barišić, S.; Labbé, J.; Leman, G.

Fritzsche, H. See Hudgens, S.

Fröhlich, H., Pelzer, H., and Zienau, S. (1950), Phil. Mag., **41** (221). 325

Froyen, S., and Harrison, W. A. (1979), *Phys. Rev.*, **B** (in press). 48f, 408–410

Froyen, S. See Harrison, W. A.

Fuchs, K. (1935), *Proc. Roy. Soc. A*, **151** (585). 349

Fuchs, K. (1936), *Proc. Roy. Soc. A*, **153** (622). [Corrected values are quoted by Mott and Jones (1936), (149)]. 349

Gadzuk, J. W. (1974), *Phys. Rev.*, **B10** (5030). 105

Gale, S. J., and Pettifor, D. G. (1977), *Solid State Commun.*, **24** (175). 455

Garrett, D. G., and Swihart, J. C. (1976), *J. Phys.*, **F6** (1781). 389

Gaspari, G. D. See Shyu, W.-M.

Gavini, A. A., and Cardona, M. (1969), *Phys. Rev.*, **177** (1351). 116

Gay, J. G., Smith, J. R., and Arlinghaus, F. J. (1977), *Phys. Rev. Letters*, **38** (561). 342

Gay, J. G. See Smith, J. R.

Gayton, W. R. See Hickernell, F. S.

Gelatt, C. D., Jr., Ehrenreich, H., and Watson, R. E. (1977), *Phys. Rev.*, **B15** (1613). 507

Gelatt, C. D., Jr. See Esposito, E.

Gelbart, W. M. See Heller, D. F.

Gell-Mann, M., and Brueckner, K. A. (1957), *Phys. Rev.*, **106** (364). 540

Gielisse, P. J., Mitra, S. S., Plendl, J. N. Griffis, R. D., Mansur, L. C., Marshall, R.,

and Pascoe, E. A. (1967), *Phys. Rev.*, **155** (1039). 115

Gilat, G., and Bharatiya, N. R. (1975), *Phys. Rev.*, **B12** (3479). 56

Gilat, G. See Raubenheimer, L. J.

Gilbert, T. L. (1970), in *Sigma Molecular Orbital Theory* (p. 244), Sinanoğlu, O., and Wiberg, K. B., eds., Yale University Press, New Haven. 46

Gilbert, T. L. Stevens, W. J., Schrenk, H., Yoshimine, M., and Bagus, P. S. (1973), *Phys. Rev.*, **B8** (5977). 263

Gladstone, G., Jensen, M. A., and Schrieffer, J. R. (1969), in *Superconductivity*, vol. 2, Parks, R. D., ed., Marcel Dekker, New York. 493

Gobeli, G. W., and Allen, F. G. (1962), *Phys. Rev.*, **127** (141). 252, 254

Gobeli, G. W., and Allen, F. G. (1965), *Phys. Rev.*, **137** (A245). 254

Gobeli, G. W. See Lander, J. J.; MacRae, A. U.

Gold, A. V. (1958), *Phil. Trans. Roy. Soc. London)*, **A251** (85). 379

Goltzene, A. See Chemla, D.

Gombas, P. (1949), *Die Statistische Theories des Atoms und Ihre Anwendungen*, Springer-Verlag, Berlin. 535

Goodenough, J. B. (1971), in *Progress in Solid State Chemistry*, vol. 5 (p. 145), Reiss, H., ed., Pergamon, Oxford. 438, 440, 441, 452

Gordon, R. G., and Kim, Y. S. (1972), *J. Chem. Phys.*, **56** (3122). 292–295, 307, 308, 540ff

Gordon, R. G. See Kim, Y. S.

Goryuneva, N. A. (1965), *Chemistry of Diamond-Like Semiconductors*, Chapman and Hall, London. 115

Gout, C. See Albert, J. P.

Graebner, J. E., Greiner, E. S., and Ryden, W. D. (1976), *Phys. Rev.*, **B13** (2426). 452

Gregory P. See Chye, P. W.

Greiner, E. S. See Graebner, J. E.

Griffing, V. See Padgett, A. A.

Griffis, R. D. See Gielisse, P. J.

Grindlay, J., and Howard, R. (1965), *Lattice Dynamics*, p. 129, Wallis, R. F., ed., Pergamon Press, Oxford. 295

Grobman, W. D., Eastman, D. E., and Freeouf, J. L. (1975), *Phys. Rev.*, **B12** (4405). 79

Groves, W. O. See Weil, R.

Gudat, W. See Cardona, M.

Gupta, K. P. See Cheng, C. H.

Gurney, R. W. See Mott, N. F.

Gurskii, Z. A. See Krasko, G. L.

Hafner, J. (1976), *J. Phys.*, **F6** (1243). 344

Herman, F., and Skillman, S. (1963), *Atomic Structure Calculations*, Prentice Hall, Englewood Cliffs, N.J. 15, 50f, 53f, 109, 540f

Herman, F., Kortum, R. L., Kuglin, C. D. (1966), *Int. J. Quant. Chem.*, **IS** (533). 138, 420

Herman, F., Kortum, R. L., and Kuglin, C. D. (1967), *Int. J. Quant. Chem.*, **IS** (533). 162

Herman, F., Kortum, R. L., Kuglin, C. D., Van Dyke, J. P., and Skillman, S. (1968), *Methods in Computational Physics*, **8** (193). 138, 152, 161, 162

Herring, C. (1940), *Phys. Rev.*, **57** (1169). 138, 343, 543

Herring, C. (1953), in *Structure and Properties of Solid Surfaces*, p. 5, Gomer, R., and Smith, C. S., eds., University of Chicago Press, Chicago. 231f

Herring, C. (1966), in *Magnetism*, Rado, G. T., and Suhl, H., eds., Academic Press, New York. 433

Herzfeld, K. F. *See* Lyddane, R. H.

Hickernell, F. S., and Gayton, W. R. (1966), *J. Appl. Phys.*, **37** (462). 196

Higginbotham, C. W., Cardona, M., and Pollak, F. H., (1969), *Phys. Rev.*, **184** (821). 116

Himpsel, F. J., and Steinmann, W. (1978), *Phys. Rev.* **B17** (2537). 324

Hite, G. E., Marple, D.T.F., Aven, M., and Segall, B. (1967), *Phys. Rev.*, **156** (850). 157

Ho, K. M. *See* Larsen, P. K.

Hodges, L., and Ehrenreich, H. (1965), *Phys. Letters*, **16** (203). 479

Hodges, L., Watson, R. E., and Ehrenreich, H. (1972), *Phys. Rev.*, **B5** (3953). 507

Hodges, L. *See* Watson, R. E.

Hoffmann, R. (1963), *J. Chem. Phys.*, **39** (1397). 46

Hoffmann, R. *See* Woodward, R. B.

Hohenberg, P., and Kohn, W. (1964), *Phys. Rev.*, **136** (B864). 540

Holas, A. *See* Brovman, E. G.

Hopfield, J. J., and Thomas, D. G. (1961), *Phys. Rev.*, **122** (35). 157

Howard, R. *See* Grindlay, J.

Howe, J. B. *See* Czyzak, S. J.

Huang, C., Moriarty, J. A., Sher, A., and Breckenridge, R. A. (1975), *Phys. Rev.*, **B12** (5395). 69

Huang, C., Moriarty, J. A., and Sher, A. (1976), *Phys. Rev.*, **B14** (2539). 69

Huang, K. *See* Born, M.

Hudgens, S. (1973), *Phys. Rev.*, **B7** (2481). 133, 135

Hudgens, S., Kastner, M., and Fritzsche, H. (1974), *Phys. Rev. Letters*, **33** (1552). 133, 135, 136

Hughes, A. J. *See* Callaway, J.

Huggins, R. A. (1975), in *Diffusion in Solids*, p. 445, Nowick, A. S., and Burton, J. J., eds., Academic Press, New York. 316

Hund, F. (1925), *Z. Physik*, **34** (833). 314

Huntington, H. B. (1958), *Solid State Physics*, vol. 7, Seitz, F., and Turnbull, D., Academic Press, New York. 193, 196

Ibach, H., and Rowe, J. E. (1974a), *Phys. Rev.*, **B9** (1951). 247

Ibach, H., and Rowe, J. E. (1974b), *Surface Science*, **43** (481). 247

Ibach, H., and Rowe, J. E. (1974c), *Phys. Rev.*, B10 (710). 271

International Critical Tables, Vol. III (1928), p. 46. McGraw-Hill, New York. 356

Inoguchi, T., Okamoto, T., and Koba, M. (1969), *Sharp Techn. J.*, **12** (59). 196

Ishiguro, E. *See* Kotani, M.

Iyengar, P. K. *See* Roy, A. P.

Jaffe, H. *See* Berlincourt, D.

Jahn, H. A., and Teller, E. (1937), *Proc. Roy. Soc. A*, **161** (220). 234

James, R. W. (1950), *Optical Principles of the Diffraction of X-rays*, chapter 5, Bell, London. 390

Janak, J. F. *See* Moruzzi, V. L.

Jayaraman, A. *See* McSkimin, H. J.

Jenkin, J. G. *See* Poole, R. T.

Jennison, D. R. and Kunz, A. B. (1976), *Phys. Rev.*, **B13** (5597). 331

Jensen, M. A. *See* Gladstone, G.

Jepsen, D. W. *See* Shih, H. D.; Strozier, J. A.

Jepsen, O., and Andersen, O. K. (1971), *Solid State Commun.*, **9** (1763). 55

Jepsen, O., Andersen, O. K., and Mackintosh, A. R. (1975), *Phys. Rev.*, **B12** (3084). 490, 492

Jepsen, O. *See* Andersen, O. K.

Joannopoulos, J. D., and Pollard, W. B. (1976), *Solid State Commun.*, **20** (947). 279

Joannopoulos, J. D., Schlüter, M. A. and Cohen, M. L. (1974), in *Proceedings of the Twelfth International Conference on the Physics of Semiconductors (Stuttgart)*, p. 1304, Pilkun, M. H., ed., Teubner, Stuttgart. 92

Joannopoulos, J. D. *See* Laughlin, R. B.; Yndurain, F.

Johansson, B., and Rosengren, A. (1975), *Phys. Rev.*, **B11** (2836). 500

Johansson, B. *See* Skriver, H. L.

Johnson, E. J. (1967), *Phys. Rev. Letters*, **19** (352). 157

Johnson, E. J. *See* Dickey, D. H.

Johnson, K. H., Norman, J. G., Jr., and Connolly, J. W. D. (1973), in *Computational Methods for Large Molecules and Localized States in Solids*, p. 161, Herman, F., McLean, A. D., and Nesbet, R. K., eds., Plenum Press, New York. 138

Johnson, Q.C., and Templeton, D.H. (1961), *J. Chem. Phys.*, **34** (2004). 304, 305

Jona, F. *See* Shih, H. D.; Strozier, J. A.

Jones, H. *See* Mott, N. F.

Jouain, D. *See* Albert, J. P.

Kagan, Yu. *See* Brovman, E. G.

Kaldis, E. *See* Batlogg, B.

Kaminsky, A. *See* Pokrovsky, Ya.

Kanazawa, K. K., and Brown, F. C. (1964), *Phys. Rev.*, **135** (A1757). 157

Kane, A. B. *See* Kane, E. O.

Kane, E. O. (1966), *Phys. Rev.*, **146** (558). 106

Kane, E. O., and Kane, A. B. (1978), *Phys. Rev.* **B17** (2691). 87, 88

Känzig, W. *See* Castner, T.

Kaplan, R., Kinch, M. A., and Scott, W. C. (1969), *Solid State Commun.*, **7** (883). 157

Karim, D. P. *See* Crabtree, G. W.

Kastner, M. (1972), *Phys. Rev.*, **B6** (2273). 116

Kastner, M. *See* Hudgens, S.

Kaufmann, U., and Schneider, J. (1974), in *Festkörperprobleme XIV, Advances in Solid State Physics*, p. 229, Pergamon (Vieweg), Braunschweig. 131

Kayama, K. *See* Kotani, M.

Keating, P. N. (1966), *Phys. Rev.*, **145** (637). 194

Kellermann, E. W. (1940), *Phil. Trans. Roy. Soc.* (London), **A238** (513). 312

Ketterson, J. B. *See* Crabtree, G. W.

Keyes, R. W. (1962), *J. Appl. Phys.*, **33** (3371). 186

Kim, Y. S., and Gordon, R. G. (1974a), *Phys. Rev.*, **B9** (3548). 309, 540

Kim, Y. S., and Gordon, R. G. (1974b), *J. Chem. Phys.*, **60** (1842). 309, 540

Kim, Y. S., and Gordon, R. G. (1974c), *J. Chem. Phys.*, **60** (4323). 309

Kim, Y. S. *See* Gordon, R. G.

Kinch, M. A. *See* Kaplan, R.

King, J. S. *See* Feldkamp, L. A.

Kirkpatrick, S. *See* Velický, B.

Kittel, C. (1963), *Quantum Theory of Solids*, John Wiley and Sons, New York. 158, 325, 533, 542

Kittel, C. (1967), *Introduction to Solid State Physics*, 3rd. ed., John Wiley and Sons, New York. 176, 183, 188, 219, 291, 295, 309, 311, 315f, 326, 424

Kittel, C. (1971), *Introduction to Solid State Physics*, 4th ed., John Wiley and Sons, New York. 171, 249, 292, 296, 434

Kittel, C. (1976), *Introduction to Solid State Physics*, 5th ed., John Wiley and Sons, New York. 99, 295, 357

Kittel, C. *See* Ruderman, M. A.

Klein, G., and Chun, H. U. (1972), *Physica Status Solidi* **(b),49** (167). 272

Kleinman, D. A., and Spitzer, W. G. (1962), *Phys. Rev.*, **125** (16). 279ff, 284f

Kleinman, L. (1962), *Phys. Rev.*, **128** (2614). 199

Kleinman, L. *See* Phillips, J. C.

Klose, W. *See* Andersen, O. K.

Koba, M. *See* Inoguchi, T.

Koda, T. *See* Langer, D. W.

Koelling, D. D. *See* Freeman, A. J.; Crabtree, G. W.

Kohn, S. E. *See* de Alvarez, C. W.

Kohn, W. (1959), *Phys. Rev. Letters*, **2** (393). 395f

Kohn, W., and Sham, L. J. (1965), *Phys. Rev.*, **140** (A1133). 535, 540

Kohn, W. *See* Hohenberg, P.; Lang, N. D.

Koiller, B., and Falicov, L. M. (1974), *J. Phys.*, **C7** (299). 433–438

Kokes, R. J. *See* Andrews, D. H.

Kondo, K., and Moritani, A. (1977), *Phys. Rev.*, **B15** (812). 107

Kortum, R. L. *See* Herman, F.; Koster, G. F.; Slater, J. C.

Kotani, M., Mizuno, Y., Kayama, K., and Ishiguro, E., (1957), *J. Phys. Soc. Japan*, **12** (707). 28

Kroemer, H. *See* Frensley, W. R.

Krasko, G. L., and Gurskii, Z. A. (1969), *Zh. E. T. F. Pis Red.*, **9** (596) [*Soviet Phys. JETP Letters (English Transl.)*, 344 **9** (363).]

Krumhansl, J. A. (1959), *J. Appl. Phys.*, **30** (1183). 131

Kubo, R., and Nagamiya, T. (1969), *Solid State Physics*, McGraw-Hill. 234, 297

Kuglin, C. D. *See* Herman, F.

Kunc, K., Balkanski, M., and Nusimovici, M. A. (1971), *Proceedings of the International Conference on Phonons, Rennes, France*, p.

109, Nusimovici, M. A., ed., Flammarion Sciences, Paris. 211

Kunc, K. *See* Carabatos, C.

Kunz, A. B. *See* Jennison, D. R.

Kupecek, P. *See* Chemla, D.

Labré, J., and Friedel, J. (1966), *J. Phys. Radium*, **27** (153). 455

Labbé, J. *See* Barištić, S.

Lander, J. J., Gobeli, G. W., and Morrison, J. (1963), *J. Appl. Phys.*, **34** (2298). 233, 235

Lander, J. J., and Morrison, J. (1963), *J. Appl. Phys.*, **34** (1403). 249

Lander, J. J., and Morrison, J. (1964), *Surface Science*, **2** (553). 248

Landolt, H., and Börnstein, R. (1966), *Numerical Data and Functional Relationships in Science and Technology, New Series*, Hellwege, K. H., and Hellwege, A. M., eds., Springer-Verlag, Berlin. 135, 196, 313

Lang, N. D., and Kohn, W. (1970), *Phys. Rev.*, **B1** (4555). 399ff

Lang, N. D., and Kohn, W. (1971), *Phys. Rev.*, **B3** (1215). 402

Lang, N. D., and Williams, A. R. (1976), *Phys. Rev. Letters*, **37** (212). 402f

Lang, N. D. *See* Yu, K. Y.

Langer, D. W., Euwema, R. N., Era, K., and Koda, T. (1970), *Phys. Rev.*, **B2** (4005). 157

Langreth, D. C. *See* Ashcroft, N. W.

Lannoo, M. (1977), *J. Phys. (Paris)*, **38** (473). 97

Lannoo, M. *See* Decarpigny, J. N.

Lannoo, M., and Decarpigny, J. N. (1973), *Phys. Rev.* **B8** (5704). 60

Lannoo, M., and Decarpigny, J. N. (1974), *J. Phys. (Paris)*, **35C** (3). 282

Larsen, D. M. *See* Dickey, D. H.

Larsen, P. K., Smith, N. V., Schlüter, M., Farrell, H. H., Ho, K. M., and Cohen, M. L. (1978), *Phys. Rev.*, **B17** (2612). 105

Laughlin, R. B., and Joannopoulos, J. D. (1977), *Phys. Rev.*, **B16** (2942). 279

Lawaetz, P. (1971), *Phys. Rev.*, **B4** (3460). 103, 156f, 158, 253

Lax, B. *See* Reine, M.

Lax, M. (1958), *Phys. Rev. Letters*, **1** (133). 211

Leckey, R. C. G. *See* Poole, R. T.

Lee, B. W. *See* Lubinsky, A. R.

Lee, Y. T. *See* Parson, J. M.; Siska, P. E.

Lehmann, G., and Taut, M. (1972), *Physica Status Solidi*, **(b)54** (469). 55

Leman, G., and Friedel, J. (1962), *J. Appl. Phys.*, **33** (281). 60

Lekner, J. *See* Ashcroft, N. W.

Lennard-Jones, J. E. (1924), *Proc. Roy. Soc. A*, **106** (441). 294

Lennard-Jones, J. E. (1925), *Proc. Roy. Soc. A*, **109** (584). 294

Levin, A. A. (1974), *Solid State Quantum Chemistry*, McGraw-Hill, New York. 53, 149

Lewis, G. N. *See* Coulson, C. A. 59

Liesegang, J. *See* Poole, R. T.

Lightowlers, E. C. *See* Collins, A. T.

Lindgren, I. (1971), *Intern. J. Quant. Chem.*, **5** (411). 513

Lindquist, R. E., and Ewald, A. W. (1964), *Phys. Rev.*, **135** (A191). 115

Lines, M. E., and Waszczak, J. V. (1977), *J. Appl. Phys.*, **48** (1395). 131

Lloyd, P. (1967), *Proc. Phys. Soc.* **90** (207). 508

Lloyd, P., and Sholl, C. A. (1968), *J. Phys.* **C1** (1620). 344, 389

Lomer, W. M. (1962), *Proc. Phys. Soc. A*, **80** (489). 490

Louie, S. G., Chelikowsky, J. R., and Cohen, M. L. (1975), *Phys. Rev. Letters*, **34** (155). 97

Louie, S. G., and Cohen, M. L. (1975), *Phys. Rev. Letters*, **35** (866). 428

Louie, S. G., Schlüter, Chelikowsky, J. R., and Cohen, M. L. (1976), *Phys. Rev.*, **B13** (1654). 249

Lubinsky, A. R., Duke, C. B., Lee, B. W., and Mark, P. (1976), *Phys. Rev. Letters*, **36** (1058). 242

Lucovsky, G., Martin, R. M., and Burstein, E. (1971), *Phys. Rev.*, **B4** (1367). 219f, 335, 336

Lucovsky, G. *See* Burstein, E.; Martin, R. M.

Ludeke, R. *See* Sakaki, H.

Lyddane, R. H., and Herzfeld, K. F. (1938), *Phys. Rev.*, **54** (846). 218

Lyddane, R. H., Sachs, R. G., and Teller, E. (1941), *Phys. Rev.*, **59** (673). 218

McElroy, P. *See* Cardona, M.

McFee, J. H., Boyd, G. D., and Schmidt, P. H. (1970), *Appl. Phys. Letters*, **17** (57). 122

McKelvey, J. P. (1966), *Solid State and Semiconductor Physics*, p. 485, Harper & Row, New York 244, 246.

Mackintosh, A. R. (1977), *Physics Today*, **30: 6** (23). 431

Mackintosh, A. R. *See* Jepsen, O.

McLaren, R. E. (1974), Ph.D. dissertation, University of New England, Armidale, Australia. 343

McLaren, R. E., and Sholl, C. A. (1974), *J. Phys.*, **F4** (2172). 344

McMillan, W. L. (1968), *Phys. Rev.*, **167** (331). 397f, 399

MacRae, A. U., and Gobeli, G. W. (1966), in *Semiconductors, and Semimetals, vol. 2*, p. 115, Willardson, R. K., and Beer, A. C., eds., Academic Press, New York. 242

McSkimin, H. J., and Bond, W. L. (1957), *Phys. Rev.*, **105** (116). 193, 196

McSkimin, H. J., Jayaraman, A., Andreatch, P., Jr., and Bateman, T. B. (1968), *J. Appl. Phys.*, **39** (4127). 196

Madelung, E. (1909), *Gött. Nach.* (100). 304

Madelung, E. (1918), *Physik, Z.*, **19** (524). 336

Madelung, E. (1919), *Physik, Z.*, **20** (494). 336

Manabe, A., Mitsuishi, A., and Yoshinaga, H. (1967), *Japan J. Appl. Phys.*, **6** (593). 115

Manca, P. *See* Bailly, F.

Manghi, F. *See* Bertoni, C. M.

Mann, J. B. (1967), *Atomic Structure Calculations, 1: Hartree-Fock Energy Results for Elements Hydrogen to Lawrencium.* Distributed by Clearinghouse for Technical Information, Springfield, Virgina 22151. 533

Mansur, L. C. *See* Gielisse, P. J.

Marcus, P. M. *See* Shih, H. D.

Mark, P. *See* Lubinsky, A. R.

Marple, D.T.F. (1964b), *J. Appl. Phys.*, **35** (539, 1879). 115, 157

Marple, D. T. F. *See* Hite, G. E.; Segall, B.

Marshall, R. *See* Gielisse, P. J.

Martin, R. M. (1970), *Phys. Rev.*, **B1** (4005). 190, 195, 199, 200

Martin, R. M. (1972a), *Phys. Rev.*, **B5** (1607). 220, 224

Martin, R. M. (1972b), *Phys. Rev.*, **B6** (4546). 190, 194, 211

Martin, R. M., Lucovsky, G., and Helliwell, K. (1976), *Phys. Rev.*, **B13** (1383). 92, 94

Martin, R. M., and Chadi, D. J. (1976), in *Proceedings of the Thirteenth International Conference on the Physics of Semiconductors* (Rome), p. 187, Fumi, F. G., ed. 211

Martin, R. M. *See* Chadi, D. J.; Lucovsky, G.; Van Vechten, J. A.

Mattheiss, L. F. (1964), *Phys. Rev.*, **134** (A970). 477f, 485, 486

Mattheiss, L. F. (1972a), *Phys. Rev.*, **B5** (290, 306). 433, 434, 435

Mattheiss, L. F. (1972b), *Phys. Rev.*, **B6** (4718). 439–444, 449–452

Mattheiss, L. F. (1975), *Phys. Rev.*, **B12** (2161). 455

Mattheiss, L. F. (1976), *Phys. Rev.*, **B13** (2433). 452

Meijer, H. J. G., and Polder, D. (1953), *Physica*, **19** (255). 225

Meserve, R. (1975), *Ph.D. dissertation*, Stanford University, Stanford, California. 547f

Messiah, A. (1962), *Quantum Mechanics*, John Wiley and Sons, New York. 134, 509

Messmer, R. P., and Watkins, G. D. (1973), *Phys. Rev.*, **7** (2568). 249

Messmer, R. P. *See* Watkins, G. D.

Miklosz, J. C., and Wheeler, R. G. (1967), *Phys. Rev.*, **153** (913). 157

Miler, M. (1968), *Czech. J. Phys.*, **B18** (354). 286

Miller, J. N. *See* Yu, K. Y.

Milnes, A. G., and Feucht, D. L. (1972), *Heterojunctions and Metal-Semiconductor Junctions*, Academic Press, New York. 255

Mitchell, D. L. *See* Pidgeon, C. R.

Mitra, S. S. *See* Gielisse, P. J.; Tsay, Y. F.

Mitsuishi, A. *See* Manabe, A.

Mizuno, Y. *See* Kotani, M.

Moore, C. E. (1949), *Atomic Energy Levels*, vol. 1, National Bureau of Standards, Washington. 296, 434

Moore, C. E. (1952), *Atomic Energy Levels*, vol. 2, National Bureau of Standards, Washington. 296, 434

Moos, H. W. *See* Soref, R. A.

Mooser, E., and Pearson, W. B. (1959), *Acta Crystallogr.*, **A12** (1015). 89

Moriarty, J. A. (1970), *Phys. Rev.*, **B1** (1363). 512, 513, 520

Moriarty, J. A. (1972a), *Phys. Rev.*, **B5** (2066). 511, 512

Moriarty, J. A. (1972b), *Phys. Rev.*, **B6** (1239). 511, 512, 513, 520

Moriarty, J. A. (1972c), *Phys. Rev.*, **B6** (4445). 512, 520

Moriarty, J. A. (1975), *J. Phys.*, **F5** (873). 514

Moriarty, J. A. *See* Huang, C.

Moritani, A. *See* Kondo, K.

Morel, P., and Anderson, P. W. (1962), *Phys. Rev.*, **125** (1263). 399

Morrison, J. *See* Lander, J. J.

Moruzzi, V. L., Williams, A. R., and Janak, J. F., (1977), *Phys. Rev.*, **B15** (2854). 494

Moruzzi, V. L., Janak, J. F., and Williams, A. R. (1978), Pergamon Press, New York. 477, 494

Mott, N. F., and Jones, H. (1936), *Theory of the Properties of Metals and Alloys*, Clarendon Press, Oxford. 395

Mott, N. F., and Gurney, R. W. (1953), *Electronic Processes in Ionic Crystals*, Clarendon Press, Oxford. 297

Mott, N. F., and Jones, H. (1958), *Theory of the Properties of Metals and Alloys*, Dover, New York; first published by Clarendon Press, Oxford (1936). 60, 343, 412f

Mott, N. F. (1974), *Metal-Insulator Transitions*, Taylor and Francis, London. 438

Moussa, F. *See* Carabatos, C.; Hennion, B,; Prevot, B.

Mueller, F. M. (1967), *Phys. Rev.*, **153** (659). 479

Musgrave, M. J. P., and Pople, J. A. (1962), *Proc. Roy. Soc. A*, **268** (474). 194

Nagamiya, T. *See* Kubo, R.

Nahory, R. E., and Fan, H. Y. (1966), *Phys. Rev. Letters*, **17** (251). 157

Neyer, H. R. *See* Segmüller, A.

Ngoc, T. C. *See* Poppendieck, T. D.

Nikulin, V. K. (1971), *Zh. Tekhn. Fiz.*, XLI (41). [*Sov. Phys.-Tech. Phys.* (*English Transl.*), **16** (28).] 541

Nikulin, V. K., and Tsarev Yu. N. (1975), *Chem. Phys.* (*Netherlands*), **10** (433). 540

Nizzoli, F. *See* Bertoni, C. M.

Nohl, H. *See* Anderson, O. K.

Norman, J. G., Jr. *See* Johnson, K. H.

Nowotny, H. *See* Hafner, J.

Nusimovici, M. A. *See* Kunc, K.

Nye, J. F. (1957), *Physical Properties of Crystals, Their Representation by Tensors and Matrices*, Clarendon, Oxford. 198

Ockman, N. *See* Braunstein, R.

Okamoto, T. *See* Inoguchi, T.

Overhauser, A. W. *See* Dick, B. G., Jr.

Paasch, G., Felber, B., and Eschrig, H. (1972), *Physica Status Solidi*, (b)**54** (K51). 344

Padgett, A. A., and Griffing, V. (1959), *J. Chem. Phys.*, **30** (1286). 28

Painter, G. S., and Ellis, D. E. (1970), *Phys. Rev.*, **1** (4747). 95, 165

Palik, E. D., and Wallis, R. F. (1961), *Phys. Rev.*, **123** (131). 157

Pandey, K. C. (1976), *Phys. Rev.*, **B14** (1557). 53

Pantelides, S. T. (1975a), *Phys. Rev. Letters*, **35** (250). 326ff

Pantelides, S. T. (1975b), *Phys. Rev.* **B11** (2391). 324ff

Pantelides, S. T. (1975c), *Phys. Rev.*, **B11** (5082). 296, 297, 320ff

Pantelides, S. T., and Harrison, W. A. (1975), *Phys. Rev.*, **B11** (3006). 44, 46, 47, 81, 142–151

Pantelides, S. T., and Harrison, W. A. (1976), *Phys. Rev.*, **B13** (2667). 264–287, 306

Pantelides, S. T. *See* Harrison, W. A.

Parker, V. B. *See* Wagman, D. D.

Parson, J. M., and Lee, Y. T. (1971), in *Third International Symposium on Molecular Beams, Cannes.* 293

Parson, J. M. *See* Siska, P. E.

Pascoe, E. A. *See* Gielisse, P. J.

Patel, C. K. N. (1966), *Phys. Rev. Letters*, **16** (613). 122

Paul, W., and Waschauer, D. M. (1963), *Solids under Pressure*, McGraw-Hill, New York. 226

Paul, W. *See* Zallen, R.

Pauling, L. (1938), *Phys. Rev.*, **54** (899). 525

Pauling, L., (1960), *The Nature of the Chemical Bond*, Cornell University Press, Ithaca, New York. 43, 126

Pearson, W. B. *See* Mooser, E.

Pelzer, H. *See* Fröhlich, H.

Percus, J. K. (1962), *Phys. Rev. Letters*, **8** (462). 404

Percus, J. K., and Yevick, G. J. (1958), *Phys. Rev.*, **110** (1). 404

Pertosa, P., and Michel-Calendini, F. M. (1978), *Phys. Rev.*, **B17** (2011). 439

Petroff, Y. *See* de Alvarez, C. V.

Pettifor, D. G. (1969), *J. Phys.*, **C2** (1051). 514

Pettifor, D. G. (1977a), *J. Phys.*, **F7** 613). 487, 493, 495f, 499f, 503, 506f

Pettifor, D. G. (1977b), *J. Phys.*, **F7** (1009). 493, 495, 500, 507

Pettifor, D. G. *See* Duthie, J. C.; Gale, S. J.

Pettit, G. D. *See* Turner, W. J.

Philipp, H. R., and Ehrenreich, H. (1963), *Phys. Rev.*, **129** (1550). 101

Phillips, J. C. (1970), *Rev. Mod. Phys.*, **42** (317). 43, 47, 111

Phillips, J. C. (1973a), *Bonds and Bands in Semiconductors*, Academic Press, New York. 54, 60, 103, 117, 126, 162, 173, 253

Phillips, J. C. (1973b), *Surface Science*, **40** (459). 240

Phillips, J. C. (1975), *Phys. Rev. Letters*, **34** (1196). 102f, 111

Phillips, J. C., and Kleinman, L. (1959), *Phys. Rev.*, **116** (287). 343, 408

Phillips, J. C., and Van Vechten, J. A. (1970), *Phys. Rev.*, **B2** (2147). 60, 177, 178

Phillips, J. Christopher. *See* Harrison, W. A.

Pickett, W. E. *See* Chakraborty, B.

Pidgeon, C. R., Mitchell, D. L., and Brown, R. N. (1967), *Phys. Rev.*, **154** (737). 157

Pietronero, L. (1978), *Phys. Rev.*, **B17** (3946). 337

Pikus, G. E. *See* Bir, G. L.

Platzöder, K. (1968), *Physica Status Solidi*, **29** (k63). 270

Plendl, J. N. *See* Gielisse, D. J.

Pokrovsky, Ya., Kaminsky, A., and Svistunova, K. (1970), in *Proceedings of the Tenth International Conference on the Physics of Semiconductors, Cambridge, Mass.* p. 504, Keller, S. P., Hensel, J. C., and Stern, F., eds. 164

Polder, D. *See* Meijer, H. J. G.

Pollak, F. H. *See* Cardona, M.; Higginbotham, C. W.; Shileika, A. Yu.; Yu, P. Y.

Pollard, W. B. *See* Joannopoulos, J. D.

Poole, R. T., Jenkin, J. G., Liesegang, J., and Leckey, R. C. G., (1975), *Phys. Rev.*, **B11** (5179). 320, 321, 322, 324

Poole, R. T., Liesegang, J., Leckey, R. C. G., and Jenkin, J. G. (1975), *Phys. Rev.*, **B11** (5190). 320, 323, 324

Pople, J. A., and Beveridge, D. L. (1970), *Approximate Molecular Orbital Theory*, McGraw-Hill, New York. 47

Pople, J. A. *See* Mugrave, M. J. P.

Poppendieck, T. D., Ngoc, T. C., and Webb, M. B. (1978), *Surface Science*, **75** (287). 241f

Pratt, G. W. *See* Zeiger, H. J.,

Prevot, B., Carabatos, C., Schwab, C., Hennion, B., and Moussa, F. (1973), *Solid State Commun.*, **13** (1725). 208

Prevot, B. *See* Hennion, B.

Price, D. L., and Rowe, J. M. (1969), *Solid State Commun.*, **7** (1433). 208

Quinn, J. J. (1960), in *The Fermi Surface*, p. 58, Harrison, W. A., and Webb, M. B., eds., John Wiley and Sons, New York. 398

Ransil, B. J. (1960), *Rev. Mod. Phys.*, **32** (245). 28

Rath, J., and Callaway, J. (1973), *Phys. Rev.*, **B8** (5398). 342, 488, 490f

Rath, J., and Freeman, A. J. (1975), *Phys. Rev.*, **B11** (2109). 56

Raubenheimer, L. J., and Gilat, G. (1966), *Phys. Rev.*, **144** (390). 55f, 80, 216

Rèdei, L. R. *See* Coulson, C. A.

Reese, W. E. *See* Turner, W. J.

Reilly, M. H. (1970), *J. Phys. Chem. Sol.*, **31** (1041). 263

Reine, M., Aggarwal, R. L., and Lax, B. (1970a), *Solid State Commun.*, **8** (35). 157

Reine, M., Aggarwal, R. L., Lax, B., and Wolfe, C. M. (1970b), *Phys. Rev.*, **B2** (458). 157

Ren, Shang-Yuan (1979—to be published). 106

Ressler, N. W. *See* Wang, C. C.

Reynolds, D. C., and Collins, T. C. (1969), *Phys. Rev.*, **185** (1099). 157

Riccius, H. D., and Turner, R. (1968), *J. Phys. Chem. Solids*, **29** (15). 157

Rose, M. E. (1957), *Elementary Theory of Angular Momentum*, John Wiley and Sons, New York. 460

Rosengren, A. *See* Johansson, B.

Rössler, U., and Schütz, O. (1973), *Physica Status Solidi*, **(b)56** (483). 296

Roth, L. M. *See* Bennett, A. J.

Rowe, J. E. *See* Ibach, H.; Price, D. L.

Rowe, J. M. *See* Price, D. L.

Roy, A. P., Dasannacharya, B. A., Thaper, C. L., and Lyengar, P. K. (1973), *Phys. Rev. Letters*, **30** (906) and (1278, E). 389

Ruderman, M. A., and Kittel, C. (1954), *Phys. Rev.*, **96** (99). 432f, 528f

Ruffa, A. R. (1968), *Physica Status Solidi*, **(b)29** (605). 263

Ruffa, A. R. (1970), *Phys. Rev. Letters*, **25** (650). 263

Ryden, W. D. *See* Graebner, J. E.

Sachs, R. G. *See* Lyddane, R. H.

Saffren, M. (1960), in *The Fermi Surface*, Harrison, W. A., and Webb, M. B., eds., John Wiley and Sons, New York. 479, 483

Sai-Halasz, G. *See* Sakaki, H.

Sakaki, H., Chang, L. L., Ludeke, R., Chang, Chin-An, Sai-Halasz, G., and Esaki, L. (1977), *App. Phys. Lett.*, **31** (211). 255

Saksena, B. D. (1940), *Proc. Indian Acad. Sci.*, A12 (93). 279

Salvan, F. *See* Benoit á la Guillaume. C.

Samara, G. A. (1976), *Phys. Rev.*, **B13** (4529). 336

Sanchez, C., and Cardona, M. (1971), in *Proceedings of the Conference on Vacuum UV Radiation Physics*, (Tokyo), Nakai, Y., ed., Physics Society of Japan, Tokyo. 116

Schafer, T. P. *See* Siska, P. E.

Schiferl, D. (1974), *Phys. Rev.*, **B10** (3316). 424f

Schiff, L. I. (1968), *Quantum Mechanics*, 3rd ed., McGraw-Hill, New York. 10, 29, 98, 99, 380, 509, 511, 514, 517, 518, 519

Schlegel, A. *See* Batlogg, B.

Schlier, R. E., and Farnsworth, H. E. (1959), *J. Chem. Phys.*, **30** (917). 233, 240

Schlüter, M. A. *See* Chelikowsky, J. R.; Joannoupoulos, J. D.; Larsen, P. K.; Louie, S. G.

Schmidt, P. H. *See* McFee, J. H.

Schneider, J. *See* Kaufmann, U.

Schrenk, H. *See* Gilbert, T. L.

Schrieffer, J. R. *See* Gladstone, G.

Schultz, T. D. (1956), *Tech. Report 9*, Solid-State and Molecular Theory Group, Massachusetts Institute of Technology, Cambridge, Mass. 325

Schütz, O. *See* Rössler, U.

Schwab, C. *See* Carabatos, C.; Chemla, D.; Hennion, B.; Prevot, B.

Schwartz, C. *See* Chemla, D.

Schwarz, K. (1977), *J. Phys.*, **C10** (195). 455

Scoles, G. *See* Cavallini, M.

Scott, W. C. *See* Kaplan, R.

Segall, B., and Marple, D. T. F. (1967), in *Physics and Chemistry of II–VI Compounds*, chapter VII, Aven, M., and Prener, J., eds., North-Holland, Amsterdam. 157

Segall, B. *See* Hite, G. E.

Següller, A., and Neyer, H. R. (1965), *Physik Kondensierten Materie*, **4** (63). 199

Seitz, F. (1940), *Modern Theory of Solids*, McGraw-Hill, New York. 218, 336

Seitz, F. *See* Wigner, E. P.

Selwood, P. (1956), *Magnetochemistry*, Interscience, New York. 135

Sen, P. N., and Thorpe, M. F. (1977), *Phys. Rev.*, **B15** (4030). 279

Shaklee, K. L. *See* Cardona, M.

Sham, L. J. (1965), *Proc. Roy. Soc. A*, **283** (33). 393

Sham, L. J. *See* Austin, B. J.; Hanke, W.; Kohn, W.

Sharma, R. R. (1979), *Phys. Rev.*, **B19** (2813). 481

Shaw, R. W., Jr. (1968), *Phys. Rev.*, **174** (769). 343

Shaw, R. W., and Harrison, W. A. (1967), *Phys. Rev.*, **163** (604). 545

Shay, J. L., Wagner, S., and Phillips, J. C. (1976), *Appl. Phys. Lett.*, **28** (31). 255

Shen, Y. R. *See* de Alvarez, C. V.

Sher, A. *See* Huang, C.

Shevchik, N. J. *See* Tejeda, J.

Shih, H. D., Jona, F., Jepsen, D. W., and Marcus, P. M. (1976), *Phys. Rev. Letters*, **36** (798). 403

Shileika, A. Yu, Cardona, M., and Pollak, F. H. (1969), *Solid State Commun.*, **7** (1113). 116

Shiozawa, L. R. *See* Berlincourt, D.

Shockley, W. (1950), *Electrons and Holes in Semiconductors*, Van Nostrand, New York. 226

Sholl, C. A. *See* Lloyd, P.; McLaren, R. E.

Shumm, R. H. *See* Wagman, D. D.

Shyu, W.-M., and Gaspari, G. D. (1967), *Phys.*

Rev., **163** (667). 344

Shyu, W.-M., and Gaspari, G. D. (1968), *Phys. Rev.*, **170** (687). 344

Siska, P. E., Parson, J. M., Schafer, T. P., and Lee, Y. T. (1971), *J. Chem. Phys.*, **55** (5762). 293

Skillman, S. *See* Herman, F.

Skriver, H. L., Andersen, O. K., and Johansson, B. (1978), *Phys. Rev. Letters*, **41** (42). 431

Slater, J. C. (1937), *J. Appl. Phys.*, **8** (385). 525

Slater, J. C. (1951), *Phys. Rev.*, **81** (385). 535

Slater, J. C. (1968), *Quantum Theory of Matter*, 2nd. ed., McGraw-Hill, New York. 22, 25, 28

Slater, J. C., and Koster, G. F. (1954), *Phys. Rev.*, **94** (1498). 11, 48, 480f

Smith, J. R., and Gay, J. G. (1975), *Phys. Rev.*, **B12** (4238). 342

Smith, J. R. *See* Gay, J. G.

Smith, N. V. *See* Larsen, P. K.

Smith, S. D. *See* Summers, C. J.

Smitt, F. T. (1971), *Seventh International Conference on The Physics of Electronic and Atomic Collisions, Amsterdam*, Branscomb, L. M., ed., North Holland, Amsterdam. 307

Sokel, R. (1976), *Bull. Am. Phys. Soc.*, **21** (1315). 142, 143, 185ff, 193

Sokel, R. C. (1978). Thesis, Stanford University. 185–193

Sokel, R., and Harrison, W. A. (1976), *Phys. Rev. Letters*, **36** (61). 212–215

Sokel, R. *See* Harrison, W. A.

Sonntag, B. *See* Cardona, M.

Soref, R. A., and Moos, H. W. (1964), *J. Appl. Phys.*, **35** (2152). 122

Soven, P. (1967), *Phys. Rev.*, **156** (809). 508

Spicer, W. E. *See* Chye, P. W.; Wagner, L. F.; Yu, K. Y.

Spitzer, W. G. *See* Kleinman, D. A.

Staver, T. *See* Bohm, D.

Steinman, D. K. *See* Feldkamp, L. A.

Steinman, W. *See* Himpsel, F.-J.

Stevens, W. J. *See* Gilbert, T. L.

Stillman, G. E., Wolfe, C. M., and Dimmock, J. O. (1969), *Solid State Commun.*, **7** (921). 157

Stocker, D. *See* Coulson, C. A.

Strehlow, W. H., and Cook, E. L. (1973), *J. Phys. Chem. Ref. Data*, **2** (163). 321, 322

Strozier, J. A., Jr., Jepsen, D. W., and Jona, F. (1975), in *Surface Physics of Materials*, (vol. I, p. 1), Blakely, J. M., ed., Academic Press, New York. 403

Sturge, M. (1962), *Phys. Rev.*, **127** (768). 157

Sukegawa, T. *See* Chye, P. W.

Sukhatme, V. P., and Wolff, P. A. (1975), *Phys. Rev. Letters*, **35** (1369). 131

Summers, C. J., Dennis, R. B., Wherrett, B. S., Harper, P. G., and Smith, S. D. (1968), *Phys. Rev.*, **170** (755). 157

Sunami, H. *See* Chye, P. W.

Sundfors, R. K. (1974), *Phys. Rev.*, **B10** (4244). 125

Suslina, L. G. *See* Bir, G. L.

Svistunova, K. *See* Pokrovsky, Ya.

Swank, R. K. (1967), *Phys. Rev.*, **153** (844). 254

Swihart, J. C. *See* Garrett, D. G.

Taut, M. *See* Lehmann, G.

Tejeda, J., and Shevchik, N. J. (1976), *Phys. Rev.*, **B13** (2548). 538

Tejedor, C., and Vergés, J. A. (1979), *Phys. Rev.*, **B19** (2283). 87

Teller, E. *See* Jahn, H. A.; Lyddane, R. H.

Templeton, D. H. *See* Johnson, Q. C.

Thaper, C. L. *See* Roy, A. P.

Thomas, D. G. *See* Hopfield, J. J.

Thomson, J. J. (1897), *Phil. Mag.*, **44** (293). 59

Thorpe, M. F. *See* Sen, P. N.; Weaire, D.

Tinkham, M. (1964), *Group Theory and Quantum Mechanics*, McGraw-Hill, New York. 442

Tosi, M. P. (1964), *Solid State Phys.*, **16** (1). 309, 311

Tsay, Y. F., Corey, A. J., and Mitra, S. S. (1975), *Phys. Rev.*, **B12** (1354). 115

Tsarev, Yu. N. *See* Nikulin, V. K.

Turnbull, D. *See* Frank, F. C.

Turner, R. *See* Riccius, H. D.

Turner, W. J., Reese, W. E., and Pettit, G. D. (1964), *Phys. Rev.*, **136** (A1467). 157

Valbusa, U. *See* Cavallini, M.

Van Dyke, J. P. *See* Herman, F.

Van Gool, W. (1973), ed., *Fast Ion Transport in Solids*, North Holland, Amsterdam. 36

van Reuth, E. *See* Cheng, C. H.

Van Vechten, J. A. (1969a), *Phys. Rev.*, **182** (891). 60, 328

Van Vechten, J. A. (1969b), *Phys. Rev.*, **187** (1007). 60, 108

Van Vechten, J. A., Cardona, M., Aspnes, D. E., and Martin, R. M. (1970), in *Proceedings of the Tenth International Conference on the Physics of Semiconductors*, (Cambridge, Mass.), Keller, S. P., Hensel, J. C., and Stern, F., eds. 124

Van Vechten, J. A. *See* Phillips, J. C.

Varma, C. M. and Weber, W. (1977), *Phys. Rev. Letters*, **39** (1094). 473

Vagelatos, N. *See* Feldkamp, L. A.

Velický, B. Kirkpatrick, S., and Ehrenreich, H. (1968), *Phys. Rev.*, **175** (747). 508

Venkataraman, G. *See* Feldkamp, L. A.

Vérié, C. (1972), in *New Developments in Semiconductors*, p. 153, Wallace, P. R., ed., Noordhoff International Publishing, Leyden. 163

Vogl, P. (1978), *J. Phys.*, **C11** (251). 424

Voos, M. *See* Benoit à la Guillaume, C.

Wachter, P. *See* Batlogg, B.

Wagman, D. D., Evans, W. H., Halow, I., Parker, V. B., Bailey, S. M., and Shumm, R. H. (1968), *National Bureau of Standards Technical Note 270*, United States Government Printing Office, Washington, D.C. 176

Wagner, L. F., and Spicer, W. E. (1972), *Phys. Rev. Letters*, **28** (1381). 246

Wakoh, S., and Yamashita, J. (1966), *J. Phys. Soc. Japan*, **21** (1712). 521

Wallace, D. C., (1972), *Thermodynamics of Crystals*, John Wiley and Sons, New York. 311

Wallis, R. F. (1965), ed., *Lattice Dynamics* (Proceedings of International Congress, Aug. 5–9, 1963, Copenhagen), Pergamon Press, Oxford. 204, 312

Wallis, R. F. *See* Palik, E. D.

Wang, C. C., and Ressler, N. W. (1970), *Phys. Rev.*, **B2** (1827). 123

Wannier, G. H. (1937), *Phys. Rev.*, **52** (191). 87

Warren, J. L., Yarnell, J. L., Dolling, G., and Cowley, R. A. (1967), *Phys. Rev.*, **158** (805). 208

Warren, J. L. *See* Yarnell, J. L.

Waschauer, D. M. *See* Paul, W.

Watkins, G. D., and Messmer, R. P. (1973), in *Computational Methods for Large Molecules and Localized States in Solids*, p. 133, Herman, F., McLean, A. D., and Nesbet, R. K., eds., Plenum Press, N.Y. 249ff

Watkins, G. D. *See* Messmer, R. P.

Watson, R. E., Ehrenreich, H., and Hodges, L. (1970), *Phys. Rev. Letters*, **24** (829). 507

Watson, R. E., and Ehrenreich, H. (1970), *Comments Solid State Phys.*, **3** (109). 507

Watson, R. E. *See* Gelatt, C. D., Jr.; Hodges, L.

Waugh, J. L. T. *See* Dolling, G.

Weaire, D. (1971), *Phys. Rev. Letters*, **26** (1541). 251

Weaire, D., and Thorpe, M. F. (1973), in *Computational Methods for Large Molecules and Localized States in Solids*, p. 295,

Herman, F., McLean, A. D., and Nesbet, R. K., eds., Plenum Press, New York. 273

Weaire, D. *See* Heine, V.

Weast, R. C. (1965), ed., *Handbook of Chemistry and Physics*, 46th edition, The Chemical Rubber Company, Cleveland. 115

Weast, R. C. (1975), ed., *Handbook of Chemistry and Physics*, 56th edition, The Chemical Rubber Company, Cleveland. 381, 395

Webb. M. B. *See* Harrison, W. A.; Poppendieck, T. D.

Weber, W. (1973), *Phys. Rev.*, **B8** (5082). 455

Weber, W. (1974), *Phys. Rev. Letters*, **33** (371). 212

Weber, W. *See* Varma, C. M.

Weil, R., and Groves, W. O. (1968), *J. Appl. Phys.*, **39** (4049). 196

Weinreich, G. (1965), *Solids*, John Wiley and Sons, New York. 87

Weissbluth, M. (1978), *Atoms and Molecules.*, Academic Press, New York. 460

Wemple, S. H. (1977), *J. Chem. Phys.*, **67** (2151). 114, 115

Wenzel, R. G. *See* Yarnell, J. L.

Wheeler, R. G., and Dimmock, J. O. (1962), *Phys. Rev.*, **125** (1805). 157

Wheeler, R. G. *See* Miklosz, J. C.

Wherrett, B. S. *See* Summers, C. J.

White, R. M. (1974), *Phys. Rev.*, **B10** (3426). 131

White, R. M. *See* Chadi, D. J.

Wiegmann, W. *See* Dingle, R.

Wigner, E. P., and Seitz, F. (1934), *Phys. Rev.*, **46** (509). 500f, 506

Williams, A. R. *See* Appapillai, M.; Lang, N. D.; Moruzzi, V. L.; Yu, K. Y.

Williams, R. K. *See* Butler, W. H.

Wilson, J. A. (1972), *Ad. Phys.*, **21** (143). 438

Wolfe, C. M. *See* Reine, M.; Stillman, G. E.

Wolff, P. A. *See* Sukhatme, V. P.

Woods, A. D. B. *See* Cowley, R. A.

Woodward, R. B., and Hoffmann, R. (1971), *The Conservation of Orbital Symmetry*, Academic Press, New York (English edition). 39

Wulff, G. (1901), *Zeits für Kristallog.*, **34** (449). 231f

Wyckoff, R. W. G. (1963), *Crystal Structures*, 2nd. ed., vol. 1, Interscience New York. 171, 263, 302, 322

Wynne, J. J. (1969), *Phys. Rev.*, **178** (1295). 123

Wynne, J. J., and Bloembergen, N. (1969), *Phys. Rev.*, **188** (1211). 122

Yablonovitch, E. C., Flytzanis, Ch., and Bloembergen, N. (1972), *Phys. Rev. Letters*, **29** (865). 123, 124

Yamashita, J. *See* Wakoh, S.

Yarnell, J. L., Warren, J. L., Wenzel, R. G., and Dean, P. J. (1968), in *Neutron Inelastic Scattering*, p. 301, International Atomic Energy Agency, Vienna. 208

Yarnell, J. L. *See* Warren, J. L.

Yevick, G. J. *See* Percus, J. K.

Yip, K. L., and Fowler, W. B. (1974), *Phys. Rev.*, **B10** (1400). 263

Yndurain, F., and Joannopoulos, J. D. (1976), *Phys. Rev.*, **B14** (3569). 279

Yoshimine, M. *See* Gilbert, T. L.

Yoshinaga, H. *See* Manabe, A.

Yu, K. Y., Miller, J. N., Chye, P., Spicer, W.E., Lang, N. D., and Williams, A. R. (1976), *Phys. Rev.*, **B14** (1446). 403

Yu, P. Y., Cardona, M., and Pollak, F. H. (1971), *Phys. Rev.*, **B3** (340). 116

Yu, P. Y. *See* Cardona, M.

Zallen, R., and Paul, W. (1967), *Phys. Rev.*, **155** (703). 109f

Zeiger, H. J., and Pratt, G. W. (1973), *Magnetic Interactions in Solids*, Clarendon Press, Oxford. 214

Zienau, S. *See* Fröhlich, H.

Ziman, J. M. (1961), *Phil. Mag.*, **6** (1013). 404

Subject Index

the role of the central ion
450